FOOD SCIENCE AND TECHNOLOGY

FOCUS ON FOOD ENGINEERING

FOOD SCIENCE AND TECHNOLOGY

Additional books in this series can be found on Nova's website under the Series tab.

Additional E-books in this series can be found on Nova's website under the E-books tab.

FOOD SCIENCE AND TECHNOLOGY

FOCUS ON FOOD ENGINEERING

ROBERT J. SHRECK
EDITOR

Nova Science Publishers, Inc.
New York

For permission to use material from this book please contact us:
Telephone 631-231-7269; Fax 631-231-8175
Web Site: http://www.novapublishers.com

LIBRARY OF CONGRESS CATALOGING-IN-PUBLICATION DATA

Focus on food engineering / editor: Robert J. Shreck.
p. cm. -- (Food science and technology)
ISBN 978-1-61209-598-1 (hardcover)
1. Food industry and trade. I. Shreck, Robert J. II. Series: Food science and technology series (Nova Science Publishers)
TP370.F528 2011
664--dc22

2011002381

Published by Nova Science Publishers, Inc. † New York

CONTENTS

PREFACE

In the development of food engineering, one of the many challenges is to employ modern tools and knowledge to develop new products and processes. Simultaneously, improving quality, safety, and security remain critical issues in food engineering. Additionally, process control and automation regularly appear among the top priorities identified in food engineering. This book presents topical research in the study of food engineering, including: ozone technology in the food industry; current trends in drying and dehydration of foods; strategies for extending the shelf-life of foods using antimicrobial edible films; developments in high-pressure food processing; as well as tempering and polymorphism during chocolate manufacture.

Chapter 1 - Behavior of the hen's eggs under impact loading has been investigated. Two main problems have been solved.

The first one was focused on the non- destructive impact of the egg. In this part, eggs were excited by the ball impact on the blunt side, sharp side, or on the equator, and the response signals were detected by the laser-vibrometers. These sensors record the velocity of the vibration at a certain point in the direction of the laser beam. In the test, the laser beam was focused normally to the eggshell surface at a selected node on the meridian of the egg. The response wave signals were then transformed from time to frequency domain and the frequency spectrum was analyzed. The specific objectives of the research were to:

(1) analyze the response time signals and frequency signals of eggs,

(2) find the effect factors on dynamic resonance frequency, and

(3) establish relationship between the dominant frequency and the egg's physical properties.

The finite element model of the egg has been developed. The eggshell is considered as linear isotropic elastic material. Its behavior is then described by the Young modulus E and by the Poisson constant ν. The numerical simulation has been performed using LS DYNA 3D finite element code. Computed signals exhibit very good agreement with experimental ones.

The second part of the research was focused on a different type of impact loading when the egg, lying in a planar support, was loaded by the falling rod. The instrumentation of the rod enabled us to obtain the time history of the force at the point of the bar impact. The velocity of the rod was gradually increased up to a certain critical value at which the eggshell failure starts. Numerical simulation of these experiments enabled us to obtain the stress at which the eggshell fracture occurs. This stress represents the eggshell's strength. This strength is dependent on the egg shape as well as on the eggshell thickness. It seems that this strength is an intrinsic material parameter which may be affected by the eggshell microstructure, by its chemical composition and by structural elements distribution. Achieved results have been used for the study of the hen's eggshell behavior at the impact on a rigid plate. Numerical results are in a reasonable agreement with records of the high speed camera.

Chapter 2 - The development and production of new packaging materials, which are friendlier with the environment, is actually being studied with the purpose of minimizing the environmental pollution that is produced by the use of traditional, non-biodegradable packaging. In the framework of this interest, the study of the use of biopolymers to produce "edible films" has considerably progressed in the last decade. Starches, proteins, cellulose and derivatives, gums, chitosan, among other hydrocolloids, have been used for producing this kind of films. The presence of a plasticizer agent is always required to minimize brittle structure and antimicrobials or other additives can be included in the film formulation. Antimicrobials will provide the film with specific functional properties in addition to their inherent barrier properties to the water vapor and oxygen and, in this case, the edible films can be thought of as an active packaging material since they are able to support and, eventually, release the food preservatives. The films will perform as an additional microbial stress factor in order to protect the food from the external contamination and, therefore, will contribute to produce shelf life extension.

The object of the present study was the production of tapioca starch - glycerol based edible films containing the preservatives potassium sorbate (KS) or nisin. Physicochemical properties of films such as crystalline fraction, solubility in water, sorptional behavior and color attributes were studied. In order to optimize the film functionality, the influence of soy oil addition to the film formulation, the use of sodium trimetaphosphate-chemically cross-linked-tapioca starch and the use of different filmmaking techniques, were evaluated. The study of the effect of the film composition on the physicochemical properties and antimicrobial activity behavior will help to predict the potential usefulness of the film for a particular food system.

Chapter 3 - This chapter reports the developments made in the processing of foods using high pressure which over the past two decades. Consumers these days generally expect the food to be of a high quality, minimally processed, natural, additive-free, high in nutritional value as well as safe to eat. High Pressure processing is an alternative to thermal processing which can destroy harmful microorganisms rendering the food safe to eat. As a way of minimally processing food, it has the potential to preserve the quality of foods in many cases and even be responsible for producing new textures and properties. The effect of high pressure on the molecular structure of food proteins is to change their functional properties in surprising and often useful ways. A pressure of ten thousand times greater than atmospheric is capable of coagulating the albumin of egg without the use of heat.

The purpose of using high pressure instead of heat is to preserve and even improve food quality in terms of taste, flavour, texture and colour. The molecular structure of many food components including sugars, oils, vitamins, lipids and pigments are able to resist the effects of high pressures. Pressure is capable of affecting only the weaker bonds and forces sufficient to alter the delicate molecular structures, as in the case of proteins.

There have been some excellent examples worldwide of commercially applying high pressure in the processing of fruits, fish and shellfish, meat and dairy products. Research continues to understand fully the remarkable effects of high pressure on the constituents of food. In general, this has been in the areas of food safety with the destruction of micro-organisms, the activation and deactivation of enzymes; the functional properties of foods components to form foams, gels and emulsions; thermodynamics with the control of phase change. The most important of these has been to establish the sterilisation properties of high pressure food processing. Many harmful microorganisms differ significantly in their ability to withstand pressure while bacteria, yeasts and moulds are readily killed with spores being only inactivated by pressure after germination.

Chapter 4 - This chapter describes and discusses some results obtained through the study of the microencapsulation of açai juice by spray drying using different carrier agents. Initially, the influence of process conditions on the moisture content, process yield and anthocyanin retention was evaluated using a central composite design. From the conditions selected in this first section (inlet air temperature of 140°C, feed flow rate of 15 g/min and 6% of carrier agent), particles were produced using four types of carrier agents: maltodextrin 10DE, maltodextrin 20DE, gum Arabic and tapioca starch. These particles were then characterized with respect to water activity, bulk and absolute density, porosity, particle size distribution and morphology. The samples produced with maltodextrin 20DE and with gum Arabic exhibited the highest water activity and the lowest particles size, while those produced with tapioca starch were the less porous, with the lowest bulk density and highest mean diameter. Then, physical stability of particles, when exposed to different relative humidities, was evaluated through the construction of sorption isotherms and determination of the glass transition temperature. The samples produced with maltodextrin 10DE exhibited the highest critical water activity, being considered as the most stable powder. Glass transition temperature decreased with increasing moisture content, confirming the plasticizant effect of water on this property. Finally, anthocyanin stability of powders stored at different temperatures and relative humidities was evaluated. The increase of both these parameters resulted in higher anthocyanin degradation. Maltodextrin 10DE was the carrier agent that showed the best pigment protection, for all the conditions studied.

Chapter 5 - Computed tomography (CT) is one of the emerging technologies of interest to food science as it permits a non-destructive characterization of food products and their control throughout processing. This work describes the history and physical basis of this technology as well as the working principles of CT. It focuses on the latest research findings related to the application of this technology to different food products; especially dry-cured ham production as well as other issues like pig carcass classification. A revision of other X-ray technologies applied to food science is also included. In dry-cured ham production, CT helps the study of the factors which affect the salting/curing processes. These processes can be monitored because salt can easily be detected due to the differences in densities of meat and salt. Using experimental models, salt and water contents can be non-destructively determined at any moment during the process thus enabling the establishment of safety and quantity criteria in order to avoid either sensory defects or the microbiological hazards common in dry-cured ham. For carcass classification purposes, CT can be used to obtain the lean content of carcasses which is of interest to the food industry as it defines the commercial value of the pig. The estimation of the lean content is usually calculated from the physical measurements of subcutaneous fat depths and muscle thicknesses in specific locations. Devices for this task need to be calibrated and therefore, the dissection is the reference method most commonly used, but this method is difficult and time consuming. CT is an excellent tool for this task as it easily distinguishes the differences between lean, fat and bone.

Chapter 6 - High Pressure Processing (HPP) allows decontamination of foods with minimal impact on their nutritional and sensory features. The use of HPP to reduce microbial loads has shown great potential in the muscle-derived food industry. HPP has proven to be a promising technology and industrial applications have grown rapidly, especially in the stabilization of ready-to-eat meats and dry-cured products, satisfying the demands of regulatory agencies such as the United States Department of Agriculture-Food Safety and Inspection Services (USDA-FSIS). Applications also extend to seafood products and HPP has been used in a wide range of operations, from non-thermal decontamination of acid foods to combined pressure-heating treatments to inactivate pathogenic bacteria, pressure supported freezing and thawing, texturization, and removal of meat from shellfish and crustaceans. Research has also been conducted on the impact of the technology on quality features. Processing-dependent changes in muscle foods include changes in colour, texture and water-holding capacity, with endogenous enzymes playing a major role in the phenomena. This review summarizes the current approaches to the use of high hydrostatic pressure processing, focusing mainly on meat, meat products and seafood. Recent findings on the microbiological, chemical and molecular aspects, along with commercial and research applications, are described.

Chapter 7 - Mathematical modeling represents an effective support to design and control industrial processes. Different approaches can be used to develop reliable models aimed at investigating how system responses may change, with time, under the influence of both external disturbances and manipulated variables. In the present chapter, it will be shown how various kinds of advanced models, based either on fundamental, or on artificial neural networks or on hybrid modeling, can be used to predict the behavior of a typical industrial process: the convective drying of food. The main aims of convective drying are a decrease of food water content and an increase of its temperature; both the above effects improve food preservation since microbial spoilage is favored by low temperature and high moisture content. Drying does actually preserve foods by decreasing water activity, thus stopping the micro-organisms growth. If food drying is performed in uncontrolled conditions, the level of water content could be not sufficiently low to stop the activity of micro-organisms, which continue proliferating. The starting point of any kind of automatic control is definitely represented by the availability of a predictive model of the process under study.

Chapter 8 - Cheese whey is a by-product that recently has gained attention because it represents a source of food ingredients of high nutritive value. It is generally processed by ultra filtration and spray drying to produce whey powder and protein concentrates. However, the concentrated fluid obtained from the membrane process could be used directly in the formulation of foods if it maintains a proper microbial stability, avoiding the drying process which can be rather expensive and affects the nutritional and functional properties.

Food-processing methods have been developed to interfere with bacterial homeostasis, prevent growth, or kill food-borne pathogens. The growing demand for fresh, minimally processed foods by consumers had led to the need for natural food preservation methods such as the use of natural antimicrobials and combination with other hurdles, without adverse effects on the consumer or the food itself. Additionally, natural antimicrobials can be used alone or in combination with other non-thermal technologies, but it has also been demonstrated that microorganisms can display tolerance to these factors when they are applied in sub-lethal concentrations. Thus, it may be useful to assess whether pre-treatments with natural antimicrobials like green tea extract, Microgard™ or nisin, would enhance sensitivity or resistance of food-borne bacteria to different stress conditions, such as acidity, heat, hydrogen peroxide or salt addition, which are commonly applied in food-processing methods. For this purpose different techniques were used in order to evaluate possible membrane damages produced by the combined treatments. The results of this study revealed that the bacterial reduction when applying stress conditions was improved in most of the cases after exposure to natural antimicrobials in liquid cheese whey. Synergistic effects are important to minimize the concentration of additives required to achieve a particular antibacterial effect without adversely affecting the sensorial acceptability. This knowledge about the interactions between natural antimicrobials and stressors has important implications for the food industry when considering which food preservation regime is the best to adopt.

Chapter 9 - This study is aimed at characterizing the non-linear viscoelastic behavior of hard cheese during ageing. Edam block (corresponding to the Gouda cheese) and Brie were used as examples. The rheological properties of these cheeses were evaluated using different experimental techniques:

The compression tests, including stress relaxation, were performed. The material data from monotonic compression and stress relaxation tests were used for creation of a constitutive equation describing the cheese viscoleastic properties.

Viscoelastic properties were also obtained from indentation tests.

In the next step, the behavior of the tested cheeses under dynamic loading was studied during ageing processes. The block of tested cheese was loaded by the impact of an aluminum bar. The force between the bar and cheese was recorded. The surface displacements, as well as surface velocities, were obtained at different points from the position of bar impact. The laser vibrometers were used to record the displacements. Response functions were evaluated both in the time and frequency domains. It was found that the degree of the cheese ageing process is characterized well by the reduction of the surface displacement maximum. This process of ageing is also by the maximum of the impact force. The spectral analysis of the response functions revealed that there was a dominant frequency, which depends only on the degree of the cheese ageing process. The developed method represents a promising procedure for the continuous recording of cheese ageing.

Chapter 10 - This paper reviews the factors affecting the rennet paste composition and highlights strengths and weaknesses in the production of this type of coagulant and in its use for cheesemaking. Rennet paste is used almost exclusively in the manufacture of PDO cheese from ovine and goat milk and of some pasta filata cheeses as Caciocavallo.

New strategies for improving enzymatic and hygienic quality of rennet paste are presented as the incorporation of selected probiotic strains into rennet. Innovative lamb rennet paste containing probiotic is able to transfer microbioal cells into the curd matrix during the milk coagulation step. The study of ovine cheese produced using lamb rennet paste containing *Lactobacillus acidophilus* and a mix of *Bifodobacterium lactis* and *Bifidobacterium longum* evidenced enhanced nutritional and health properties and accelerated ripening process in terms of proteolysis and lipolysis, with cheese maintaining acceptable organoleptic characteristics. In particular, *Lb. acidophilus* and bifidobacteria evidenced the ability to liberate *9c,11t*-CLA, *9t,11t*-CLA, and *10t,12c*-CLa in the ovine cheese as an outcome of the peculiar metabolic pathway associated to microbial strains. The current consumer market of dairy products requires a continuous effort in terms of quality and innovation; dairy compartment offers the chance of developing innovative products beginning from traditional cheeses which exert ameliorated healthfulness and beneficial properties.

Chapter 11 - Texture and color are major quality attributes of plant-based foods. Texture lies in a “mechanical unit” whose components are cell wall, cellular membrane or plasmalemma and middle lamella. Color lies in the presence of pigments compartmentalized into the cells, which selectively absorb certain wavelengths of light while reflecting others. The application of different stress factors for food preservation purposes can alter enzymatic activity and can affect texture and color, compromising consumer acceptability of resultant food products.

In this chapter, the effect on the activity of some enzymes, textural behavior and color changes, of the osmotic treatment of cucumber (*Cucumis sativus* L.) and butternut (*Cucurbita moschata*, Duch. ex. Poiret) or of the gamma irradiation of red beet (*Beta vulgaris* L. var. *conditiva*) or of the blanching of kiwifruit (*Actinidia deliciosa*, A. Chev.), is analyzed.

In: Focus on Food Engineering
Editor: Robert J. Shreck

ISBN: 978-1-61209-598-1

Chapter 1

BEHAVIOR OF HEN'S EGGS AT IMPACT LOADING

Šárka Nedomová[1], Jan Trnka[2], Libor Severa[1], Pavla Stoklasová[2], and Jaroslav Buchar[1]

[1] Mendel University in Brno, Zemědělská 1, 613 00 Brno, Czech Republic

[2] Institute of Thermomechanics, Czech Academy of Science, Dolejškova 5, 182 00, Praha 8, Czech Republic

ABSTRACT

Behavior of the hen's eggs under impact loading has been investigated. Two main problems have been solved.

The first one was focused on the non- destructive impact of the egg. In this part, eggs were excited by the ball impact on the blunt side, sharp side, or on the equator, and the response signals were detected by the laser-vibrometers. These sensors record the velocity of the vibration at a certain point in the direction of the laser beam. In the test, the laser beam was focused normally to the eggshell surface at a selected node on the meridian of the egg. The response wave signals were then transformed from time to frequency domain and the frequency spectrum was analyzed. The specific objectives of the research were to:

(1) analyze the response time signals and frequency signals of eggs,
(2) find the effect factors on dynamic resonance frequency, and
(3) establish relationship between the dominant frequency and the egg's physical properties.

The finite element model of the egg has been developed. The eggshell is considered as linear isotropic elastic material. Its behavior is then described by the Young modulus E and by the Poisson constant ν. The numerical simulation has been performed using LS DYNA 3D finite element code. Computed signals exhibit very good agreement with experimental ones.

The second part of the research was focused on a different type of impact loading when the egg, lying in a planar support, was loaded by the falling rod. The instrumentation of the rod enabled us to obtain the time history of the force at the point of the bar impact. The velocity of the rod was gradually increased up to a certain critical value at which the eggshell failure starts. Numerical simulation of these experiments enabled us to obtain the stress at which the eggshell fracture occurs. This stress represents the eggshell's strength. This strength is dependent on the egg shape as well as on the eggshell thickness. It seems that this strength is an intrinsic material parameter which may be affected by the eggshell microstructure, by its chemical composition and by structural elements distribution. Achieved results have been used for the study of the hen's eggshell behavior at the impact on a rigid plate. Numerical results are in a reasonable agreement with records of the high speed camera.

1. INTRODUCTION

The eggshell is the natural packing material for the egg's contents, and as a result, it is important to obtain high shell strength to resist all impacts an egg is subjected to during the production chain (Bain, 1992). Broken eggs cause economic damage in two ways: they cannot be sold as first-quality eggs, and the occurrence of hair cracks raises the risk for bacterial contamination of the broken egg and of other eggs when leaking, creating problems with internal and external quality and food safety. Eggshell strength is generally measured using either direct tests, such as nondestructive deformation (Voisey and Hunt, 1974) or destructive fracture force (Voisey and Hunt, 1967a, b) of an egg under quasistatic compression between two parallel plates, or indirect tests, such as the measurement of eggshell thickness (Brooks and Hale, 1955; Voisey and Hamilton, 1976; Ar and Rahn 1980; Thompson et al., 1981; Bell, 1984; Hunton, 1993) or specific gravity (Olsson, 1934). Many of these methods, however, are destructive, slow, or subject to environmental influences and, hence, are regarded as being unpractical. Coucke (1998) presented a fast, objective, and nondestructive method for the determination of the eggshell strength, based on acoustic resonance analysis. This research revealed that the acoustic response of the eggs to the impulse loading could be used for the identification of many physical properties of the eggs, including the crack detection (Coucke, 1998; Coucke et al., 1999; De Ketelaere et al., 2000, 2002, 2003; Dunn et al., 2005 a, b). Old poultry farmers checked the mechanical integrity of hatching eggs by placing two eggs in one hand and spinning the eggs around each other while gently impacting both eggs several times. The characteristic sound produced by this impact contains information about the presence of hairline cracks in the shell. Eggs with cracked shells sound dull and highly damped and are consequently rejected from incubation. Intact eggs produce a typical sound, which is consistent when impacting on several places around the equator. In fact, by impacting the eggs in this way, they start to vibrate at one of their resonant frequencies, producing the typical sound. This test gave rise to the idea that the vibration response of eggs after impact excitation could be used to gain knowledge about the structural integrity of the eggshell. The study of the acoustic vibrations as an indicator of egg and eggshell quality is limited and relatively recent. Sinha et al. (1992) reported the use of acoustic resonant frequency analysis in chicken eggs for the detection of *Salmonella enteriditis* bacteria in the egg content. The excitation was done at the blunt side of the egg with a piezoelectric crystal and the response is measured with an accelerometer at the opposite side. Yang et al. (1995) used the vibrational behavior of a chemically treated egg as a quality detection method. Bliss (1973) and Moayeri (1997) detect the local stiffness of the eggshell by analyzing the time signal of the impactor after impacting. Multiple measurements distributed over the whole eggshell surface are required to have a global overview of the eggshell strength.

The present study was intended to better understand the vibration characteristics of an egg. For this purpose, an experimental modal analysis was performed on an intact egg. This analysis gives us the ability to visualize the spatial motion of the egg after being impacted. In order to achieve these goals, eggs were excited by the impact of the steel ball on the blunt side, and the response signals were detected by the laser vibrometers at the different points on the eggshell surface. The response wave signals were then transformed from time to frequency domain and the frequency spectrum was analyzed. The specific objectives of the research were to:

- analyze the response time signals and frequency signals of eggs
- develop a finite element model of an egg.

The next part of the presented paper deals with the problem of the eggshell strength evaluation under impact loading. A new experimental technique has been developed. Numerical simulation of this method has also been performed. This procedure enables us to determine the eggshell strength in terms of the stress. Experiments focused on the hen's eggshell behavior at the impact on a rigid plate have been used for the verification of the obtained model of an egg. Through this method, the complex behavior of the eggshell at impact loading has been described.

2. Eggshell Behavior at Non - Destructive Impact

2.1. Egg Samples

Eggs were collected from a commercial packing station. The physical properties of egg samples were determined by the following methods: linear dimensions, i.e. length (L) and width (W), were measured with a digital calliper to the nearest 0.01 mm. The geometric mean diameter of eggs was calculated using the following equation given by (Mohsenin, 1970):

$$D_g = \left(LW^2\right)^{\frac{1}{3}} \tag{1}$$

According to Mohsenin (1970), the degree of sphericity of eggs can be expressed as follows:

$$\Phi = \frac{D_g}{L} x100 \quad (\%) \tag{2}$$

The surface area of eggs was calculated using the following relationship given by (Mohsenin, 1970; Baryeh and Mangope, 2003):

$$S = \pi D_g^2 \tag{3}$$

Volume of the egg is than given as

$$V = \frac{\pi}{6} LW \tag{4}$$

Shape index of the eggs was determined using following equation:

$$SI = \frac{W}{L} x100 \quad (\%) \tag{5}$$

The main properties of the eggs are presented in Table 1.

Table 1. Main characteristics of the tested eggs

No.	Mass (g)	Width (cm)	Length (cm)	Shape index (%)	Air cell (mm)
1	60.39	44.71	54.16	82.55	2
2	60.75	45.28	53.50	84.64	1
3	66.50	43.99	61.10	72.00	2
4	65.17	45.21	56.32	80.27	1
5	61.73	44.74	55.50	80.61	1
6	64.17	44.60	57.33	77.80	2
7	65.79	45.20	57.21	79.01	2
8	63.95	44.21	57.36	77.07	2
9	68.66	46.03	57.18	80.50	2
10	64.30	45.99	54.03	85.12	2
11	67.51	45.69	57.52	79.43	1
12	66.67	45.45	57.63	78.87	2
13	65.30	45.02	57.70	78.02	1
14	60.39	43.51	58.23	74.72	2
15	63.86	44.33	58.73	75.48	2
16	63.25	44.62	55.93	79.78	2
17	60.84	43.91	56.26	78.05	2
18	62.17	45.06	54.86	82.14	2
19	62.54	43.78	58.51	74.82	2
20	63.13	44.20	58.53	75.52	2
21	62.92	44.51	56.37	78.96	2
22	64.26	45.62	54.72	83.37	2

23	65.54	44.90	58.03	77.37	2
24	67.79	46.20	56.89	81.21	2
25	61.48	44.29	55.28	80.12	1
26	66.77	45.56	56.93	80.03	2
27	63.43	44.43	56.48	78.67	2
28	68.54	45.81	57.91	79.11	2
29	64.37	44.41	58.05	76.50	2
30	66.44	45.19	57.94	77.99	2
average	64.29	44.88	56.87	78.99	1.80

Although the shape index given in Table 1 may be sufficient for the description of many problems – see e.g. (Altuntas and Seroglu, 2008), the solution of stress and strain state in the eggshell must be based on the exact description of the eggshell shape. There are many procedures which can be used for the solution of this task. One of the most popular is the mathematical equation given by Narushin (2001) for the profile of an egg. This equation (Eq. 6) adequately defined the surface profile of the egg based on the length (L) and the maximum breadth (B) of the egg:

$$y = \pm\sqrt{L^{\frac{2}{n+1}} x^{\frac{2n}{n+1}} - x^2} \tag{6}$$

where:

$$n = 1.057\left(\frac{L}{B}\right)^{2.372} \tag{7}$$

There are many other equations – see (Erdogdu et al., 2005) for a review. A more accurate determination of an egg's shape is based on the analysis of the digital egg photographs. The application required one measured dimension (e.g. the egg length, measured with sliding calliper), and allowed the user to determine any user-defined distance on the photograph from the derived number of pixels per unit length. From the dimensional measures of individual eggs, their contours could be accurately described in a user-defined Cartesian coordinate system, using a mathematical equation. For more details, the reader is referred to the procedure described by Denys et al. (2003). Three dimensional egg shapes can be then obtained by revolving the contours 180° about the axis of symmetry. The shape of the eggshell counter can be described using of the polar coordinates r,φ as:

$$x = r\cos\varphi \qquad y = \sin\varphi, \tag{8}$$

where

$$r(\varphi) = a_o + \sum_{i=0}^{\infty} a_i \cos\left(2\pi \frac{\varphi}{c_i}\right) + b_i \sin\left(2\pi \frac{\varphi}{c_i}\right) \tag{9}$$

The analysis of our data led to the conclusion that the first four or five coefficients of the Fourier series are quite sufficient for the egg's counter shape description (the correlation coefficient between measured and computed egg's profiles lies between 0.98 and 1).

In the Figure 1a, an example of the egg counters computed using of the Eq. (6) and determined from the digital photography is displayed. It can be seen that there is a difference between these two counters. The knowledge of the equation describing the eggshell counter is necessary namely for the numerical simulation of egg behavior under different mechanical loading, at the numerical simulation of different heat treatment and also for the determination of the curvature of this curve. The radius of the curvature, R, than plays meaning role at the evaluation of some egg's loading tests (compression test, etc.) – see e.g. MacLeod et al. (2006). This radius can be solved by the solving of the following system of equations:

$$\begin{aligned} &F(\varphi) = (x - x_o)^2 + (y - y_o)^2 - R^2 \\ &F(\varphi) = 0 \\ &\frac{dF(\varphi)}{d\varphi} = 0 \\ &\frac{d^2F(\varphi)}{d\varphi^2} = 0 \end{aligned} \tag{10}$$

where x_o, y_o are the coordinates of the center of the curvature of the given curve. If the curve is described as function y=f(x) – see Eq. 6, than the radius of the curvature is given as:

$$R = \frac{\left(1 + \left(\frac{dy}{dx}\right)^2\right)^{\frac{3}{2}}}{\frac{d^2y}{dx^2}} \tag{11}$$

An example of the radius of the eggshell contour curvature is displayed in the Figure 2. Examples of the parameters $a_0 - a_5$ are given in the Table 2.

Table 2. Parameters of the Fourier series

Egg No.	a_0 (mm)	a_1 (mm)	a_2 (mm)	a_3 (mm)	a_4 (mm)	a_5 (mm)
1	2.12499	-0.29335	0.34217	-0.12421	0.17387	0.09230
2	2.09077	-0.48952	0.34469	-0.17596	0.32507	0.00512
3	2.20550	-0.22044	0.35509	0.00817	0.24207	0.02732

4	2.07706	-0.53085	0.46642	-0.26738	0.17613	0.13600
5	2.10414	-0.37483	0.52265	-0.16073	0.16685	0.08387
6	2.21139	0.08756	0.41620	-0.18620	0.23424	0.33348
7	2.11313	-0.30195	0.39717	0.01323	0.18090	0.08909
8	2.25341	-0.32390	0.36839	-0.00570	0.22664	0.08596
9	2.09777	-0.56694	0.44252	-0.20463	0.24923	0.04899
10	2.10381	-0.26255	0.35309	-0.02600	0.30705	-0.00028
11	2.14153	-0.42771	0.46006	-0.24329	0.31052	0.16553
12	2.08970	-0.47290	0.46277	0.15292	0.17199	-0.16723
13	2.25533	0.43566	0.05081	0.54100	0.58856	-0.22585
14	1.98390	-0.60629	0.54410	-0.36056	0.21227	0.08132
15	2.04938	-0.30450	0.38307	-0.09057	0.15791	0.07845
16	2.02310	-0.56432	0.56001	-0.18340	0.24989	-0.01039
17	1.90091	-0.60728	0.66132	-0.47316	0.27122	0.11678
18	2.00963	-0.64351	0.39148	-0.23410	0.30908	-0.01978
19	2.26103	-0.20451	0.27660	0.01392	0.17835	-0.03563
20	2.05119	0.51333	0.64971	-0.06500	0.05849	0.22746
21	2.02030	-0.39033	0.55991	-0.11977	0.31929	0.05467
22	2.11194	-0.22980	0.38933	0.09574	0.13560	0.02557
23	2.19283	-0.23854	0.36055	-0.11654	0.18820	0.07914
24	2.17823	-0.18673	0.30867	0.01258	0.07538	0.02141
25	2.15933	-0.13679	0.38355	-0.06446	0.15137	0.13713
26	2.16613	-0.27726	0.32321	-0.34317	0.24034	0.16623
27	2.00980	-0.53561	0.55880	-0.47912	0.20549	0.22067
28	2.16542	-0.15770	0.32021	0.15656	0.25722	0.05457
29	2.07503	-0.25397	0.49153	-0.15944	0.16201	0.08691
30	2.15128	0.00970	0.30886	-0.07292	0.19626	0.13395

Figure 1 shows an example of the egg's shape obtained using Eqs. (6) and (8). A significant difference can be seen. It means that the exact evaluation of the egg's shape is needed.

Product support - During the measurements the support of the egg must be such that the distortion of its natural motion caused by this support is minimal. A teflon ring was chosen as a supporting mean.

The impact excitation method was chosen because of its fast and simple nature. The egg is excited at top of the blunt part of the egg by the impact of the steel ball. The ball (6 mm in diameter) falls from the height of 121 mm.

The egg response measurement. A laser vibrometer was used to measure the egg response to the impact. This contactless sensor adds no extra mass to the structure and does not disturb the free vibration of the egg. The laser-vibrometer measures the velocity of the vibration at a certain point in the direction of the laser beam. In the test, the laser beam is focused normally to the eggshell surface at a selected node on the meridian of the egg. The laser vibrometer is isolated from the egg supporting structure so no disturbing vibrations are introduced when performing the measurements.

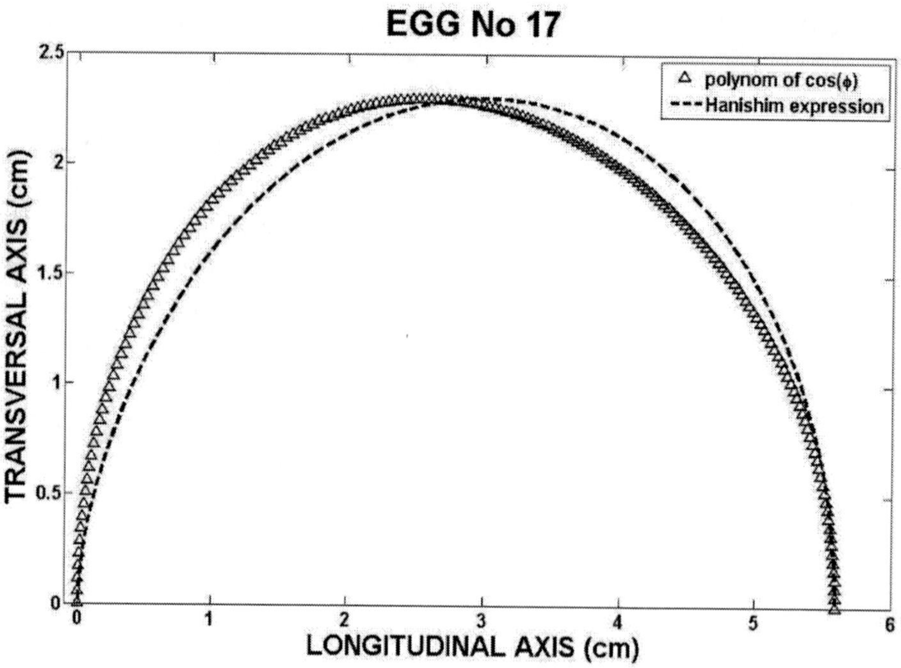

Figure 1. Egg's profile determined using Equations (6) and (8).

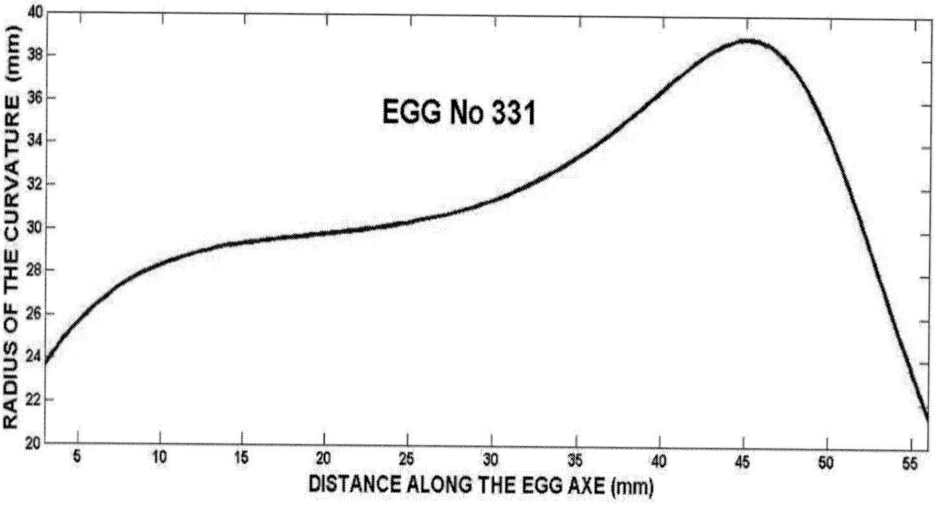

Figure 2. Variation of the radius of the curvature along the egg's profile.

2.2. Experimental Method

The measurement set-up is shown in Figure 3. The system has three major parts, namely the product support, the excitation device and the response-measuring device.

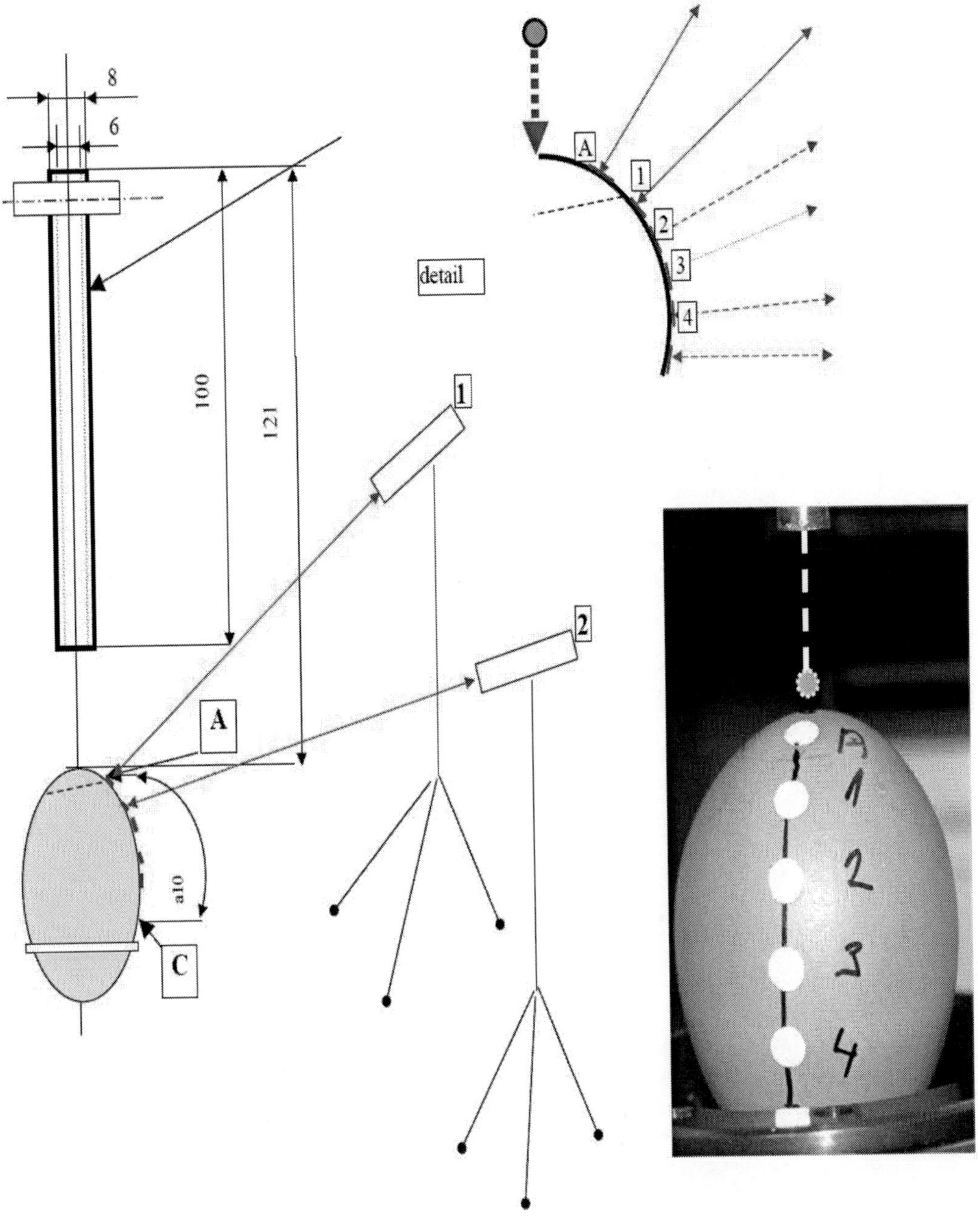

Figure 3. Experimental set up (positions A, 1, 2, 3, 4 are denoted as A0, A1, A2, A3, A4).

Data acquisition and analysis. When eggs were excited, the response acceleration signals in time domain were detected, and MATLAB computer program was used to transform the response from time to frequency domain, by means of FFT. Dynamic response curves were then gained and statistically analyzed in the time and frequency domain for all eggs. The experiments were conducted with five replicates.

2.3. Experimental Results

2.3.1. Time Domain

In the Figure 4 an example of the surface velocity of an egg surface is given. It can be seen that the reproducibility of the experiments is relatively very high. Pronounced damping of the time signal is clearly visible. The damping is also shown in the Figure 5. The positions of the single points along the meridian are given in the Table 3.

Table 3. Position of the detection points along the meridian (Egg No. 19)

Point	Distance from the blunt pole (mm)	ANGLE α (°)
A	7.4	14.45
1	18.7	37.63
2	28.7	59.74
3	38.8	18.01
4	48.4	109.46

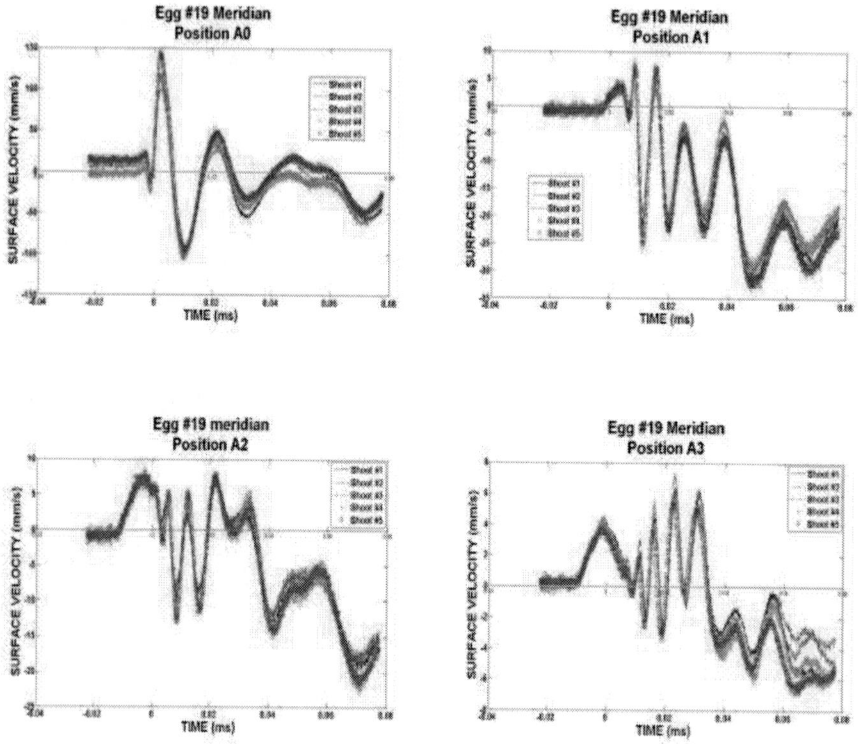

Figure 4. Experimental records of the surface velocities at the different points along the meridian.

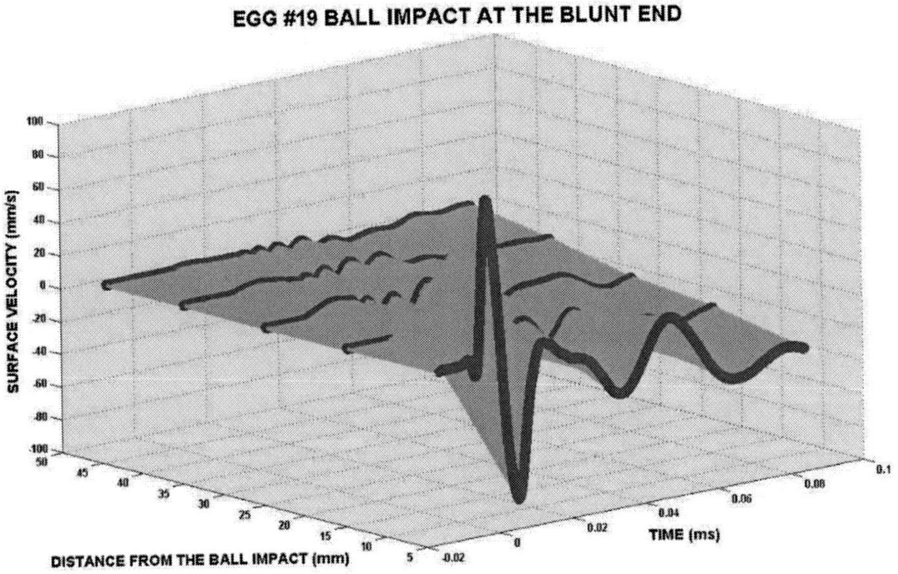

Figure 5. Development of the surface velocity along the meridian.

The damping of the surface velocity can be also expressed as the dependence of the surface velocity maximum on the distance from the blunt pole – see Figure 6.

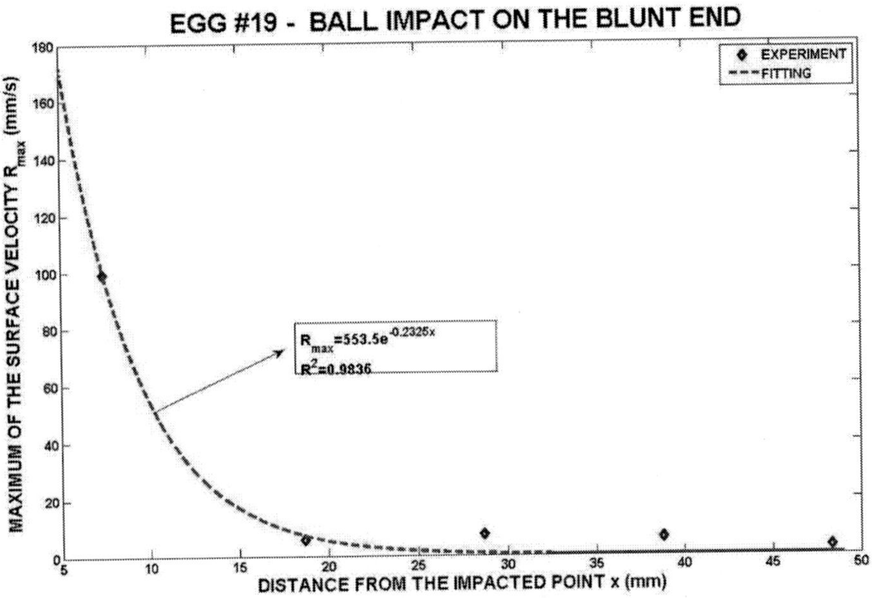

Figure 6. Damping of the surface velocity along the meridian.

The surface velocity corresponds to the surface wave which is initiated at the blunt pole by the ball impact. The velocity of this wave can be evaluated. The dependence of this velocity on the distance from the point of the ball impact is shown in the Figure 7. The decrease in this velocity is a consequence of the egg surface curvature, contact with the egg membranes, egg liquids etc. The investigation of these effects must be based on further experiments.

The remaining eggs have been tested by the same way as the egg No. 19. The response has been recorded at the position A0 and on the equator. Examples of these records are shown in Figure 8. The recorded signals are very similar. The surface wave velocities have also been evaluated. Results are given in Table 4.

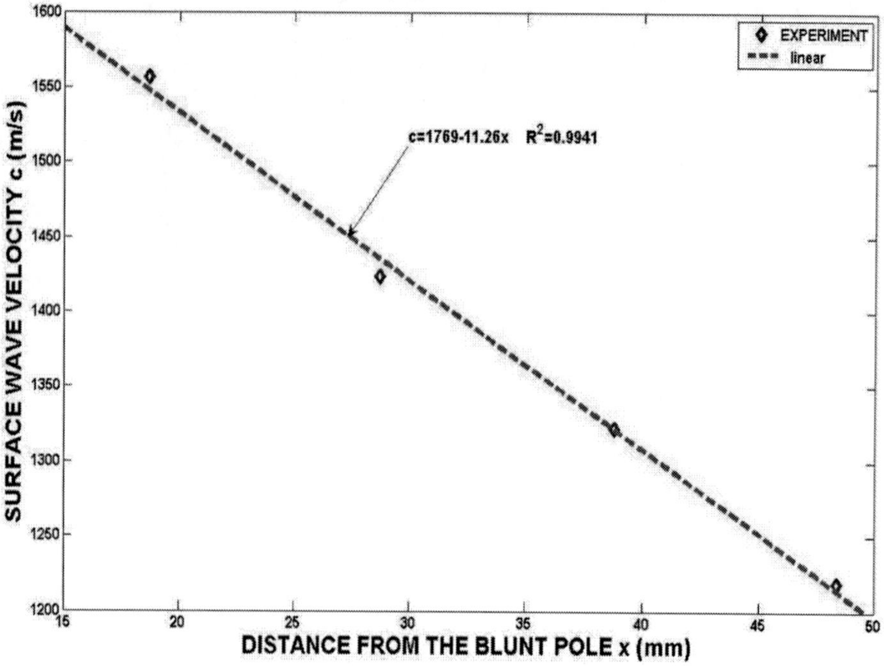

Figur 7. The decrease of the surface wave velocity with the distance from the blunt pole (Egg No. 19).

Table 4. Surface wave velocities

Egg No.	c (m/s)	Egg No.	c (m/s)	Egg No.	c (m/s)
1	1219.3	11	1226.3	21	1221,6
2	1222.5	12	1221.6	22	1218.3
3	1223.8	13	1217.6	23	1219.1
4	1218.6	14	1224.3	24	1222.7
5	1217.3	15	1227.1	25	1225.7
6	1219.1	16	1215,8	26	1218.4
7	1222.4	17	1216,3	27	1219.5
8	1220.4	18	1214,2	28	1221.7
9	12222.6	19	1218,2	29	1223.4
10	1219.3	20	1225.4	30	1224.1

The distribution of these velocities is displaced in the Figure 8.

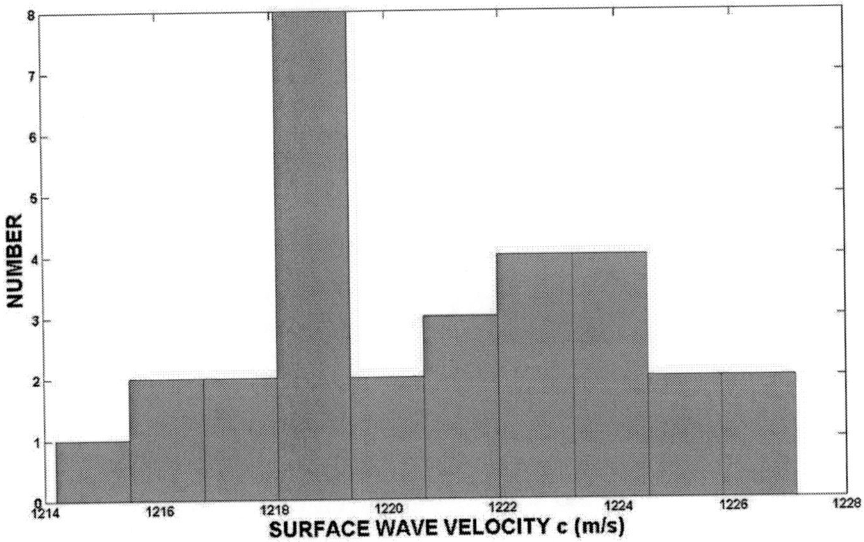

Figure 8. Distribution of the surface wave velocities.

2.3.2. Frequency Domain

The transition from the time to the frequency domain has been performed using the fast Fourier transform (FFT). This transform enables to express a time function f(t) as:

$$f(t)=\frac{1}{\pi}\int_{=\infty}^{\infty} S(\omega)e^{i\omega t}d\omega \qquad S(\omega)=\int_{-\infty}^{\infty} f(t)e^{i\omega t}dt, \tag{12}$$

where S(ω) is the spectral function and ω denotes the angular frequency. The transform into the frequency domain will be a complex valued function, that is, with magnitude and phase. The fast Fourier technique (FFT) has been used for the evaluation of the magnitude and phase. This algorithm is a part of the MATLAB software. Examples of the amplitudes of the spectral functions of the surface velocities are shown in the Figs. 9-11. The amplitudes are significant namely for a lower frequencies. It is evident that the maximum of the spectral function amplitude decreases with the transition from the point A0 (near of the blunt end) to the equator. This decrease can be expressed using of the transfer function T(ω) which is defined as ratio:

$$T(\omega)=\frac{S(\omega)|x=EQUATOR}{S(\omega)|x=A0} \tag{13}$$

Examples of the amplitudes of the transfer functions are displaced in the Figure 12.

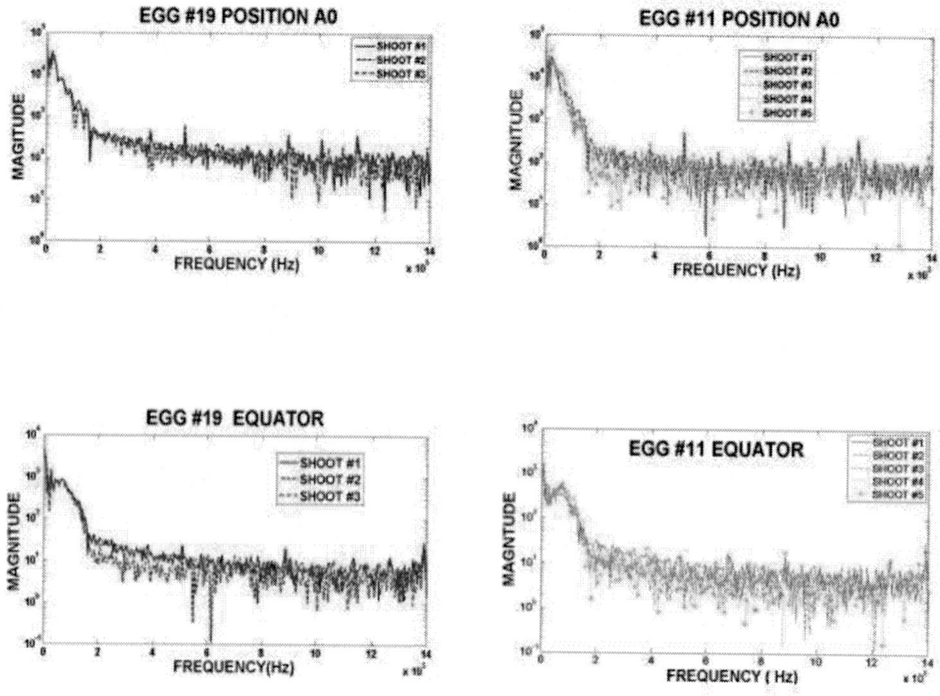

Figure 9. Amplitudes of the spectral functions.

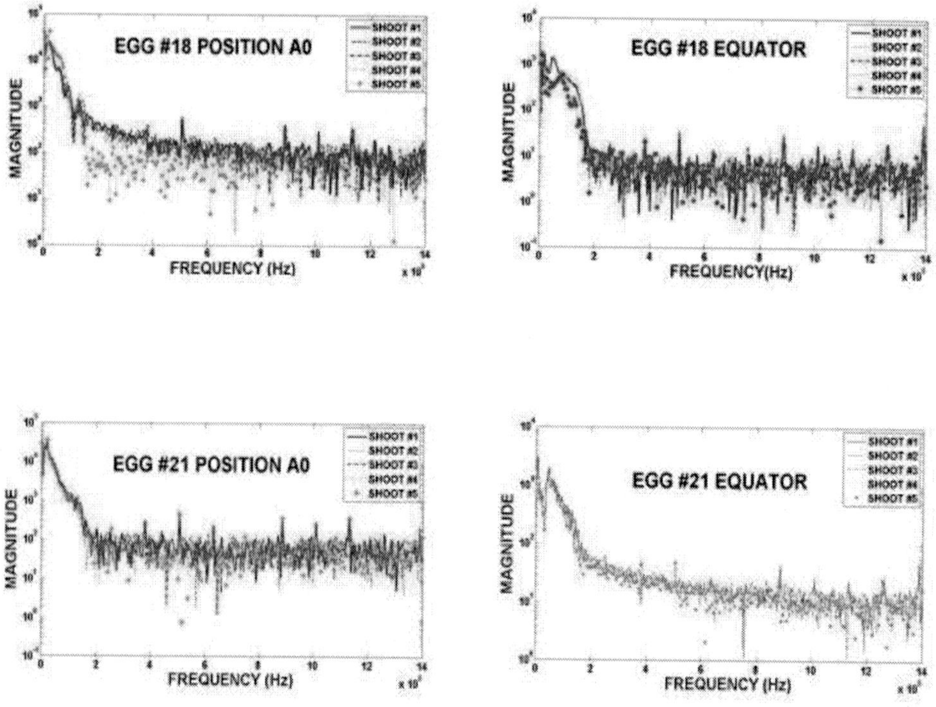

Figure 10. Amplitudes of the spectral functions.

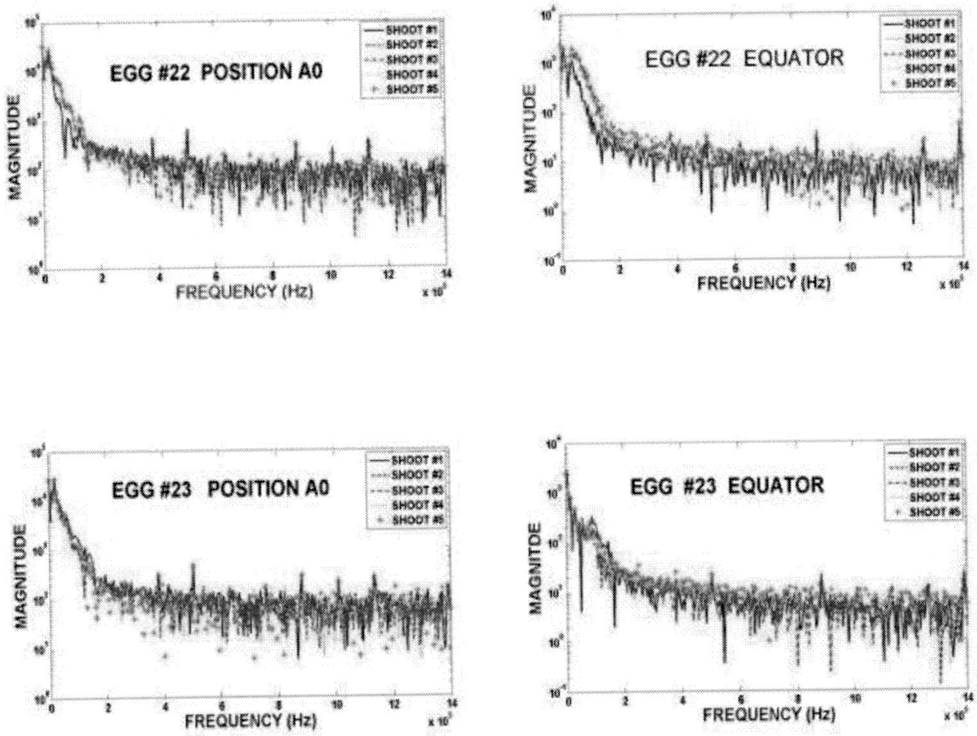

Figure 11. Amplitudes of the spectral functions.

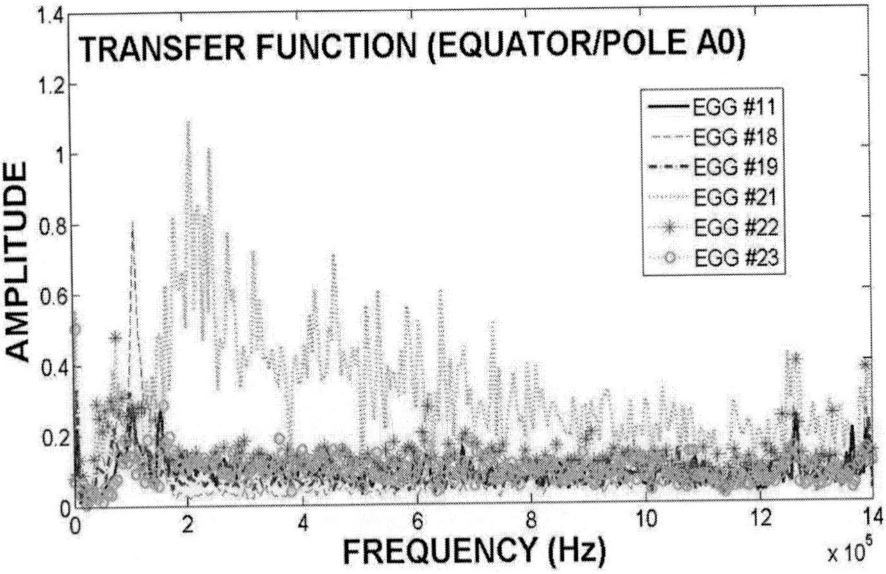

Figure 12. Examples of the transfer functions.

More detailed view on the changes of the spectral functions in the direction of the surface wave propagation can be seen in the Figure 13.

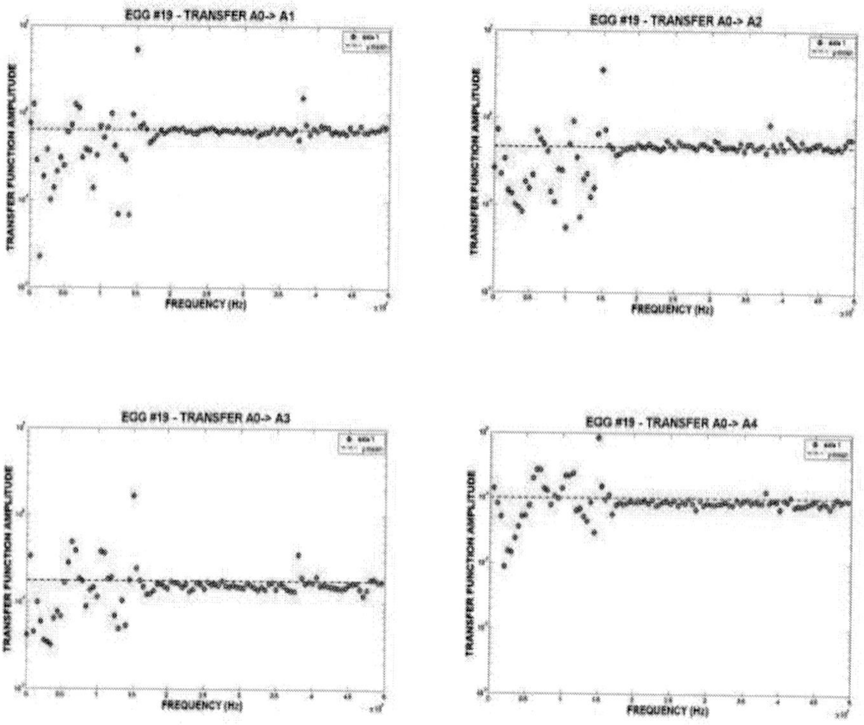

Figure 13. Transfer function amplitude around the meridian – egg No. 19.

General view on the transfer functions along the meridian is displaced in the Figure 14. There are certain frequencies at which some amplification of the spectral functions can be observed. In these figures the average values of the transfer functions amplitude are also shown. The average amplitude is decreasing function of the distance from the blunt pole – see Figure 15.

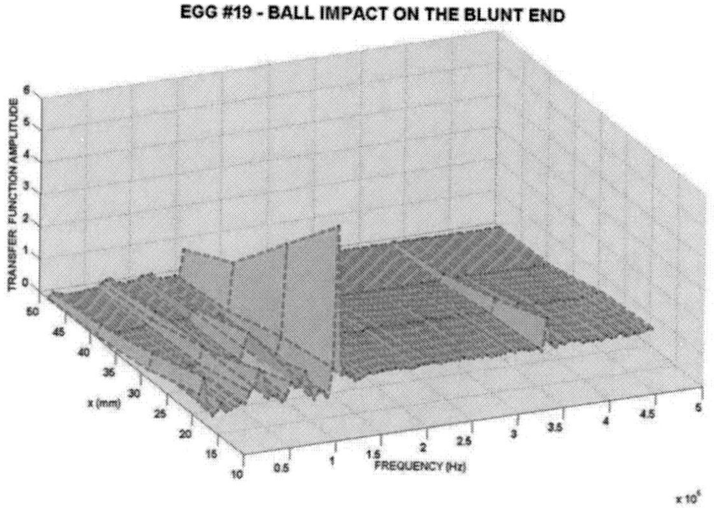

Figure 14. Transfer function amplitudes.

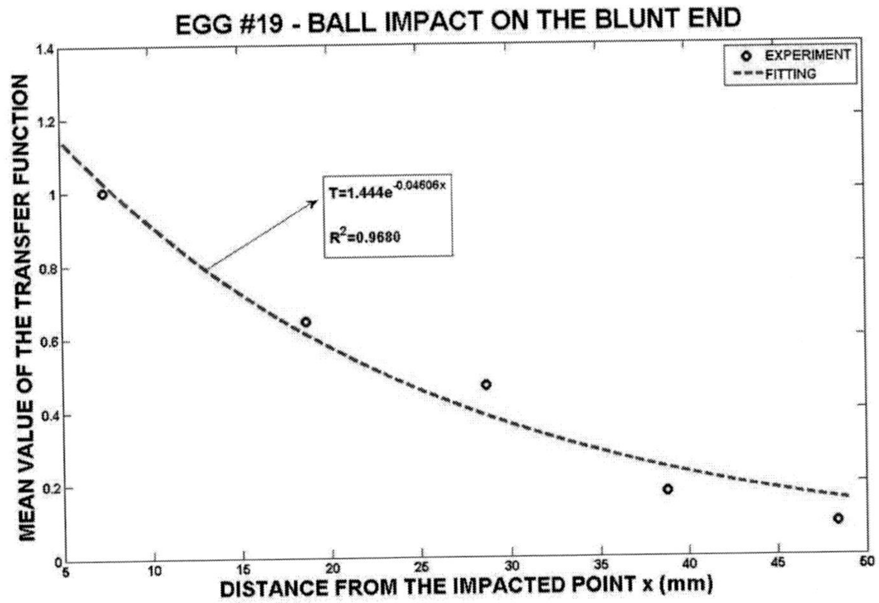

Figure 15. Average values of the transfer functions amplitudes.

The average values for all tested eggs are given in the Table 5.

Table 5. Average values of the transfer functions (blunt pole - equator) T

Egg No.	Shape Index (%)	T
1	82.55	0.16532
2	84.64	0.18112
3	72.00	0.04892
4	80.27	0.09021
5	80.61	0.09032
6	77.80	0.06451
7	79.01	0.08695
8	77.07	0.06594
9	80.50	0.08976
10	85.12	0.18640
11	79.43	0.07983
12	78.87	0.07234
13	78.02	0.07562
14	74.72	0.05631
15	75.48	0.08204
16	79.78	0.08917
17	78.05	0.08192

Table 5. (Continued)

Egg No.	Shape Index (%)	T
18	82.14	0.16000
19	74.82	0.10850
20	75.52	0.07146
21	78.96	0.08361
22	83.37	0.17231
23	77.37	0.05983
24	81.21	0.09235
25	80.12	0.08967
26	80.03	0.09023
27	78.67	0.08236
28	79.11	0.08712
29	76.50	0.07153
30	77.99	0.06611

The average value of the transfer function amplitude increases with the increase of the shape index. This tendency is shown in the Figure 16. It means that approach to the spherical shape leads to improvement of the transfer capability of the eggshell.

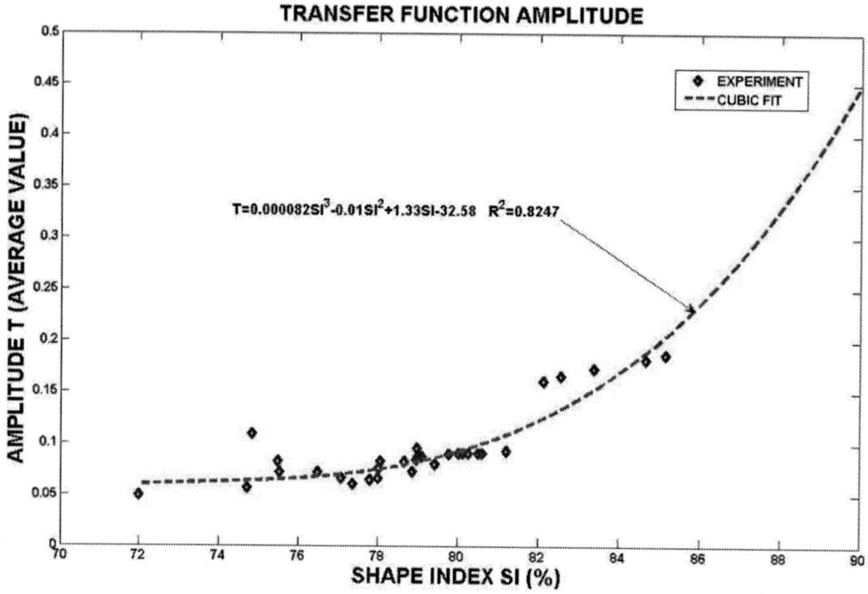

Figure 16. The dependence of the transfer function amplitude on the egg's shape.

Obtained results show that the best position for the egg excitation is the blunt pole. The response function of the egg should be detected namely in the area of the air cell.

2.4. Numerical Simulation

The numerical model should reflect the main features of the egg structure. Schematic of the hen's egg is shown e.g. in Sugino et al. (1997). The detail description of the single elements of this structure is given e.g. in Wells (1968), Rodrigez-Navarro et al. (2002). The finite element model has been developed using the following assumption:

a) Eggshell is homogeneous isotropic linear elastic material. The properties of such material are described by the Young modulus E, Poisson ratio ν and material density ρ.
b) Membranes are also taken as linear elastic material. No difference between membranes has been considered.
c) Air is considered as an ideal gas.
d) Egg yolk and egg white are considered as compressible liquids.

Cross section of the used model is shown in the Figure 17. The numerical model is shown in the Figure 18.

Parameters of the model are:

- Total number of nodes 141249
- Total number of solid elements 88050
- Total number of shell elements 46710

Numerical analysis has been performed by LS DYNA 3D finite element code.

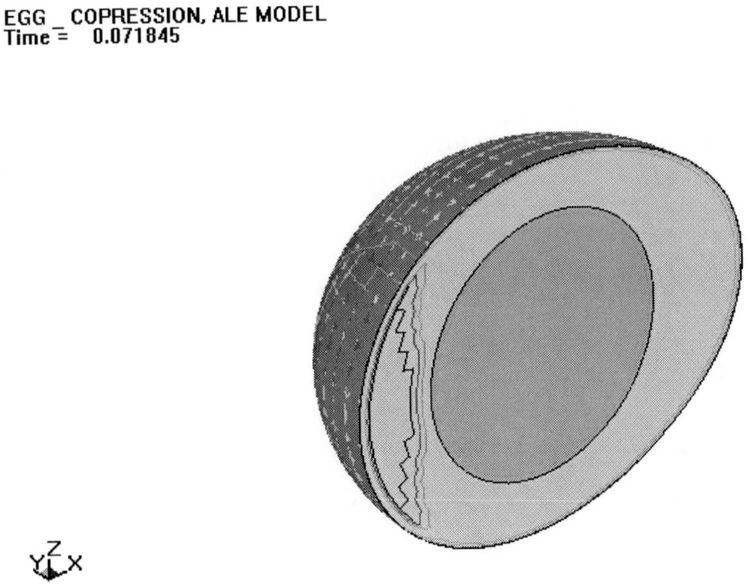

Figure 17. Schematic of the egg model used for the numerical simulation.

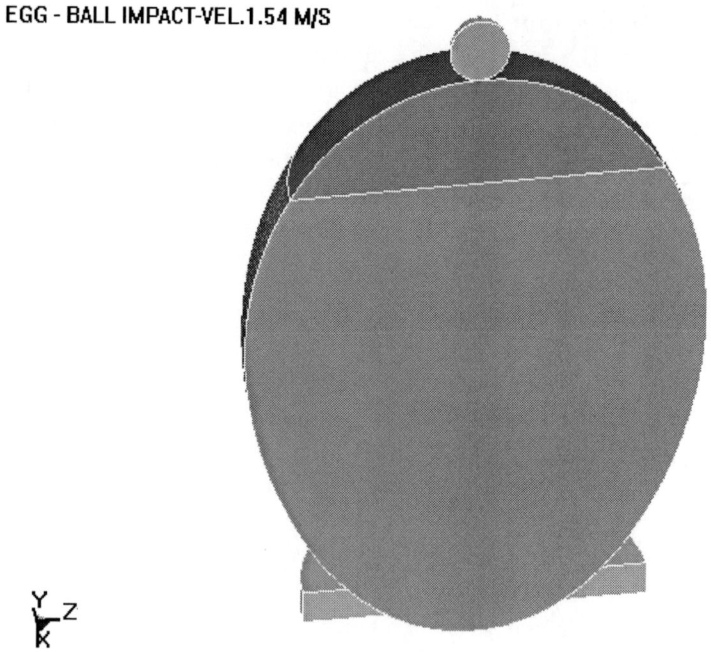

Figure 18. Numerical model of the experiment.

The elastic properties of the eggshell have been obtained using of the procedure developed in Buchar and Simeonovová (2001). Elastic properties of the membranes were determined by the method described in Bing Feng Ju et al. (2002). The compressibility of the egg liquids was taken from the study Chung and Stadelman (1965). The elastic properties are given in Table 6 and 7.

Table 6. Elastic properties of the egg parts

Egg part	ρ (kg/m^3)	E (GPa)	ν
eggshell	2140	20.8	0.37
membrane	1005	0.0035	0.45

Table 7. The properties of the egg liquids. (K- bulk modulus)

Egg liquid	ρ (kg/m^3)	K (GPa)
white	2140	2.0
yolk	1005	1.8

The surface velocities have been evaluated at different nodes corresponding to the points where the experimental data have been reported. A schematic of these nodes is shown in the Figure 19.

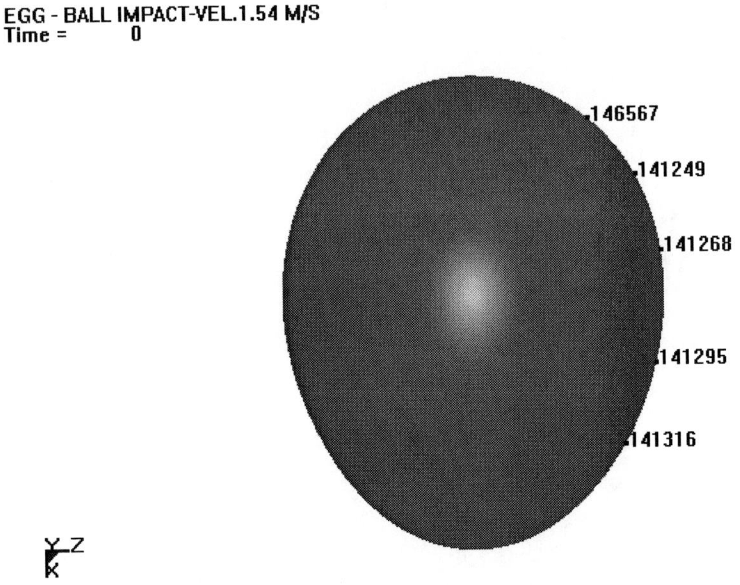

Figure 19. Nodes where the velocities have been computed. Single nodes correspond to the point A0, A1, A2, A3, and A4.

The surface velocity is computed in the x and y directions. The velocity normal to the eggshell surface has been evaluated by the procedure outlined in Figure 20a-e. The five versions have been considered. Examples of the numerical computations are displayed in the following figures.

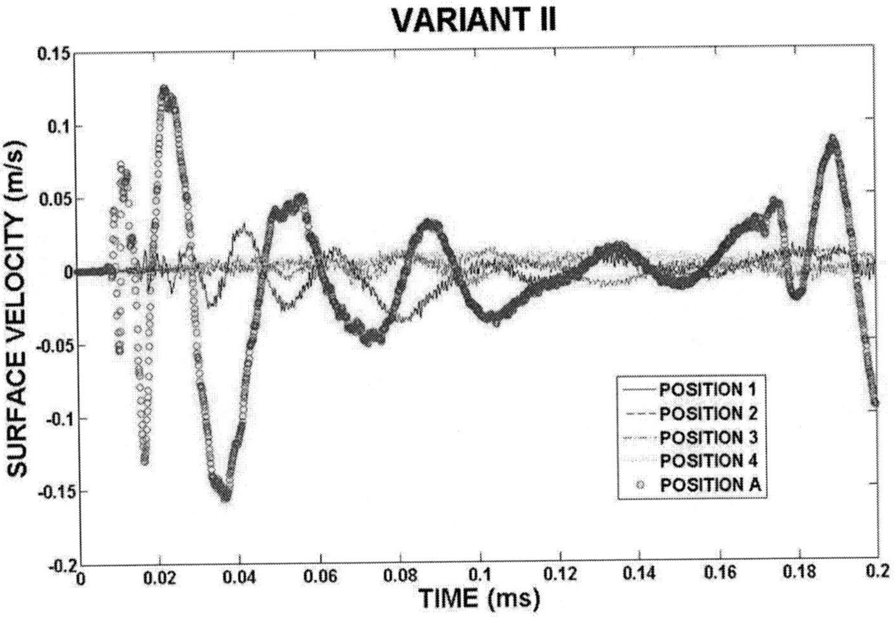

Figure 20a. Schematic of the surface velocity evaluation. Variant I. Egg is freely supported by the teflon ring.

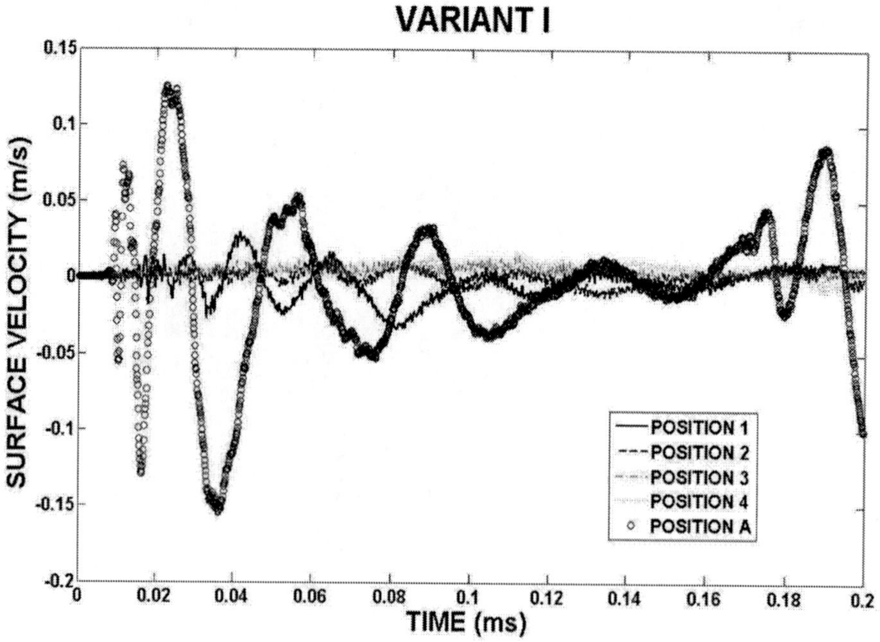

Figure 20b. Schematic of the surface velocity evaluation. Variant II. Egg is connected with the teflon ring.

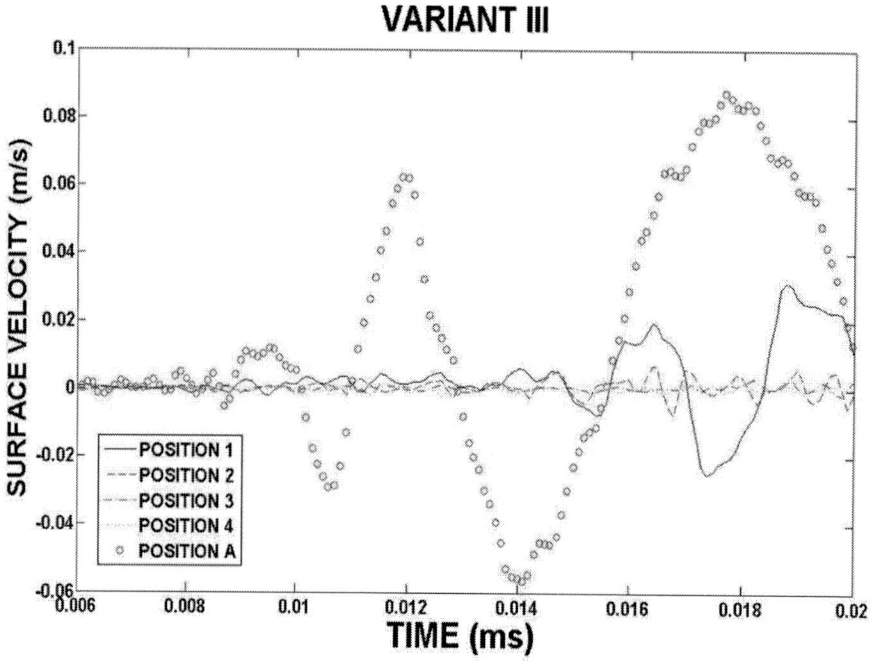

Figure 20c. Schematic of the surface velocity evaluation. Variant III. Egg is in the contact with a rigid plate.

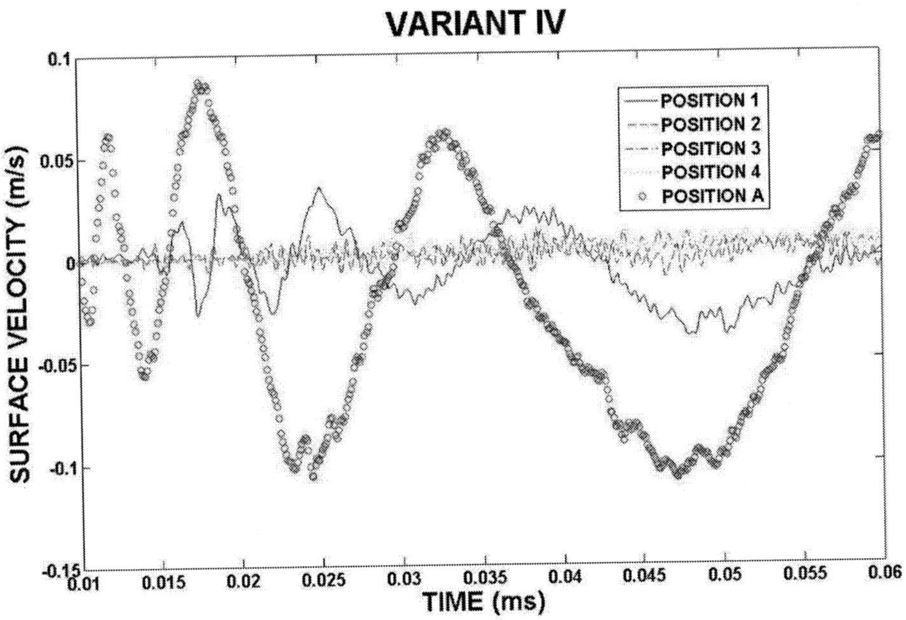

Figure 20d. Schematic of the surface velocity evaluation. Variant IV. No ring is considered.

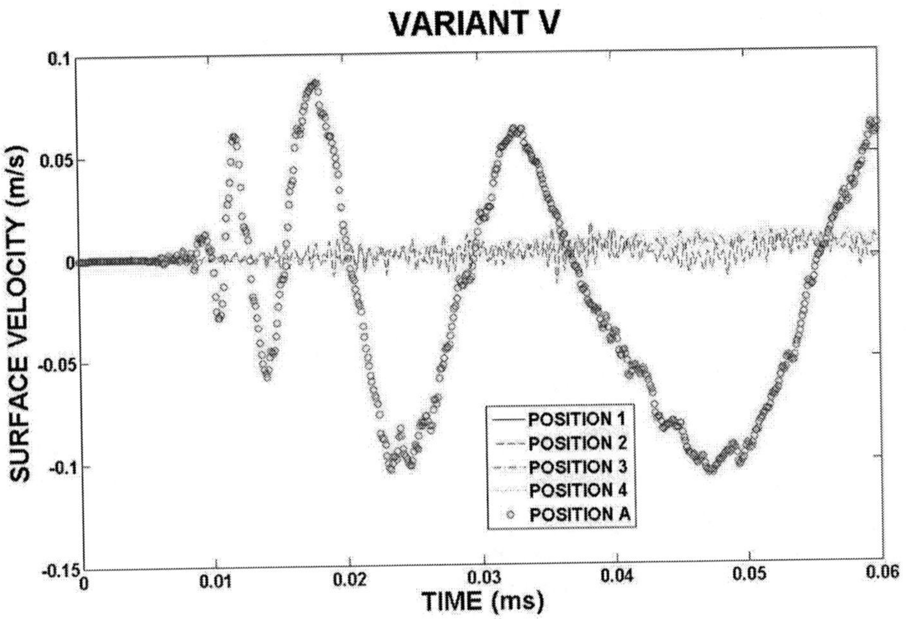

Figure 20e. Schematic of the surface velocity evaluation. Variant V. Egg is only supported at the sharp end.

It seems that the best agreement with the experimental results was obtained in the case of variant I. It is evident that the boundary value conditions play meaningful role. The process of the impact loading cannot be considered as a local event. It means it is necessary to use more exact description of the egg liquid's behavior. The reasonable agreement between numerical and experimental response functions approves using of the model developed in the given paper for solution of impact problems.

The results obtained during experiments on the non-destructive impact show that the acoustic methods can be considered as very effective tool for the study of the egg's behavior under different loading conditions. In the next step the new method of the dynamic strength of the eggshell has been developed.

3. Eggshell Strength Under Impact Loading

To study the egg resistance to impact, many methods have been developed - see e.g. Tyler and Geake (1963), Voisey and Hunt (1967, 1968). These methods use ball or rod, which are dropped on the eggshell and the height of fall, the size of ball or a number of blows are used to estimate shell strength. Within the work presented in this chapter, the new experimental method of the dynamic strength evaluation has been developed. Loading has been performed by the impact of the free falling rod. The record of the force at the point of rod-eggshell contact enables to evaluate the rupture force at a definite impact velocity. Obtained results have been compared with results of the static compression of the eggs in order to find possible evidence of the loading rate influence.

The numerical simulation of the impact experiments have been used for evaluation of stresses at the moment of the egg's break.

3.1. Experimental Details

Eggs (*Hisex Brown* strain) were collected from a commercial packing station. Typically, eggs were maximal 2 days old when they arrived at the grading station. The main characteristics of the eggs have been evaluated. These characteristics are given in the Table 1.

The experimental set-up is very similar to that shown in Figure 3 – see Figure 21. It also consists of three major components; they are the egg support, the loading device and the response-measuring device.

1) The egg support is a cube made of soft polyurethane foam. The stiffness of this foam is significantly lower than the eggshell stiffness; therefore there is very little influence of this foam on the dynamic behavior of the egg.
2) A bar of the circular cross-section with strain gauges (semi conducting, 3 mm in length) is used as a loading device. The bar is made from aluminum alloy. Its length is 200 mm, diameter is 6 mm. The bar is allowed to fall freely from a pre-selected height. The instrumentation of the bar by the strain gauges enables to record time history of the force at the area of bar-eggshell contact.
3) The response of the egg to the impact loading, described above, has been measured using the laser vibrometer. This device enables to obtain the time history of the eggshell surface displacement.

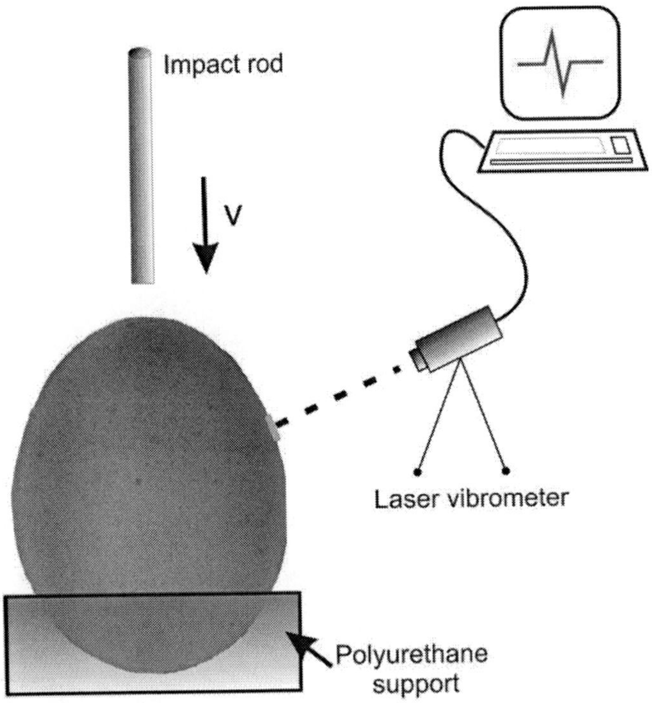

Figure 21. Schematic of the impact loading of the egg.

The eggs have been impacted on the sharp end, on the blunt end, and on the equator. The height of the bar fall has been increased up to value at which the eggshell damage has been observed. The displacement has been recorded on the equator of the egg. The displacement has been measured in normal direction to the eggshell surface.

The static compression has been performed using the test device (TIRATEST 27025, Germany) which has three main components: a stationary and a moving platform, and a data acquisition system. Compression force was measured by the data acquisition system. The egg was placed at the block of polyurethane foam positioned on the stationary plate - as it is shown in the Figure 21. The egg has been loaded by the moving rod (6 mm in diameter) at a speed of 20 mm/min. Two compression axes (*X* and *Z*) for the egg were used in order to determine the rupture force and deformation. The *X*-axis was the loading axis through the length dimension, and the *Z*-axis was the transverse axis containing the width dimension. Along the X - axis two other orientations have been considered. The eggs have been loaded at the sharp end and at the blunt end. For each orientation 30 eggs have been tested.

3.2. Experimental Results

3.2.1. Static Compression

In Figure 22 the examples of experimental records of the force vs. rod displacements for the different loading axes are shown.

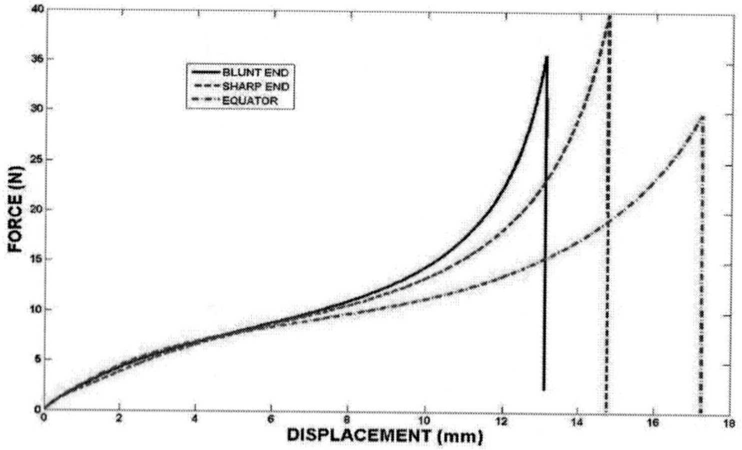

Figure 22. Experimental records of the force – displacements for the different loading orientation.

It can be seen that the shape of these curves is different from those obtained at the eggs compression between two plates - see Altuntas and Sekeroglu (2008). The observed dependences can be fitted by the polynomial:

$$F = a_1x^4 + a_2x^3 + a_3x^2 + a_4x + a_5 \tag{14}$$

where x is the rod displacement in millimeters. For all the experiments the correlation between this fit and experimental record is better than 0.98.

The egg stiffness at this type of loading can be expressed using the dF/dx function. Its course is shown in Figure 23. One can see that this stiffness exhibits more or less significant dependence on the rod displacement. The stiffness at the egg compression between two plates is nearly constant. The maximum of the force has been taken as the rupture force. This force can be considered as the strength of the eggshell.

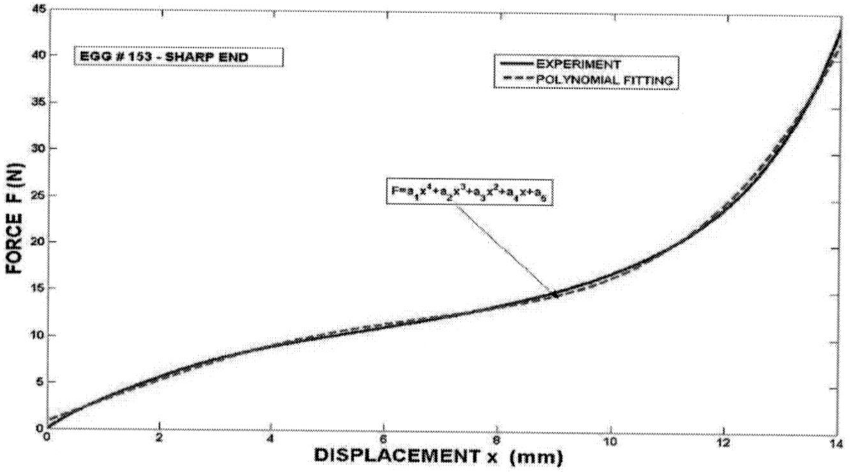

Figure 23. Stiffness of the egg loaded along the X – axis (sharp end).

These forces are given in the Tables 8 - 10. In these tables the values of the displacement at the fracture x_{max} and energy Q absorbed by the eggshell during the compression are also given. The values of Q are computed as

$$Q = \int_0^{x_{max}} F(x)dx \tag{15}$$

The distribution of these quantities is shown in the Figs. 24 – 26. The t-test of the given data revealed that there is a significant difference between values of the rupture function obtained for the egg loaded along X and Z axes. The rupture forces obtained for the egg loaded along the X-axis cannot be considered as different. The use of Pearson test led to the conclusion that the variability in the egg shape, eggshell thickness and, eggshell mass did not have any influence on the rupture force. The dependence of the rupture force on compression axes agree with results reported by Altuntas and Sekeroglu (2008). These authors also reported some other works supporting these results. The rupture force obtained at the compression test between two plates exhibited significant dependence on the egg shape.

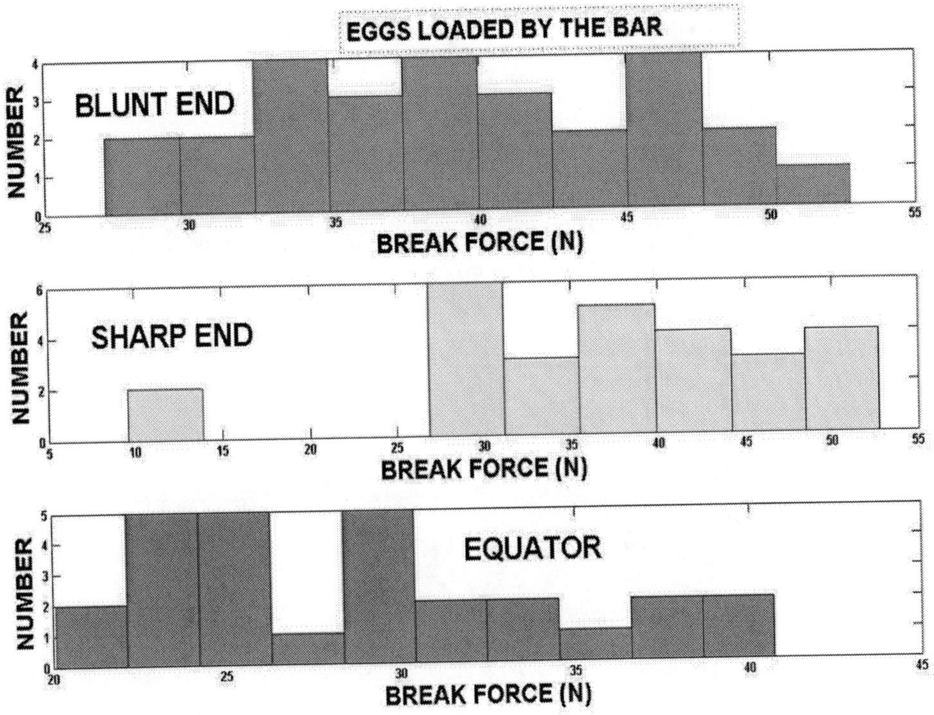

Figure 24. Distribution of the eggshell strength.

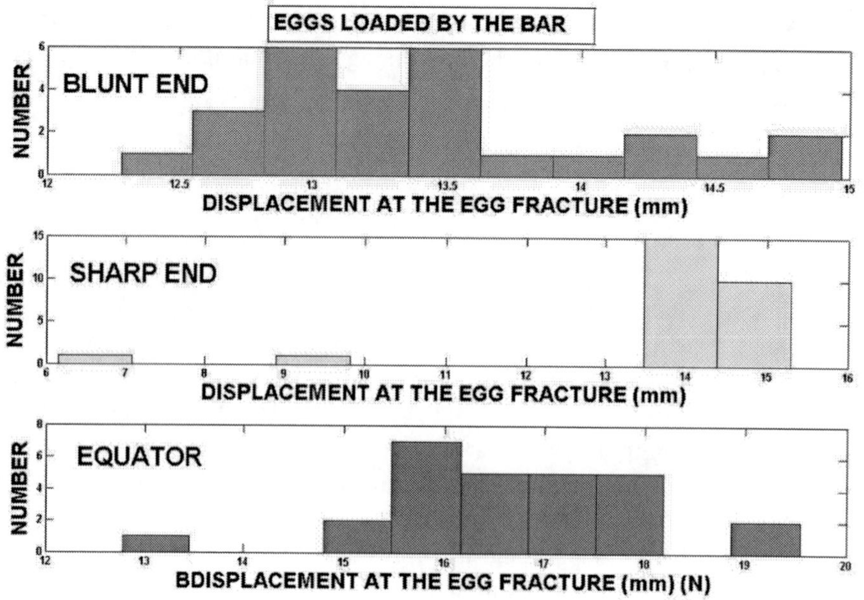

Figure 25. Distribution of the displacements at the eggshell break.

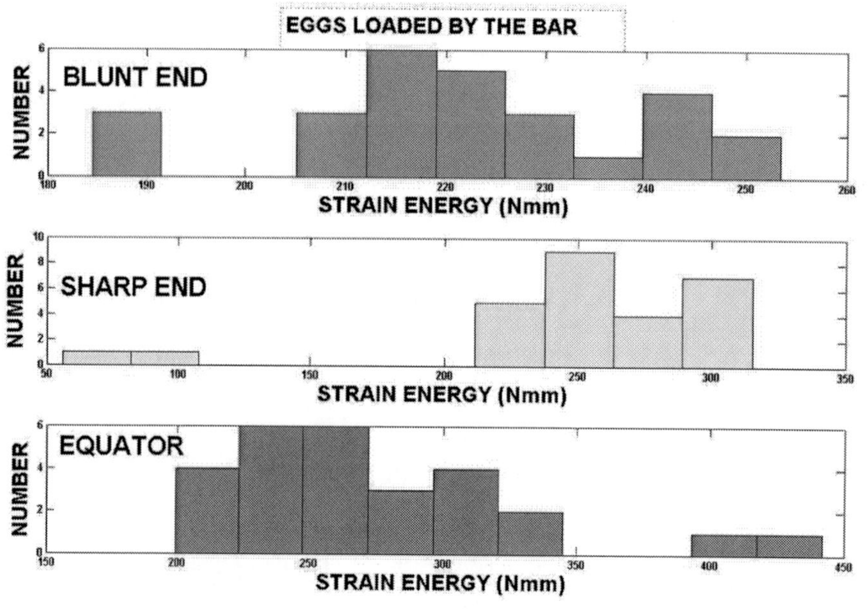

Figure 26. Distribution of the energy absorbed by the eggshell.

The independence reported in this chapter may be a consequence of the local character of loading. In order to verify these results, additional experiments on the eggs exhibiting higher variability in their shapes are needed. The obtained data serve namely for the comparison with data obtained at the impact loading.

Table 8. Static compression of the eggs – blunt end

Egg No.	Height (mm)	Width (mm)	Mass (g)	SI (%)	Strenght (N)	Displacement (mm)	Energy (Nmm)
151	57.6	44.4	64.0	77.08	52.16	17.23	500.88
152	55.8	44.0	61.2	78.85	30.45	14.04	262.95
153	56.7	42.5	57.5	74.96	46.72	14.21	311.13
154	60.4	46.1	72.0	76.32	36.70	14.08	274.46
155	57.4	43.5	61.7	75.78	30.70	13.51	236.54
156	60.9	46.0	72.3	75.53	42.31	14.96	298.73
157	58.2	44.4	65.2	76.29	49.30	15.01	315.39
159	58.0	44.5	64.1	76.72	33.98	13.79	247.56
160	59.3	43.9	64.6	74.03	29.93	13.59	232.49
161	56.8	44.8	65.6	78.87	26.90	13.57	222.56
162	57.2	43.2	61.0	75.52	36.52	14.20	261.18
163	61.7	44.2	67.7	71.64	9.57	6.16	56.16
164	58.9	46.0	71.2	78.10	35.96	15.24	279.92
165	62.0	45.3	72.3	73.06	30.43	13.63	240.79
166	57.2	44.4	63.0	77.62	42.63	14.75	278.04
167	56.0	41.9	56.1	74.82	41.42	14.03	254.23
168	56.4	43.9	61.7	77.84	33.30	13.66	240.53
169	63.5	45.2	74.5	71.18	51.21	14.40	295.92
170	59.3	44.9	67.3	75.72	41.91	14.55	271.73
171	56.8	43.7	61.1	76.94	52.76	14.53	291.33
172	59.3	45.0	68.2	75.89	39.75	14.75	262.67
173	58.4	43.8	62.8	75.00	34.54	13.66	234.50
174	54.4	42.4	55.5	77.94	36.34	14.10	244.45
175	57.3	44.1	62.9	76.96	48.85	15.30	291.56
176	60.4	46.4	72.7	76.82	30.07	14.04	226.62
177	58.0	42.8	58.9	73.79	45.69	14.15	254.95
178	57.7	44.5	64.8	77.12	12.07	9.16	97.28
253	58.4	44.4	64.8	76.03	46.72	14.45	297.18
263	59.0	43.1	62.1	73.05	49.39	14.89	288.85
268	63.2	45.6	71.2	72.15	35.80	13.36	225.35
Average	58.5	44.3	64.9	75.72	37.80	13.90	259.86
std	2.10	1.10	4.95	1.98	10.18	1.84	68.79

Table 9. Static compression of the eggs – sharp end

Egg No	Height (mm)	Width (mm)	Mass (g)	SI (%)	Strenght (N)	Displacement (mm)	Energy (Nmm)
181	62.4	44.6	69.4	71.47	41.06	14.56	213.827
183	59.8	45.9	70.1	76.76	47.60	14.96	212.758
186	57.2	44.9	65.2	78.50	49.80	14.18	184.571
187	62.7	46.3	75.1	73.84	37.46	12.74	242.296
188	57.6	44.4	65.2	77.08	39.55	13.44	223.207
189	59.6	43.8	64.5	73.49	35.71	13.12	245.775
190	55.4	44.1	61.9	79.60	52.74	14.97	253.508
191	57.8	44.4	63.9	76.82	42.70	13.22	213.827
192	57.3	43.9	62.1	76.61	41.01	13.55	212.758
193	56.2	42.9	58.1	76.33	33.80	13.05	184.571
194	59.2	44.2	65.2	74.66	43.33	13.07	211.524
195	59.4	43.9	65.9	73.91	46.07	13.24	229.463
196	56.4	45.6	66.2	80.85	49.17	14.10	232.312
197	55.8	43.0	58.8	77.06	32.61	13.53	236.303
198	59.7	44.9	67.6	75.21	39.28	12.91	208.455
199	61.4	46.4	74.5	75.57	40.56	13.42	231.407
200	59.2	43.7	64.5	73.82	39.66	12.86	216.893
201	59.7	44.6	66.3	74.71	33.39	12.85	213.827
202	57.6	45.0	66.1	78.13	27.12	12.71	212.758
203	57.7	43.8	62.2	75.91	36.54	12.28	184.571
204	58.1	44.2	63.3	76.08	46.54	12.60	211.524
206	58.0	44.3	65.2	76.38	45.66	13.68	242.296
207	57.6	42.9	60.6	74.48	29.44	13.19	223.207
208	58.2	45.2	67.7	77.66	36.63	14.66	245.775
209	56.9	45.6	66.7	80.14	30.07	14.37	253.508
244	59.4	46.1	71.4	77.61	33.21	13.08	220.627
249	57.7	45.3	66.8	78.51	31.71	13.55	223.140
254	60.3	45.2	69.7	74.96	42.40	13.55	221.423
259	57.4	43.0	60.8	74.91	35.42	13.47	231.337
264	57.4	44.6	64.5	77.70	35.10	13.27	219.914
Average	58.4	44.6	65.7	76.29	39.2	13.5	221.912
std	1.7	1.0	3.9	2.05	6.32	0.67	17.580

Table 10. Static compression of the eggs – equator

Egg No	Height (mm)	Width (mm)	Mass (g)	SI (%)	Strenght (N)	Displacement (mm)	Energy (Nmm)
245	57.5	44.9	65.3	78.09	28.16	16.91	259.548
255	59.4	45.7	69.2	76.94	25.33	16.15	248.346
260	56.3	42.9	58.7	76.20	33.62	17.90	302.310
269	61.4	47.8	78.4	77.85	38.11	12.78	341.391
271	57.7	43.7	61.6	75.74	31.86	19.54	441.994
272	59.1	44.8	67.1	75.80	38.97	19.16	396.437
274	58.8	44.2	64.8	75.17	30.25	17.59	306.032
275	59.8	45.3	69.2	75.75	29.82	17.24	298.157
276	59.2	45.6	69.4	77.03	31.73	17.26	293.975
277	59.0	46.3	71.2	78.47	33.33	17.99	312.254
278	56.0	45.0	63.8	80.36	28.79	17.06	267.554
281	57.8	42.7	59.4	73.88	24.76	16.46	249.745
282	60.7	44.4	67.8	73.15	23.10	16.48	249.323
284	55.6	43.3	58.6	77.88	23.84	16.14	225.622
286	61.6	44.9	69.1	72.89	22.61	15.99	236.632
287	56.7	45.5	67.1	80.25	28.38	15.76	221.703
289	60.6	47.4	75.4	78.22	34.54	16.98	267.447
291	58.7	45.4	68.3	77.34	40.70	17.51	292.045
292	59.4	45.4	69.7	76.43	24.36	18.11	325.213
293	56.7	44.7	63.0	78.84	27.19	16.04	225.622
294	58.7	46.6	70.8	79.39	26.25	17.55	284.695
295	58.7	46.4	71.4	79.05	22.94	16.32	241.275
296	57.9	44.8	66.0	77.37	29.91	15.70	218.574
297	58.1	44.8	66.8	77.11	20.85	16.70	263.377
298	56.1	43.7	60.3	77.90	20.09	15.21	199.260
299	61.1	46.6	73.2	76.27	25.55	14.91	203.383
300	59.5	45.6	68.0	76.64	22.88	15.79	233.454
325	62.8	45.9	75.3	73.09	37.93	16.20	234.750
326	55.2	44.4	61.8	80.43	39.53	18.21	314.900
329	63.1	46.5	76.4	73.69	29.80	18.17	337.841
Average	58.8	45.2	67.6	76.91	29.17	16.79	276.429
std	2.0	1.2	5.1	2.06	5.67	1.29	53.773

3.2.2. Eggs Loaded by the Rod Impact

The main geometric characteristics of the eggs which have been tested by the procedure shown in the Figure 21 are given in the Table 11. Eggs have been tested by the rod impact in the three directions shown in the Figure 27.

Table 11. Geometry of the tested eggs (W – width, L – length, m – mass)

Egg No	L (mm)	W(mm)	m (g)	SI (%)
331	58.70	43.80	63.60	74.62
332	58.00	44.40	65.20	76.55
333	58.10	44.40	66.50	76.42
334	58.80	43.70	63.90	74.32
335	60.40	44.20	66.50	73.18
336	58.50	42.40	60.50	72.48
337	59.60	43.30	64.00	72.65
338	60.50	44.90	67.90	74.21
339	63.70	44.70	71.50	70.17
340	57.20	45.00	63.60	78.67
341	61.30	45.50	71.40	74.23
342	57.20	45.70	66.90	79.90
343	58.40	45.90	70.00	78.60
344	58.00	44.30	63.70	76.38
345	61.00	45.40	70.50	74.43
346	57.90	45.20	65.10	78.07
347	58.40	43.80	62.40	75.00
348	58.20	45.40	67.80	78.01
349	61.40	46.50	74.60	75.73
350	60.40	44.80	67.70	74.17
351	56.30	44.30	61.60	78.69
352	60.00	44.40	66.00	74.00
353	59.30	43.70	62.80	73.69
354	58.80	43.50	63.60	73.98
355	58.60	43.40	62.20	74.06
356	58.00	44.90	65.80	77.41
357	58.40	45.10	65.70	77.23
358	56.70	45.00	64.20	79.37
359	58.80	43.80	63.90	74.49
360	58.30	43.10	60.50	73.93

In Figure 28 an example of the experimental record of the forces at the point of contact between bar and egg is shown. It has been found that the shape of the force-time function reflects the eggshell damage. If the eggshell is not damaged the shape of this function is nearly „half-sine“. The origin of the eggshell damage is connected with an abrupt in this dependence. The dependence of the force maximum on the height of the fall of the rod is shown in Figure 29. The time history of the eggshell surface displacement is shown in Figure 30. The origin of the damage leads to significant increase in peak value of this displacement. The same qualitative features, both force and surface displacement, have been observed for the remaining eggs.

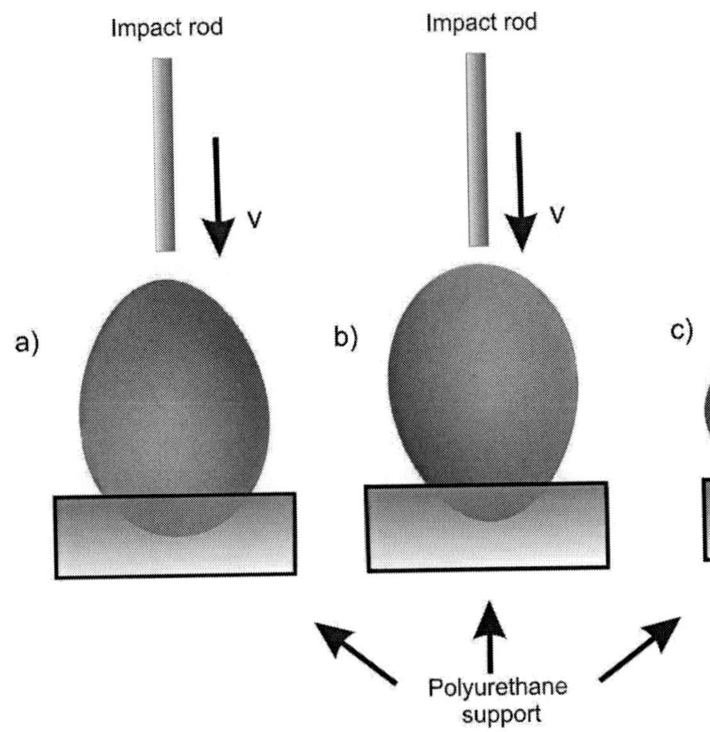

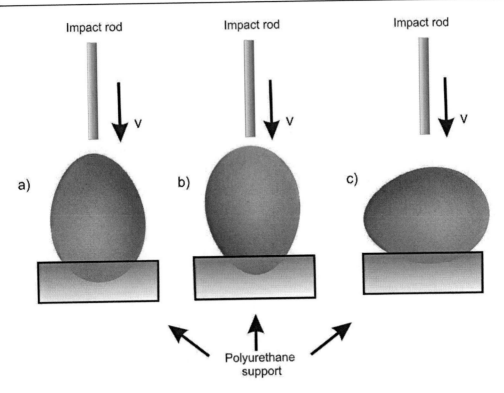

Figure 27. Schematic of the eggs loading.

The force corresponding to the eggshell strength has been chosen as the average between the force at the intersection of two lines (see Figure 29) and the highest value of the force at which no eggshell damage has been detected. The exact evaluation of the eggshell strength should be determined using the continuous increasing height of the rod fall.

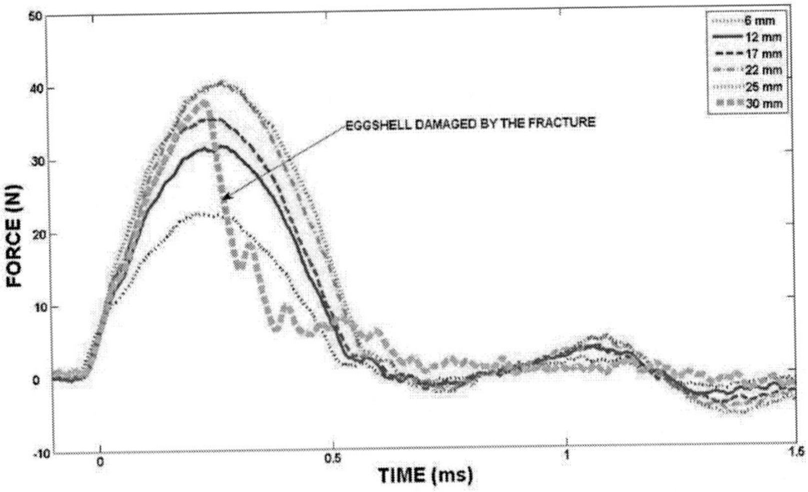

Figure 28. Experimental records of the time history of the force at the rod impact. Egg No. 331 – Rod impacted the blunt end of the egg. The displacement of the egg surface has been detected on the

meridian at the distance of 36 mm from the impacted end. The different values of heights h are given in the upper right corner.

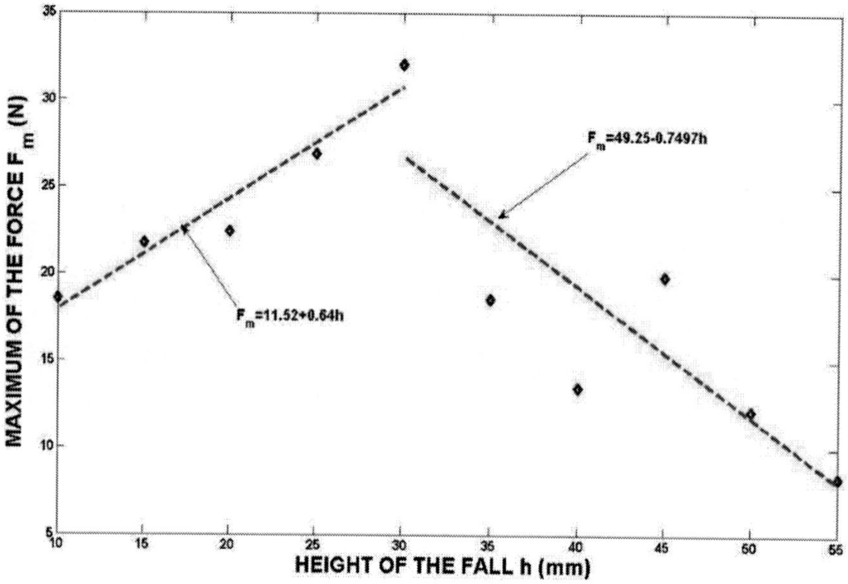

Figure 29. The dependence of the maximum of the force on the height of the rod fall.

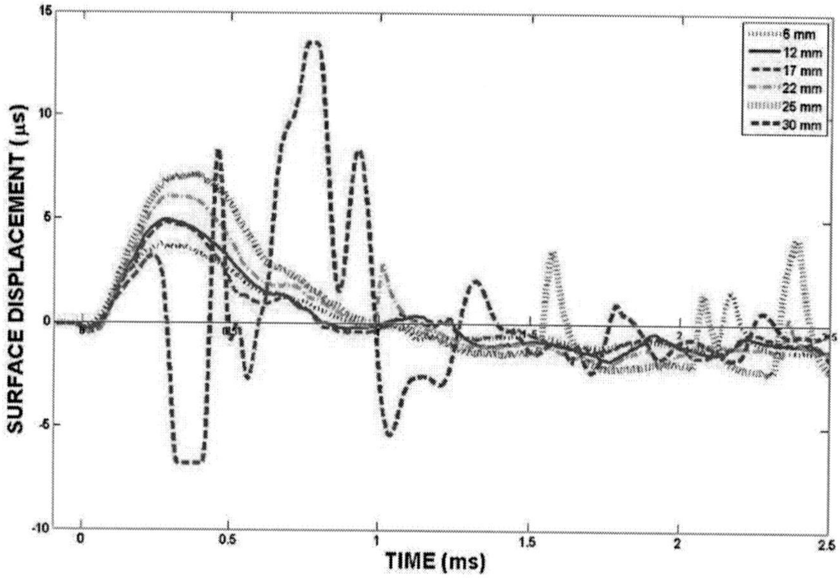

Figure 30. Surface displacement – time curves. Egg No. 331.

Experimental results can be also evaluated in the frequency domain using the Fast Fourier Transform – see chapter 2. The example of the amplitude of the spectral function of the force is shown in the Figure 31.

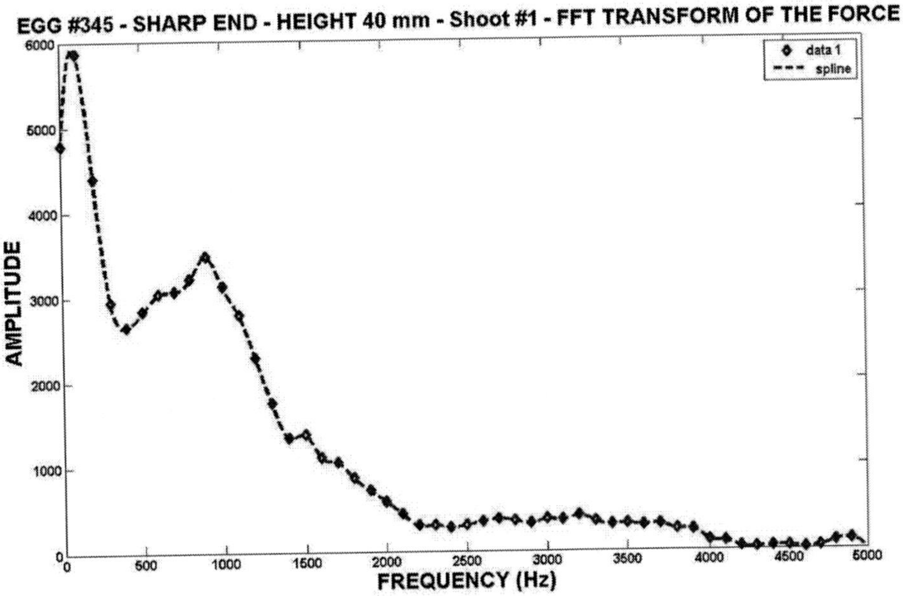

Figure 31. Frequency dependence of the spectral function.

Example of the spectral function for the displacement is shown in the Figure 32.

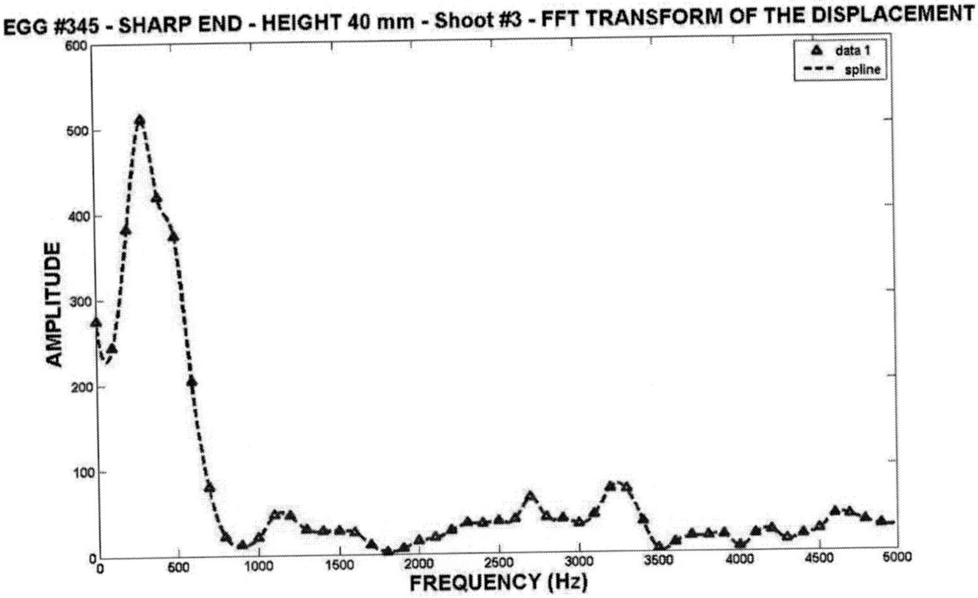

Figure 32. Frequency dependence of the spectral function.

The knowledge of the spectral functions both, force and amplitude, enables designing of the transfer function. Example of this function is given in Figure 33.

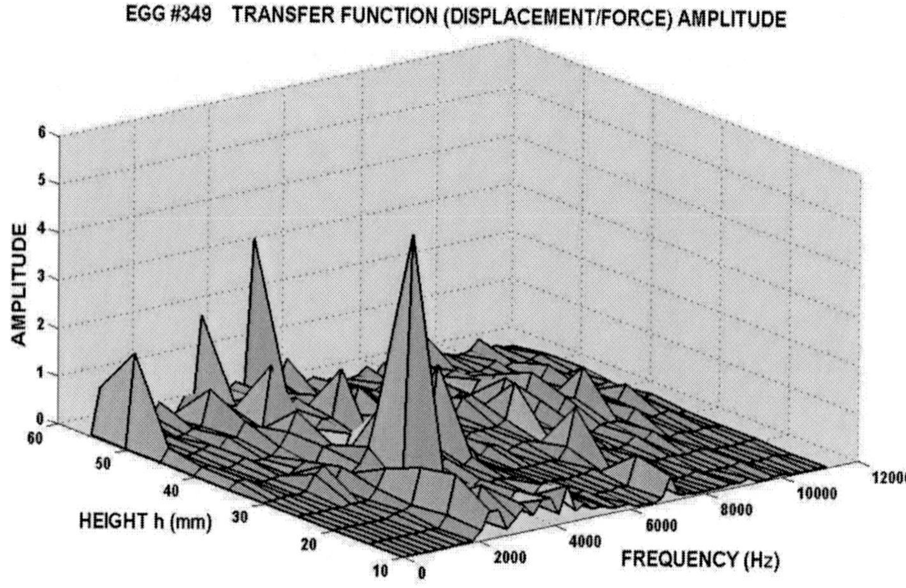

Figure 33. Amplitude of the transfer function.

There are some frequencies which are characterized by an amplification of this amplitude. The frequency dependence of the transfer amplitude for the eggs without any damage is presented in the Figure 34 there.

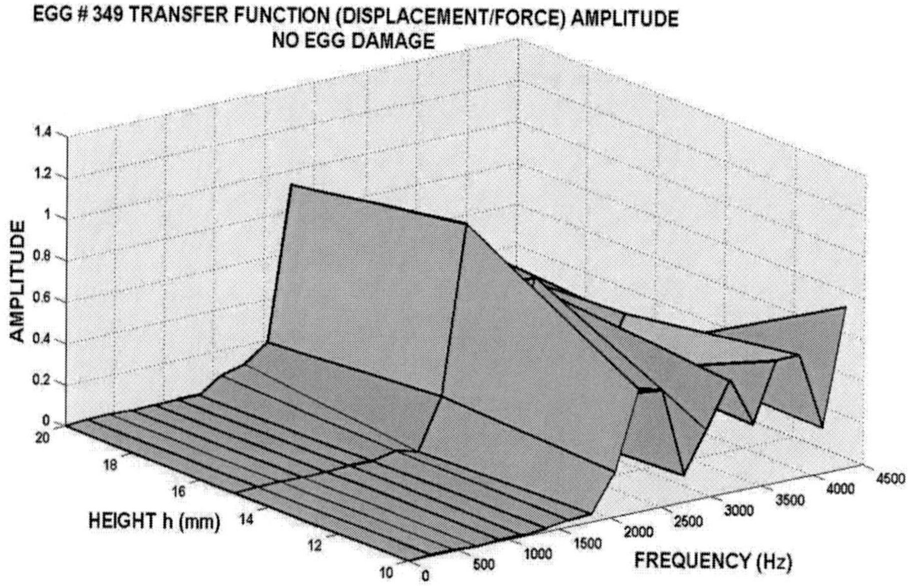

Figure 34. Amplitude of the transfer function.

The influence of the egg damage is shown in the Figure 35.

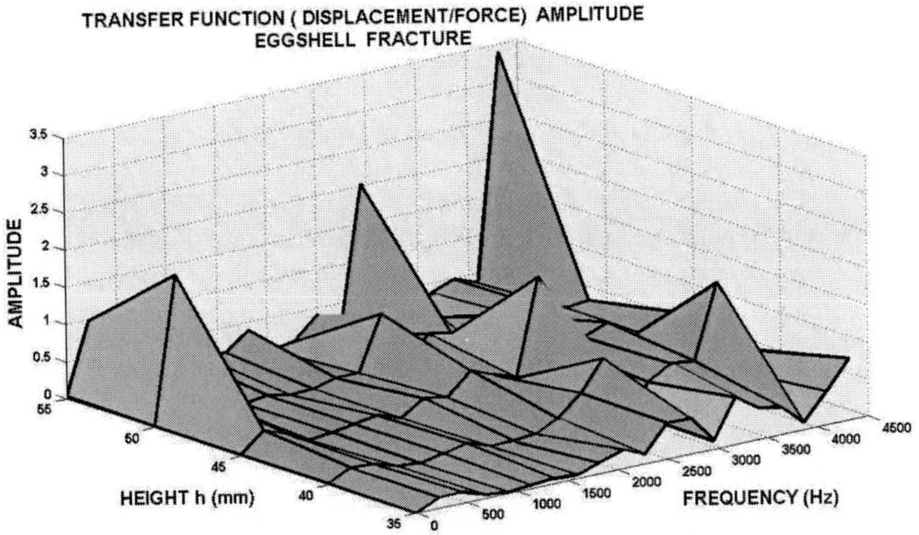

Figure 35. Amplitude of the transfer function.

Mean values of the transfer function amplitude are displayed in Figure 36.

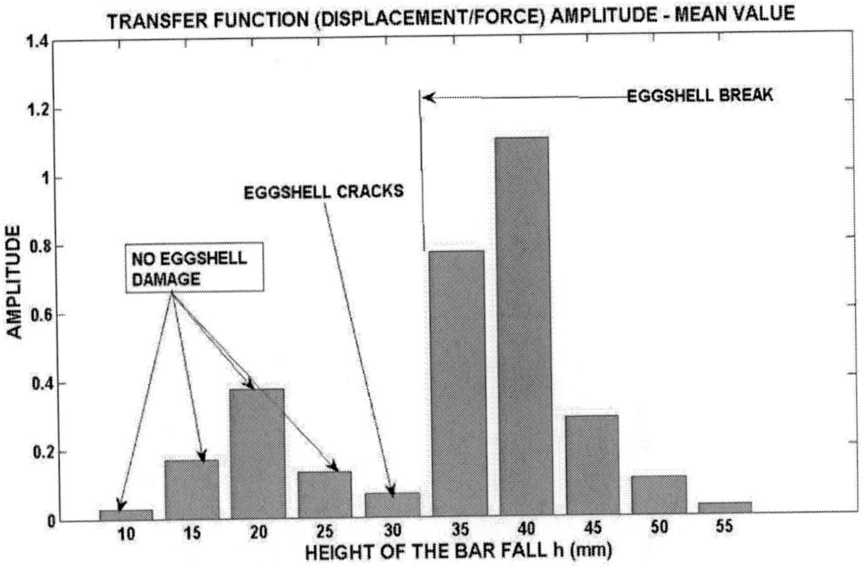

Figure 36. Mean values of the transfer function amplitude.

One of the objections to this method can consists in the possible initiation of the micro-defects during subsequent loading of the eggshell. Such micro-cracks can seriously affect the strength. In order to study this problem, an experiment with repeating loading of the egg has been performed. The height of the rod fall has been constant (h = 22 mm). Ten repeated impacts have been performed. The eggshell has not exhibited any visible damage. The shape of the force - time curve is typical for non-destructive impact. The maximum values of the force are presented in Figure 37. It can be seen that these values are very close to the strength of the egg. There is an exception at the 6th and 7th impacts. This effect may be a consequence of certain changes of the impact conditions (lower height). In this figure the higher amplitudes of the impact force are also plotted.

The values of the eggshell strength are shown in Figure 38. Ten eggs have been used for the evaluation of the dynamic strength for all three impact orientations (sharp end, blunt end, and equator). The basic statistics of the given data is given in the Table 12 and Table 13. The maximum of the dynamic strength can be observed for the impact of the rod on the sharp end. The minimum value of strength can be found for the rod impact on the equator.

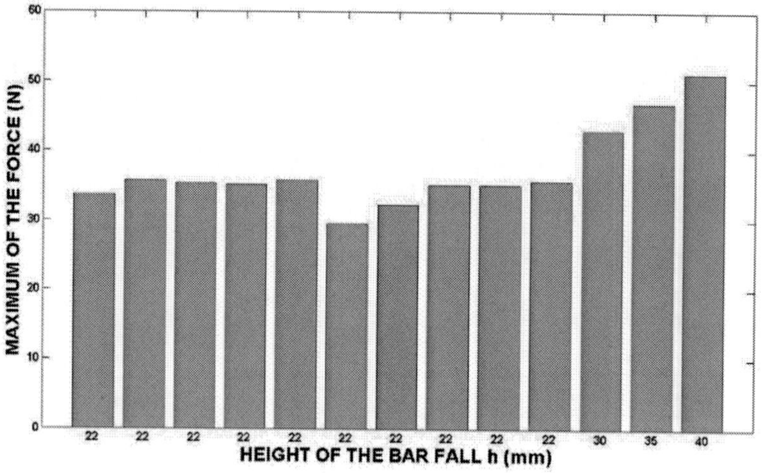

Figure 37. Maximum of the force during rod impact. Egg No. 339. Height of the fall h = 22 mm. Rod impacted the blunt end of the egg.

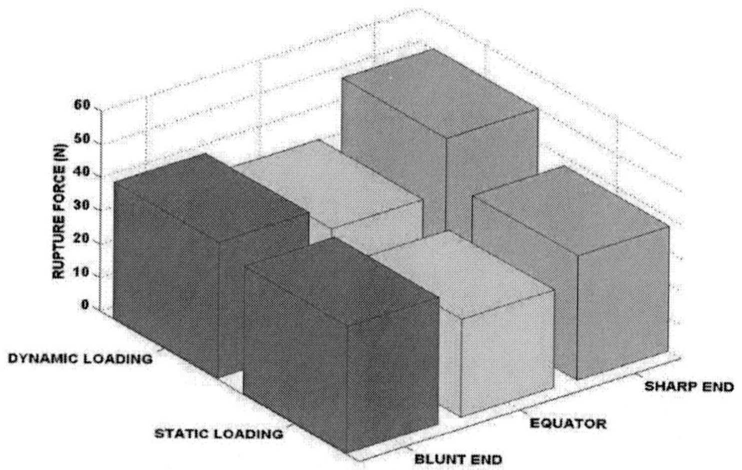

Figure 38. Dynamic and static strength of the eggshells.

These results qualitatively agree with the results obtained at the static loading. Contrary to the results of the static loading tests, there is a significant difference between the rupture force at the sharp and blunt end. The results also suggest that dynamic rupture force is higher than that obtained at the static loading. This can be taken as an evidence of the influence of the loading rate. This phenomenon has been reported in some papers - see e.g. Altuntas and Sekeroglu (2008). In order to obtain a more detailed insight into this problem some other experiments are needed. It is particularly necessary to obtain results for different speeds of the rod at the static loading.

Table 12. Average values of the static rupture force F_r

Point of loading	Rupture force F_r (N)	Standard deviation (N)
Blunt end	38.2	± 7.9
Equator	29.3	± 6.2
Sharp end	37.4	± 10.2

Table 13. Average values of the dynamic rupture force F_r

Point of the rod impact	Rupture force F_r (N)	Standard deviation (N)
Blunt end	40.9	± 3.0
Equator	33.9	± 3.6
Sharp end	50.1	± 5.2

As it has been mentioned the values of the eggshell strength in terms of the rupture force are affected by many factors (egg specific gravity, egg mass, egg volume, egg surface area, egg thickness, shell weight, shape and shell percentage). In order to avoid the simultaneous influence of these factors the numerical simulation of these experiments should be performed. Some preliminary results in this field are involved in the following paragraph.

3.2.3. Numerical Simulation of the Rod Impact

For the numerical simulation of the experiments described in the previous paragraph, the LS DYNA 3D finite element code has been used, as in the case of simulation of the ball impact. The same model of the egg has been used.

The numerical model of the experiment is shown in Figure 39. The following problems have been solved:

1) Impact of the rod on the blunt end of the egg (Egg 331 has been used)
2) Impact on the egg equator (Egg 349)
3) Impact on the sharp end of the egg (Egg 347)

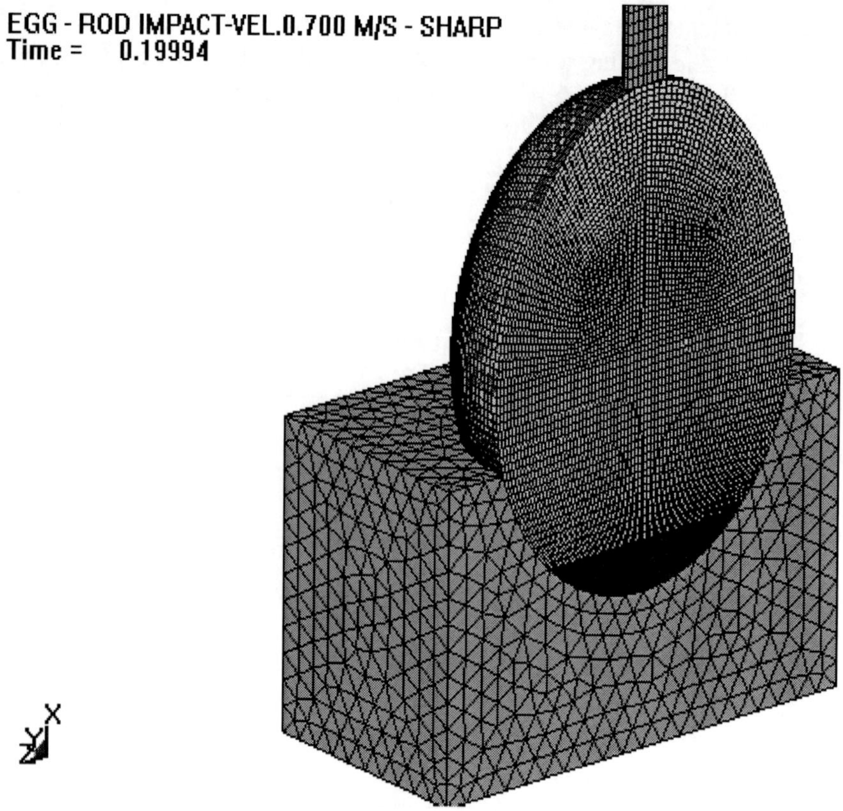

Figure 39. Finite element model of the solved problem.

Table 14. Main characteristics of the eggs used in the numerical simulation of the impact loading

Egg No.	Mass m (g)	Shape index (%)	E (GPa)	Eggshell thickness (mm)
331	63.6	74.62	73	0.230
347	61.4	75.73	47	0.235
349	62.4	75.00	47	0.240
339	71.5	70.17	45	0.235

The main characteristics of the used eggs are given in the Table 14. The height of the rod fall has been kept at 25 mm. The impact velocity of the rod is than 0.7 m/s. In order to verify the validity of the model of the egg, the time histories of the forces and surface displacements have been evaluated. These data can be compared with experimental ones. In Figs. 40 and 41, the numerical and experimental records obtained for the egg No. 331 are presented. Results for the remaining problems exhibit the same qualitative features. The computed peak values of the force agree well with experimental ones. There are some differences in the time course of both functions. Owing to some problems with the force record (finite length of the strain gauges etc.) the observed discrepancy seems be acceptable. Very reasonable agreement has been exhibited between computed and experimentally recorded time histories of the surface displacements.

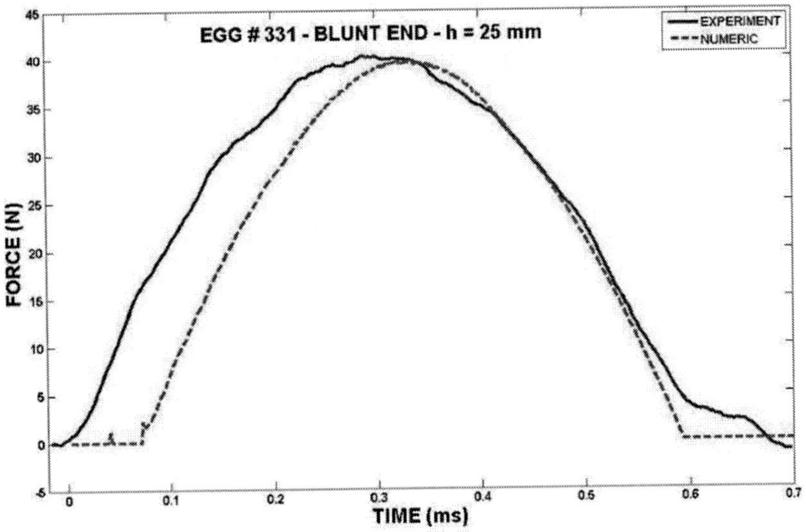

Figure 40. Experimental and numerical time histories of the force at the contact between the rod and the eggshell.

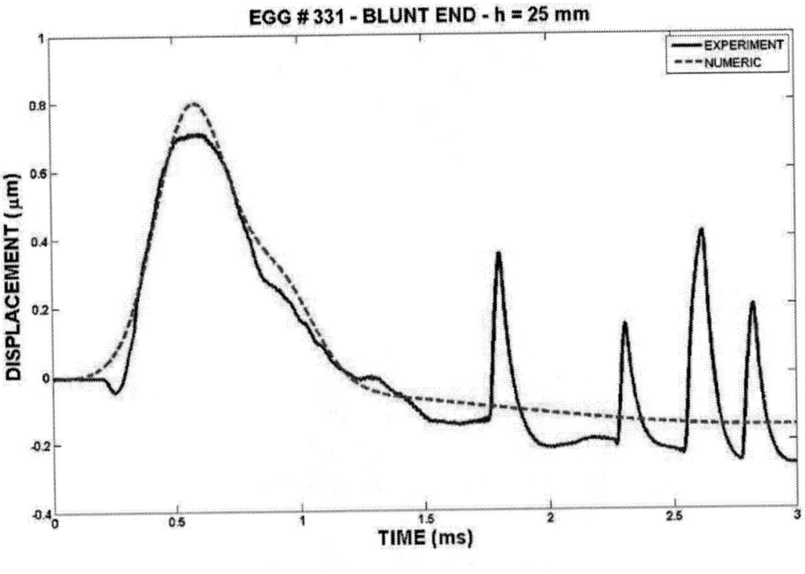

Figure 41. The dependence of the surface displacement on time. Comparison between numerical and experimental results.

The numerical results did not respect some experimentally observed peaks. These peaks are probably a consequence of possible transient phenomena in the recording system. Their occurrence has been detected only for a limited number of experiments. If we take into account some assumptions, namely the assumption on the egg liquids behavior, the agreement between numerical simulation and experiment seems to be more than satisfactory. Owing to this fact the next results of the numerical computations can be accepted as relatively reliable. The development of the stress state after the rod impact is documented in the Figure 42a-c.

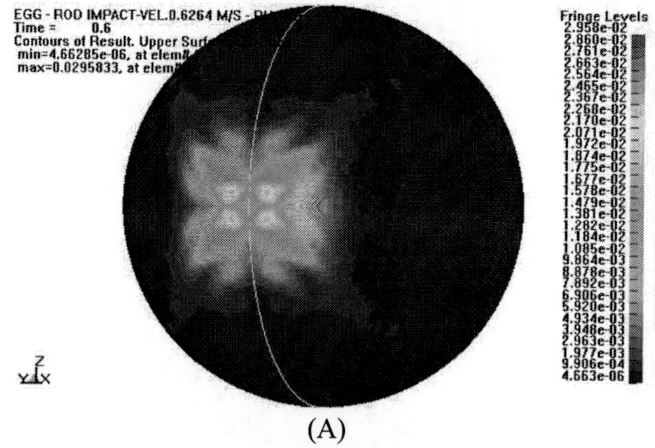

(A)

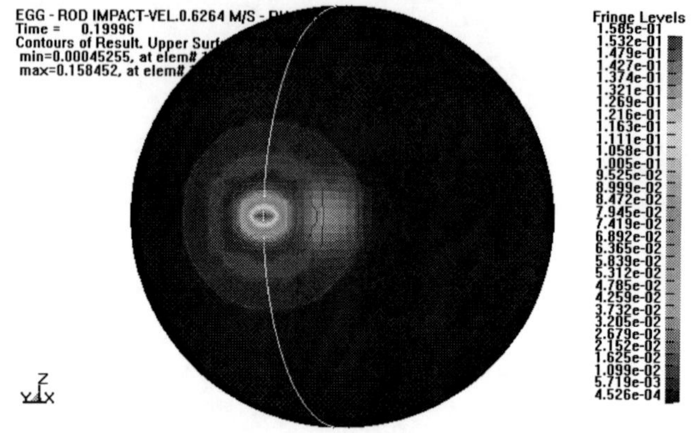

(B)

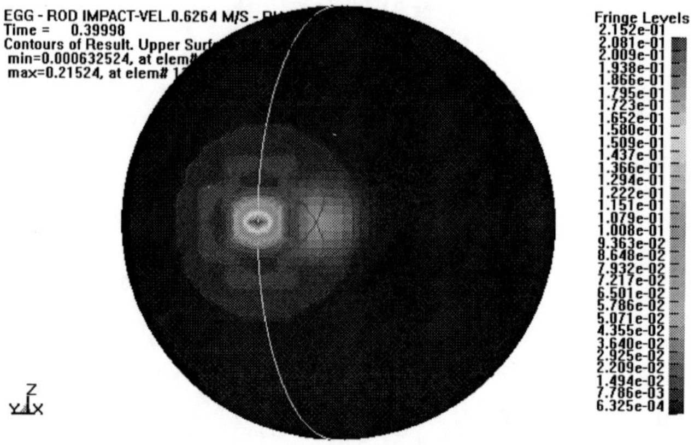

(C)

Figure 42 a-c. The development of the equivalent stress in the eggshell. Egg No. 331 - blunt end. a) time = 0.19996 ms; b) time = 0.39998 ms; c) time = 0.6 ms.

In these figures the time development of the equivalent stress is displayed. It can be seen that the stress is localized on a relatively small area around point of the contact between rod and eggshell. In the Figure 43 the computed displacements at the point of impact are shown.

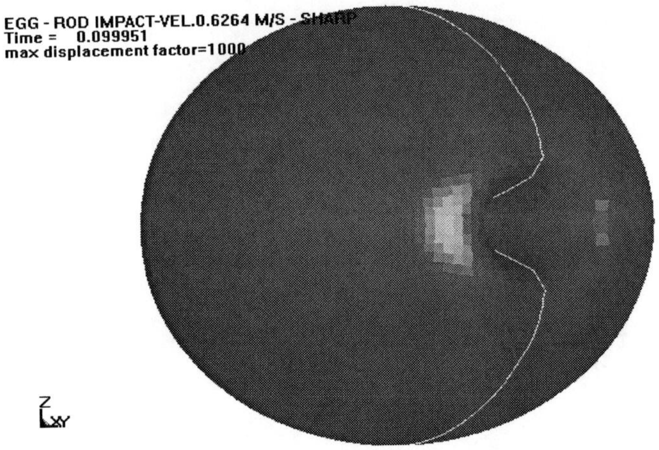

Figure 43. Computed displacement. Impact on the sharp end. Impact velocity: 0.6264 m/s.

In the next step the stress development in the eggshells has been modeled. This computation has been performed for the impact conditions at which the eggshell rupture has been observed (impact velocity = 0.741 m/s). The distribution of stresses through the eggshell thickness is the most appropriately examined using the solid element model. Directly under the rod impact on the inner surface of the shell, an equi-biaxial stress distribution develops in which the hoop and meridional stress are equal. The development of the stress on the outer and inner surfaces of the eggshell is shown in Figs. 44a, b. More detailed analysis revealed that the compressive stress decreases with the distance from the point of the rod impact and it changes to the tensile stress. The stress at the inner surface is only tensile. Very similar features of the stress distribution in the eggshells have been also reported for the numerical simulation of the quasi-static compressive loading of the eggs - see MacLeod et al. (2006). If we take the maximum of the tensile stress as a measure of the eggshell strength, we obtain results given in the Table 14. It seems that these stresses are independent on the position of the point of the rod impact.

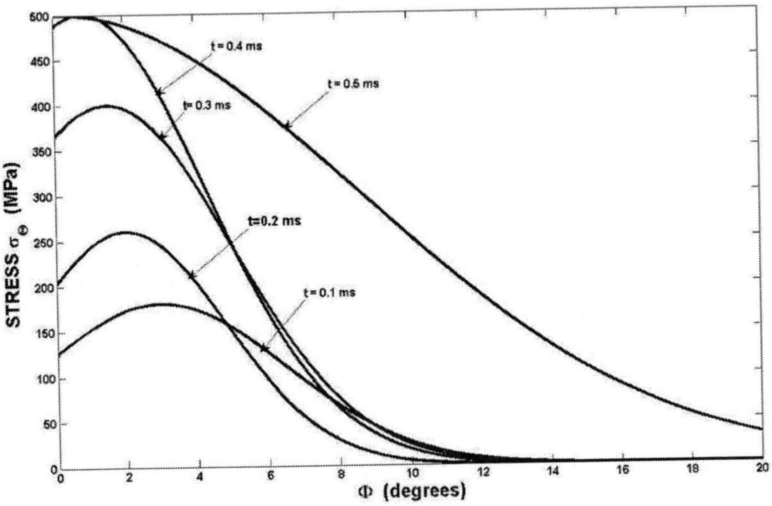

Figure 44a. Time development of the stress at the inner surface of the eggshell.

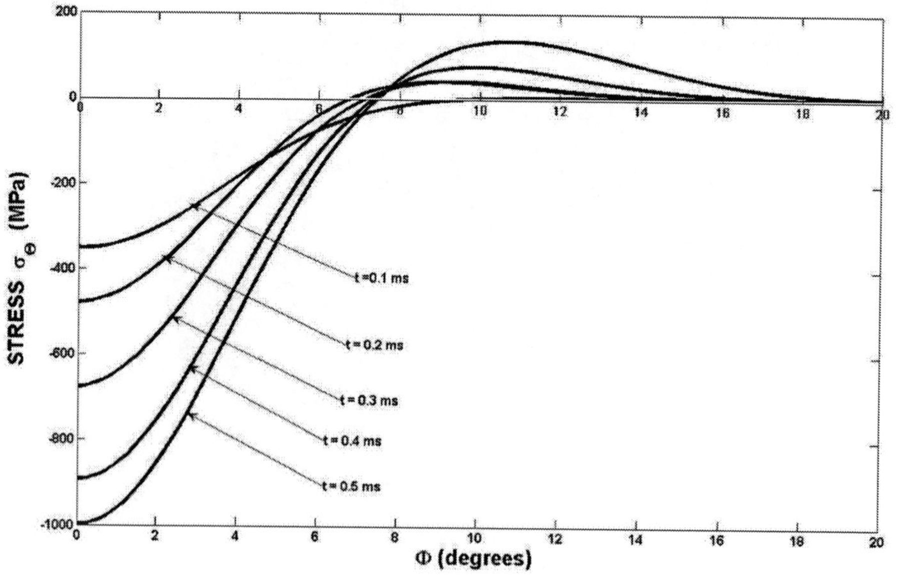

Figure 44b. Time development of the stress at the inner surface of the eggshell.

Although there is still a lack of experimental and namely the numerical results, the obtained results suggest that the value of the maximum of the tensile stress at which the eggshell rupture starts can be independent of many factors affecting the value of the experimentally found rupture force. This stress can be considered as an intrinsic strength property of the eggshell material. In order to verify this hypothesis many other experiments are needed.

4. Hen's Eggs Impact Against the Elastic Obstacle

Breakage occurs whenever the stress at some point in the shell exceeds the ultimate strength of shell material at given point. When attempting to predict when breakage may occur two questions arise:

a) What is the maximum external force upon the shell?
b) What is the maximum consequential stress within the shell?

These problems must be solved for all kinds of eggs loading. When breakage occurs under dynamic conditions question *(a)* is concerned with ballistics and the elasticity of shell material. The ballistic aspects are complicated by two factors: the shape of the egg and the nature of its contents. At most points on the shell of an egg-shaped body a line drawn normal to the surface does not pass through the centre of mass, so any impact results in rotation as well as translatory movement. Because the egg contents include yolk and thick white floating in thin white (but moored to the poles by chalazae), rotation about the long axis may involve little more than the shell, whereas rotation about a transverse axis must involve the greater part of the egg. The distribution of the energy acquired at impact into three components corresponding with translation, rotation about the long axis and rotation about a transverse axis cannot be predicted accurately without further information about the properties of the egg contents, but it seems likely that in all situations where shell breakage may occur the energy of rotation about the long axis will be small compared with the sum of the other two components. It is therefore disregarded in what follows. Conversely, for rotation about a transverse axis the egg contents are treated as though they were solidly anchored to the shell. Given these assumptions and the further assumption that eggs behave as elastic bodies, their change of motion at impact can be predicted from their motions before impact and their masses and moments of inertia.

Question *(b)* relates the external force, *P,* on the shell to the location and magnitude of the maximum stress, *Stm,* within it. This problem has been solved namely for the eggs compressed between two flat plates at quasistatic loading conditions. The shell theory of engineers, as developed for shallow spherical domes (Reissner, 1946) and for flat plates loaded by a concentrated force (Timoshenko and Woinowsky-Krieger, 1959), has been applied to the case of an egg shell loaded at its poles. The parameters of the theoretical model shell are its thickness, *T,* and radius, *R,* the radius, *r,* of the area over which it is loaded and Young's Modulus, *E,* Poisson's Ratio, ν, and the ultimate tensile strength, *Stu,* of the material of which the shell is made. With this model the stress is expected to be maximal at the inner surface of the shell below the centre of the loaded area and the expected values of *Pm* can be calculated in terms of *T, R, r, E,* ν and *Stu.* However, the values of *Pm* observed experimentally were greater, by a factor of 20, than predicted values based on the best available estimates of *E,* ν and *Stu* (Voisey and Hunt, 1967a). Later work in which eggs were compressed between flat plates (Tung et al., 1969) likewise led to estimates of failure stress that were consistent but considerably higher than those obtained previously by methods that measure membrane stresses (Hammerle and Mohsenin, 1967; Sluka et al., 1967). These findings call into question the suitability of the model. In one respect, at least, it is clearly unsuitable; it does not take into account the existence, discovered subsequently, of an inner layer of shell that is weak in tension (Carter, 1970, 1971). A quite different model, in which the shell is treated as a set of incompressible radial prisms held together by elastic material, leads to the conclusion that the stress at the inner surface of an area of shell flattened by an external load is independent of the load: increasing *P* merely increases the contact area, in such a way that *r/P* remains constant. Data of Brooks and Hale (1960) v show that this is the relationship between *r* and *P.* Since experience nevertheless teaches that the probability of shell fracture is related to *P,* this model implies that the maximal stress, and therefore the stress that initiates fracture, must occur not within the flattened area but at some point outside it. The photographs of Voisey and Hunt (1967a), who compressed eggs that had been covered with a strain detecting brittle coating, point to the same conclusion; the cracks that were formed first - the continuous "main roads" that are joined but not crossed by the "side roads" - do not run through the contact area but run outwards from points near it. This implies that they were formed as a result of circumferential tensile stresses at some distance from the contact area. In that case the relevant shell thickness is *Te,* the thickness that is effective in respect of tensile strength. The difference, ε, between *T* and *Te* depends on the strain and identity of the hen; strain mean values range from about 85 to 130 (Carter, 1970c, 1971). Time is another parameter that may have to be taken into account, since the external force that will just crack an egg shell has been shown experimentally by Voisey and Hunt (1969) to depend on compression speed, at least if the speed is below about 10 mm/s. As these workers also showed that the stiffness of an egg shell is nearly constant at such speeds, *Stu* must be a function of compression speed, at least for speeds up to about 10 mm/s. It is not clear whether or not *Stu* depends on compression speed at the much higher speeds, usually in excess of 200 mm/s, at which shell cracking may occur at impact. The possibility of such dependence must be borne in mind, though if the energy-absorbing capability of egg shell is limited, as the same workers suggested, the rate of change of *Stu* with impact speed may be low. A consistent feature of egg shell strength is its high variability under apparently similar experimental conditions. It is due in part to the fact that egg shell is brittle material, in the sense in which engineers use the term. However, this is not the whole explanation, since an appreciable part of the residual variation in egg shell strength, after overall shell thickness and curvatures have been taken into account, is usually associated with the identity of the hen. This suggests that between-hen variation in *Stu* or ε may play an important role.

The theories mentioned above have been used by Carter (1976) for the solution of many problems of the hen's egg impact. The theory predicts the maximum of the force exerted on an egg at impact. As an example, let us consider the impact of an egg against an obstacle with the mass and the stiffness which are much smaller than those of the egg. This example represents e.g. the impact of the egg against an elastic bar. The maximum force P_m is given by:

$$P_m = \left(\frac{1}{2} M v_o^2 \right)^{\frac{1}{2}} S^{\frac{1}{2}}, \qquad (16)$$

where M is the mass of the egg, v_o is its impact velocity and S is the eggshell stiffness. Although this theory exhibited a reasonable agreement with experimental data its use is limited only to some very simple impact conditions. The most effective way how to describe the impact problems consists in the use of the numerical simulation. This procedure has been very successful for the analysis of the quasi static compression of the eggs (MacLeod et al., 2006) and also for the explanation of some phenomena connected with the dynamic loading (Nedomová et al., 2009). The verification of the results of the numerical simulation must be based on the reliable experiments. These experiments are described in the following section.

4.1. Experimental Arrangement

The research of egg impact loading was performed with use of specially developed testing method – see schematic in Figure 45. The egg falls from selected height on the round section bar (diameter of 50 mm) made of PMMA. The egg is (during its fall) guided in the hollow cylinder, in order to prevent its swinging. The egg falls either on its sharp or blunt end. The impacted bar is deformed purely elastically. Thus determination of the force in the contact point between egg and bar is possible. The force is quantified by use of strain gauge attached to the bar.

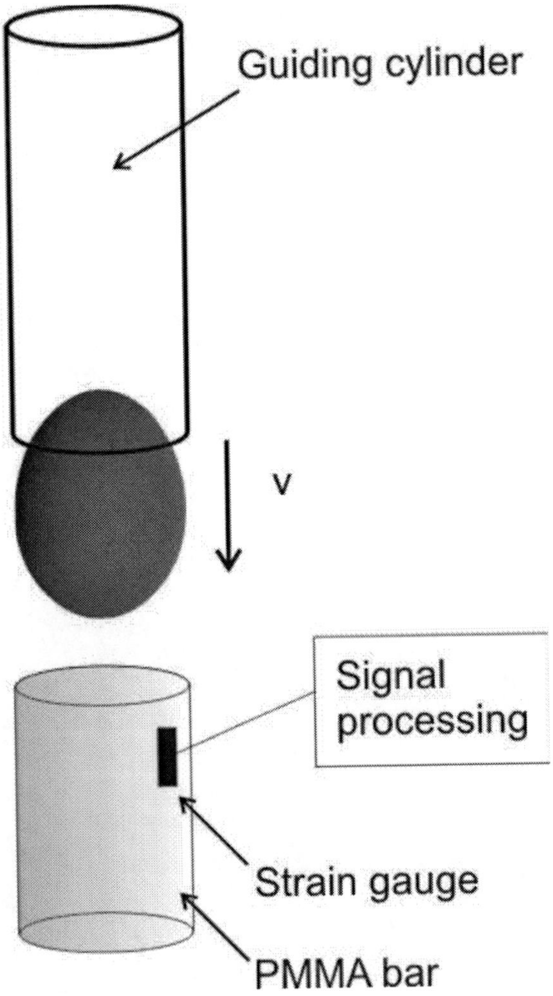

Figure 45. Schematic of the experimental arrangement.

4.2. Experimental Results

The hen's eggs were dropped from different heights ranging from 100 to 780 mm. The time dependencies of the force acting in the impact point were determined. In order to evaluate the influence of egg liquids on the impact behavior, the raw as well as boiled eggs were examined. The eggs were kept in boiling water for 10 minutes. The data were evaluated and processed in time as well as frequency domain. Regarding the response and sensitivity of force sensing, only those experiments were relevant, where the eggs were dropped from 105 (and more) mm. In this case, the breakage of the eggshell always occurred. The character of the breakage is shown on Figs 46a – d.

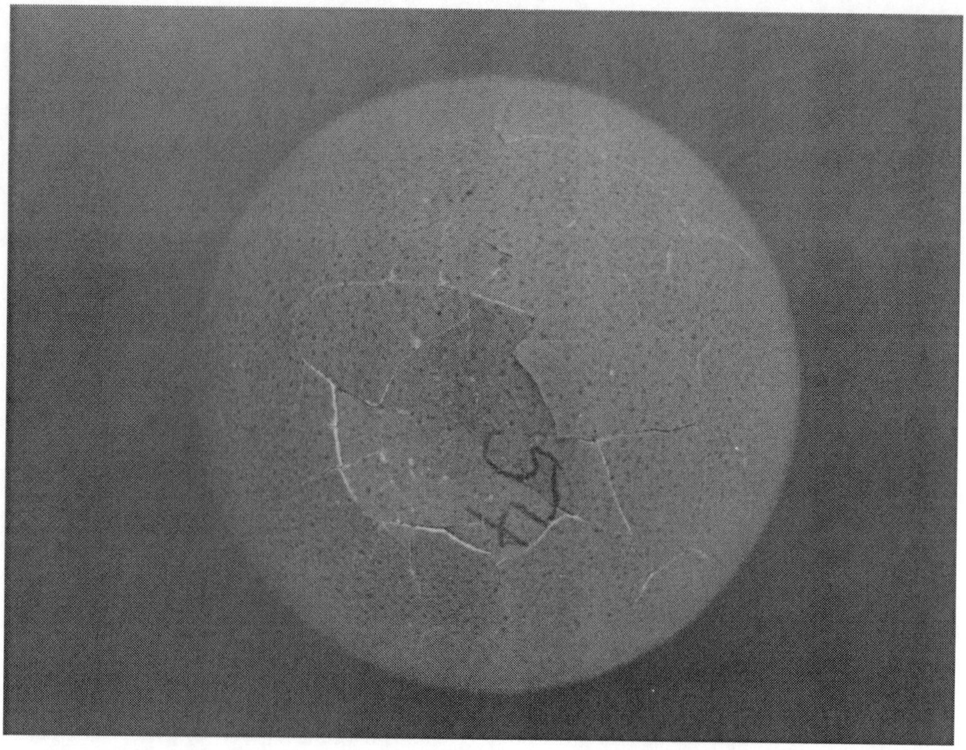

Figure 46a. Boiled eg, fall height 200 mm – blunt end.

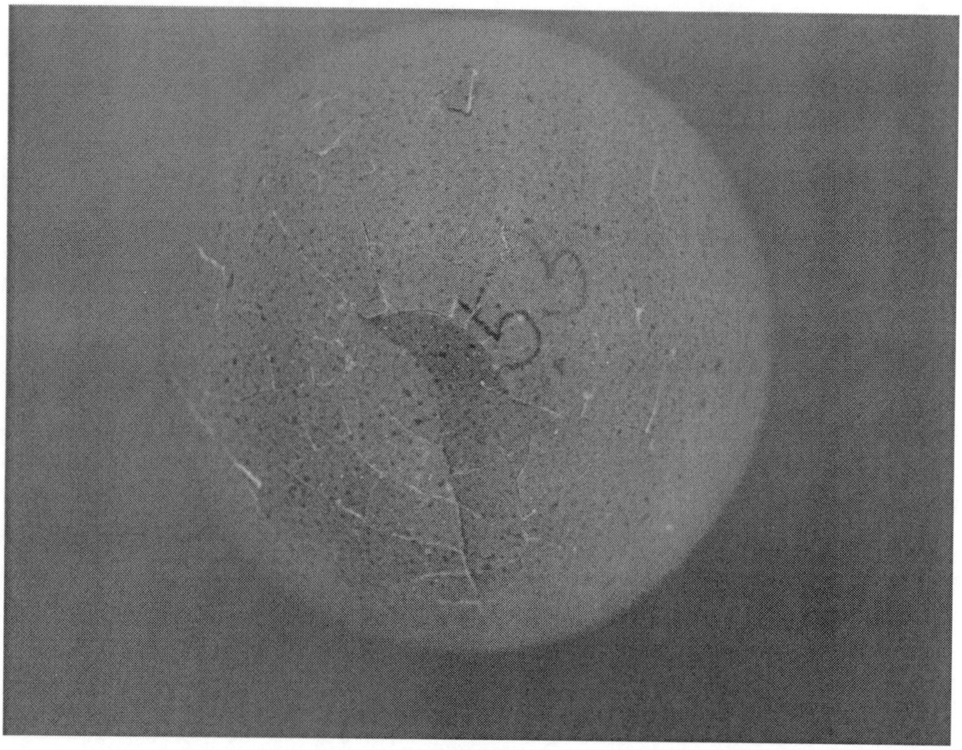

Figure 46b. Boiled egg, fall height 300 mm – blunt end.

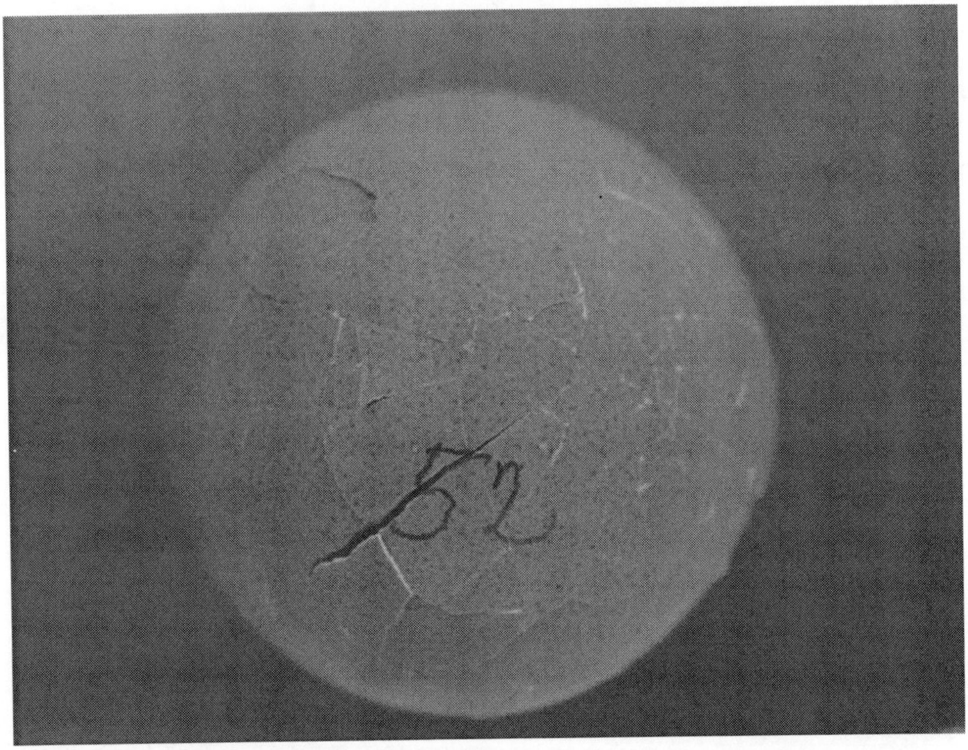

Figure 46c. Boiled egg, fall height 400 mm – blunt end.

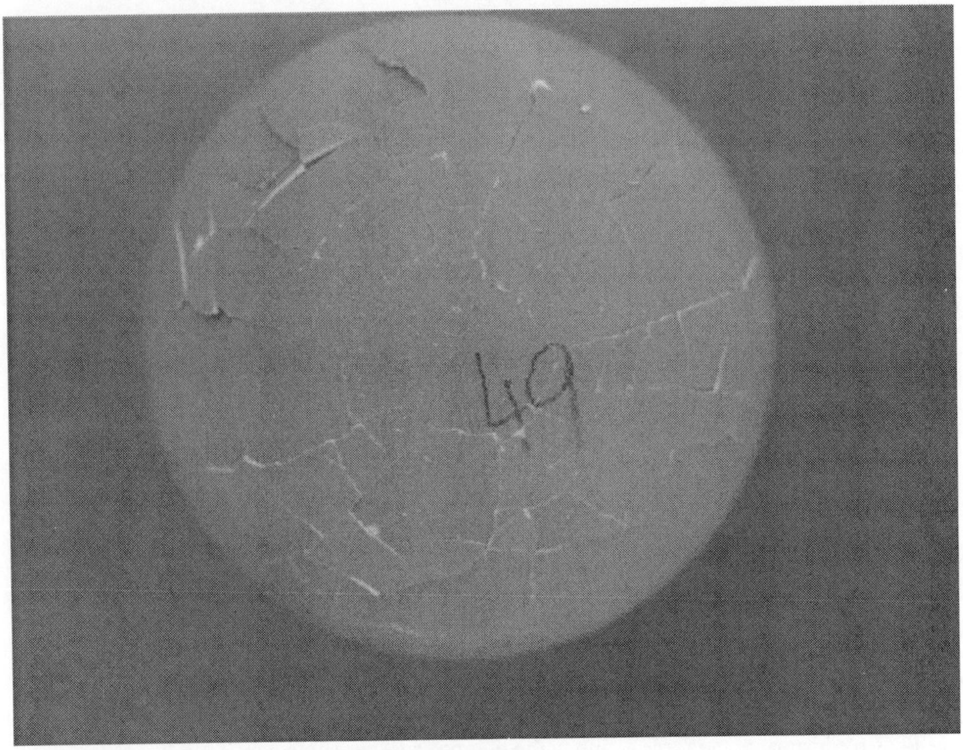

Figure 46d. Boiled egg, fall height 500 mm – blunt end.

4.2.1. Raw Eggs

Time histories of the force acting in the point of egg-bar contact are presented in the Figs 47 – 52. The different fall heights (or impact velocities) are considered.

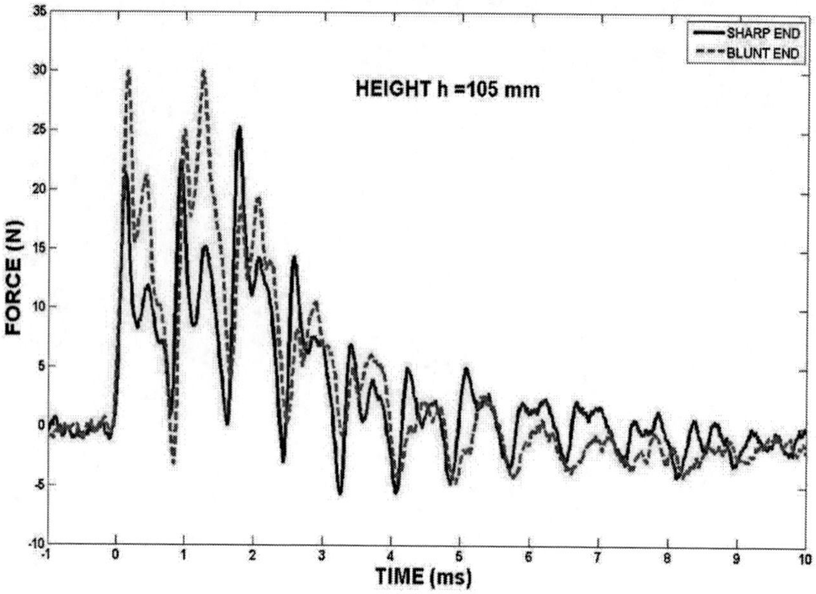

Figure 47. Time history of the force acting in the point of egg-bar contact.

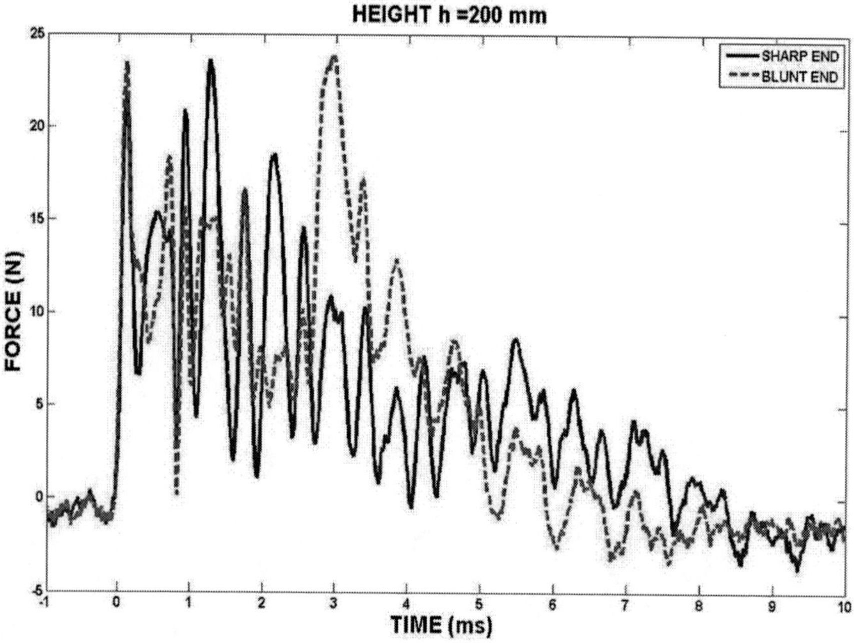

Figure 48. Time history of the force acting in the point of egg-bar contact.

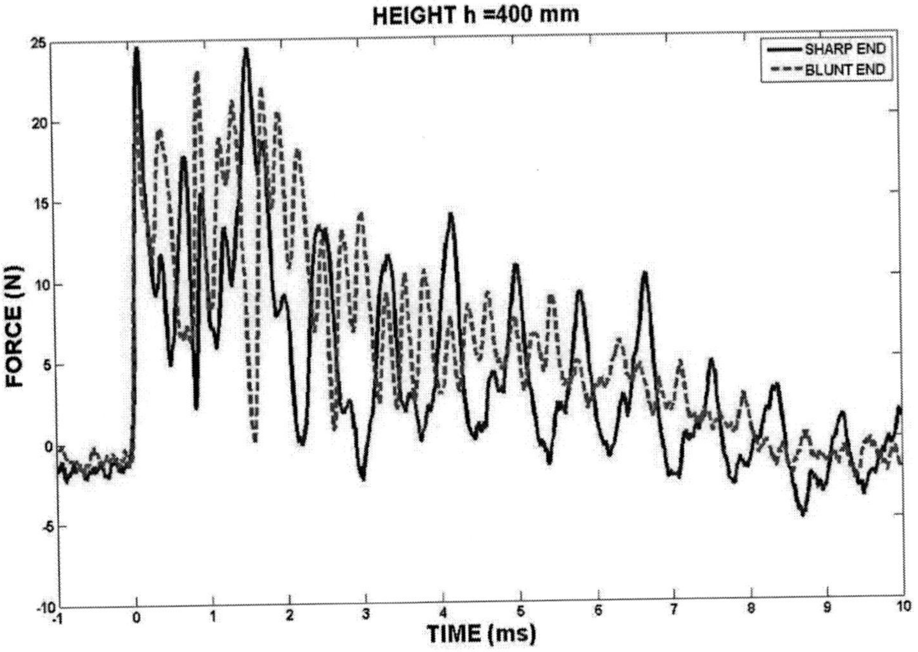

Figure 49. Time history of the force acting in the point of egg-bar contact.

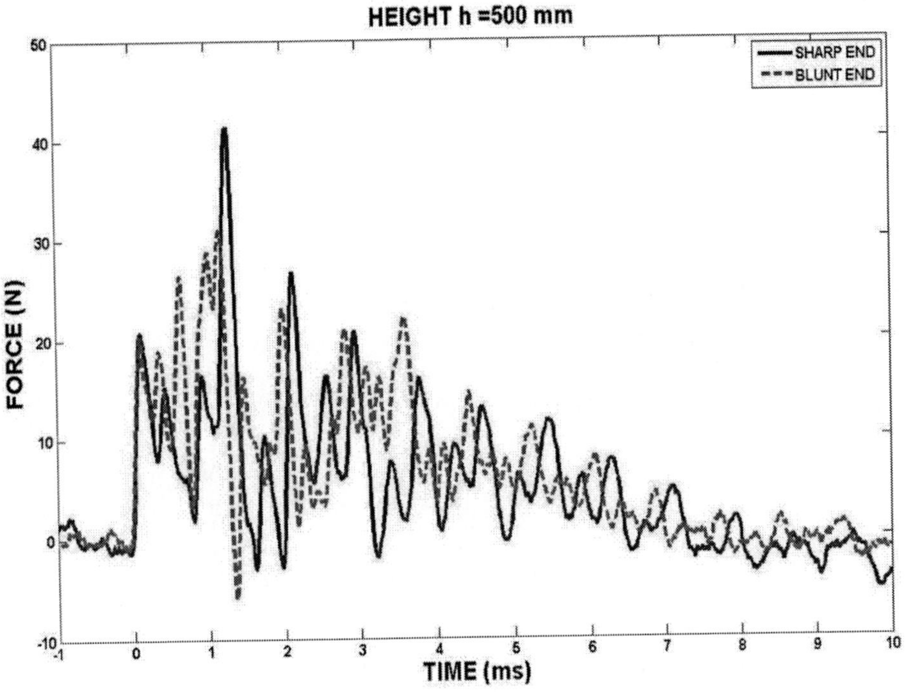

Figure 50. Time history of the force acting in the point of egg-bar contact.

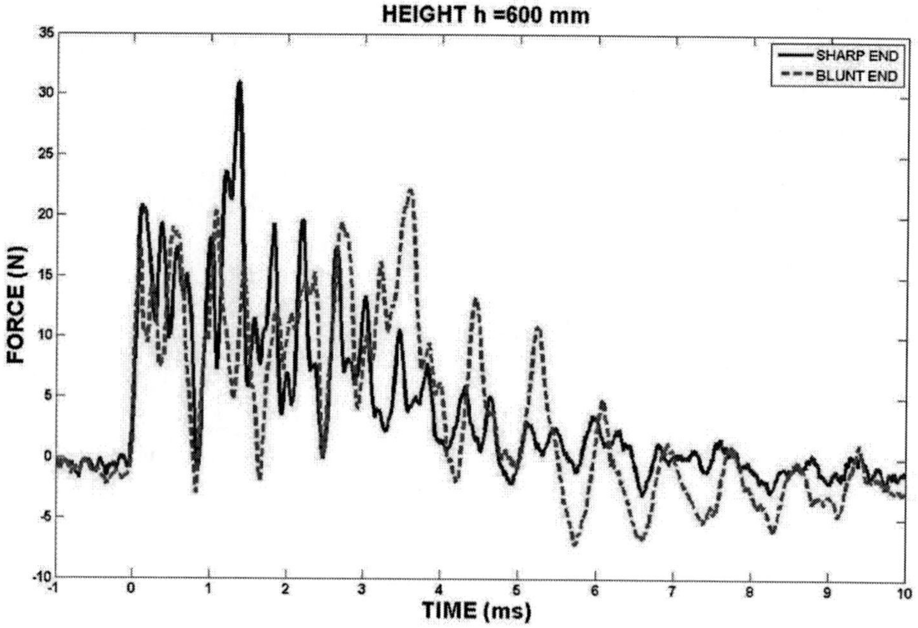

Figure 51. Time history of the force acting in the point of egg-bar contact.

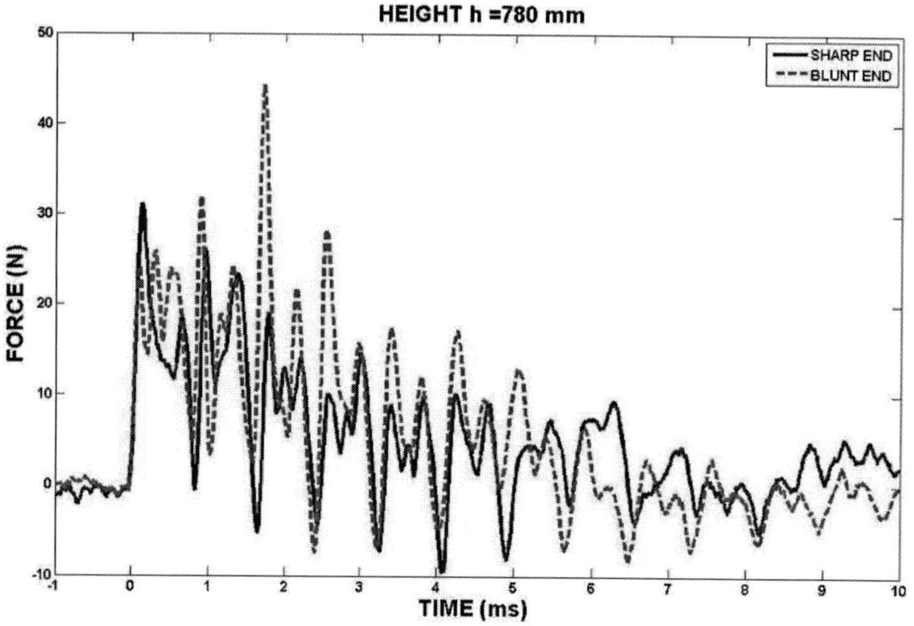

Figure 52. Time history of the force acting in the point of egg-bar contact.

The course of contact force is generally connected with numerous oscillations, which is valid for all fall heights and both egg ends. The amplitudes of individual oscillations are characterized by similar size. The values of maximum forces are visualized in Fig 53.

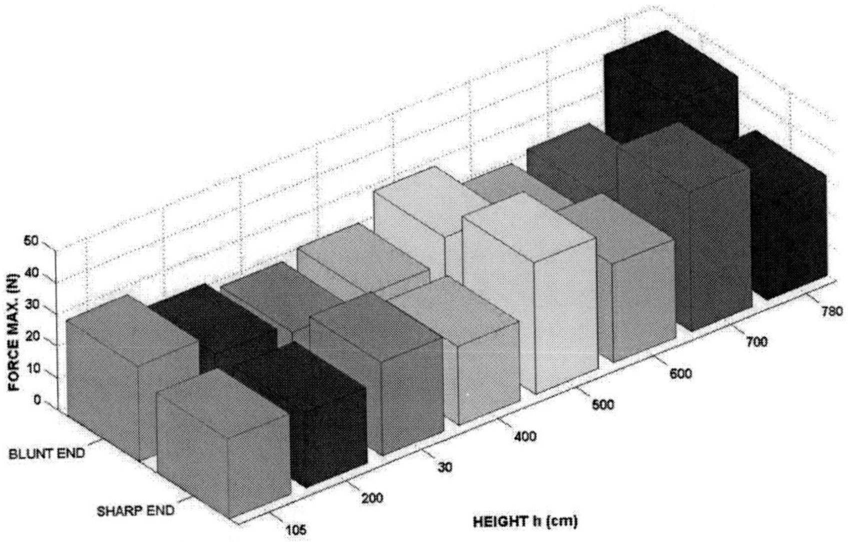

Figure 53. Maximum contact forces.

It is evident that maximum of the contact force is independent on impact velocity and it is not possible to find the relevant difference between force values during fall on either sharp or blunt end. These results are visualized in Figure 54.

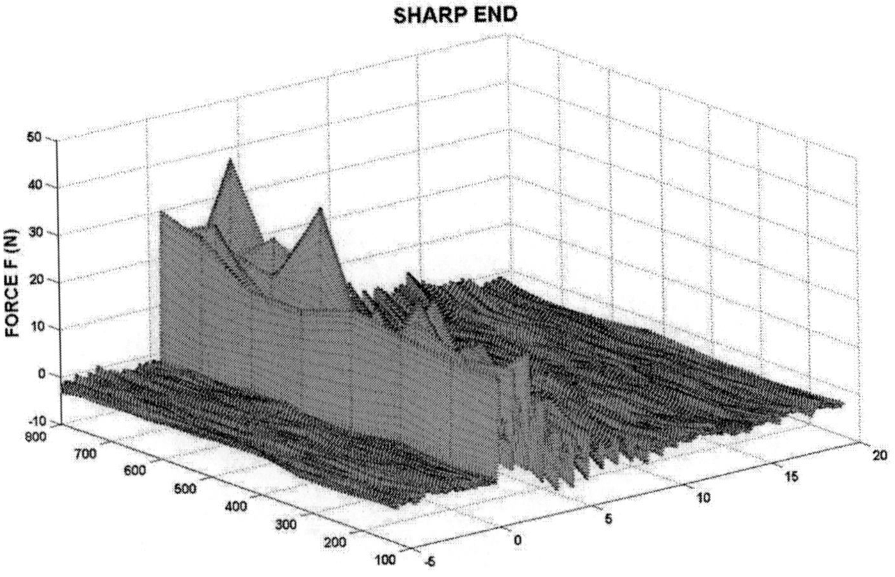

Figure 54. Influence of fall height on the force time history.

Considering rather complicated time history of the contact force *F(t)*, use of impulse of the force seems to be more efficient:

$$I = \int F(t)dt \tag{17}$$

The example of time history of this value is shown in Figure 55.

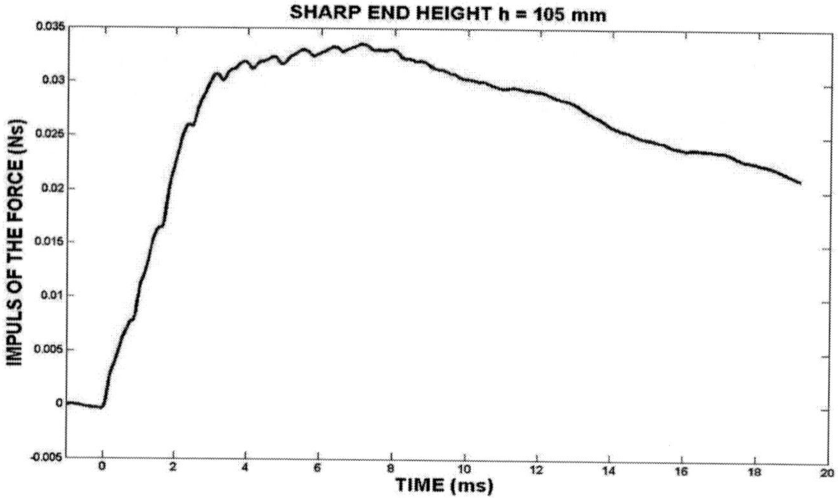

Figure 55. Time history of the impulse of the force.

The decrease of the variable is connected with negative values of oscillating force. Fall height dependence of this variable is presented in Figs 56 and 57.

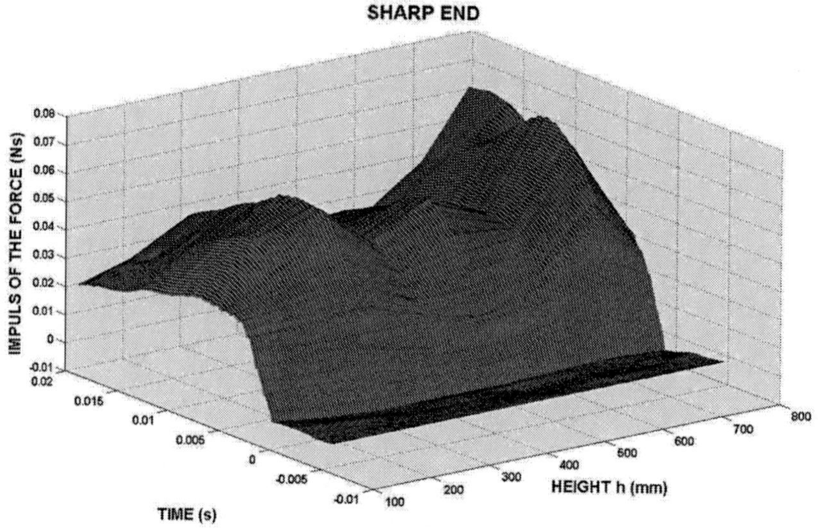

Figure 56. Influence of the fall height on the time history of the impulse of the force.

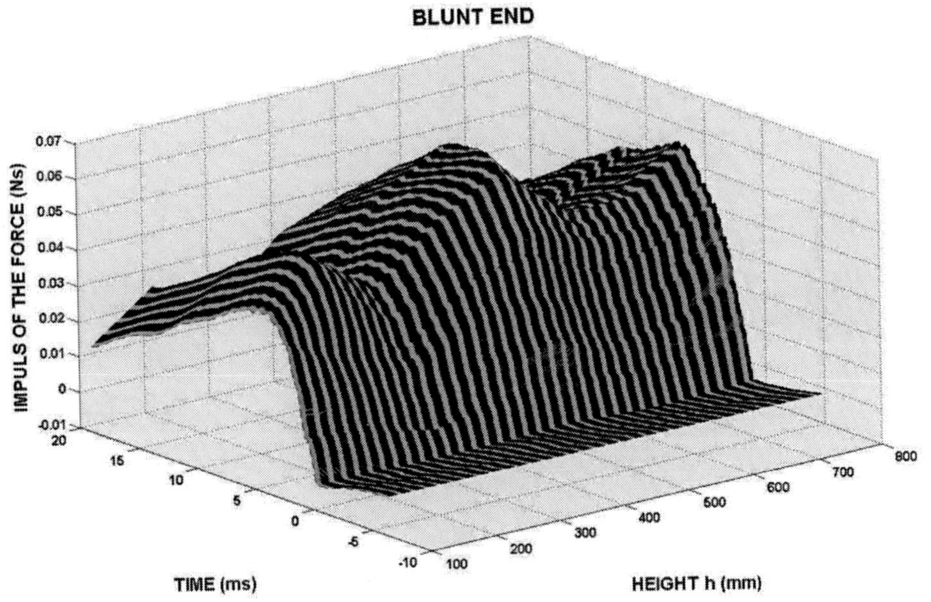

Figure 57. Influence of the fall height on the time history of the impulse of the force.

The next information on the force can be obtained in the Frequency domain. The analysis in the frequency domain is based on the Fourier transform (see e.g. Stein et al., 2003)

For a continuous function of one variable *f(t)*, the Fourier Transform *F(ω)* is defined as:

$$F(\omega)=\int_{-\infty}^{+\infty} f(t)e^{-i\omega t}dt \tag{18}$$

And the inverse transform as:

$$f(t)=\int_{-\infty}^{+\infty} F(\omega)e^{i\omega t}d\omega \tag{19}$$

,

where F is the spectral function and ω is the angular frequency.

The same procedure can be used for the Fourier transform of a series *x(k)* with N samples. This procedure is termed as the discrete Fourier Transform (DFT). A special kind of this transform is Fast Fourier Transform (FFT). This procedure is a part of the most of software packages dealing with the signal processing.

This algorithm is a part of the MATLAB software. An example of the spectral function amplitude is shown in the Figure 58. This function is characterized by a peak value at relatively low frequency. For the whole spectrum, the momentum M_o (Eq. (20), the momentum M_1 (Eq. (21) the central frequency *CF* (Eq. (22) and the variance *Var* (Eq. (23) were calculated (Oppenhein and Schafer, 1989).

$$M_0 = \sum F(\omega)\Delta\omega \tag{20}$$

$$M_1 = \sum F(\omega)\omega\Delta\omega \tag{21}$$

$$CF = \frac{M_1}{M_0} \tag{22}$$

$$Var = \frac{\sum(\omega - CF)F(\omega)}{\sum F(\omega)} \tag{23}$$

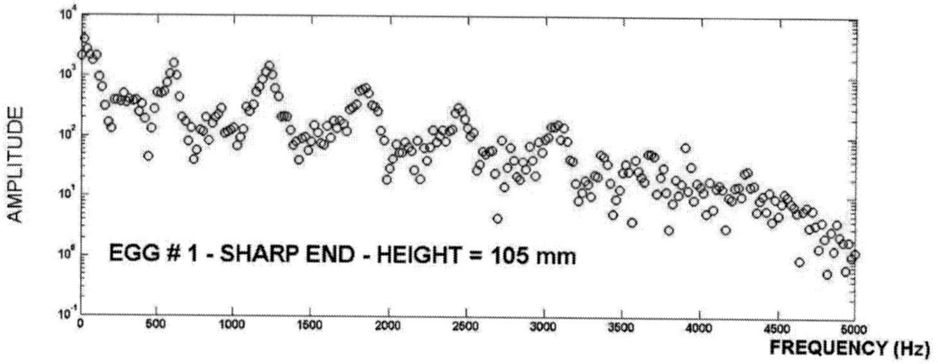

Figure 58. Fourier analysis (FFT) of the force.

The parameters M_o, M_1 and CF are given in the Table 15 together with the frequency at which the maximum on the amplitude – force dependence occurs.

Table 15. Characteristics of the amplitude – frequency spectrum – raw eggs

	Egg No	ω (s^{-1})	M_o	M_1	CF=$M_1/M_o(s^{-1})$	h (mm)
Sharp end	1	20	2230137	1.11E+11	49980	105
	2	20	2447438	1.36E+11	49980	200
	3	20	2717143	1.36E+11	49980	300
	4	20	2611760	1.31E+11	49980	400
	6	20	3067572	1.53E+11	49980	500
	7	20	2800472	1.4E+11	49980	600
	9	20	3325453	1.66E+11	49980	700
	10	20	3106422	1.55E+11	49980	780
	11	20	3489978	1.74E+11	49980	780
	12	20	3556307	1.78E+11	49980	780
	13	20	2826837	1.41E+11	49980	700
	14	20	2722207	1.36E+11	49980	600

Blunt end	15	20	3124508	1.56E+11	49980	500
	16	20	2591775	1.3E+11	49980	400
	17	20	2484916	1.24E+11	49980	300
	18	20	2496437	1.25E+11	49980	200
	19	20	2737091	1.37E+11	49980	105

It is obvious that frequency at which the maximum in contact force frequency spectrum occurs is independent on impact velocity as well as egg position (direction of loading). The same rule is valid for central frequency.

4.2.2. Boiled Eggs

Examples of force-time dependencies in the egg-rod contact point are shown in Figs 59-60. Different fall heights and/or impact velocities are considered. Similar dependencies were obtained for other impact velocities. Concerning qualitative course of the dependencies, the courses are simile as those received for raw eggs.

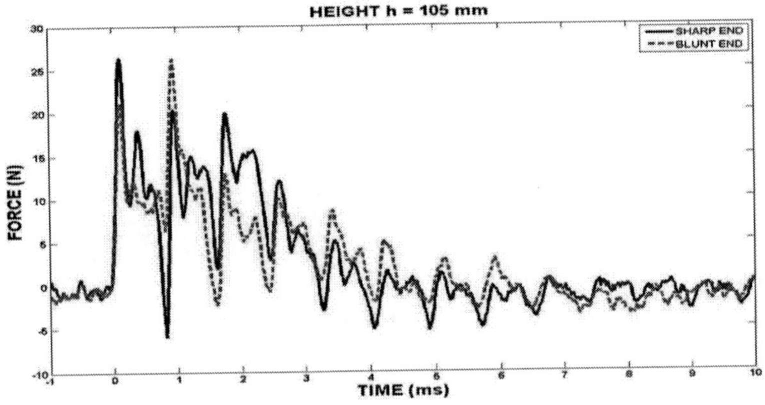

Figure 59. Force-time dependence in the egg-rod contact point.

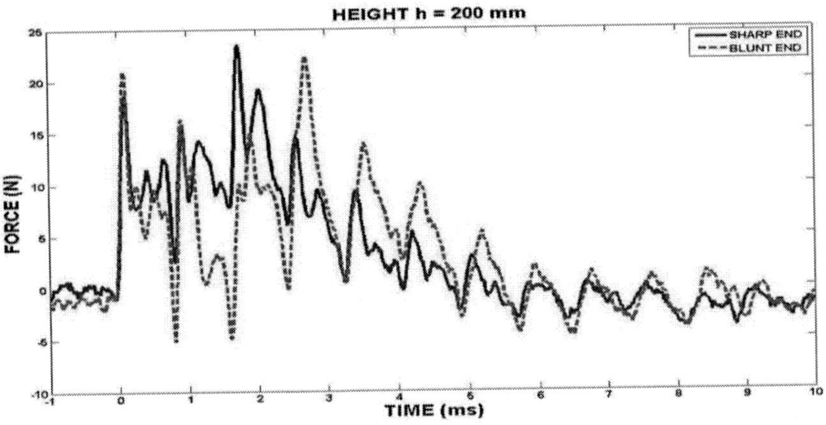

Figure 60. Force-time dependence in the egg-rod contact point.

The force courses in the egg-rod contact point are qualitatively similar as ones recorded for raw eggs – see examples given in Figs 61-62.

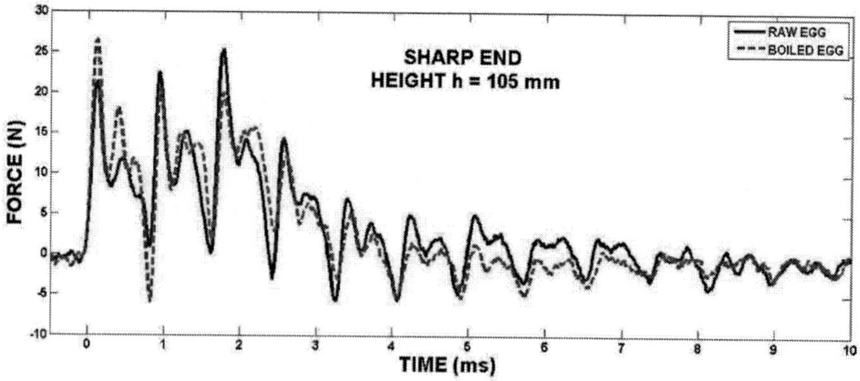

Figure 61. Force-time dependence in the egg-rod contact point.

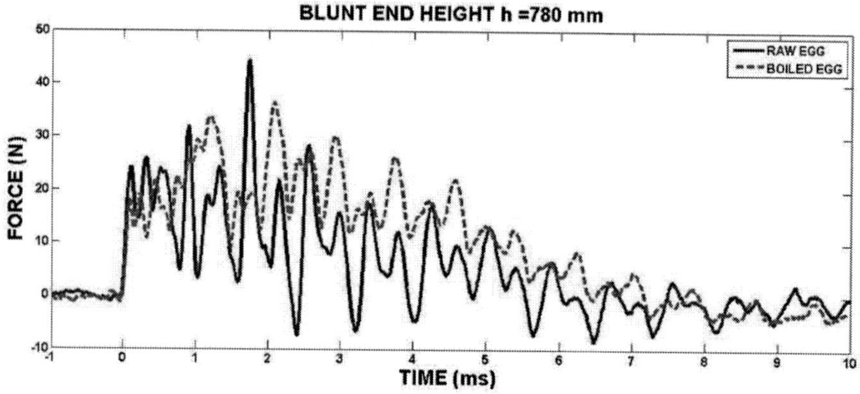

Figure 62. Force-time dependence in the egg-rod contact point

Following pictures 63 and 64 show the values of contact force impulse. Neither these values indicate considerable influence of thermal treatment on their strain behavior of the eggs.

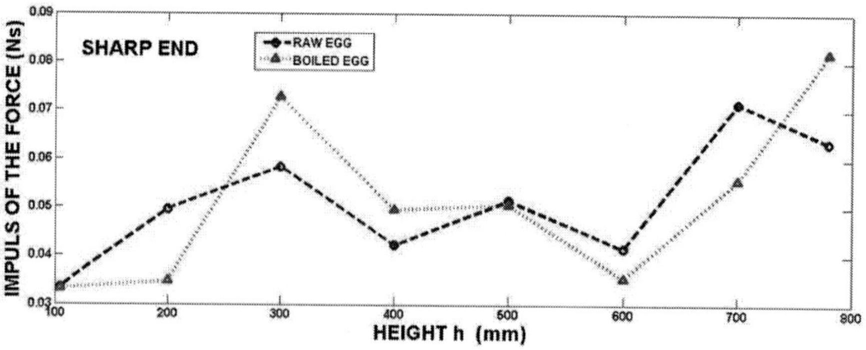

Figure 63. Contact force impuls.

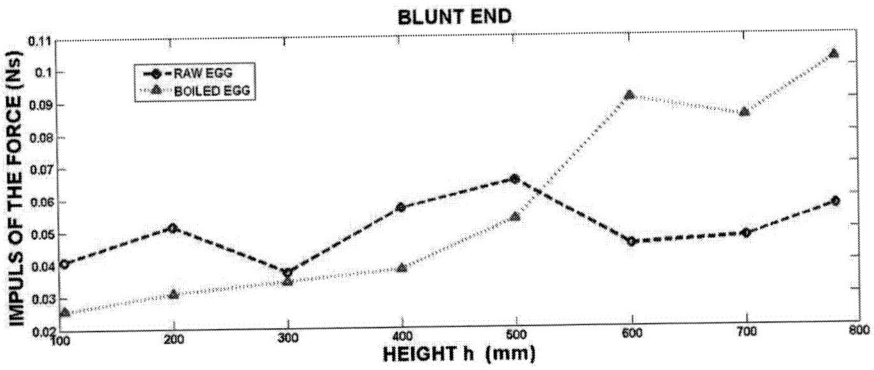

Figure 64. Contact force impulse.

Similar results arise from data obtained in frequency spectrum – see Table 16.

Table 16. Characteristics of the amplitude – frequency spectrum – boiled eggs

	h (mm)	ω_{max} (s-1)	M_0	M_1	CF (Hz)
Sharp end	105	20	2462049	1.23E+11	49980
	200	20	2025868	1.01E+11	49980
	300	20	2721826	1.36E+11	49980
	400	20	2714301	1.36E+11	49980
	500	20	2395353	1.2E+11	49980
	600	20	2382926	1.19E+11	49980
	700	20	3395568	1.7E+11	49980
	780	20	3256423	1.63E+11	49980
Blunt end	105	20	2135069	1.07E+11	49980
	200	20	2303722	1.15E+11	49980
	300	20	2696738	1.35E+11	49980
	400	20	2557428	1.28E+11	49980
	500	20	3283512	1.64E+11	49980
	600	20	3723301	1.86E+11	49980
	600	20	3608312	1.8E+11	49980
	700	20	3843944	1.92E+11	49980
	780	20	3309206	1.65E+11	49980

4.2.3. Peeled Eggs

The eggs were peeled and dropped from different heights in such way that they impacted either on their sharp or blunt end. These impact experiments were performed also for extracted yolk itself. Example record of the force course in the peeled egg-rod contact point is shown in Figure 65. It can be seen that force course includes less oscillations than in the case of egg with shell. Also the oscillation amplitudes are considerably lower. The same results are valid for force courses of falling egg yolk – see Figure 66.

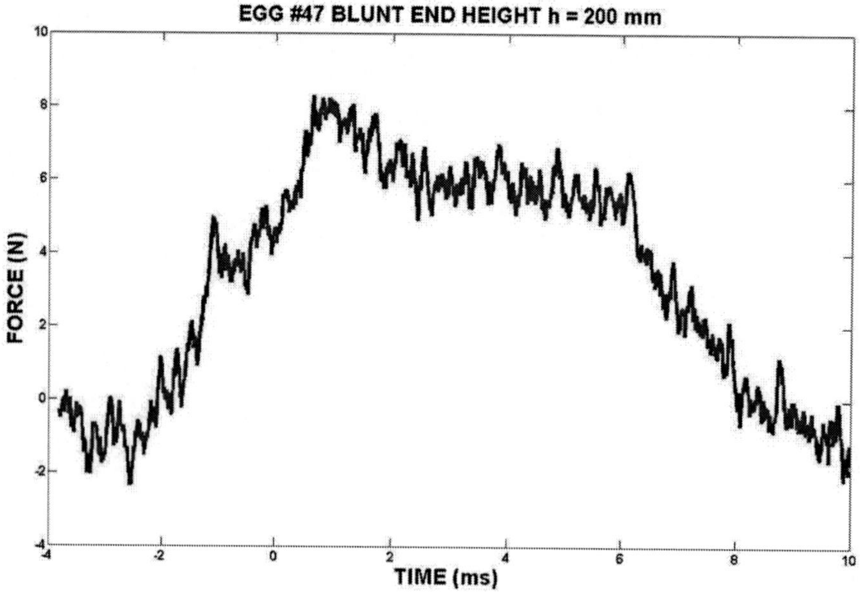

Figure 65. Contact force acting after fall of peeled egg.

Figure 67 shows a comparison of different force records for different fall combinations. A relevant qualitative as well as quantitative difference can be seen for strain behavior of the egg with and/or without eggshell, raw egg and egg yolk. It is obvious that observed oscillations are connected with wave propagation in the eggshell.

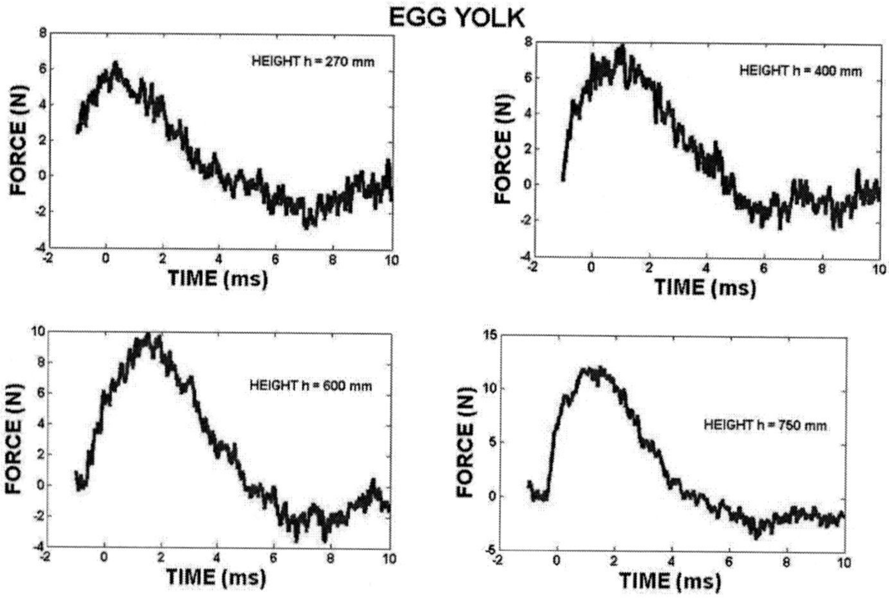

Figure 66. Contact force acting after fall of egg yolk.

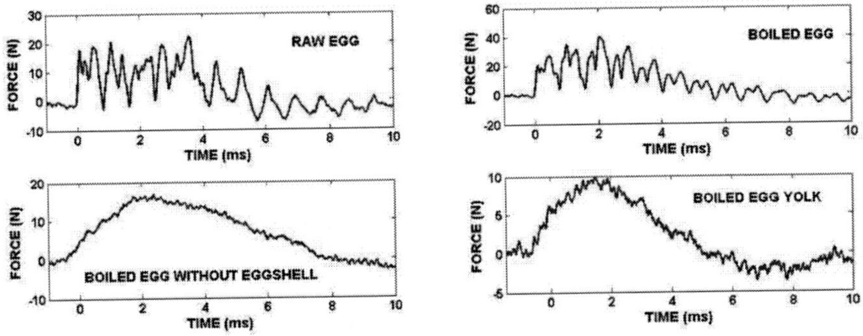

Figure 67. Contact forces acting after fall of raw, boiled, peeled egg, and egg yolk. Height h = 600 mm. Blunt end.

For further description of the problem, a frequency analysis using FFT was performed. Tables 17-19 contain results and basic characteristics obtained from amplitude spectrum of the contact force.

Table 17. Characteristics of the amplitude – frequency spectrum – boiled eggs without eggshell – blunt end

Egg No	h (mm)	ωmax (-1)	M_0	M_1	CF (Hz)
47	200	20	1046892	5.23E+10	49980
48	300	20	1053188	5.26E+10	49980
51	400	20	1394342	6.97E+10	49980
59	600	20	1466786	7.33E+10	49980

Table 18. Characteristics of the amplitude – frequency spectrum – boiled eggs without eggshell – sharp end

Egg No	h (mm)	ωmax (-1)	M_0	M_1	CF(Hz)
60	105	20	958730.5	4.79E+10	49980
57	300	20	1199991	6E+10	49980
56	600	20	1253331	6.26E+10	49980

Table 19. Characteristics of the amplitude – frequency spectrum – boiled yolk

Egg No	h (mm)	ωmax (-1)	M_0	M_1	CF (Hz)
-	270	20	936597	4.68E+10	49980
-	400	20	1033762	5.17E+10	49980
-	600	20	1126646	5.63E+10	49980
-	750	20	1278810	6.39E+10	49980

It is evident that there is no change in frequency, which is the amplitude maximum reached in. These spectra are shown in Figs. 68-70.

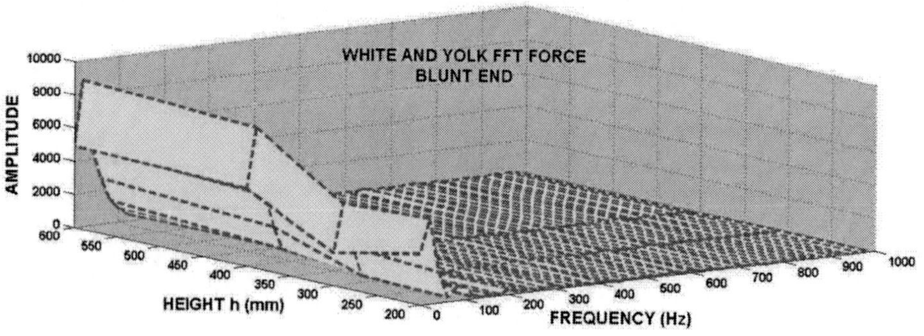

Figure 68. Amplitude frequency spectrum of the contact force.

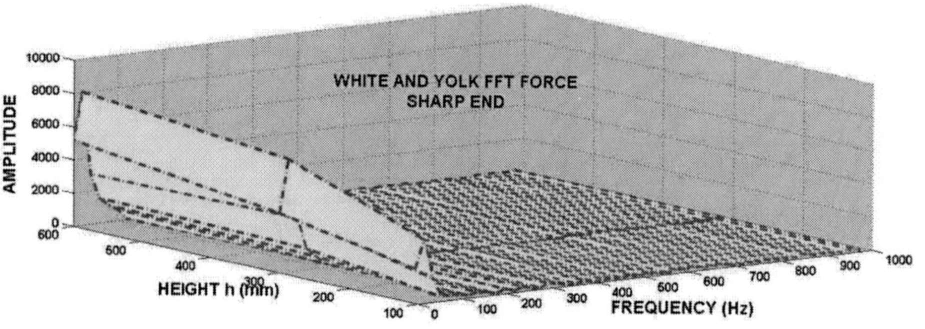

Figure 69. Amplitude frequency spectrum of the contact force.

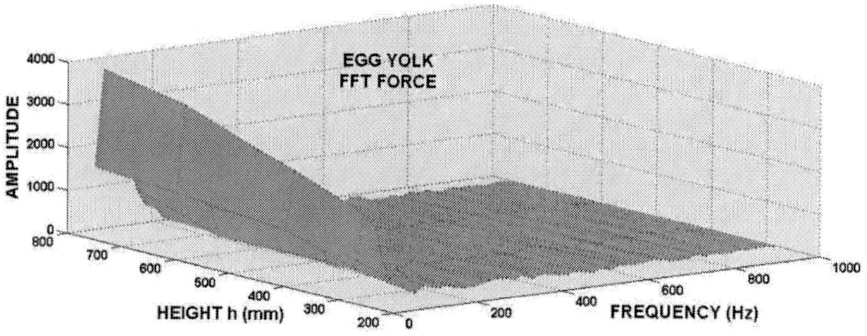

Figure 70. Amplitude frequency spectrum of the contact force.

The differences are largely notable especially in case of M_0 and M_1 – see Figs. 71-72.

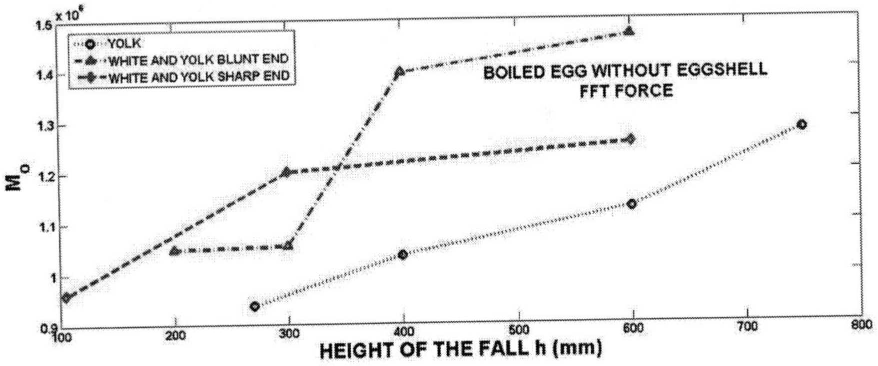

Figure 71. Size of moment M_0.

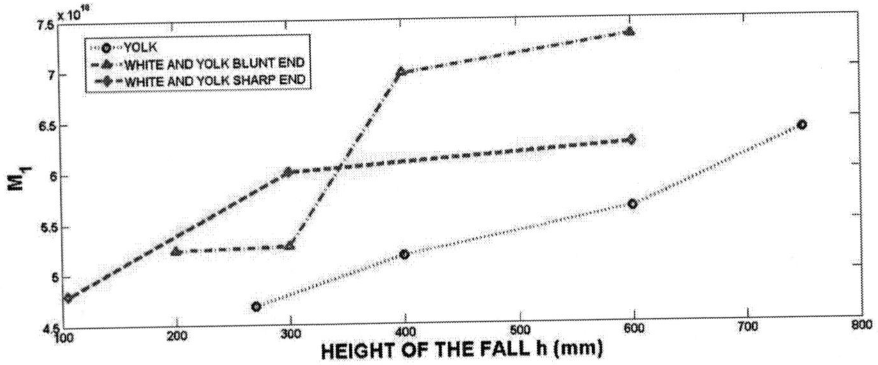

Figure 72. Size of moment M_1.

More detailed analyses of the interaction between egg and rod (or stop block generally) are conditioned by gaining more data, e.g. by use of high-speed camera. Such data would complete the information on impact behavior of eggs and could serve as an input for evaluation of numerical methods reliability and creating of numerical simulations of given problems. Such approaches will be the subject of future research.

5. Conclusion

This presented chapter contains an overview of the basic results received within the research focused on impact loading of the hen's eggs. The following problems have been solved:

1) Non-destructive ball impact on the eggshell resulting in data collecting for evaluation of non-destructing eggshell behavior.
2) Loading of the eggs by instrumented bar resulting in determination of the eggshell strength.

3) Fall of the egg on the rigid stop, which was represented by rod enabling recording of the contact forces. This research was aimed at collecting the data for reliable evaluation of given problem numerical simulation applicability. If the applicability of numerical simulation will be confirmed, generalization of the egg impacting would be possible and different conditions such as transport or storing manipulation could be simulated.

The following results have been obtained for above listed research problems:

Ad 1. The detail analysis of the eggshell response to the non-destructive impact of the ball led to the obtaining of a large amount of new knowledge concerning egg acoustic behavior. The egg dynamic resonance frequency was detected. It was obtained through the analysis of the dynamically measured frequency response of an excited egg. The effect of excitation point, detected point, impact intensity, excitation material, different egg mass, density, shell stiffness, and shell crack on the dominant frequency were assessed. The basic results are summarized below:

- The dominant frequency was significantly affected by the shell crack, shell stiffness, egg mass and egg density.
- The excitation point, detected point, excitation velocity, and the impacting material did not significantly affect the dominant frequency.
- The dominant frequency increased with increase of the shell stiffness and/or the density and decreased with increase of the egg mass.

The numerical model of the egg has also been proposed. Although this model uses many simplified assumptions, the good agreement between experimental and numerical results suggests that there is a good chance to obtain reliable model. The proposed numerical model can minimize the number of experiments but it can also serve as a tool for describing of many other dynamic loading conditions.

Ad 2. The experimental method of the free falling rod has been developed for the measurement of the eggshell strength under dynamic loading conditions. This method enables simultaneous evaluation of the eggshell strength and acoustic response of the tested egg. The effort in this study was limited on determination of the dynamic strength value. It has been found that the force at which the eggshell rupture started was significantly dependent on the position of the rod impact. Contrary to the results of the static loading tests, there is a significant difference between the rupture force at the sharp and blunt end. This force is also higher than that obtained at the static loading. This phenomenon, which is common for most metallic materials, polymers and also some brittle materials, must be verified and explained in terms of the eggshell microstructure.

Because the values of the eggshell strength expressed in terms of the rupture force are affected by many factors (egg specific gravity, egg mass, egg volume, egg surface area, egg thickness, shell weight, shape and shell percentage), it is necessary to perform the numerical simulation of these experiments in order to avoid the simultaneous influence of these factors. The numerical model of the given experiment has been proposed. The model respects the true shape of the eggshell. Preliminary results of the numerical simulation show reasonable agreement with experimental data. The tensile stresses which correspond to the impact conditions under which the eggshell damage starts have been determined. It seems that this quantity may represent an intrinsic property of the eggshell. Verification of this hypothesis needs both further experiments and numerical computations and some other experiments with rods of different diameters and rod tip shapes (conic, ball etc.). Instead of further research focused on numerical analysis, some other problems should be solved in order to improve capabilities of the experimental method described in this chapter. The main problem consists of the use of microscopic method for the eggshell defects observation.

Ad 3. In this part of the research, experimental method enabling detection of time history of contact force between an egg and rigid stop was implemented. The rigid stop was represented by the elastic rod of a circular cross-section equipped with the strain gauges used for force detecting. The advantage of this approach consists of the possibility of testing different rigid stop materials. In this case, the rod was made of PMMA material. Both, the raw and boiled eggs were examined within this research. The following results have been made:

- Time course of the contact force is connected with numerous oscillations of large amplitude, which are comparable to force maximum. The qualitative character of this value's time history is analogous, for both raw and boiled eggs fall on the sharp as well as blunt end, and it is also dependent on impact velocity. The above-mentioned variation reflects the size of the moment, which is the area under force-time curve.
- The frequency spectrum of the force pulse is characteristic with its marked and noticeable maximum. The maximum is reached in the frequency which is independent on impact velocity and egg orientation during the impact (sharp or blunt end). The frequency is the same for both raw and boiled eggs.
- The fall of the peeled egg is connected with much smaller oscillations of egg-rod contact forces and relatively low oscillation amplitudes. The value of contact force is obviously lower than in the case of an unpeeled egg. The same result is valid for the case of yolk fall. It is evident that the presence of the oscillations is connected with the eggshell. The oscillations are probably not developed as a consequence of the egg fluid's movement, because qualitatively similar results were obtained for both raw and boiled eggs. To a certain extent it is also possible to assume that the development of these oscillations is not connected with crack creation and/or gradual eggshell breaking. This result is based on the fact that oscillation character is not dependent on impact velocity and, thus, damage extent. It is evidently the consequence of wave propagation and interference in the material.

The following result can be pronounced as a general rule: Vibrational characteristics during non-destructive egg impact loading are connected with egg geometry and dimensions, egg mass, eggshell thickness and strength properties. The determination of these characteristics is considerably significant for both, practical applications (e.g. automatic egg sorting) and basic research. The aforementioned values are dependent on eggshell structure and can be used as a datum for evaluating of feed quality effect and/or cracks presence evaluation.

The importance of the fall experiments exists especially in the possibility of a reliable verification of the egg models in the different numerical software. Development of the numerical simulation is essential for evaluation of wide range of problems connected with egg mechanical loading (egg collecting, transport, storing etc.). It is obvious that experiments alone cannot cover and implicate all possible variations of the aforementioned influences – for practical, time consuming and economical reasons. One of the main goals of the work presented in this chapter was to point out and refer to potential perspectives in this problematic.

ACKNOWLEDGMENTS

The research was supported by the Grant Agency of the Czech Academy of Science under contract No IAA 201990701.

REFERENCES

Altuntas, E. & Sekeroglu, A. (2008). Effect of shape index on mechanical properties of chicken eggs. *J. Food Eng., 85,* 606-612.

Ar, A. & Rahn, H. (1980). Water in the avian egg: overall budget of incubation. *American Zoologist, 20,* 373-384.

Bain, M. M. (1990). *Eggshell strength - a mechanical/ultrastructuralevaluation*. PhD Thesis, University of Glasgow.

Baryeh, E. A. & Mangope, B. K. (2003). Some physical properties of QP- 38 variety pigeon pea. *Journal of Food Engineering, 56,* 59-65.

Bell., D. (1984). Egg breakage - From the hen to the consumer. *Calif. Poult. Lett.,* 2-6.

Bliss, G. N. (1973). *Crack detector*, U.S. Patent 3744299.

Brooks, J. & Hale, H. P. (1955). Strength of the shell of the hen's egg. *Nature*, 175, 848-849.

Buchar, J. & Simeonovova, J. (2001). On the identification of the eggshell elastic properties under quasistatic compression. In *19th CAD – FEM USERS MEETING 2001- International Congress on FEM Technology.* 17-19. October, Potsdam, Berlin Germany, Vol. 2, pp 1-8. Published by CAD – Fem GmbH, Munchen.

Carter, T. C. (1970). The hen's egg: factors affecting the shearing strength of shell material. *Br. Poult. Sci., 11,* 433-449.

Carter, T. C. (1971). The hen's egg: variation in tensile strength of shell material and its relationship with shearing strength. *Br. Poult. Sci., 12,* 57-76.

Carter, T. C. (1976). The hen's egg: Shell forces at impal and quasi–static compression, *Br. Poult. Sci., 17,* 199-214.

Coucke, P. (1998). *Assessment of some physical egg quality parameters based on vibration analysis.* PhD Thesis. Katholieke Universiteit Leuven. Belgium.

Coucke, P., Dewil, E., Decuypere, E., & De Baerdemaeker, J. (1999). Measuring the mechanical stiffness of an eggshell using resonant frequency analysis. *Br. Poult. Sci., 40,* 227-232.

Chung, R. A. & Stadelman, W. J. (1965). A study of variations in the structure of the hens egg. *British Poultry Science, 6,* 277-282.

De Ketelaere, B., Coucke, P., & De Baerdemaeker, J. (2000). Eggshell crack detection based on acoustic resonance frequency analysis. *J. Agric. Eng. Res., 76,* 157-163.

De Ketelaere, B., Govaerts, T., Coucke, P., Dewil, E., Visscher, J., Decuypere, E., & De Baerdemaeker, J. (2002). Measuring the eggshell strength of 6 different genetic strains of laying hens: techniques and comparisons. *British Poultry Science, 43,* 238-244.

De Ketelaere, B., Vanhoutte, H., & De Baerdemaeker, J. (2003). Parameter estimation and multivariable model building for the non-destructive on-line determination of the eggshell strength. *Journal of Sound and Vibration, 266,* 699-709.

Denys, S., Pieters, J. G., & Dewettinck, K. (2003). Combined CFD and experimental approach for determination of the surface heat transfer coefficient during thermal processing of eggs. *Journal of Food Science, 68,* 943-951.

Dunn, I. C., Bain, M., Edmond, A., Wilson, P. W., Joseph, N., Solomon, S., De Ketelaere, B., De Baerdemaeker, J., Schmutz, M., Preisinger, R., & Waddington, D. (2005a). Heritability and genetic correlation of measurements derived from acoustic resonance frequency analysis; a novel method of determining eggshell quality in domestic hens. *British Poultry Science, 46,* 280-286.

Dunn, I. C., Bain, M., Edmond, A., Wilson, P. W., Joseph, N., Solomon, S., De Ketelaere, B., De Baerdemaeker, J., Schmutz, M., Preisinger, R., & Waddington, D. (2005b). Dynamic stiffness (K_{dyn}) as a predictor of eggshell damage and its potential for genetic selection. In *Proc. XIth European Symposium on the Quality of Eggs and Egg Products*. Doorwerth. The Netherlands. 23-26 May 2005.

Erdogdu, F., Sarkar, A., & Singh, R. P. (2005). Mathematical modeling of air-impingement cooling of finite slab shaped objects and effect of spatial variation of heat transfer coefficient. *Journal of Food Engineering, 71,* 287-294.

Erdogdu, F., Ferrua, M., Singh, S. K., & Singh, R. P. (2007). Air-impingement cooling of boiled eggs: Analysis of flow visualization and heat transfer. *Journal of Food Engineering, 79,* 920-928.

Hammmerle, J. R. & Mohsenin, N. N. (1967). Determination and analysis of failure stresses in egg shells. *J. agric. Engng Res., 12,* 13-21.

Hunton, P. (1993). Understanding the architecture of the eggshell. In *Proc. 5th Eur. Symp. Qual. Eggs and Egg Prod*, pp 141–147, Tours, France. World's Poult. Sci. Assoc., Cedex, France.

Ju, B. B., Liu, K. K., Ling, S. F., & NG, W. H. (2002). A novel technique for characterizing elastic properties of thin biological membrane. *Mechanics of Materials, 34,* 749-754.

MacLeod, N., Bain, M. M., & Hancock, J. W. (2006). The mechanics and mechanisms of failure of hens' eggs. *Int. J. Fracture, 142,* 29-41.

Moayeri, A. (1997). *Probe, device and method for testing eggs.* U.S. Patent 5728939.

Mohsenin, N. N. (1970). *Physical properties of plant and animal materials.* New York: Gordon and Breach Science Publishers.

Narushin, V. G. (2001). Shape geometry of the avian egg. *Journal of Agricultural Engineering Research, 79,* 441-448.

Nedomová, Š., Severa, L., & Buchar, J. (2009). Influence of hen egg shape on eggshell compressive strength. *Int. Agrophysics, 23,* 249-256.

Olsson, N. (1934). *Studies on Specific Gravity of Hen's Egg.* A Method for Determining the Percentage of Shell on Hen's Eggs. Otto Harrassowitz. Leipzig. Germany.

Oppenhein, A. V. & Schafer, R. W. (1989). *Discrete-time signal processing.* New Jersey: Prentice Hall International, Inc.

Reissner, E. (1946). Stresses and small displacements of shallow spherical shells. *Int. J. Math. Phys., 25,* 80-85.

Rodriguez-Navarro, A., Kalin, O., Nys, Y., & Garcia-Ruiz, J. M. (2002). Influence of the microstructure and crystallographic texture on the fracture strength of hen's eggshells. *Br. Poult. Sci, 43,* 395-403.

Sinha, D. N., Johnston, R. G., Grace, W. K., & Lemanski, C. L. (1992). Acoustic resonances in chicken eggs. *Biotechnology Progress, 8,* 240-243.

Sluka, S. J., Besch, E. L., & Smith, A. H. (1967). Stresses in impacted egg shells. *Trans. Am. Soc. Agric. Engrs, 10,* 364-369.

Sugino, H., Nitoda, T., & Juneja, L. R. (1997). General chemical composition of hen eggs. In Yammamoti, T., Juneja, L. R., Hatta, H. and Kim, M. (Eds). *Hen eggs.* Their basic and applied science. CRC Press.

Thompson, B. K., Hamilton, R. M. G., & Voisey, P. W. (1981). Relationships among various traits relating to shell strength, among and within five avian species. *Poult. Sci., 60,* 2388-2394.

Timoshenko, S. P. & Woinowsky-Krieger, S. (1959). *Theory of Plates and Shells.* New York, McGraw-Hill Book Company, Inc.

Tung, M. A., Staley, L. M., & Richards, J. F. (1969). Estimation of Young's modulus and failure stresses in the hen's egg shell. *Can. Agric. Engng, 11,* 3-5.

Tyler, C. & Geake, F. H. (1963). A study of various impact and crushing methods used for measuring shell strength. *Br. Poult. Sci., 4,* 49-61.

Voisey, P. W. & Hamilton, R. M. G. (1976). Factors affecting the non-destructive and destructive methods of measuring eggshell strength by the quasi-static compression test. *Br. Poult. Sci,. 17,* 103-124.

Voisey, P. W. & Hunt, J. R. (1967a). Physical properties of egg shells. Stress distribution in the shell. *Br. Poult. Sci. 8,* 263-271.

Voisey, P. W. & Hunt, J. R. (1967b). Relationship between applied force, deformation of eggshells and fracture force. *J. Agric. Eng. Res., 12,* 1-4.

Voisey, P. W. & Hunt, J. R. (1969). Effect of compression speed on the behaviour of eggshells. *J. agric. Engng Res., 14,* 40-46.

Voisey, P. W. & Hunt, J. R. (1974). Measurement of eggshell strength. *J. Texture Stud., 5,* 135-182.

Wells, R.G. (1968). Egg quality Characteristics. In T.C. Carter, ed., *Egg Quality: A Study of the Hens Eg.* (pp. 214- 225), Edinburgh: Oliver and Boyd.

Yang, S. H., Hsi-Lung, C., & CHUNGHWA, W. (1995). Quality control on thousand year egg by vibration identification, In *Proceedings of the 13th International Modal Analysis Conference*, Nashville, TN, pp. 1242-1247.

In: Focus on Food Engineering
Editor: Robert J. Shreck

ISBN: 978-1-61209-598-1

Chapter 2

STRATEGIES FOR EXTENDING SHELF LIFE OF FOODS USING ANTIMICROBIAL EDIBLE FILMS

***Silvia K. Flores*[*1,2], *Lía N. Gerschenson*[1,2], *Rosa J. Jagus*[3], and *Karen J. Sanjurjo*[3]**

[1]Departamento de Industrias,
Facultad de Ciencias Exactas y Naturales (FCEN), Universidad de Buenos Aires (UBA)
Ciudad Universitaria. Intendente Güiraldes 2620, (1428)
Ciudad Autónoma de Buenos Aires, Argentina.
[2]Research Member of the National Scientific and Technical Research Council of Argentina (CONICET).
[3]Departamento de Ingeniería Química,
Facultad de Ingeniería (FI), Universidad de Buenos Aires (UBA), Ciudad Universitaria,
Intendente Güiraldes 2620, (1428) Ciudad Autónoma de Buenos Aires, Argentina

ABSTRACT

The development and production of new packaging materials, which are friendlier with the environment, is actually being studied with the purpose of minimizing the environmental pollution that is produced by the use of traditional, non-biodegradable packaging. In the framework of this interest, the study of the use of biopolymers to produce "edible films" has considerably progressed in the last decade. Starches, proteins, cellulose and derivatives, gums, chitosan, among other hydrocolloids, have been used for producing this kind of films. The presence of a plasticizer agent is always required to minimize brittle structure and antimicrobials or other additives can be included in the film formulation. Antimicrobials will provide the film with specific functional properties in addition to their inherent barrier properties to the water vapor and oxygen and, in this case, the edible films can be

*Corresponding author: Silvia Karina Flores
e-mail: sflores@di.fcen.uba.ar
1 Phone: 54 – 11 – 4576- 3366
Fax number: 54 – 11 – 4576- 3366

thought of as an active packaging material since they are able to support and, eventually, release the food preservatives. The films will perform as an additional microbial stress factor in order to protect the food from the external contamination and, therefore, will contribute to produce shelf life extension.

The object of the present study was the production of tapioca starch - glycerol based edible films containing the preservatives potassium sorbate (KS) or nisin. Physicochemical properties of films such as crystalline fraction, solubility in water, sorptional behavior and color attributes were studied. In order to optimize the film functionality, the influence of soy oil addition to the film formulation, the use of sodium trimetaphosphate-chemically cross-linked-tapioca starch and the use of different filmmaking techniques, were evaluated. The study of the effect of the film composition on the physicochemical properties and antimicrobial activity behavior will help to predict the potential usefulness of the film for a particular food system.

INTRODUCTION

During the last twenty years, one of the main goals of the research in the area of food science and technology has been the development of new technologies in preservation. These new technologies, which might be an option to the traditional ones, shall be less aggressive in order to obtain processed foods with better nutritional and organoleptical properties as well as safe and stable. Among the proposed methodologies, the use of techniques that combine different factors to attain microbial stress can be mentioned. Slight heat treatment, pH and a_w control, authorized chemical preservatives addition and irradiation, often coupled with special packaging and refrigeration, are some of the factors that are combined in these techniques tending towards the optimization of the food quality (Leistner, 1995; Zeuthen & Bøgh-Sørensen, 2003).

Packaging materials have traditionally aimed the main objective of protecting foods from external chemical and microbiological contamination as well as, from the deteriorative action of environmental agents such as oxygen, water vapor and light (Fabech et al, 2000). Food packaging area is undergoing a very significant progress mainly due to increased demands on product safety, nutrition, shelf-life extension, cost-efficiency, environmental issues, and consumer convenience (Ahvenainen, 2003). Many studies have been carried out in relation to new packaging materials and the possible interactions between the packaging and the food (EUFIC, 2002). In addition, due to the increasing consciousness about the environmental pollution, the study of the negative impact of the current packaging used was performed. The raw material employed for the elaboration of traditional food packaging (i.e.: polyethylene, polyvinyl chloride, polypropylene, polystyrene, polyethylene terephthalate) comes from non renewable sources (petroleum derivatives), which are not biodegradable and, therefore, represents a massive contamination medium (Piermaria et al., 2009). However, because of their mechanical resistance and water vapor, oxygen and/or carbon dioxide barrier properties, those materials could not be easily replaced by others with similar characteristics but with reduced environmental consequences. For that reason, the development of new materials with potential application in food packaging is an interesting research area that is continuously making progress.

Following that trend, one of the emerging technologies that has generated a lot of expectations is the use of biodegradable or edible films to protect food. The edible films can provide specific functional properties for food and constitute an interesting alternative to the traditional packaging technology (Dutta et al., 2009; Cutter, 2006). Edible films offer a number of advantages that have not been completely explored. They have potential to increase the food shelf life, especially considering film application as a stress factor used in combination with others. The possible actions that the edible films can exert on the food to preserve it are: moisture or gas (O_2, CO_2) passage retardation, lipids and solute transport retardation, addition of structural integrity, volatile flavor components retention, food additives support (Flores, et al., 2007a; Buonocore et al., 2003a). The edible films can act in an additional or cooperative way with other factors, optimizing the food global quality, protecting against external microbial contamination, extending food shelf life and, probably, improving the efficiency of the other packaging materials (Kester & Fennema, 1986). The use of biopolymers from available and renewable sources such as starches, cellulose derivatives, gums, pectins, chitosan, whey or soy proteins, has been proposed (Phan The et al., 2009; Chillo et al., 2008; Flores et al., 2007b; León & Rojas, 2007; Han & Krochta, 2007; García et al., 2000). According to Moura et al. (2009), the use of these biopolymers to obtain biodegradable films with specific functionality, means an innovative contribution to help to solve the problem of pollution.

The use of a biopolymer such as native starch is interesting because this polymer is abundant and quite economical. Starches are polymers that naturally occur in a variety of botanical sources such as wheat, corn, potatoes and tapioca or cassava. They are a renewable resource that is widely available and can be obtained from different by-products from harvesting and raw material industrialization (García et al., 2008).

The formation of starch edible film involves gelatinization of starch granules by heating in excess of water. This procedure results in granule swelling and disruption as well as leaching of soluble components (amylose) from the granule. Amylose is the starch component responsible for the film-forming capacity. A viscous mass is obtained consisting of a continuous phase constituted basically by solubilized amylose and a discontinuous phase of remnant granules, mainly based on amylopectin (Zobel, 1994). The cooling of the hot paste, results in a viscoelastic gel. The formation of the junction zones (polymer molecules joined by covalent bonds, hydrogen bonding and/or Van der Waal forces) of a gel can be considered the first stage for starch crystallization. The collective processes that take part in the reduction of the solubility of dissolved starch are called retrogradation and involve both constituent polymers, amylose and amylopectin, with amylose undergoing retrogradation at a much more rapid rate than that of amylopectin. Gelation and retrogradation can be interpreted as the result of double helices forming a network of physically cross-linked molecules. As initial juncture points grow into helical segments and then aggregate into A–B-type crystallites, gels or retrograded materials become more rigid and difficult to disperse (Flores et al., 2007a).

Grown in tropical areas (Latin America, Asia and Southern Africa) of the world, tapioca (cassava) is used in Latin America as a meal, as animal fodder, or cooked and eaten as a vegetable. Tapioca starch is used to a lesser extent than other starches, like corn, in the food industry. The Food and Agriculture Organization (FAO) highlighted that tapioca is a good commercial cash crop and a major source of food security, and that it needs a competitive edge to thrive in the global starch market. Due to the shortage or high price of traditional starch sources, such as wheat and soybeans, tapioca starch is viewed as an alternative source by food companies for use as an ingredient (FAO, 2004).

Starch based films are reported to be transparent, odorless, tasteless, semi-permeable to CO_2 and resistant to O_2 passage (Nísperos-Carriedo, 1994). However, the application of starch based films is limited by their solubility in water as well as their poor water vapor barrier characteristics (Vásconez et al., 2009).

In order to acquire flexibility and extensibility, the incorporation of a plasticizing agent is a must. The plasticizer has to be compatible with the biopolymeric matrix and be able to disrupt the intermolecular array. Glycerol, polyethilenglicol, sorbitol are the most used plasticizers for edible film preparation (Fernández Cervera et al., 2004). Rodríguez et al. (2006) evaluated the influence of the combined presence of surfactants and plasticizers on the hydrophilicity and on the mechanical properties of starch films. The authors observed a synergistic behavior between the plasticizer and the surfactants (Tween 20).

In addition to their properties of biodegradability, edibility, low oxygen permeability, one of the most remarkable advantages of these films is the possibility of the incorporation of food additives. In that sense, antimicrobials, antioxidants, flavorings, colorants can be supported in film matrix improving the food appearance, texture, taste or storage conditions (Rojas-Graü et al., 2009).

In the particular case of antimicrobials, they can be added to film formulation to delay the growth of bacteria, molds and yeasts during food storage and distribution. Antimicrobial films can act as a barrier to the contamination or as a reserve for the preservative gradual release. Some antimicrobial agents normally added to edible films are: benzoic acid and its sodium salt, sorbic acid and its potassium salt, propionic acid and nisin (Flores et al., 2007a; Buonocore et al., 2003b; Han, 2000).

Sorbic acid and its potassium salt (sorbates) are generally recognized as safe (GRAS) additives and are active against yeast, molds, and many bacteria (Sofos, 1989). These preservatives are unstable in aqueous solution and can suffer an oxidative degradation or can be metabolized by microorganisms under certain conditions of storage (Gerschenson & Campos, 1995). The addition of sorbates to edible films has been proposed as a way of minimizing surface microbial contamination (Flores et al., 2007b). To accomplish this objective, a certain concentration of the preservative must be present at the surface of the product. The reduction of the surface level due to the diffusion in food or due to degradation of the preservative must be taken into account when designing an antimicrobial film (Flores et al., 2007b; Gliemmo et al., 2004).

One interesting natural antimicrobial is nisin. It is an antibacterial peptide produced by *Lactococcus lactis* that effectively inhibits Gram-positive bacteria and also the outgrowth of spores of bacilli and clostridia (Cleveland et al., 2001; Hurst & Hoover, 1993; Hurst, 1981). The mode of action of nisin on sensitive bacteria is based on, first, interaction between the peptide and the cell membrane (Sebti et al., 2003). Nisin is the first antimicrobial peptide with a "generally recognized as safe" status in the United States (Food and Drug Administration, 1988). Its use in various food products is allowed in several countries (Delves-Broughton, 1990), as is the case of certain cheese products in the USA (Ko et al., 2001).

According to previously mentioned facts, it is important to explore the KS and nisin behavior when they are supported in edible films in order to evaluate the usefulness of films as a hurdle for food preservation.

Regarding the active packaging definition, there are several assumptions according to the source of the literature (Ahvenainen, 2003; Fabech et al., 2000; Vermeiren et al., 1999). Active packaging can be considered as a sort of wrapping that changes the condition of the packed food to extend shelf life or to improve safety or sensory properties, while maintaining the quality of the packaged food (Ahvenainen, 2003). Bureau (1996) has stressed the advantages of the use of active material packaging, which can release authorized substances in order to contribute to the product stability. It can be of particular interest in the case of food storage at room temperature. An edible film can be classified as "active packaging" if it has or produces substances ("actives") available to migrate into the food head space or into the food itself, with the objective of getting a technological effect in the atmosphere, in the packaging or in the food. Those "actives" must be an authorized food additive (Fabech, 2000; Han, 2000). The use of edible films as a way to support antimicrobials offers the possibility to provide a highly local functional effect preventing the increase of global additive concentration in the bulk of the food (Giannakopoulos & Guilbert, 1986) or to produce the antimicrobial controlled release into food (Flores et al., 2007c). There are several studies that deal with the use of edible films that support preservatives such as natamicin and potasium sorbate (Flores et al., 2007 a and b; Franssen et al., 2002), for the slow release of lisozime, nisin and propyl parabene (Sanjurjo et al., 2006; Buonocore et al., 2003a; Chung et al., 2001). Considering that the surface microbial growth is the main cause of spoilage for many food products, it is important to remark that the food quality can be affected by diffusion ability of the preservative from the surface to the whole product. According to Guillard et al. (2009) and Guilbert (1988), protective edible coatings can be used to ensure the retention of additives, control the preservative surface concentration and to prevent that the additives freely diffuse into food.

In order to improve the starch based edible film properties, different strategies have been proposed. As an example, it can be mentioned the blending of the starch with other biopolymers such as chitosan (Vasconez et al., 2009; Chillo et al., 2008), gelatin (Arvanitoyannis et al., 1997), gums and the addition of cross-linking agents like sodium trimetaphosphate (Chaisawang & Suphantharika, 2005) or citric acid (Coma et al., 2003). Some reports indicate that the different cross-linking degree of the polymeric matrix can control the relaxation degree of the films in high water activity media and, therefore, it could be possible to contribute to the active substances control release from films (Buonocore et al., 2003b; Rhim, 2004). Another possibility is the incorporation of lipids into the film formulation. It has been reported that oil presence generates a film that combines the advantages of both lipid and hydrocolloid components. The lipid component in the coating formulation can serve as a good barrier to water vapor while the hydrocolloid can provide a selective barrier to oxygen and carbon dioxide and the necessary supporting matrix (García et al., 2000). It has also been reported the diffusive mechanism of KS through edible films elaborated with whey proteins (Ozdemir & Floros, 2001), κ-carragenane (Choi et al., 2005), or also through synthetic materials (Han, 2000). However, systematic information about the sorbate behavior supported in starch based films is not available.

The object of this research was the production of tapioca starch - glycerol based edible films containing the preservatives KS or nisin. Physicochemical properties of films such as crystalline fraction, solubility in water, sorptional behavior and color attributes were studied. In order to optimize the film functionality, the effect on film physicochemical properties and antimicrobial action of: i) soy oil addition to the film formulation or ii) the use of sodium trimetaphosphate-chemically crosslinked-tapioca starch and iii) the influence of the filmmaking technique, were evaluated.

The results will allow obtaining information in relation to the influence of the composition and biopolymer macromolecular structure on the physicochemical and antimicrobial properties of the film in order to optimize its activity and performance as well as the global quality of the food at which the film can be applied. This information will contribute to broaden the knowledge concerning the development of biopolymer based films which can be applied as an additional stress factor to extend food shelf life.

EXPERIMENTAL SECTION

Materials

The films were prepared using the following materials and chemicals:

- Biopolymer and plasticizer: In the basic formulation, tapioca starch used was provided by Bernesa S.A. (Argentina) and the plasticizer was glycerol (Mallickrodt, Argentina) of analytical grade. In the case of lipid containing films, commercial soy oil (Sojola, Molinos S.A., Argentina) was incorporated into the basic film formulation. In the case of the use of modified starch as biopolymer, cross-linked tapioca starch was obtained by the reaction of native tapioca starch with sodium trimetaphosphate (STMP, Sigma, St Louis, Missouri) of analytical grade.
- Antimicrobial agents: potassium sorbate (Sigma, St Louis, Missouri) or a stock solution of nisin ($1x10^5$ IU/ml) prepared by dissolving Nisaplin™ (Aplin & Barret Ltd., Dotset, U.K.) in sterile distilled water; the pH was adjusted to 2.0 with 0.1N HCl to ensure high bacteriocin solubility and solution was stored at - 20ºC until used.

Film Preparation

Native Starch Based Edible Films

Film forming solutions were mixtures of native tapioca starch, glycerol and water (5.0:2.5:92.5 in weight) or native starch, glycerol, antimicrobial agent and water. In the case of films containing KS, 0.3 g of water were replaced by this preservative and in the case of nisin, 15 g of water were replaced by a solution of nisin of a concentration such that each milliliter of final system contained 5000 IU/ml of the antimicrobial.

300 g of film forming solution were heated on a magnetic stirrer with a hot plate and gelatinization was performed at a constant rate of 1.8°C/min for ≅ 30 min. The temperature of gelatinization was ≅ 70 - 75 °C. The sample final temperature was ≅ 82°C. After gelatinization, vacuum was applied to remove air from the systems before casting. The mixtures were casted onto glass plates and dried at 50°C for two hours. The drying was finished at 25 °C over calcium chloride. The films without antimicrobials were denominated **F1** and the films with KS or nisin were called **F1KS** or **F1NIS** respectively.

In the case of systems containing nisin, the antimicrobial was added after gelatinization to preclude affecting nisin activity. The addition was accomplished under agitation to assure system homogeneity.

In order to study the influence of filmmaking techniques on film properties, a different drying process was also assayed for films made from native tapioca starch and containing KS. The drying process was performed in a temperature controlled chamber (Velp, Italy) at 25°C and RH: 80–90% for a week. These films were denominated **F2KS**; a control named **F2** and without sorbate was also made.

Cross-Linked Starch Based Edible Films

In order to obtain cross-linked tapioca starch, 5g of STMP were suspended in 30 mL of distilled water and the pH was adjusted to 10.5 by the addition of solid NaOH (Anedra, Argentina) under stirring. 100 g of native tapioca starch were incorporated into the suspension and the cross-linking reaction was allowed to be carried out for 3.5 hours at 25°C. Then, the pH of the reaction batch was adjusted to 6.5 by the addition of HCl (Anedra, Argentina) 0.2 N. The suspension was filtered under vacuum using a Buchner funnel and glass fiber filter paper. The cross-linked tapioca starch was washed with distilled water in several steps. The total volume of water used for the washing was around 2L (Atichokudomchai & Varavinit, 2002). The cross-linked tapioca starch was freeze-dried for 24 hours to obtain a dry product, packed in polyvinyl – polyvinyliden chloride (Cryovac, Grace, USA) bags, sealed under vacuum and stored at -18°C.

The filmmaking method applied to elaborate cross-linked starch based films was similar to the one described for native starch based films, with the exception that the drying process was completed in a temperature controlled chamber (Velp, Italy) at 25°C and RH: 80–90% for a week. These films were denominated **F2CS**; a control named **F2CS WS** and without sorbate was also made.

Lipid Containing Edible Films

Once the native tapioca starch gel was obtained as described previously and, at a system temperature of 82°C, 0.6 g of soy oil was added to the starch gel leading to an oil concentration of 0.2% (w/w) in the initial suspension. Immediately after the lipid addition, an oil / water emulsion was generated using an Ultra Turrax device (dispersion tool S25N – 18G, Staufen, Germany). The rate selected was 13500 RPM and it was applied for 5 min. Air-bubbles were eliminated under vacuum for 15 min and, then, the suspension was casted onto glass dishes of 5 cm diameter. The drying process was similar to the one described for cross-linked tapioca starch based films. These films were denominated **F2LIP**; a control named **F2LIP WS** and without sorbate was also made.

Film Stabilization

Once constituted, all the films were peeled off from the glass plates and conditioned at 25°C, over a saturated solution of NaBr (water activity, $a_W \cong 0.575$) for 7 days previous to film testing.

Physical Characterization of Edible Films

X-ray Diffraction Analysis

A Philips X-ray diffractometer with vertical goniometer was used (Cu Kα radiation λ = 1.542 Å). The operation was performed at 40 kV and 30 mA. The samples mounted on a glass and conditioned at an a_W of 0.575 were attached to the equipment holder and the X-ray intensity was recorded with a scintillation counter in a scattering angle (2θ) range of 6–33° with a scanning speed of 1 °/min. Distances between the planes of the crystals *d* (Å) where calculated from the diffraction angles (°) obtained in the X-ray pattern, according to Bragg's law:

$$n\lambda = 2d\sin(\theta)$$

where λ is the wavelength of the X-ray beam and *n* is the order of reflection. From the scattering spectrum, the effective percent crystallinity of films was determined, according to Koksel et al. (1993) as the ratio of the integrated crystalline intensity to the total intensity. The crystalline area was evaluated on the basis of the area of the main peaks (main *d*-spacing). Because of the complexity of the system, the calculated crystallinities are not taken as absolute, but are rather used for comparative purposes.

Moisture Determination

The samples were dried in a vacuum oven at 70ºC until constant weight. The determination was performed on five film specimens of each formulation and the average was reported.

The moisture content of the films was determined as:

$$Moisture\ content\ (\%) = \frac{(wet\ mass - dry\ mass)}{dry\ mass} \times 100$$

Solubility in Water

Solubility is defined (Gontard et al., 1992) as the percentage of film dry matter solubilized after 24 h of immersion in distilled water. The initial percentage of dry matter was determined by drying 2 cm diameter disks in a vacuum oven at 100 ºC for 24 h. Disks were cut, weighed and immersed in 50 mL of distilled water, with periodic stirring, for 24 h at 25 ºC. The non-solubilized films were taken out and dried (at 100 ºC for 24 h) to determine the final weight of dry matter. The solubility is reported as the difference between the initial and final dry matter with respect to the initial dry matter as follows:

$$Solubility\ in\ water\ (\%) = \frac{(weight\ of\ initial\ dry\ matter - weight\ of\ dry\ matter\ not\ solubilized)}{weight\ of\ initial\ dry\ matter} \times 100$$

Moisture Sorption Isotherms

Aliquots of 0.3-0.4 g of each sample were distributed in glass containers (30 mm-diameter and 20 mm-height). For each system, three samples were assayed. The samples were equilibrated at 25°C in desiccators containing $CaCl_2$ (Anedra, Argentina).

The moisture sorption isotherms of the films were determined at 25ºC according to the standard gravimetric method (Mathlouthi, 2001). Ten different humidity conditions (11.3, 22.5, 32.8, 43.1, 57.6, 75.3, 84.3, 90.1, 94.3 and 97.3 % R.H.) were obtained by using saturated salt solutions of LiCl, $KC_2H_3O_2$, $MgCl_2 . 6H_2O$, K_2CO_3, BrNa, ClNa, KCl, $BaCl_2$ and K2SO_4 respectively (Greenspan, 1977). The films were allowed to reach equilibrium, evaluated as less than 0.2% change in the sample mass. The sample weights were measured to the nearest 0.0001g at 25 °C for each relative humidity specified above.

The equilibrium moisture content was determined by drying samples in a vacuum oven at 70ºC until constant weight.

The experimental moisture sorption values were averaged (n=3) and fitted to the Oswin empirical model:

$$m = a\left(\frac{a_w}{1-a_w}\right)^n$$

where m is the moisture content of the films (g / 100g dry basis, db), a_w is the water activity (RH/100), a and n, are the fitting parameters. This equation fits adequately curves with a sigmoidal shape (Chen, 1990).

Color Evaluation

Film disks of an appropriate diameter were rested on white background standard (Trezza & Krochta, 2000). The measurements were performed in a Minolta colorimeter (Minolta CM-508d, Tokyo, Japan) using an aperture of 1.5 cm-diameter. The exposed area was sufficiently great relative to the illuminated area to avoid any light trapping effect. The Hunter parameters: *L*, *a* and *b*, and the yellow index (*YI*) were measured according to a standard test method (ASTM E1925, 1988) in, at least, five positions randomly selected for each sample. The color parameters ranged from $L = 0$ (black) to $L = 100$ (white), $-a$ (greenness) to $+a$ (redness), and $-b$ (blueness) to $+b$ (yellowness).The standard values considered were those of the white background. Calculations were made for D-65 illuminant and 2º observer.

Water Vapor Permeability

The water vapor permeability (WVP) of films was determined gravimetrically, using a modified ASTM E96-00 (2000) procedure. The permeation cell (acrylic cups) had an internal diameter of 4.4 cm, an external diameter of 8.4 cm and a depth of 3.5 cm (exposed area: 15.2 cm^2). They contained $CaCl_2$ (Anedra S.A., Argentina) to generate a 0% relative humidity (0% RH) inside the cell. The film was located between the cell and its acrylic ring shaped cover which was adjusted to the cup with four screws. A 10 mm air gap was left between the film and the $CaCl_2$ layer. A rubber O-ring and vacuum grease helped to assure a good seal. The covered cell was placed in a temperature and RH controlled chamber (Ibertest, Madrid, Spain) maintaining a temperature of 30 ºC and an RH of 70%. After 20–24 h, a stationary water vapor transmission rate was attained and, from that moment on, changes in the weight of the cell (to the nearest 0.1 mg) were recorded daily over a 6-day period. All the tests were conducted in triplicate, the WVP values were calculated using the WVP Correction Method described by Gennadios et al. (1994) and the means were reported.

Evaluation of Antimicrobial Activity of Edible Films

For the purpose of determining the performance of KS supported in films, the effectiveness of the antimicrobial released into a liquid medium or acting as a barrier to yeast contamination of high a_w products was studied.

As stated in previous papers, filmmaking techniques have no influence on the film antimicrobial performance (Flores et al., 2007b). Consequently, the **F2** series film methodology (gelatinization rate: 1.8 ºC/min and drying at 25 ºC for a week) was selected to obtain the cross-linked tapioca starch based edible films or with lipid addition, in order to test the influence of the formulation changes on the antimicrobial activity of the film.

The film's antimicrobial effectiveness was evaluated using as an indicator *Zygosaccharomyces bailii.* The spoilage yeast *Z. bailli,* is known for its resistance to several stress factors commonly used in food elaboration such as the decrease in pH, the incorporation of high levels of sugar, pasteurization and the addition of lipophilic preservatives (Warth, 1977). These particular characteristics of *Z. bailii* make it very important to study the conditions to minimize its growth in order to ensure the proper quality of foods.

Inoculum Preparation

A *Zygosaccharomyces bailii* NRRL 7256 inoculum was prepared in Sabouraud broth (Biokar Diagnostics, Beauvais, France) at 25°C until an early stationary phase was achieved (24 h).

Effectiveness for Controlling Microbial Growth of Sorbates Released into a Liquid Medium

In order to evaluate the performance of films, the preservative release and its protective action against yeast in a liquid medium at pH 3.0 or 4.5 was evaluated. Twenty-two discs (1.3–1.4 cm diameter) of each type of edible film, weighing approximately 1.3 g, were introduced into 250 mL glass flasks containing 100 mL of Saboureaud broth with the pH being adjusted to 3.0, 4.5 with citric acid (50 % w/w) and inoculated with 3–5×10^6 colony forming units per mL (CFU/mL) of *Z. bailii.* The flasks were shaken at 150 rpm by means of an orbital shaker (Shaker Pro, Vicking S.A., Buenos Aires, Argentina) for 7 days at 25ºC and, at selected times (0, 5, 10, 22, 26, 30, 48, 54, 72, 96 and 144 h), the samples were collected and the microbial growth was evaluated. Analogous assays were performed using films without sorbates (**F2**, **F2CS WS**, **F2LIP WS**) to test the effects of other components of the film on the microbial growth.

Z. bailii growth curves were modeled using the Gompertz equation (Gliemmo et al., 2006):

$$y = N_0 + C\left\{\exp\left[-\exp\left(\frac{2{,}78\mu}{C}\right)(lag - t) + 1\right]\right\}$$

This equation expresses the change in the microorganism population (y = log CFU/mL) as a time function (t). The biological growth parameters of the yeast are: the specific growth rate (μ), delay before yeast growth or latent phase (*lag*) and the maximum population reached in the stationary phase. The latter can be calculated as (N_0 + C), being C the difference between the final and initial N_0 cell count.

In the case of the death of microorganisms, the count reduction rate was determined applying a linear fitting to the experimental data.

Effectiveness of Sorbate Containing Films as Barriers to Yeast Contamination

In order to study the performance of the films to prevent the microbial contamination of a high a_w product, Sabouraud agar with a_w depressed to 0.980 by the addition of glucose and pH adjusted to 4.5 with citric acid (50 % w/w) was formulated to resemble that kind of products. Disks of 1 cm diameter were aseptically cut from the films and applied onto the surface of the agar. Then, 10 µL of a culture of *Z. bailii* containing approximately 7×10^6 CFU/mL was seeded onto the film disks. The samples were incubated at 25°C for 48 h.

At selected times, two disks were sampled and suspended, each one, in 1 mL of peptone water (Biokar Diagnostics, Beauvais, France) placed into a short glass tube (16 x 100 mm) and shaken for 2 min at 2500 rpm with a vortex, previous to enumerating *Z. bailii* populations. The results were expressed as antimicrobial activity which is defined as the log of the ratio between (CFUt / g film) and (CFUt_0 / g film), being t the time at which the disks were sampled and t_0 the initial time.

Enumeration of Z. bailii

For all assays performed, the *Z. bailii* population was enumerated by surface plating on Sabouraud agar and incubation at 25°C for 5 days prior to counting.

Potassium Sorbate Content

The KS content of the films and broths was measured according to the Association of Official Analytical Chemists (AOAC 1990) oxidation method which includes steam distillation followed by oxidation to malonaldehyde and measurement at 532 nm of the pigment formed between the malonaldehyde and thiobarbituric acid. The determinations were performed in duplicate.

Statistical Analysis of Data

The physicochemical and microbiological results are reported on the basis of their average and standard deviation ($n \geq 2$). A nonlinear or linear regression analysis was applied to model water sorption isotherms and kinetics of microbial growth. The parameters obtained were analyzed through the analysis of variance (ANOVA, α: 0.05) and the Tukey's or least significant difference (LSD) post tests (Sokal & Rohlf, 2000) was applied to establish significant differences between the parameters.

The linear and nonlinear regression and statistical analysis were performed using the Statgraphics Plus program for Windows, version 3.0, 1997 (Manugistics, Inc., Rockville, Maryland, U.S.A.).

RESULTS AND DISCUSSION

Physical Characterization of Edible Films

X-ray Diffraction Analysis

Starch granules from tubercles typically have a B-type crystalline structure and this kind of array presents peaks at 15.8–16.0 Å (4.0 ° 2 θ); 5.9 Å (15.0 ° 2 θ); 5.2 Å (higher intensity) and a medium intensity doublet at 4.0 y 3.7 Å (21.9 ° y 24.0 ° 2 θ respectively). It is also observed, in starches that have suffered a gelatinization process, the V-type crystalline structure which is a result of a stable amylose simplex helix being complexed with substances such as aliphatic fatty acids or related compounds (Liu, 2005). Hydrated V-conformation has spatial-d peaks at 12.0; 6.8 y 4.4 Å (8.9 °, 13.0 ° y 20.1 ° 2 θ, respectively) and the dehydrated V-type crystal structure presents peaks at 11.3; 6.5 y 4.3 Å (Manzocco et al., 2003; Fanta et al., 1999).

The molecular spatial order in starch based edible films is a consequence of the interactions between the amylose and amylopectin helices. During gelatinization, starch molecules have the possibility of increasing their mobility, especially by solvent assistance, and as a result the original granule structure is disrupted. As starch gel is cooled and dried, regular arrays of the helices emerge and, once drying is completed, a semicrystalline network appears which can be analyzed by X-ray diffraction. From this study, it is possible to analyze the influence of the formulation on the film crystallinity. Crystallinity is one of the most important properties since it affects film physicochemical properties such as mechanical and sorptional behavior.

The X-ray diffraction patterns of the films studied can be seen in **Figure 1**. Panel A shows the X-ray diffraction pattern of films without antimicrobials (**F1**) and with nisin (**F1NIS**) or KS (**F1KS**). For these samples, the most intense peaks were identified and the distances (*d*) between the planes of the crystallites (Å) were calculated from the diffraction angle. The films with the preservatives nisin or KS showed a B–V pattern and a crystallinity of 4.5% and 6.4% respectively, and spatial-*d* of ≅ 3.6, 4.5, 5.2 and 5.9 Å for the KS containing films and 4.5 and 5.2 Å for nisin containing films could be observed. The films without antimicrobials showed more and sharper diffraction peaks and therefore greater crystallinity (16.2%); a B–V-type crystal structure was also observed (Manzocco et al., 2003; Fanta et al., 1999), probably due to the presence of a plasticizer (glycerol), that diffracts for the following main spatial-*d* (arranged in an ascending order) ≅ 3.6, 4.0, 4.5, 5.2, 5.9 Å. According to Zobel (1994), when amylose is in the presence of polar lipids, V-structures could result from gelatinization, both during heating and cooling.

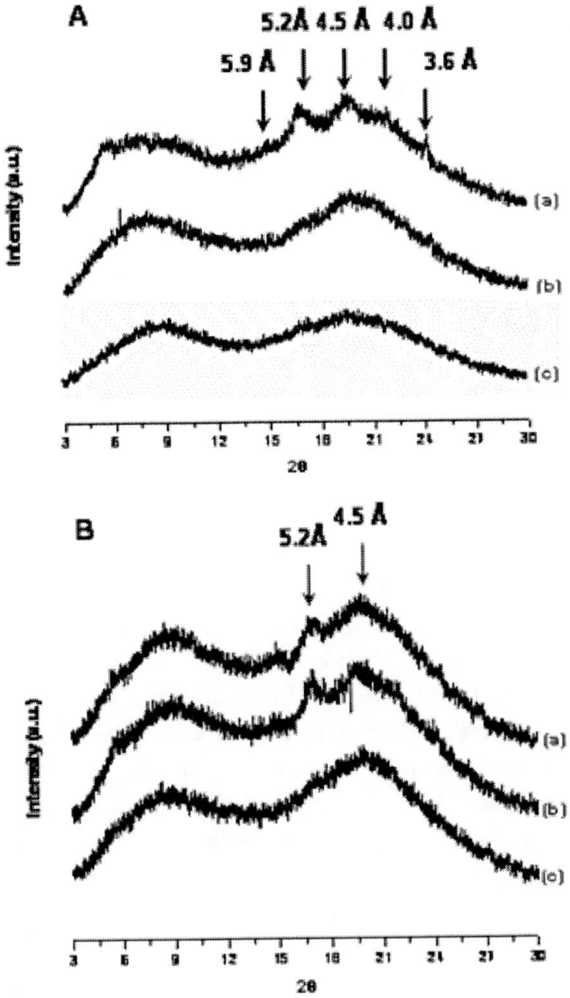

Figure 1. X-ray diffraction pattern of different films studied. **Panel A**: (a): **F1** (without antimicrobials); (b): **F1KS** (with sorbate); (c): **F1NIS** (with nisin). **Panel B**: (a): **F2KS** (with KS and native starch); (b): **F2CS** (with KS and cross-linked starch); (c): **F2LIP** (with KS and soy oil).

The results suggest that the antimicrobials acted as a plasticizer agent disturbing the helix arrays between polymeric molecules, a fact that is reflected in the reduction of the number and the intensity of the diffraction peaks and in a lower crystallinity of the films.

Nisin seems to produce a more important plasticizer effect since these antimicrobial films turned out to be practically amorphous. It could be a consequence of its more extended spatial conformation, as a result of its voluminous size in comparison to KS. The type of plasticizer used in film formulation has a strong influence in the final physical properties. As an example, Zhang & Han (2006) made pea starch films plasticized with monosaccharides and polyols. The authors concluded that the molecular size, configuration, total number of functional hydroxyl groups of the plasticizer as well as its compatibility with the polymer could affect the interactions between the plasticizers and starch molecules and, consequently, the effectiveness of plasticization.

It could be also mentioned that the films **F1KS** showed a lower degree of crystallinity than the **F2KS** (**Figure 1**, panel B) as a consequence of the faster drying rate involved in the elaboration of the **F1KS** films. Flores et al. (2007a) concluded that the gelatinization technique and drying method used to obtain edible films affected network characteristics determining changes in physical properties, such as the ratio of crystalline / amorphous zones.

Figure 1, panel B shows the X-ray diffraction pattern of films supporting KS and elaborated with native or cross-linked tapioca starch (**F2KS** and **F2CS**) or with native starch with lipid addition (**F2LIP**). For these films, only two spatial–*d* means of 4.5 and 5.2Å, corresponding to a B-V type pattern were established. According to Xie et al. (2005) and Dumoulin et al. (1998) an increase in starch cross-linking degree could produce a reduction of the crystallinity because the linking points could limit the chain mobility impairing, consequently, the formation of double helix ordered structures. However, Del Ville et al. (2003) reported that the starch retrogradation could not be totally prevented and crystalline zones were observed when wheat starch based films were subjected to a photo-cross-linking process. In addition, Atichokudomchai & Varavinit (2002) observed that neither the crystallinity nor the fusion enthalpy suffered a significant change when comparing tablets of native tapioca starch with tablets of STMP cross-linked tapioca starch. Finally, it could be mentioned that Le Tien et al. (2000) reported that the gamma irradiation cross-linking of films made from whey proteins, conferred a more stable and ordered structure since sharper peaks were observed in the diffraction X-ray pattern of treated films. The films **F2CS** and **F2KS** showed similar crystalline fraction (~8.6 %). On the contrary, the films formulated with lipid were practically amorphous (crystalline fraction 2.6%). In this case, the filmmaking process involved a homogenization step (emulsion generation) including a very intense shear. As a result, the gel viscosity was drastically reduced, in part because of the breaking of the gelatinized starch granules, fact that might produce a reduction of the order that can be attained during retrogradation. On the other hand, soy oil can act as a plasticizer agent, blocking interactions between chains and, therefore, restricting molecular ordering. Sebti et al. (2002) reported that a hydroxypropyl methylcellulose (HPMC) based film with a fifteen percent (w/w) addition of stearic acid improved the moisture barrier, whereas the film mechanical resistance was reduced, attributing this to the partial replacement of the polymer by the lipids, in the film matrix, creating discontinuities within the HPMC network, favoring the disruption of the network and, consequently, producing a decrease in the quality of mechanical properties.

Table 1 shows the moisture content of the films studied. It can be observed that the moisture content (% db) for all the samples casted by the rapid method (**F1, F1KS, F1NIS**) did not differ significantly. It is normally accepted that amorphous zones allow holding a relative high quantity of water through hydrogen bond interactions, while crystalline zones restrict the amount of water according to the conformation of crystal lattice (Zobel, 1994). Rindlav et al. (1997) reported that the water content of potato starch films increased with increasing B-type crystallinity, which was partly explained by the higher water content of the B-type crystalline areas as compared with amorphous areas. However, the results obtained in the present work suggest that, probably, the antimicrobials might interfere with amylose packing through the development of polymer-preservative hydrogen bonds which replaced polymer–polymer interactions (Yang & Paulson, 2000) and might inhibit the formation of polymer–water hydrogen bonds in the amorphous areas. It is also possible that the water incorporated in the B-type crystalline structure present in a higher proportion in films without antimicrobials has compensated the diminishing of water sorbed in the decreased amorphous structure. It is interesting to mention the work published by Godbillot et al. (2005) who reported that below 44% equilibrium relative humidity, glycerol-plasticized wheat starch films were less hygroscopic than unplasticized starch films. This fact could reveal the capacity of certain substances, such as glycerol, to block the entrance of water to the network due to the glycerol - starch interaction.

Table 1. Effect of the formulation on solubility and moisture content of tapioca starch based edible films

Film	Solubility (%)	Moisture content (g / 100 g db)
F1	20 ± 2	47 ± 1 [b]
F1KS	33 ± 3 [a, d]	49 ± 4 [b]
F1NIS	32.0 ± 0.4 [a]	46 ± 5 [b]
F2KS	30 ± 2 [d]	42 ± 1
F2CS	26 ± 1	38 ± 1 [c]
F2LIP	21 ± 4	37.0 ± 0.5 [c]

Values are the average ± standard deviation (n=3) of the measurements.
Values, in each column, followed by the same letter are not significantly different (α:0.05).
db: dry basis.

Regarding filmmaking technique, **F1KS** films moisture content was higher than **F2KS**. Flores et al. (2007a) observed that for the different techniques of fabrication assayed, moisture content of tapioca starch films was dependent of the filmmaking methodology but moisture did not change in the absence or presence of sorbate. However, the crystalline degree and Young's modulus diminished when the antimicrobial was present. On the other hand, Che & Rhee (2002) reported a slight increase in moisture for soy protein films when they were plasticized using a 0.3, 0.5 or 0.7 g plasticizer/g protein.

It is important to remark that, water molecules present in film structure could affect its physical properties, such as mechanical resistance and gas permeability. In general, water produces a weakening of the film matrix, reducing its performance in tensile and gas barrier tests. However, the relation of crystalline / amorphous regions in the film seems to rule its mechanical properties more than its moisture content.

Table 1 shows that native starch based edible films containing KS (**F2KS**) presented higher moisture content than films made from cross-linked starch (**F2CS**). The STMP cross-linking reaction produces ester groups between a phosphate group of STMP and two HO- of starch polymeric chains (distarch phosphate). This reaction involves the formation of covalent bonds which reinforce the starch three-dimensional structure and, at the same time, reduces the number of hydroxyl groups available to interact with water molecules, allowing in this way to reduce the moisture sorption.

Lipid containing films (**F2LIP**) also showed reduced moisture content in relation with **F2KS** films (**Table 1**). Although oil addition exerted an important plasticizer effect and, therefore, gave origin to a more amorphous structure, the results obtained indicate that the lipid presence reduced the amount of starch-water interactions determining a decrease of the moisture content. Probably, the hydrophobic nature of the soy oil generated a reduction of the film hydrophilicity and, accordingly, a lower affinity for water molecules (Yang & Paulson, 2000) being this one the prevailing effect.

Solubility in Water

Another important film characteristic is its solubility, because it might condition the uses of the films for technological situations. **Table 1** shows the solubility in water of the films with and without antimicrobials. It can be observed that the presence of preservatives in the film formulation increases significantly (α: 0.05) the solubility. As was previously mentioned, the incorporation of nisin or KS into the films produced an increment of amorphous zones, enhancing interactions with water; this can increase solubility in water and decrease film integrity in contact with the solvent. Antimicrobial-free films showed a restricted interaction of water with HO- groups of starch since those groups are involved in the crystalline lattice. No significant differences in solubility were observed in relation to filmmaking technique (**F1KS** vs. **F2KS**) or when comparing films supporting nisin or KS (**F1KS vs. F1NIS**).

The solubility values for **F2CS** and **F2LIP** films can also be seen in **Table 1**. For those formulations, solubilities in water were significantly lower than those observed for films **F2KS**. This interesting result shows that the interaction of the film matrix with water can be reduced by changes in the formulation, improving the film resistance to water action. In the case of **F2CS** films, the cross-linking degree reinforces the starch matrix effectively reducing the water-polymer interaction points capable of promoting the film swelling and further solubilization. Demirgöz et al. (2000) obtained composite films using corn starch cross-linked with STMP and cellulose acetate. These films had lower water absorption capacity and degradation rate in an aqueous medium as a result of the cross-linking process. Furthermore, Seker & Hanna (2006) reported that the water absorption index was inferior when extruded corn starch was put in contact with increasing levels of STMP. For the **F2LIP** system, the addition of lipids to the formulation helped to reduce the film hydrophilicity and, therefore, films resulted less affected by the immersion in water. Kim & Unsutol (2001) reported that lipid incorporation reduced the solubilities of whey protein / lipid-emulsion edible films plasticized with glycerol or sorbitol, as a consequence of the decrease of the hydrophilicity of film components.

Moisture Sorption Isotherms

Figure 2 describes the sorptional behavior of the studied films.

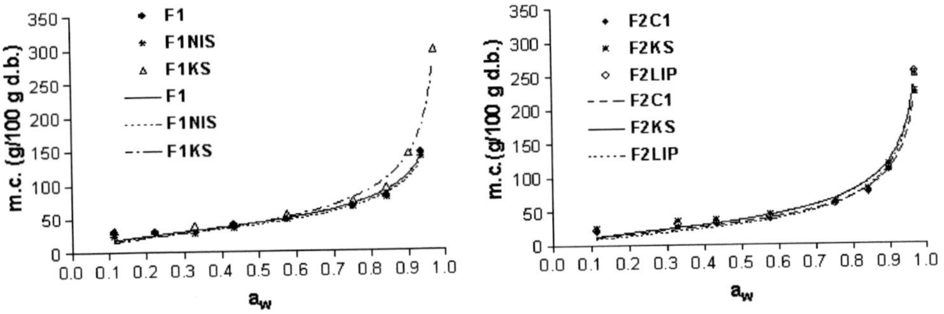

Figure 2. Water sorption isotherms of starch based films. Panel A: films containing KS (**F1KS**) or nisin (**F1NIS**). Panel B: films made with a different technique and containing KS: Native tapioca starch (**F2KS**), cross-linked tapioca starch (**F2CS**), native starch and soy oil (**F2LIP**). Experimental moisture values reported are the average of three measurements (n=3). Oswin model was used for the fitting of experimental data.

As can be observed in **Figure 2**, Panel A, in general, the curves revealed a slight increment in the moisture value with the increase of a_w, until a value of 0.6 approximately, for all films evaluated. Then, the films moisture content rose sharply between 0.75 and 0.97 a_w values, specially for **F1KS** films, suggesting a major affinity for water vapor molecules for this film. Data could be properly fitted to the Oswin equation which has been used to describe the sorptional behavior of protein and starchy foods (Chen, 1990). The Oswin parameters can be observed in **Table 2**. The *a* parameter for rapid method films ranked from 41.0 to 44.3 %db being these values statistically similar (α: 0.05). On the other hand, the Oswin *n* parameter was significantly higher for **F1KS** films suggesting that the addition of KS to film formulation generated a matrix with a greater affinity for water, in particular at high a_w values, and in comparison with the antimicrobial-free film and with nisin supporting film.

Table 2. Oswin parameters for sorptional behavior of tapioca starch based edible films

Film	Oswin Parameters		
	a (% db)	*n*	R^2
F1	43.9 ± 2.7^a	0.429 ± 0.028^b	0.9398
F1KS	44.3 ± 2.3^a	0.528 ± 0.017^e	0.9826
F1NIS	41.0 ± 1.7^a	0.441 ± 0.021^b	0.9692
F2KS	37.5 ± 1.9	$0.541 \pm 0{,}012^{d,e}$	0.9932
F2CS	33.5 ± 1.2^c	$0.540 \pm 0{,}017^d$	0.9838
F2LIP	30.3 ± 1.9^c	$0.609 \pm 0{,}020$	0.9981

Best values of fitting are informed ± standard error.

Values, in each column, followed by the same letter are not significantly different (α: 0.05).

R^2: the goodness of the fit.

db: dry basis.

With respect to the effect of the filmmaking method, it can be observed in **Table 2** that the **F1KS** film presented a significantly higher *a* value in comparison with **F2KS** films. This trend can be explained considering the higher proportion of amorphous regions present in the first film as was indicated by the cristallinity analysis performed with the X-ray diffraction technique. It is important to remark that the moisture content of the **F1KS** films was higher than the one of the **F2KS** films for all the range of a_w assayed. No significant differences (α: 0.05) in *n* parameter were observed for these films.

Figure 2, panel B, shows that the lipid containing (**F2LIP**) and the cross-linked starch (**F2CS**) films, presented lower moisture content in the a_w range 0.11 – 0.84 than the film **F2KS**. After this point, the water content increased rapidly attaining values similar to those of **F2KS** film. The *a* parameter of **F2KS** was significantly higher than the one of **F2LIP** while the *n* parameter of **F2LIP** is the highest one for this filmmaking technique (**Table 2**).

Color Evaluation

Table 3 shows the color parameters for the films studied. In a direct naked-eye examination, it was possible to appreciate that films with nisin (**F1NIS**) were darker, brown colored and opaque. These trends were reflected in their significantly higher *a* and *b* parameter values and in the lower *L* value in comparison with those values for the films without nisin. The Hunter *a* parameter was negative in the absence of nisin and turned to a positive value in its presence. Coincidently, the *YI* value was significantly higher for nisin containing films fact that might compromise the organoleptical acceptability. The presence of sorbate increased *YI* but in a lesser degree. Free antimicrobial films were the less colored systems rendering the lowest *b* and *YI* values. According to these results, the presence of preservatives in formulation produces darker films, being this effect less significant when KS was added. For the **F1KS** films, sorbate degradation would be the responsible for the browning development (Gliemmo et al., 2004).

Table 3. Color parameters of tapioca-starch edible films

Films	a[1]	b[1]	L[1]	YI[2]
F1	-1.30 ± 0.02	4.2 ± 0.3	86.2 ± 1.1	8.08 ± 0.70
F1KS	-0.69 ± 0.04	5.6 ± 0.1[c]	83.2 ± 0.3[d]	11.9 ± 0.2[e]
F1NIS	0.52± 0.13	12.1 ± 0.4	78.7 ± 0.7	29.1 ± 1.2
F2KS	-0.82 ± 0.03	5.6 ± 0.2[a, c]	83.0 ± 0.5[d]	12.0 ± 0.4[b, e]
F2CS	-1.43 ± 0.03	4.7 ± 0.1	86.5 ± 0.3	8.9 ± 0.2
F2LIP	-1.36 ± 0.05	5.8 ± 0.2[a]	85.7 ± 0.4	11.6 ± 0.5[b]

Values are the average ± standard deviation (n=5) of the measurements.
Values, in each column, followed by the same letter are not significantly different (α: 0.05).
[1]: Hunter parameters.
[2]: Yellow Index.

When cross-linked tapioca starch was used for film elaboration, the films with the lowest *YI* and *b* values and, at the same time, with the highest luminosity (*L*) were obtained (**Table 3**). It can be observed that those parameters were similar to the ones of the **F1** films.

Probably, the lower water uptake of the cross-linked starch reduced the KS degradation rate and exalted the transparency of the films and determined the decrease of yellow color.

The filmmaking technique did not affect the color parameters since there were no significant differences (α: 0.05) in *b*, *L* and *YI* of **F1KS** and **F2KS** samples (**Table 3**). The difference in *a* value was small.

Edible films containing soy oil presented similar *YI* and *b* parameters to those of the **F2KS** films, indicating that lipid incorporation did not introduce significant changes in film color.

Water Vapor Permeability

Another relevant characteristic of edible films is their capacity to act as a gas barrier. In the present work, the effect of formulation on the WVP of the **F2** series films was studied.

Several studies have revealed that lipid incorporation into the formulation of hydrophilic films, could improve their water barrier properties (García et al., 2000; Kester & Fennema, 1986; Anker et al., 2001; Ayranci & Tunc, 2000). However, according to our results, all the systems analyzed showed similar (α: 0.05) WVP values: 19.1 – 20.6 x 10^{-10} g / s m Pa. Probably, the high proportion of the amorphous phase of the **F2LIP** film counterbalance the increase of hydrophobic character of the films containing lipids. Moreover, the cross-linking of the tapioca starch did not affect the WVP characteristics of the films.

Evaluation of Antimicrobial Activity of Edible Films

Effectiveness for Controlling Microbial Growth of Sorbates Released into a Liquid Medium

The ability to control the *Z. bailii* growth of the KS incorporated into films and released into Sabouraud broth adjusted to pH 4.5 or 3.0, can be appreciated in **Figure 3**. The Gompertz or linear fitting is included in the figure. As can be appreciated in panel A, when sorbates came from the **F2LIP** films and the receiving medium had a pH value of 4.5, the lag phase was approximately 15 hours and the stationary phase was reached around 72 hours. However, for free-lipid formulation, the lag phase was extended to 30 hours and the stationary phase of growth was observed for 96 hours of incubation. For the pH 4.5 broth, the control system without preservatives did not inhibit the yeast growth and the stationary phase was reached after 48 hours, approximately.

The results suggest that the antimicrobial released from the films and into the pH 4.5 broth, produced a concentration lower than the minimal inhibitory level needed to prevent yeast growth in the medium, allowing the microorganism to develop. The KS concentration determined in the broth for lipid containing the edible films **F2LIP** was 0.44 g/L at pH 4.5. This value did not differ significantly from that founded for the **F2KS** films. Therefore, the reduction observed in the capacity of inhibition of sorbates released in the presence of soy oil, reveals an antagonistic effect between these components.

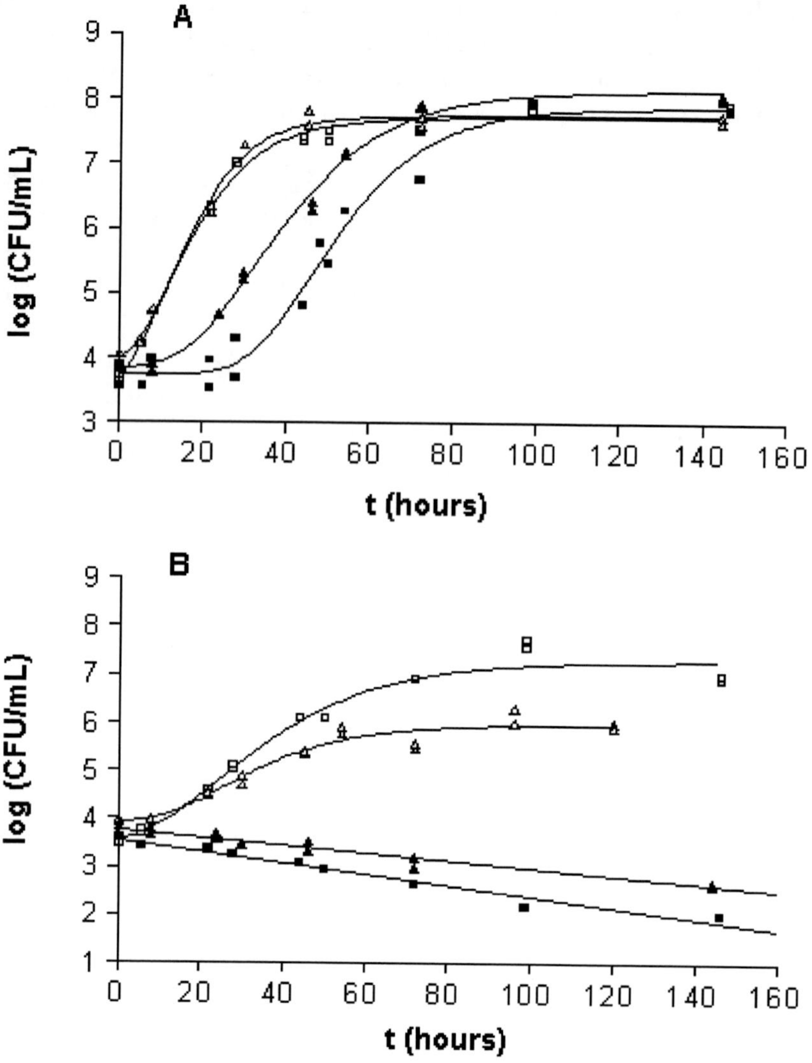

Figure 3. *Z. bailii* growth in Sabouraud broth adjusted to different pH values. CFU/ml: colony forming units per ml at any time. Panel A: pH 4.5; panel B: pH 3.0. ▲, Δ, broth with **F2LIP** films; ■, □ broth with **F2KS** films. Full symbols: Films supporting KS. Empty symbols: films without sorbate. —— Gompertz or linear fitting. Data for **F2KS** is adapted from Flores et al., 2007b.

The **F2LIP** films did not maintain their integrity when in contact with the liquid medium under constant stirring. As a consequence, the lipid could diffuse to the receiving broth. Castro et al. (2003) reported that the sorbate distribution between the aqueous and oil phase of the emulsion systems decreased the KS antimicrobial effect. As a consequence, it can be concluded that the lipid contained in the film might affect the sorbate efficiency as an antimicrobial.

The **F2LIP** and **F2KS** films produced a reduction of *Z. bailii* growth of 2 log cycles, approximately, after 140 hours of incubation when the pH broth was 3.0 (**Figure 3**, panel B). On the contrary, the yeast could grow, after a lag phase of around 12 hours, when the antimicrobial-free films were present. The rate of growth was lower than the one observed for pH 4.5.

The effectiveness to control the *Z. bailii* growth of KS supported in films made from cross-linked tapioca starch (**F2CS**) and released into a liquid medium can be seen in **Figure 4**, panels A and B. For comparison, the results obtained for the films obtained from native tapioca starch and containing KS, were added to the figure (**F2KS**). The Gompertz curves or linear tendency were also included in the figure. As can be appreciated, the KS effectiveness was similar for both films analyzed. The yeast lag phase in pH 4.5 broth was extended to 30 hours and the stationary phase was reached after approximately 96 hours of incubation. The control system, without preservative, showed a similar behavior for both formulations suggesting that no additional inhibitory effect was exerted by the cross-linking: the yeast could develop from the beginning of the incubation at a higher rate than the one observed for systems containing KS, and the stationary phase was reached after 30 hours of incubation. As was previously mentioned, the microorganism development in pH 4.5 broth can be explained considering that the antimicrobial amount released into the medium was around 0.46 g/L, determining a KS concentration lower than the minimum inhibitory concentration (MIC) needed to prevent the yeast growth in this medium.

Figure 4, panel B shows the performance of KS released into a liquid medium of pH 3.0. When KS was present, a reduction of the population of around 2 log cycles after 140 hours of incubation was observed. In this case, the amount of preservative present in the broth was 0.45 g/L which is higher than the MIC for the medium and the pH used (MIC: 0.30 g/L according to Gliemmo et al., 2004) and, consequently, the yeast growth was inhibited. On the contrary, when the KS-free films were immersed in the pH 3.0 Sabouraud broth, the inhibition of the yeast growth was not observed since after a lag phase of approximately 16 hours, the *Z. bailii* grew. The growth rate was lower than the one observed at pH 4.5, reaching a smaller population level (log CFU/mL $\cong$ 6.9).

According to the results obtained, the KS antimicrobial activity was similar when the films were constituted from cross-linked or native starch indicating that the modification of starch did not affect the KS performance.

In **Figure 5**, the Gompertz or linear parameters obtained from the growth data fitting for systems **F2KS**, **F2CS**, and **F2LIP** are summarized. It can be observed that, in general, the performance of sorbate contained in the studied films was similar for pH 4.5 or pH 3.0 receiving media: the growth rate was higher at pH 4.5 than pH 3.0 and the presence of KS in the film formulation delayed the *Z. bailii* development in pH 4.5 Sabouraud broth and promoted the yeast death in the pH 3.0 medium (**Figure 5**, panel A).

The **F2LIP** films immersed in pH 4.5 broth determined a reduction of lag phase in comparison with the **F2CS** and **F2KS** films, probably because the KS distribution between oil and aqueous phases diminished its antimicrobial effectiveness. The maximum yeast population was, in general, the same for all of the studied systems (**Figure 5**, panel C). The KS-free films showed a faster *Z. bailii* development at pH 4.5 than at pH 3.0, indicating the existence of an inhibitor effect exerted *per se* by the pH level on the yeast growth.

Therefore, we can conclude that the starch modification (cross-linking with STMP) did not affect the KS antimicrobial activity in relation with the inactivation of *Z. bailii* inoculated into the liquid medium of pH 4.5 or pH 3.0. However, the lipid addition presented an antagonistic effect (shorter lag phase) when a broth of pH 4.5 was used.

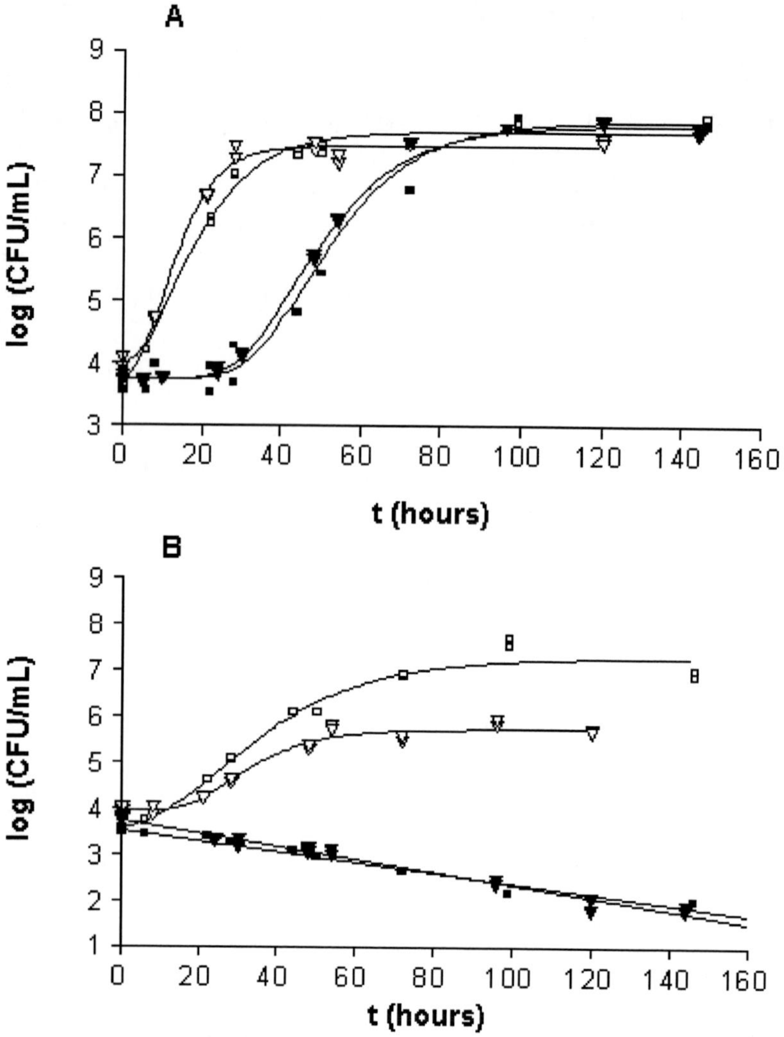

Figure 4. *Z. bailii* growth in Sabouraud broth adjusted to different pH values. CFU/ml: colony forming units per ml at any time. Panel A: pH 4.5; panel B: pH 3.0. ▼, ∇ broth with **F2CS** films; ■,□ broth with **F2KS** films. Full symbols: films with KS. Empty symbols: films without KS. —— Gompertz or linear fitting. Data for **F2KS** is adapted from Flores et al., 2007b.

The performance of nisin supported in tapioca starch based films was studied by Sanjurjo et al. (2006). The authors established that the initial release of nisin into a Tryptic Soy broth enriched with 0.6% (w/v) yeast extract (TSBYE), with different pH values (5.5, 6.5 or 7.2) adjusted with citric acid (50% w/w), and with an initial count of 10^4 CFU/mL of *L. innocua*, produced an important bacterial inhibition, specially when the medium had a pH 5.5. The decrease in microorganism growth was greater (1.5 log cycle) at 1-2 hours of contact and after that, the cell count remained approximately constant (≅ 300 CFU/mL) throughout the experiment (24 hours of incubation at 28 °C). This trend seems to indicate a high initial release of preservative suggesting that the nisin released could inhibit the microbial growth at the early stage of storage. The authors also observed that although the nisin released from films suffered a loss of activity, the gradual delivery of nisin still contained in the film, controlled growth of *L. innocua*, helping to reduce contamination during storage. It was concluded that the gradual release of the antimicrobial from the edible film can help to preclude microorganism proliferation better than nisin directly added because it seems to counterbalance, at least partially, the inactivation of nisin.

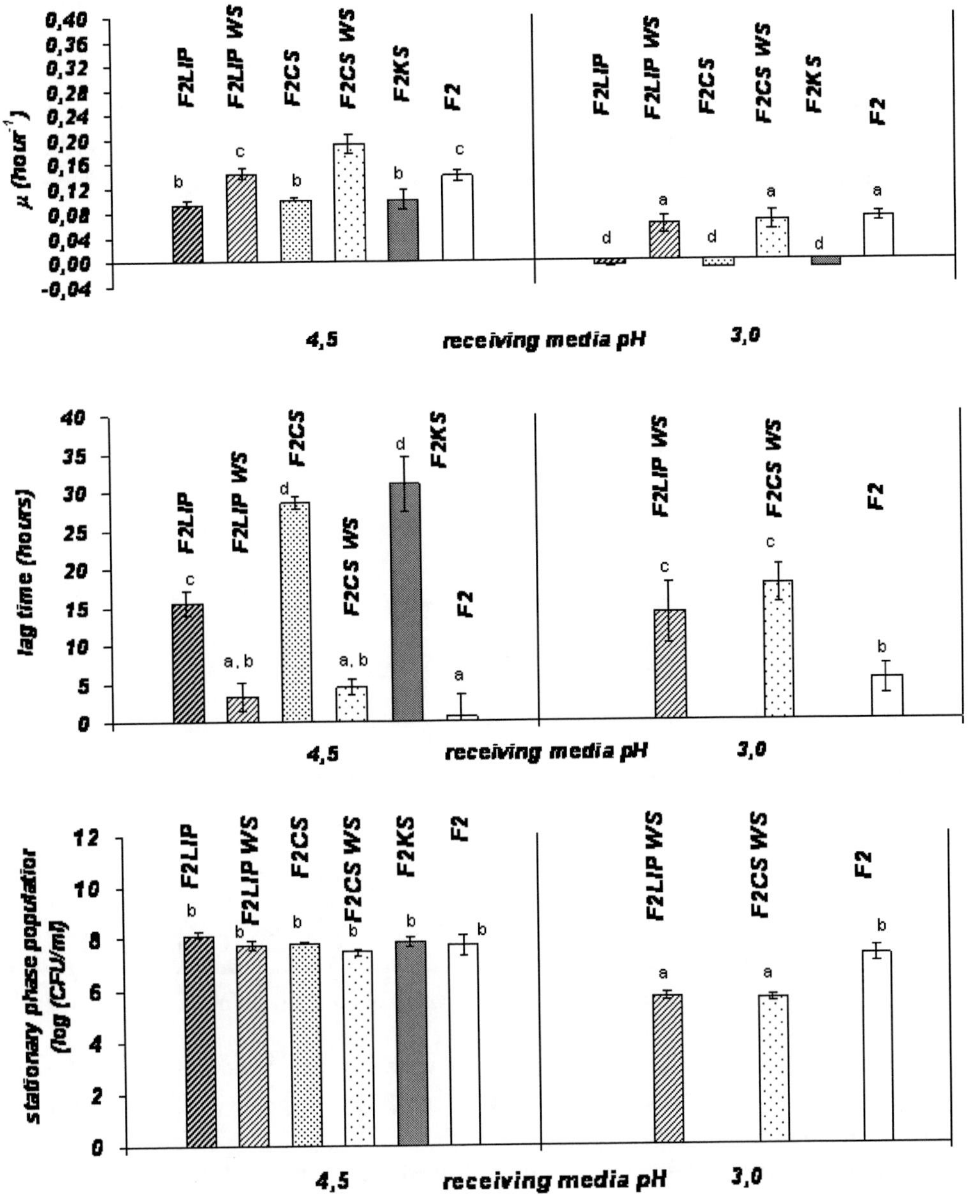

Figure 5. Gompertz and linear parameters obtained by the fitting of *Z. bailii* growth data in Sabouraud broth, at different pHs and containing edible films. Panel A: specific growth rate; Panel B: lag time; Panel C: stationary phase population. Bars with the same letter are not significantly different (α: 0.05). **F2KS**: Film made with native starch and KS; **F2CS**: Film made with cross-linked starch and KS; **F2LIP**: Film made with native starch, soy oil and KS; **F2**: Film made with native starch and without KS; **F2CS WS**: Film made with cross-linked starch and without KS; **F2LIP WS**: Film made with native starch, soy oil and without KS.

Effectiveness of Potassium Sorbate Containing Films as Barriers to Yeast Contamination

For all the studied systems, the antimicrobial effect was also evaluated in semisolid media of a_w and pH controlled. **Figure 6** shows the behavior of films containing or not soy oil and with or without KS. *Z. bailii* growth was delayed for approximately 25 hours when the **F2LIP** film was used as a barrier to yeast contamination. This delay is shorter that the one observed for the films without lipids (**F2KS**) which took a value of 50 hours. On the other hand, the yeast population was 1 log cycle higher for the film **F2LIP** when compared with **F2KS** after 48 hours of incubation. The control films (without SK) of both formulations (**F2LIP WS** or **F2**) did not inhibit the yeast development and the cell population increased 3 log cycles after 48 hours of incubation.

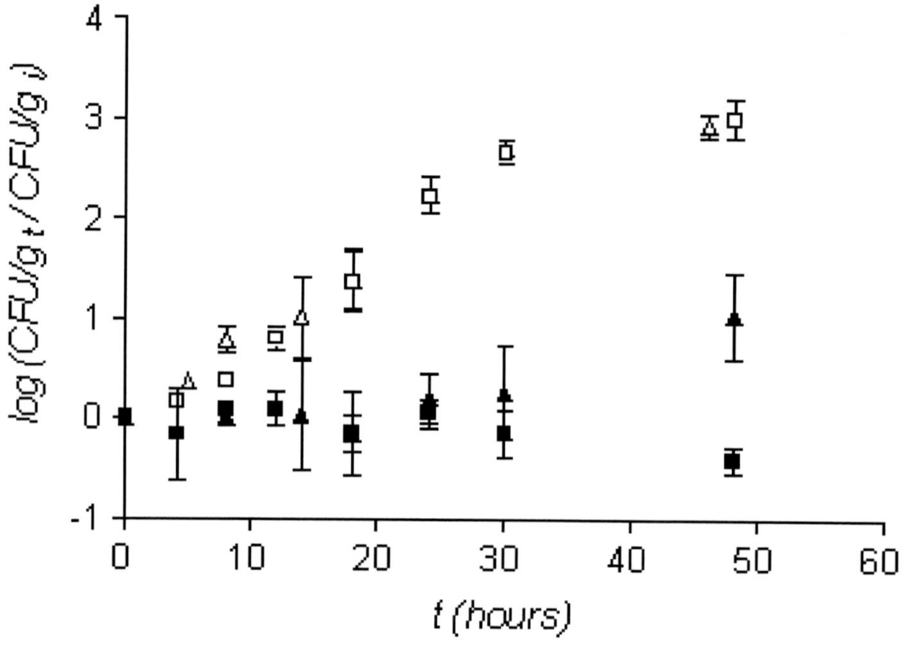

Figure 6. Edible films as barrier against external contamination. CFU / g $_t$: Colony forming units per gram at any time. CFU / g $_i$: Colony forming units per gram at initial time. ▲ **F2LIP**; △ **F2LIP WS**; ■ **F2KS**; □ **F2**. Vertical bars represent the standard deviation of the average (n=3).

For all the studied systems, the antimicrobial levels were evaluated in the agars, resulting in values of 0.126 g/kg and 0.108 g/kg for **F2LIP** and **F2KS** respectively, indicating that KS was released at levels of 90 and 87%, approximately.

When studying the performance of cross-linked starch based films for 48 hours of incubation, it was observed that the *Z. bailii* population was approximately constant, showing a similar performance of these films to that shown by native starch-based films supporting KS (**Figure 7**). On the other hand, when preservative-free films (**F2CS** or **F2**) were used, the agar system suffered a yeast count increase of 3 log cycle after the same time of incubation. The amount of antimicrobial released into the agar from the **F2CS** films was around 0.139 g/kg, amount that represents a 93% of the KS content of the film.

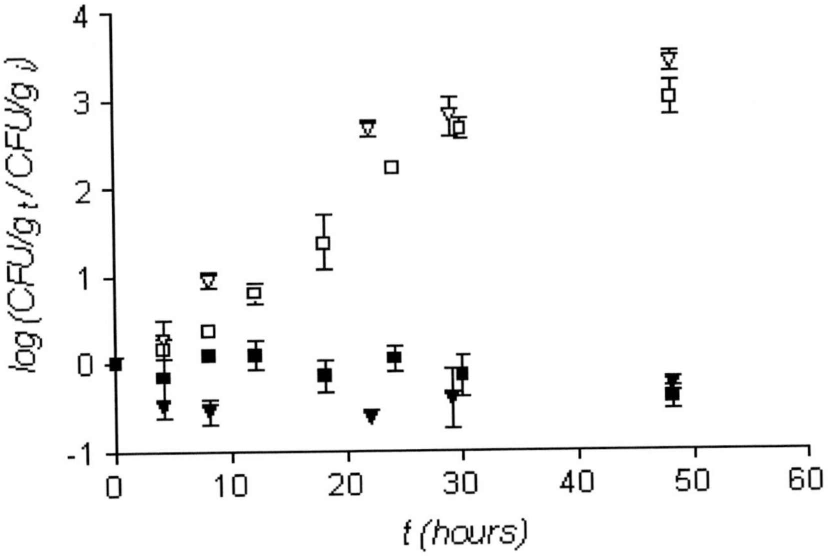

Figure 7. Edible films as barrier against external contamination. CFU / g t: Colony forming units per gram at any time. CFU / g i: Colony forming units per gram at initial time. ▼ **F2CS**; ▽ **F2CS WS**; ■ **F2KS**; □ **F2**. Vertical bars represent the standard deviation of the average (n=3).

Sanjurjo et al. (2006) also studied the performance of starch-based edible films supporting nisin and evaluated them as a barrier against *L .innocua* contamination. The assay was performed using Tryptic Soy Broth Yeast Extract (TSBYE) agar, with a_w reduced to 0.94 with dextrose and pH 5.0, adjusted with citric acid (5.2 mol/L). 1 cm diameter disks were brought in contact with the surface of the agar. Then, 10 μl of a culture of *L. innocua* containing $2 \cdot 10^9$ CFU/ml were dispensed onto the disks. The samples were incubated at 28 °C for 4 hours and the bacterial response was periodically tested. The authors reported a rapid reduction of *L. innocua* viable counts (3 log cycles) at the beginning of the assay followed by an additional but slower inactivation. After 240 hours, the systems attained a count reduction of 4 log cycle when using a film with 5000 IU nisin/ml. On the other hand, the films without the natural antimicrobial did not change the initial counts of studied bacteria.

CONCLUSION

It was observed that antimicrobial presence in the film formulation reduced the crystalline degree, increased the solubility in water but did not change the moisture content in comparison with the films without preservatives. Particularly, nisin presented an important plasticizing effect and enhanced the browning development.

The filmmaking technique applied to obtain KS containing films, affected the crystalline / amorphous ratio since the faster drying method originated a more amorphous starch matrix. Such structural characteristics increased moisture sorption. The Hunter color parameters were not influenced by the technique used to elaborate the films.

The use of cross-linked tapioca starch in film production reduced the solubility and moisture content in comparison with native tapioca starch based films. A reduction of the yellow index and a higher luminosity for this formulation were also observed.

The lipid addition resulted in almost totally amorphous films. However, a lower solubility and moisture sorption was determined as a consequence of the increase of film hydrophobicity when soy oil was incorporated. The water vapor permeability was not influenced by the changes in the formulation of films containing KS and was always high.

Regarding the antimicrobial performance of KS present in studied films, it was observed that the microbial growth was delayed or reduced (depending on the pH of the system) when films were in contact with a liquid medium. In general, films were effective to control the external contamination. The antagonism between soy oil and KS detected affected antimicrobial capacity of films containing lipids.

It can be concluded that studied films have a possible application as a stress factor to control or inhibit microbial growth. By means of changes in formulation, the physicochemical properties can be modified according to the technological application needed.

Acknowledgments

The authors acknowledge the financial support of the National Agency for Scientific and Technological Promotion (ANPCyT), the National Scientific and Technical Research Council of Argentina (CONICET) and the University of Buenos Aires.

References

Ahvenainen, R. (2003). Novel food packaging techniques. Boca Ratón, FL: CRC Press.

Anker, M., Berntsen, J., Hermansson, A. M., & Stading, M. (2001). Improved water vapor barrier of whey protein films by addition of an acetylated monoglyceride. *Innovative Food Science and Emerging Technologies*, 3, 81-92.

AOAC (1990). Official Methods of Analysis (13th). Washington, DC: Association of Official Analytical Chemists.

Arvanitoyannis, I., Psomiadou, E., Nakayama A., Aiba, S., & Yamamoto, N. (1997). Edible films made from gelatin, soluble starch and polyols, Part 3. *Food Chemistry*, 60, 593–604.

ASTM E96-00. (2000). Standard Test Method for water vapor transmission of Materials. Philadelphia: American Society for Testing and Materials.

ASTM E1925. (1988). Standard Test Method for yellowness index of plastics. Philadelphia: American Society for Testing and Materials.

Atichokudomchai, & N., Varavinit S. (2002). Characterization and utilization of acid-modified cross-linked Tapioca Starch in pharmaceutical tablets. *Carbohydrate Polymers*, 53, 263-270.

Ayranci, E., & Tunc, S. (2003). A method for the measurement of the oxygen permeability and the development of edible films to reduce the rate of oxidative reactions in fresh foods. *Food Chemistry*, 80, 423-431.

Buonocore, G. G., Del Nobile, M. A., Panizza, A., Bove, S., Battaglia, G., & Nicolais, L. (2003a). Modeling the lysozyme release kinetics from antimicrobial films intended for food packaging applications. *Journal of Food Science*, 68. 4, 1365-1370.

Buonocore, G.G., Del Nobile, M. A., Panizza, A., Corbo, M. R., & Nicolais, L. (2003b). A general approach to describe the antimicrobial agent release from highly swellable films intended for food packaging applications. *Journal of Controlled Release*, 90, 97-107.

Bureau, G. (1996). Packaging and microbiology. In G. Bureau, & J.L. Multon (Eds.), *Food packaging technology* (1°, pp. 47-56). New York: Wiley-VCH Inc.

Castro, M. P., Garro, O., Gerschenson, L. N., & Campos, C. A. (2003). Interactions between potassium sorbate, oil and Twin 20: its effect on the growth and inhibition of *Z. bailii* in model salad dressings. *Journal of Food Safety*, 23, 47-59.

Chaisawang, M., & Suphantharika, M. (2005). Effects of guar gum and xanthan gum additions on physical and rheological properties of cationic tapioca starch. *Carbohydrate Polymers*, 61, 288-195.

Che, S.Y., & Rhee, C. (2002). Sorption characteristics of soy protein films and their relative to mechanical properties. *Lebensmittel Wissenschaft und Technologie*, 35, 151-157.

Chen, C.C. (1990). Modification of Oswin EMC / ERH equation. *Journal of Agricultural Research of China*, 39, 367-376.

Chillo S., Flores S., M. Mastromatteo M., Conte A., Gerschenson, L., & Del Nobile, M.A. (2008). Influence of glycerol and chitosan on tapioca starch-based edible film properties. *Journal of Food Engineering*, 88, 159–168.

Choi, J. H., Choi, W. Y., Chinnan, M. J., Park, H. J., Lee, D. S., & Park, J. M. (2005). Diffusivity of potassium sorbate in κ-carrageenan based antimicrobial film. *Lebensmittel Wissenschaft und Technologie*, 38, 417-423.

Chung, D., Chikindas, M., & Yan, K. (2001). Inhibition of *Saccharomyces cerevisiae* by slow release of propyl paraben from a polymer coating. *Journal of Food Protection*, 64 (9), 1420-1424.

Cleveland, J., Montville, T. J., Nes, I. F., & Chikindas, M. L. (2001). Bacteriocins: safe, natural antimicrobials for food preservation. *International Journal of Food Microbiology*, 71, 1-20.

Coma V., Sebtia I., Pardonb P., Pichavantb F. H., & Deschamps A. (2003). Film properties from crosslinking of cellulosic derivatives with a polyfunctional carboxylic acid. *Carbohydrate Polymers*, 51, 265–271.

Cutter, C. N. (2006). Opportunities for bio-based packaging technologies to improve the quality and safety of fresh and further processed muscle foods. *Meat Science*, 74, 131–142.

Delves-Broughton, J. (1990). Review: nisin and its application as a food preservative. *Journal of the Society of Dairy Technology*, 43, 73-76.

Delville J., Joly C., Dole P., Biliard C. (2003). Influence of photocrosslinking on the retrogradation of wheat starch based films. *Carbohydrate Polymers*, 53, 373-381.

Demirgöz, D., Elvira, C., Mano J., Cunha, A. M., Piskin, E., & Reis, R. L. (2000). Chemical modification of starch based biodegradable polymeric blends: effects on water uptake, degradation behavior and mechanical properties. *Polymer degradation and stability*, 70, 161-170.

de Moura M. R., Aouada F. A., Avena-Bustillos R. J, McHugh T. H., Krochta J. M., & Mattoso L. H. C. (2009). Improved barrier and mechanical properties of novel hydroxypropyl methylcellulose edible films with chitosan/tripolyphosphate nanoparticles. *Journal of Food Engineering*, 92, 448–453.

Dumoulin Y., Serge A., Szabo P., Cartilier L., & Mateescu M. A. (1998). Cross-linked amylose as matrix for drug controlled release. X-ray and FT-IR structural analysis. *Carbohydrate Polymer*, 37, 361-370.

Dutta P.K., Tripathi S., Mehrotra G.K., & Dutta J. (2009). Perspectives for chitosan based antimicrobial films in food applications. *Food Chemistry*, 114, 1173-1182.

EUFIC, (2002). European Food Information Council, Bulletin Food today, 33, 4.

Fabech, B., Hellstrøm, T., Henrysdotter, G., Hjulmand-Lassen, M., Nilsson, J., Rüdinger, L., Sipiläinen-Malm, T., Solli, E., Svensson, K., Thorkelsson, A., & Tuomaala, V. (2000). Active and Intelligent Food Packaging. A Nordic report on the legislative aspects. Nordic co-operation. Available from: http://www.norden.org/pub/ebook/2000-584.pdf

Fanta, G. F., Shogren, R. L., & Salch, J. H. (1999). Steam jet cooking oh high amylose starch-fatty acid mixtures. An investigation of complex formation. *Carbohydrate polymers*, 38, 1-6.

FAO (2004). Proceedings of the validation forum on the global cassava development strategy. Global cassava market study business opportunities for the use of cassava. Rome.

Fernández Cervera, M., Karjalainen, M., Airaksinen, S., Rantanen, J., Krogars, K., Heinamaki, J., Iraizoz Colarte, A., & Yliruusi, J. (2004). Physical stability and moisture sorption of aqueous chitosan–amylose starch films plasticized with polyols. *European Journal of Pharmaceutics and Biopharmaceutics*, 58, 69-76.

Flores, S., Famá, L., Rojas, A. M., Goyanes, S., & Gerschenson, L. (2007a). Physical properties of tapioca-starch edible films: influence of filmmaking and potassium sorbate. *Food Research international*, 40, 257-265.

Flores, S., Haedo, A. S, Campos, C., & Gerschenson, L. (2007b). Antimicrobial performance of sorbates supported in a tapioca starch edible film. *European Food Research and Technology*, 225, 3-4, 375-384.

Flores, S., Conte A., Campos, C., Gerschenson, L., & Del Nobile, M.A. (2007c). Mass transport properties of tapioca-based active edible films. Journal of Food Engineering, 81, 580–586.

Food and Drug Administration. (1998). Nisin preparation: Affirmation of GRAS status as a direct human food ingredient. FDA, Federal Regulation, 53, 11247.

Franssen, L., Rumsey, T., & Krochta, J. M. (2002). Modeling of natamycin and potassium sorbate diffusion in whey protein isolate films to application to cheddar cheese. 2002 Annual Meeting and Food Expo. Anaheim, California. Session 28. Available from: http://ift.confex.com/ift/2002/techprogram/paper_11523.htm.

García, M., Bifani, V., Campos, C., Martino, M. N., Sobral, P., Flores, S., Ferrero, C., Bertola, N., Zaritzky, N. E., Gerschenson, L., Ramírez, C., Silva, A., Ihl, M., & Menegalli, F. (2008). Edible coating as an oil barrier or active system. In Food *Engineering: Integrated Approaches*, (1°, pp. 225-241. New York: Ed. Springer.

Garcia, M., Martino, M., & Zaritzky, N. (2000). Lipid addition to improve barrier properties of edible starch-based films and coatings. *Journal of Food Science*, 65, 941-947.

Gennadios, A., Weller, C. L., & Gooding, C. H. (1994). Measurement errors in water vapor permeability of highly permeable, hydrophilic edible films. *Journal of Food Engineering*, 21, 395–409.

Gerschenson, L.N., & Campos, C.A. (1995). Sorbic Acid Stability during Processing and Storage of High Moisture Foods. In: G. Barbosa Canovas, & Welti Chanes J (Eds). *Food reservation by moisture control. Fundamentals and applications*. (1°, pp.761-790). Lancaster, PA: Technomic Publishing.

Giannakopoulos, A., & Guilbert, S. (1986). Determination of sorbic acid diffusivity in model food gels. *Journal of food technology*. 21, 339-353.

Gliemmo, M. F., Campos, C. A., & Gerschenson, L. N. (2006). Effect of several humectants and potassium sorbate on the growth of *Zygosaccharomyces bailii* in model aqueous systems resembling low sugar products. *Journal of Food Engineering*, 77, 761-770.

Gliemmo, M. F., Campos, C. A., & Gerschenson, L. N. (2004). Effect of sweet humectants on stability and antimicrobial action of sorbates. *Journal of Food Science*, 69, 39-44.

Godbillot, L., Dole, P., Joly, C., Roge, B., & Mathlouthi, M. (2005). Analysis of water binding in starch plasticized films. *Food Chemistry*, 96, 380-386.

Gontard, N., Guilbert, S., & Cuq, J. L. (1992). Edible wheat gluten films: Influence of the main process variables on film properties using response surface methodology. *Journal of Food Science*, 57, 190-199.

Greenspan, L. (1977). Humidity fixed points of binary saturated aqueous solutions. *Journal Research of the National Bureau of Standards. A- Physics and Chemistry*, 81, 89-96.

Guilbert S. (1988). Use of superficial edible layer to protect intermediate moisture foods: Application to the protection of tropical fruit dehydrated by osmosis. In C. C. Seow (Ed.). *Food Preservation by Moisture Control*. (1°, pp. 199-219). New York: Elsevier Applied Science Publishers, Ltd.

Guillard, V., Issoupov, V., Redl, A., & Gontard, N. (2009). Food preservative content reduction by controlling sorbic acid release from a superficial coating. *Innovative Food Science & Emerging Technologies*, 10, 108-115.

Han, J. H., & Krochta, J. M. (2007). Physical Properties of Whey Protein Coating Physical Properties of Whey Protein Coating Solutions and Films Containing Antioxidants. *Journal of Food Science*, 72, 308-314.

Han, J. H. (2000). Antimicrobial food packaging. *Food technology*, 54, 56-65.

Hurst, A. (1981). Nisin. *Advances in Applied Microbiology*, 27, 85-123.

Hurst, A., & Hoover, D. G. (1993). Nisin. In P. M. Davidson, & A. L. Branen (Eds.), *Antimicrobials in food* (1°, pp. 369–394). New York: Marcel Dekker.

Kester, J. J., & Fennema O. R. (1986). Edible films and coatings: a review. *Food Technology*, December, 47-59.

Kim, S. J., & Ustunol, Z. (2001). Solubility and moisture sorption isotherms of whey-protein-based edible films as influenced by lipid and plasticizer incorporation. *Journal of Agricultural and Food Chemistry, 49, 4388-4391.*

Ko S., Janes, M. E., Hettiarachchy, N. S., & Johnson, M. G. (2001). Physical and chemical properties of edible films containing nisin and their action against *Listeria monocytogenes*. *Journal of Food Science*, 66, 1006-1011.

Koksel, H., Sahbaz, F., & Ozboy, O. (1993). Influence of wheat-drying temperatures on the birefringence and X-ray diffraction patterns of wet-harvested wheat starch. *Cereal Chemistry*, 70, 481–483.

Le Tien, C., Letendre, M., Ispas-Szabo, P., Mateescu, M. A., Delmas-Petterson, G., Yu H.. L., & Lacroix, M. (2000). Development of biodegradable films from whey proteins by cross-linking and entrapment in cellulose. *Journal of Agricultural and Food Chemistry*, 48, 5566-5575.

Leistner, L. (1995). Use of hurdle technology in food processing: recent advances. In G. Barbosa Cánovas, G., Welti Chanes (Eds.), *J. Food Preservation by Moisture Control. Fundamentals and Applications.* (pp. 377-396). Lancaster, Pennsylvania: Economic Publishing Co.

León P.G., & Rojas A.M. (2007). Gellan gum films as carriers of L-(+)-ascorbic acid. *Food Research International*, 40, 5, 565-575

Liu, Q. (2005). Understanding Starches and Their Role in Foods. In *Food Carbohydrates: Chemistry, Physical Properties and Applications*. CRC Press, Boca Raton, FL, USA.

Manzocco, L., Nicoli, M.C., & Labuza, T. (2003). Study of bread staling by X-ray diffraction analysis. *Italian Food Technology*, XII, 17-23.

Mathlouthi, M. (2001). Water content, water activity, water structure and stability of foodstuffs. *Food Control*, 12, 409-417.

Nísperos-Carriedo, M. O. (1994). Edible coatings and films base on polysaccharides. In J. M. Krochta, E. A. Baldwin, M. O. Nisperos-Carriedo (Eds.), *Edible coatings and films to improve food quality* (1°, pp. 305-335). Lancaster, Pennsylvania: Technomic Publishing Co., Inc.

Ozdemir, M., & Floros, J. D. (2001). Analysis and modeling of potassium sorbate diffusion through edible whey protein films. *Journal of Food Engineering*, 47, 149-155.

Phan The, D., Debeaufort, F., Voilley, A., & Luu, D. (2009). Biopolymer interactions affect the functional properties of edible films based on agar, cassava starch and arabinoxylan blends. *Journal of Food Engineering*, 90, 548–558.

Piermaria, J., Pinotti, A., Garcia, M. A., & Abraham, A. G. (2009). Films based on kefiran, an exopolysaccharide obtained from kefir grain: Development and characterization. *Food Hydrocolloids*, 23, 684–690.

Rhim, J. W. (2004). Physical and mechanical properties of water resistant sodium alginate films. *Lebensmittel Wissenschaft und Technologie*, 37, 323–330.

Rindlav, Å., Hulleman, S. H. D., & Gatenholm, P. (1997). Formation of starch films with varying crystallinity. *Carbohydrate polymers*, 34, 25-30.

Rodríguez, M., Osés, J., Ziani, K., & Maté J. I. (2006). Combined effect of plasticizers and surfactants on the physical properties of starch based edible films. *Food Research International*, 39, 840–846.

Rojas-Graü, M. A., Soliva-Fortuny R. C., & Martín-Belloso, O. (2009). Edible coatings to incorporate active ingredients to fresh cut fruits: a review. *Trends in Food Science & Technology*, 20, 438-447.

Sanjurjo, K., Flores, S., Gerschenson, L., & Jagus, R. (2006). Study of the performance of nisin supported in edible films. *Food Research International*, 39, 749-754.

Sebti, I., Delves-Broughton, J., & Coma, V. R. (2003). Physicochemical properties and bioactivity of nisin-containing cross-linked hydroxypropyl methylcellulose films. *Journal of Agricultural and Food Chemistry*, 51, 6468-6474.

Sebti, I., Ham-Pichavant, F.R., & Coma, V. (2002). Edible bioactive fatty acid-cellulosic derivative composites used in food-packaging applications. Journal of Agricultural and Food Chemistry, 50, 4290-4294.

Seker, M., & Hanna, M. (2006). Sodium hydroxide and trimetaphosphate levels affect properties of starch extrudates. *Industrial crops and products*, 23, 249-255.

Sofos, J. N. (1989). Sorbate Food Preservatives (1°). Florida: Academic Press.

Sokal R. R., & Rohlf J. B. (2000). Biometry. The principles and practice of statistics in biological research (1°). San Francisco, CA: W. H. Freeman.

Trezza, T. A., & Krochta, J. M. (2000). Color stability of edible coatings during prolonged storage. *Journal of Food Science*, 65, 1166-1169.

Vásconez, M. B., Flores, S., Campos, C., Alvarado, J., & Gerschenson, L. (2009). Antimicrobial activity and physical properties of chitosan-tapioca starch based edible films and coatings. *Food Research International*, 42, 762-769.

Vermeiren, L., Devlieghere, F., Van Beest, M., de Kruijf, N., & Debevere, J. (1999). Developments in the active packaging of foods. *Trends in Food Science and Technology*. 10, 77–86.

Warth, A. D. (1977). Mechanism of resistance of *Saccharomyces bailii* to benzoic, sorbic and other weak acids used as food preservatives. *Journal of Applied Bacteriology*, 43, 215-230.

Xie, S. X., Liu, Q., & Cui, S. W. (2005). Starch modification and applications. In S. Cui (Ed.), *Food Carbohydrates. Chemistry, physical properties and applications* (1°, pp. 285-322). Boca Ratón, FL: CRC Press.

Yang, L., & Paulson, A. T. (2000). Mechanical and water vapour barrier properties of edible gellan films. *Food Research International*, 33, 563-570.

Zhang Y., & Han J. H. (2006). Mechanical and thermal characteristics of pea starch films plasticized with monosaccharides and polyols. *Journal of Food Science*, 71, 109-118.

Zobel, H. F. (1994). Starch granule structure. In R. J. Alexander, & H. F. Zobel (Eds.), *Developments in carbohydrate chemistry* (1°, pp. 1-36). St Paul, Minnesota: The American Association of Cereal Chemists.

Zeuthen, P., & Bøgh-Sørensen, L. (2003). *Food preservation techniques* (1°). Boca Ratón, FL: CRC Press.

In: Focus on Food Engineering
Editor: Robert J. Shreck

ISBN: 978-1-61209-598-1

Chapter 3

DEVELOPMENTS IN HIGH-PRESSURE FOOD PROCESSING

Carl J. Schaschke*

Department of Chemical & Process Engineering, University of Strathclyde, Glasgow G1 1XJ, Scotland, UK

ABSTRACT

This chapter reports the developments made in the processing of foods using high pressure which over the past two decades. Consumers these days generally expect the food to be of a high quality, minimally processed, natural, additive-free, high in nutritional value as well as safe to eat. High Pressure processing is an alternative to thermal processing which can destroy harmful microorganisms rendering the food safe to eat. As a way of minimally processing food, it has the potential to preserve the quality of foods in many cases and even be responsible for producing new textures and properties. The effect of high pressure on the molecular structure of food proteins is to change their functional properties in surprising and often useful ways. A pressure of ten thousand times greater than atmospheric is capable of coagulating the albumin of egg without the use of heat.

The purpose of using high pressure instead of heat is to preserve and even improve food quality in terms of taste, flavour, texture and colour. The molecular structure of many food components including sugars, oils, vitamins, lipids and pigments are able to resist the effects of high pressures. Pressure is capable of affecting only the weaker bonds and forces sufficient to alter the delicate molecular structures, as in the case of proteins.

There have been some excellent examples worldwide of commercially applying high pressure in the processing of fruits, fish and shellfish, meat and dairy products. Research continues to understand fully the remarkable effects of high pressure on the constituents of food. In general, this has been in the areas of food safety with the destruction of micro-organisms, the activation and deactivation of enzymes; the functional properties of foods components to form foams, gels and emulsions; thermodynamics with the control of phase change. The most important of these has been to establish the sterilisation properties of high pressure food processing. Many harmful microorganisms differ

* Tel: 0141 548 2371
Email: carl.schaschke@.strath.ac.uk

significantly in their ability to withstand pressure while bacteria, yeasts and moulds are readily killed with spores being only inactivated by pressure after germination.

INTRODUCTION

Subjecting food to pressures far in excess of those found even at the bottom of the Earth's deepest oceans may seem a strange thing to do, but that is exactly what is now starting to be done by some of the world's food manufacturers. The effect of applying massive and crushing pressures on the molecular structure of foods is to change their functional properties in surprising and useful ways. Under a pressure thousands of times greater than atmospheric, it is possible, for example, to coagulate or stiffen the albumin of egg without using heat. A number of foods are already available worldwide sale which have been processed by pressure and many more are in the stages of development.

Since time immemorial, humans have sought the best way to extend the freshness of their food, to keep it safe and palatable. There is evidence that the earliest cave dwellers dried their meat to make it last presumably for lean times when food may not necessarily have been so plentiful. In South America, Inca Indians found ways of freeze-drying their food. Salt and smoke were also used to preserve food. Centuries later, 15th, 16th 17th and 18th century mariners and ocean explorers including Vasco da Gama, Ferdinand Magellan and Captain James Cook were well accustomed to long sea voyages and the necessity of stocking up on dried and salted foods for the long months at sea that lay ahead. Low temperatures and air refrigeration was also well known to slow the rate of deterioration before commercial refrigeration was developed by Clarence Birdseye and others in the early twentieth century. Many wealthy stately home owners in the United Kingdom used subterranean ice houses which were filled with winter snow and ice into which food would be packed and stored for the warmer months ahead.

From those early days, food producers pioneered many thermal and non-thermal methods such as heat pasteurisation, the application of biological and chemical agents such as bacteriocins, carbon dioxide, ozone, pulsed electric and magnetic fields, and the more controversial use of irradiation to destroy or inactivate food spoilage micro-organisms thereby extending considerably shelf life enabling food to be transported greater distances, remaining on the shelves of supermarkets for sale or consumed at a later and more convenient date (Mertens and Knorr, 1992).

A common feature of all food preservation methods is that after treatment, food has neither the same appearance, tastes the same nor has the same nutritional quality as freshly prepared food. Clarence Birdseye examined extensively the relationship between the rate of freezing and thawing on the quality of foods. Pasteurisation and commercial sterilisation which are both widely used both have a cooking effect on foods altering colour, taste and appearance. Drying and smoking, by definition, alter the state of the food yet interestingly dried fish and meats are today a delicacy in their own right.

A relatively new approach to extending the shelf life of foods is to subject them to high pressure. High pressure processing, also called high hydrostatic pressure or high isostatic pressure (or perhaps more aptly called ‘pascalisation’), provides the food industry with a plethora of new product development opportunities that is able to exploit the functional properties of ingredients of many foods such as hydrocolloids and proteins.

The effects of high pressure on certain foods is not new and, in fact, has long been known. Bert Hite, at the end of the nineteenth century, published findings of his research carried out at the West Virginia Experimental Station which reported that high pressure could effectively preserve milk (Earnshaw, 1996). The application of high pressure up to 600 MPa had been first shown to have effects on the inactivation of bacteria some 15 years earlier by H. Royer. By 1914, the effect of protein coagulation was well understood by Nobel Prize-winning physicist Percy Bridgman.

The application of high pressure *per se* is not new and has long been used in a number of industrialised areas including the production of plastics, ceramics, metal-forming and pharmaceutical tablet manufacture. While food processing typically takes effect at around 300 MPa and proceeds upwards to between typically between 600 MPa and 1000 MPa, it is worth noting, as a comparison that the RMS Titanic lies in Arctic waters at a depth of 4 kilometres at a pressure of just 40 MPa. While the coagulation of proteins with pressure was first demonstrated a century ago, with the exception of marine biology and deep sea diving physiology, there has been little interest in the effects of pressure on life's delicate biochemistry. It is only within the last two decades though that the technology reached a stage which allows the better understanding and use of the strange effects of high pressure.

The application of high pressure in food processing has been in four main areas: the destruction of micro-organisms, activation and deactivation of enzymes; the inactivation kinetics of both vegetative and pathogenic microorganisms; the change of functional properties of biopolymers such as proteins and polysaccharides used in foams, gels and emulsions; the control of phase change such as fat solidification and ice melting point (Cheftel, 1991; Knorr, 2002). The effectiveness of sterilisation of foods by high pressure has been compared to that of heat treatment. Like heat, microorganisms differ significantly in their ability to withstand pressure. Bacteria, yeasts and moulds are readily killed by high pressure while bacterial spores and some viruses are particularly resistant; spores being only inactivated by high pressure after germination.

Today, the high pressure processing of foods is considerably more sophisticated than that employed by the early pioneers. Yet comparatively during the intervening period, little work was carried out from the pioneering work a century ago until the 1980s, following significant developments in engineering design, fabrication and equipment reliability. Interest in high pressure processing applied to food processing re-surfaced particularly in light of the ever-growing demand by consumers for high quality foods which are seen as being safe, additive-free and minimally processed (Galazka and Ledward, 1995). Principally under the influence of Professor Hayashi together with significant financial support led to a consortium involving the Japanese government, academics and industry to invest in research and development. By the late 80s, a number of the high pressure processed products such as fruit juices rice cakes and some fish products including raw squid, appeared on the markets in Japan (Hayashi, 1992).

In Europe, meanwhile, high pressure processing remained largely at the research and pilot plant stage. The first recorded high pressure meeting in Europe took place in 1992 in Montpellier, France. It was here that the European Union released a mandate for research in high pressure processing. Two major EU projects began under the plan of 'Planta de Technologia dels Aliments' of the Universitat Autònoma of Barcelona, Spain and focused on milk and liquid foods (Trujillo et al. 2000). A few years later, EU legislation classified high pressure processed foods as being "novel" and the EC Novel Food regulation, EC 258/97, introduced a statutory pre-market approval system for such novel foods.

In the US, certain foods including shellfish and avocadoes proved themselves to benefit from high pressure processing. The Food and Drug Administration (FDA) classified high pressure processing as a nonconventional process. This requires that careful scrutiny and proof of food safety must be demonstrated before any process can be approved for commercial use. In spite of the growing number of applications, the US regulatory agencies have yet to approve sterilisation of low-acid foods by high pressure.

Consumers these days increasingly expect the food that they buy and consume to be of a high quality, be minimally processed, be natural, additive-free and have a high in nutritional value. The unique effect of pressure appears to be able to meet these requirements and demands. In just two decades, high-pressure processing has become now a routine procedure worldwide for the commercial processing of a wide range of foodstuffs.

The principal purpose of high pressure processing, which typically uses pressures in the order of hundreds of MPa, is to render foods microbiologically safe for human consumption. Using high isostatic pressures, the functional properties of food constituents are able to be retained and in some cases improved in terms of colour, flavour, texture, sensorial and nutritional quality as well as bringing about the necessary level of processing in terms of microbial deactivation, sterilisation, extended shelf-life and molecular alteration. High pressure can not only ensure that food is rendered microbiologically safe but ensures that the food can last longer without compromising its naturally fresh or just prepared characteristics.

By using pressure instead of heat (as in conventional cooking), the overall effect is to render the food safe as well as preserving and even improving food quality. At the molecular level, the covalent bonds of the various food components including saccharides, vitamins, lipids and pigments are able to resist the effects of high pressures in contrast to the highly damaging effects of heat. Molecular compression is generally only able to affect the weaker bonds and forces sufficient to alter the delicate molecular structures, as in the case of proteins.

High pressure food processing also has the advantage over conventional heat, microwave and irradiation by being able to transmit pressure both uniformly and rapidly throughout the body of the food undergoing treatment. Being an isostatic pressure, the pressure is applied equally from all directions. Without applied shear forces, there is consequently no distortion or misshaping of the food. In general, any food with a moisture content greater than 40% is suitable for being exposed to ultra high pressure without being crushed. While conventional thermal processing involves heat which is conducted through the exterior of foods to penetrate the interior by way of thermal diffusion, which takes time often with over-cooking of the surface, the same it not true of high pressure processing. One the other hand, unlike heat-treated food, high pressure processing is more dependent on the quality of the raw food material. High quality raw materials will, however, lead to a high quality product.

The processing of foods by high pressure offers the creation of many entirely new and exciting food textures. Pressurised beef muscle has been shown to have a texture like raw ham but without a change in taste. Fish and pork muscle become more glossy, transparent, dense, smooth and soft. Fruit-based jams, jellies, purées and juices are described as having an exceptional "just squeezed" flavour together with remarkable and striking natural colour. Protein from soya, milk and eggs can form soft gels that can be used to make new types of desserts and yoghurts. The foaming and emulsifying properties of egg white albumins can also be influenced by pressure by careful control of the molecular unfolding of the protein structure.

From its early beginnings, revenues from high pressure processed foods today is now valued in excess $2 billion annually. The range of commercially available foods today includes various fruit juices and meats in Europe, guacamole and oysters in the US, and an extensive range of products in Japan which includes fish, surimi, rice cakes and fruit juices amongst a growing number of others. Japan's Meida-ya Food Company became the world's first producer of high-pressure processed foods in 1991 and still continues to produce a range of jams, jellies and sauces. More recently in France, GEC Alstom has developed its horizontal *Hyperbar* equipment used for producing exotic fruit juices. The equipment is also being used by Spanish company Espuna to produce meat products with a production rate of about 600 kg/hour. This has proved effective in extending the shelf life of ham to 60 days, which is around three times longer than conventionally refrigerated ham.

The principal success of high pressure processing is the ability to kill pathogenic microorganisms including *Listeria*, *E.coli 0157*, *Salmonella* and *Vibrio* when subjected to pressures from 300 MPa to 700 MPa. The pressure necessary to kill these bacteria has a comparatively low energy so does not promote the formation of harmful new chemical compounds, radiolytic by-products or free radicals. Significantly, vitamins texture and flavour remain unaffected. Using water as the pressure transmitting medium which has a low compressibility ensures that the amount of energy needed to compress food is relatively low. High pressure is, in fact, more energy efficient than many high temperature food production methods, and requires only relatively small amounts of electricity and water and does not produce harmful emissions.

FOOD SAFETY AND CONTROL

The highest priority of the food industry is in ensuring that the food processed is microbiologically safe to eat. There has been much publicity in recent times regarding major food issues such as BSE in beef, genetically modified crops, nitrates in water, benzene contaminated water, *Listeria* in blue cheese and *E. coli* in cooked meats to name but a few. A major cause of illness, however, is foods contaminated by microorganisms either through poor processing conditions, poor sanitation and working practises, or poor or inadequate packaging.

There are two ways in which food quality can be adversely affected by microorganisms. Food spoilage microorganisms such as yeast and moulds can cause food to deteriorate and reduce shelf life. Secondly, food poisoning microorganisms such as *Escherichia coli* 0157 *Campylobacter jejuni* and *Listeria monocytogenes* can result in a serious health and hygiene hazard.

Rendering foods safe to consume or extending the shelf life of foods requires destroying or inactivating the microorganisms responsible for causing damage to the foodstuff. When these microorganisms are capable of producing disease in humans this becomes even more important (Chilton et al., 1997). High pressure processing will inactivate many species of microorganism but the extent of inactivation depends on a number of factors. This includes the magnitude and duration of pressure, microbial species, processing temperature, and type of substrate or food product. In general, Gram-positive bacteria are more resistant to pressure than Gram-negative, moulds and yeasts. It is important to note that a small but significant number of microorganisms, especially spores, can survive high pressure treatment. Combining high pressure processing with thermal processing can, however, increase the effectiveness (Knorr, 1993).

The type of substrate or food can affect the degree to which microorganisms are affected by high pressure. Acidic foods such as fruit juices can sensitise bacteria to high pressure. Microbial cells, which are actively multiplying, are also more sensitive to high pressure compared to those that are not. Spore-forming bacteria such as *Bacilli* can make sterilisation of foods difficult using high pressure. Combining high pressure with temperature and cycling high and low pressures between 50 to 400 MPa and 60 to 90°C can overcome this problem. While some spores of yeasts and moulds are sensitive to this combined treatment, others are resistant such as *Byssochlamys nivea* (Butz et al., 1996).

In addition to both temperature and pressure being essential parameters influencing the effectiveness of sterilisation or pasteurisation in foods, processing time is also a significant factor. The required combination of pressure, temperature and time required varies markedly between food and contaminating microorganism and many studies focus on producing data. Since the number of contaminating microorganism initially present within a foodstuff is not known for certain, a treatment has to be adopted that will reduce the highest likely number encountered to a negligible level or acceptably low level.

Microbial inactivation at high pressure is a first order process. Termed the logarithmic order of death, the same approach is taken with conventional thermal processing in which the rate of inactivation by pressure (or indeed heat) occurs at a rate that is approximately proportional to the number of microorganisms present in the food. Expressed as the so-called decimal reduction time or D-value, this is the time in minutes at a specific pressure (or temperature) required to destroy 90% (or 1 log-cycle) of the microorganisms in a population. The time (or number of D-values) required depends on the processing pressure, type of microorganism(s) present in food and the physical and chemical characteristics of the food product such as pH. While a so-called thermal botulinum cook corresponds to a 12-D process at 121°C for 3 minutes, high pressures are currently being established to determine the pressure-time combinations necessary for many species of pathogenic microorganisms. The minimum pressure for microbial inactivation is 300 MPa and a D-value of at least 5 should be achieved for pathogenic microorganisms and 8D for spoilage microorganisms. While the D-value increases with pressure, there is little virtue in exceeding pressures of 700 MPa (Toledo, 1999). Many observations, however, show that there are long survival tails indicating that it is particularly difficult to destroy every pathogenic microorganism that may be present in a food undergoing high pressure processing.

While the application of high pressure is evidently effective at killing microorganisms, the precise cause of microbial death by high pressure is not fully understood. There are a number of postulated theories. The food structure itself may be responsible for providing a pressure-protective effect and so the rate of surviving microorganisms actually increases. It is also likely that modifications in cytoplasmic membrane, which is the primary site of pressure damage, is the main cause of lethal cell injury generated by pressure treatment (Yuste et al., 2001). The effect of pressure-freezing on the phospholipid membrane has also been postulated as, too, has the disruption of key enzymes in the metabolic pathway essential for life. Rapid release of pressure has also been reported to be the cause of microbial death due to destruction of the cell wall and membrane structure.

Also contributing to the effects of cell growth and death is temperature. Unless the vessel is maintained at a constant temperature, there will be a temperature rise and consequently a heating effect associated with adiabatic compression (Hoǔska et al., 2004). While this may contribute to the thermal death of microorganisms depending on the initial temperature, there may be an associated cooking effect. For water, this corresponds to a temperature rise of between 3°C to 4°C per 100 MPa (Rasanayagam et al., 2003). Temperature rises may therefore be in excess of 20°C and therefore contribute to a cooking effect. Foods with a low density and specific heat such as oils consequently have a lower adiabatic temperature rise.

FOOD QUALITY

An important aspect of high pressure processing of foods concerns the physical and chemical nature of food structure and taste. The properties and quality of foods, which affect acceptability, are referred to as organoleptic properties. Food is liked or disliked for two fundamental reasons. The first is concerned with the psychological and social attitudes towards food while the second is concerned with the properties of temperature, colour, texture, flavour and odour of the particular food, dish or product. It is, however, virtually impossible to quantify the definition of food quality because it varies with personal preference, food type, and varieties of the same food. The key determinants are nutritional value, desirable aroma, flavour, and taste, desirable appearance (size, shape, and colour), desirable texture, available variety, convenience, price and environmental impact. High pressure processing can influence each of these qualities in food in both desirable and undesirable ways.

Defined in terms of its organoleptic properties, consumers, in general, are concerned that the quality of a food or food product has a consistent standard. The quality of a processed or unprocessed product can be tested by a trained panel of experts who can detect whether a product has attained a necessary standard. It is rather expensive to use expert panels and consequently mechanical or electronic techniques are often used instead. Food quality can be tested by the use of instruments that can give an objective measurement of a particular attribute such as viscosity, colour and texture. By way of example, high pressure food processing can change the consistency of diary products such as yoghurts and the texture and colour of meats. These can be measured by rheometers, penetrometers and spectrophotometers, respectively.

An interesting effect of high pressure is the depression of the freezing point of water. Pressure induced phase transition such as pressure shift freezing or pressure assisted thawing may offer possible applications. At a pressure of 200 MPa, for example, the freezing point is reduced to -22°C. The problem with conventionally storing frozen food is that much of the textural quality is lost during the freezing or thawing often due to the damaging effects of ice crystals. A possible application of high pressure processing may be to cool food below its freezing point at high pressure and then by releasing the pressure instantly freeze the food. Tests have shown that the rapid and uniform freezing effect has been found to preserve better food texture on thawing. Pressure shift freezing may therefore reduce structural damage of frozen products or increase thawing rates as compared to conventional thawing of foods while controlling or inhibiting microbial growth (Denys et al., 2001).

Other interesting effects of high pressure are associated directly with the functionality of the constituents of foods. While covalent bonds are little affected, hydrophobic and electrostatic interactions can be altered to affect the secondary properties of proteins and thus functionality. While proteins can be denatured, lipids may solidify and biomembranes destroyed. Whey proteins, for example, has poor foaming properties but high pressure processing of β-lactoglobulin in whey improves the foaming properties making it more surface active and suitable for maintaining foam stability.

Constituents of Foods

To understand fully the effects of high pressure on foods, it is necessary to consider carefully the various constituents of food. Foods are generally complex multi-component mixtures comprising several major groups of chemicals. In addition to water, these include carbohydrates, proteins, fats and lipids. Also present but in much small quantities are flavours, vitamins, minerals, preservatives and various other additives. Not all foods contain all these groups but may feature some in combination in varying amounts.

Proteins

As well as providing nutritionally essential amino acids, proteins contribute to the acceptability of foods and many of the properties of the proteins are utilised when cooking food. Plant and animal proteins are composed of twenty amino acids, which can be combined in a variety of ways to form muscle, tendons, skin, fingernails, feathers, silk, haemoglobin, enzymes, antibodies and hormones. Proteins are therefore polyamides and the order in which amino acids are sequentially joined together in a protein molecule is called the primary structure. Not surprisingly, the word protein is derived from the Greek *proteios*, which literally means “primary”.

The shape into which a protein molecule folds its backbone is called the secondary structure. Further folding of the backbone upon itself by molecular forces to form a spherical structure is called the tertiary structure. The secondary and tertiary structures are collectively referred to as the higher structure of the protein. The functional properties of a protein are due specifically to the higher structure.

The precise shape or conformation of a protein molecule is due to weak non-covalent intermolecular forces across the higher structure. These include hydrogen bonding between side chains, disulphide cross-links, and salt bridges (ionic bonds such as $RCO_2^- \ {}^+H_3NR$ between side chains). The most stable higher structure is the one that has greatest number of stabilising interactions.

The orderly and distinguishable secondary structure consists of α helical structures and β (or pleated) sheets. Helical structures involve hydrogen bonds between one amide-carbonyl group and an NH group while the sheet arrangement consists of single protein molecules are lined up side by side and held together by hydrogen bonds between the chains.

Milk and egg white are soluble globular proteins. Their solubility is due to their tertiary structure. Polar hydrophilic side chains are positioned on the outside of their spherical structure increasing water solubility while non-polar hydrophobic side chains are arranged on the inside surface where they may be used to catalyse non-aqueous reactions. The unique surface of globular proteins enables them to recognise certain complementary organic molecules. This recognition allows enzymes to catalyse certain reactions but not others.

Protein denaturation is the loss of the higher structural features. In a high pressure treated system, it is the hydrostatic and electrostatic interactions which are most vulnerable with hydrogen bonding bonding being largely unaffected. The result is the loss or change in many of the functional properties of the protein. The mode of unfolding and subsequent aggregation of proteins differs from heat treated proteins and this may markedly affect the perceived eating and textural quality.

Some proteins are quite resistant to denaturation, while others are more susceptible. Denaturation may be reversible if a protein has been subjected to only mild denaturing conditions. Under certain conditions a protein may resume its natural higher structure in a process called renaturation. Renaturation, however, may be very slow or may not actually occur at all.

For protein denaturation to occur, it is necessary for the energy supplied to exceed the binding energy of the hydrogen bonds as well as be able to overcome hydrophobic interactions and other forces responsible for the stability of the tertiary structure. A pressure of 600 MPa is sufficient to impart the energy necessary to disrupt the structure at both the secondary and tertiary levels.

To measure and evaluate the effectiveness of high pressure to alter the conformation of proteins, complete 3-dimensional structural information of proteins can be obtained experimentally using either x-ray crystallography or nuclear spectroscopy. X-ray crystallography has been successfully used to determine the structure of many proteins. The limitation tends to be the needs to obtain good quality and purified proteins. With NMR spectroscopy, crystals are not required. Instead, NMR methods are carried out with the protein in solution. Stable and concentrated solutions are, however, required since the technique takes several days to complete.

Analytical techniques such as circular dichroism can be used to provide an overall characteristic of protein tertiary structure and protein secondary structure (Kelly and Price, 1997). The far-UV (190-260 nm) and the near-UV (260-320 nm) spectra are used to measure the secondary and tertiary structure, respectively. Secondary structure estimates (α-helix content) are obtained from the molar ellipticity of CD spectra either at a particular wavelength or by fitting observed CD spectra to combinations of 'best fit' reference spectra of α-helices, β-sheets and other elements of structure in proteins for which X-ray crystallography data are known (Provencher and Glöckner, 1981). Such procedures tend to be highly dependent on good quality noise-free data below wavelengths of 190 nm (Price, 1995).

In general, the effects of pressure on proteins are reversible and only seldom are they accompanied by aggregation or changes in covalent structure. The effect disturbing the delicate balance of stabilising-destabilising interactions causing protein unfolding is governed by Le Chatelier's principle. This states that if one of the conditions of a system in equilibrium is altered, the system will adjust itself so as to annul, or tend to annul, the effect of that change. The application of pressure shifts the equilibrium towards the state that occupies a smaller molecular volume. This is because the volume of an unfolded or denatured protein is smaller than its corresponding native state. The extent of the volume change can be estimated from the pressure dependence of the equilibrium between the folded and unfolded states. The unfolding of protein structure is accompanied by a change in their function. In this way, enzymes can be inactivated and micro-organisms killed.

Noted for its emulsifying and foaming properties, a widely used protein in the food industry is the globular bovine whey protein β-lactoglobulin. The biochemical, physicochemical and structural properties of this protein have been extensively studied, and in addition to being particularly sensitive to temperature and pH, it is sensitive to high pressure. It has, consequently, been used extensively as a model for study by high-pressure (Pittia *et al.* 1996). A growing body of literature is now to be found concerning the pressure-induced changes in structure and functionality of globular proteins. Accounting for up to half of the protein in bovine whey isolate, the high-pressure induced denaturation of β-lactoglobulin has been the focus of many studies. Denaturation involves the dissociation of dimer to monomer together with complex changes to the conformation of the polypeptide chain. The effect of pressure is to alter the balance of intramolecular and solvent-protein interactions and can result in the disruption to both internal hydrophobic bonds and salt bridges (Pittia *et al.* 1996).

It is well known that pressure-induced structural changes can cause non-reversible effects including unfolding of monomeric proteins, aggregation and formation of gel structures. Irreversible modifications in tertiary structure and surface hydrophobicity of β-lactoglobulin have been observed for pressures of between 600 and 800 MPa (Iametti *et al.* 1997). A near total reduction of the α-helix content due to pressures of 1000 MPa has been reported and considered as being more severe than temperature-induced denaturation (Hayakawa *et al.* 1996). Aggregation effects have been observed upon treatment between 200 and 600 MPa and formation of intermolecular disulphides up to 450 MPa (Funtenberg *et al.* 1995). Partial unfolding of β-LG has been detected using pressures as low as 50 MPa as well as in combination with mild thermal treatment and pH (Tedford *et al.* 1998).

Fats, Lipids and Oils

The rheological property of food solutions, suspensions, emulsions and mixtures is an important way of characterising flow behaviour (Rao, 1977). The properties of such multi-component and multiphase foods are often a function of process history and can change dramatically during the course of processing. This is due to chemical and physical alternations caused by temperature, pH, application of shear and, increasingly, through the application of high pressure. High pressure processing is well known to influence considerably the viscous behaviour of fluids and, in particular, oils.

A rather general rule is that the more complex the molecular structure of the liquid, the larger is the effect of pressure. In terms of food process engineering, it is essential to have an understanding of the nature of viscosity at high pressure since it has a profound influence in terms of plant design, quality control of both raw materials and products at different stages of the process (Mertens and Deplace, 1993; McKenna and Lyng, 2001). High pressure can also influence the degree of oxidation and other chemical reactions of lipids and oils (Severini *et al.,* 2003; Cheah and Ledward, 1995). Further, since oils and lipids are directly linked to the organoleptic properties of many foods by way of mouth-feel, viscosity is an important parameter for evaluation.

At ambient pressure, the viscosity of all vegetable oils decreases rapidly at elevated temperatures due the reduction of intermolecular forces and mobility, the converse is true with elevated pressures (Abramovič, H. and Klofutar, 1998; Santos *et al.,* 2005). While the rheological properties of the oil and fatty acids can be readily determined at ambient pressure using conventional rheometers, the high pressure viscosity measurement is more problematic.

Viscosity is measured by applying a known force to a fluid and measuring the resulting rate of deformation. The shear stress can be calculated from the magnitude of the force and the way in which the shear rate can be obtained from the rate of deformation. The design of mechanical devices either to apply a measured shear stress or to measure the resulting shear rate inside thick-walled pressure vessels presents engineering problems which have limited this approach to relatively low pressures. Without external drives due to difficulties in maintaining seals under high pressure, the use of gravitational force confines the determination of viscosity to conditions of low shear. The physical size and density of the viscometer thus determines the size of the viscometer and the magnitude of the force that can be applied (Irving and Barlow, 1970).

It is well known that the viscosity of liquids at high pressure increases considerably. There are many equations which have been proposed for the viscosity of liquids at high pressure. Various empirical expressions have been proposed relating viscosity to pressure such as the use of quadratic equations and higher polynomials (Doolittle *et al.*, 1960). However, a reliable equation based the concept of free volume of molecules has been shown to take the form

$$\mu = \mu_o \exp^{\alpha p}$$

where μ_o is the viscosity at the standard pressure. Values of α can be obtained from experimental data.

It is well known that polymerised oils possess a much higher viscosity than non-polymerised oils. All oils possess a high viscosity as a consequence of their long fatty chain structure. In the case of olive oil and other vegetable oils, viscosity increases with the chain length of triglycerides fatty acids constituents and decreases with unsaturation. Viscosity therefore increases with the degree of hydrogenation. Olive oil is composed of triglycerides of fatty acids and many other micro-components such as fat soluble vitamins, carotenes, phenolic and aroma compounds. The composition of olive oil is therefore complex and the interaction between molecules correspondingly significant. The effect of the triglyeride structure offers a greater degree of entanglement than individual fatty acids (Schaschke et al. 2006).

Various studies have shown that the viscosity of vegetable oils depends on fatty acid composition (Santos *et al.*, 2005). Since olive oil is predominantly oleic acid with a "zig-zag" configuration, the free movement of adjacent chains is inhibited due to attraction through van der Waals forces which manifests itself in the form of high viscosities. Molecules with a higher degree of unsaturation with the *cis* configuration prevents the same level of intermolecular contact, resulting in an increased capability of the fluid to flow (Abramovič and Klofutar, 1998).

Process Operation

The way in which food is pressurised processed is similar to the way food is thermally processed. Foods can be pressured treated either as liquid products in bulk semi-continuously or as packed products in batches. Liquid foods such as fruit juices are compressed directly by a piston while packed foods are sealed into flexible containers (plastic or glass with flexible foil lids) and immersed within a hydraulic fluid within the vessel. Water (and hence the term "hydrostatic") is usually preferred as being largely incompressible as well as compatible with most foods. Using three adjacent vessels in parallel to fill simultaneously, process and eject the product, liquid products can be processed at a rate of up to 1000 litres per hour per vessel with the overall rate of production being determined by the production capacity of each vessel, the number of cycles per vessel and the number of vessels in each system. Where a batch system is used which is a manual operation, the overall rate of production also includes the downtime for depressurising, discharging, cleaning, re-charging and re-pressurising.

The high pressure system consists of a pressure vessel into which the food product is held. A pressure intensifier is used to raise the pressure of the pressurising fluid to the required pressure. A hydraulic pump and piston is used to operate the intensifier and achieve the very high pressures. Using a pressure multiplication factor in which the pressure is generated from the ratio of area of the drive piston to that of the intensifier piston, pressures of up to 1000 MPa can be generated. The system is required to be operated with appropriate control of pressure and temperature, as well as pressurisation, pressure-cycling and depressurisation control.

As well as effective seals capable of withstanding the operating pressures, the vessels are required to be opened and closed easily to reduce the time of loading and unloading. Batch vessels can be loaded with pre-packaged foods loaded down vertically into the vessel. Larger vessels with operating volumes of over 300 litres and lengths of 5 m are mounted horizontally. An adequate floor space is therefore required to house the vessels as well as ability to charge and discharge.

Pre-packaged, prepared or bulk foods such as delicatessen meats, hot dogs, fruit and shellfish, can be processed as batches in which the food is first loaded into a handling basket or tray and placed into the pressure vessel. Batch equipment has a high capacity and allows for large scale production. By integrating the batch high pressure vessel with conventional filling equipment, it is possible to enable rapid change over between different types of food product. Once loaded with food packages and closed, the vessel is filled with the pressuring liquid. Water can be mixed with a small amount of soluble oil for lubrication and anti-corrosion purposes. Liquid foods, on the other hand, are placed directly into the vessel. Semi-continuous food production can be maintained using a configuration of multiple batch vessels.

Continuous food production can be achieved for liquid products such as juices, sauces and purées that can be pumped. The food products can be processed using conventional in-line production methods. The liquids are pumped into an isolator in which a separator partitions the food from the high pressure water. After pressurisation by the pressure intensifier lasting for typically between 30 seconds and 2 minutes, they are then pumped out into a clean or aseptic filling station. A variety of jams, fruit yoghurts, fruit jellies, salad dressings and fruit sauces are currently being processed by high pressure. Citrus juices can typically be made at a rate of up to 6000 litres per hour and guacamole up to 50 tonnes per day.

Following the early success of the food related high pressure science and technology with the launch of the first high pressure processed food products by Meida-ya in 1991, an increasing number of commercial products have reached the Japanese, European and US markets. A wide range of industrial and pilot scale equipment as well as research scale equipment have been developed and operated over the past twenty years. For each of these, the engineering design of high pressure presses is required to meet strict international standards. The thickness of the vessel wall is determined by the operating pressure, volume and the number of times the vessel used. It is essential that they are able to withstand the high internal pressure. No air or gas must be trapped inside. Within the European Union, the vessels are required to meet the European Pressure Equipment Directive, and American Society of Mechanical Engineers (ASME) Section VIII, Division 3 code requirements.

The vessels themselves, arguably at the heart of the high pressure process and forged to form a cylindrical vessel in low alloy steel of high tensile strength, are designed to operate at a maximum working pressure. During operation, the vessels, as well as the components including seals and valves, are exposed to extreme forces. Depending on the internal diameter, and thus volume which defines production rate, the wall thickness can be determined and are usually limited to around internal working pressures of up to 600 MPa. Where higher pressures are required, pre-stressed vessels can be fabricated as compound cylinders or through the use of metal frames which can be strengthened using a pre-stressed wire-winding. These result in a relatively compact structure that enables only compressive stresses occur. This has the benefit that the resistance to fatigue is high and the risk of critical crack propagation avoided.

A limitation of high pressure processing development is the capital cost of the equipment. The retail or unit cost of high pressure processing food is currently in excess of that of heat processed food even though high pressure processing actually consumes less energy. The energy, and consequently cost, required to reach a pressure of 400 MPa is around the same as heating from only 20°C to 25°C. The reason for the high cost is because the equipment needed is very expensive and makes up an estimated 80% of the total production costs. In spite of this, the sales of high pressure processed foods continue to rise. At the moment the improved quality to be gained by high pressure processing has found a niche market among certain consumers who are willing to pay a premium. The future may well lead to pressure processed foods eventually being more commercially competitive and affordable to everyone.

Fruits and Vegetables

High pressure processing of packaged fruit and vegetable products is noted for giving products with fresh, just-prepared characteristics. Prepared as juices, blends, purées and smoothies, fruit and vegetable products are seen as being attractive to the consumer for their evidently fresh quality. By retaining sensory qualities, texture, colour as well as nutritional content, high pressure processing is seen to present many opportunities to create exceptional value-added products. Capable of inactivating *Salmonella, E. coli* and *Listeria monocytogenes* in fruit and vegetable products, food processors have used high pressure processing within their HACCP programmes and been able to achieve the FDA requirement of the necessary 5-log reduction of pathogens in fresh juice.

Although high pressure processing can be used to deliver a microbiological stable chilled product, it is frequently enzymes which limit the acceptable shelf life of the product. Strategies for the control of enzymes in a high pressure preserved product are therefore required. The enzyme, polyphenoloxidase (PPO), is responsible for enzymatic browning of many edible plant products, especially fruits and vegetables during post-harvest handing and processing. Peroxidase (POD), is also responsible for the colour change as well as off-flavours and loss of texture. Both food-degrading enzymes are readily inactivated by sodium metabisulphite and rapid thermal treatments such as blanching. Other methods have been reported including mild thermal treatments, microwave and freezing techniques. Thermal processing, however, tends to impair the flavour and softens the texture. Blanching processes tend to compromise between enzyme inactivation and changes in flavour and texture.

Other fruit susceptible to spoilage by way of browning due to PPO include banana, apple and pear which are used in a wide number of food products including beverages, bakery and diary products. As well as being rich sources of carbohydrates and vitamin C in the diet, they are rich in a number of other vitamins. Banana, in particular, is a naturally rich source of B6, potassium and dietary fibre, and is especially ideal for baby foods as it is very low in allergens. Being virtually free of lipids, banana is also readily digested. Banana, apple and pear macerate, however, undergo rapid enzyme browning on exposure to oxygen (air) during processing as a result of cellular disruption during the peeling and slicing operations.

An early commercial success in the US in the high pressure processing of fruit susceptible to enzymatic browning was the preparation of avocado pears either as ripe avocado halves or as a guacamole preparation. In both cases, the preparation involves cutting avocadoes and mashing, respectively, thereby damaging living cells, releasing PPO, reacting with oxygen (air) resulting in dark polyphenols, and thus food spoilage and consumer unacceptability. While there has been a number of ways of reducing or preventing the colouration, high pressure processing effectively denatures and inactivates the enzyme. Further commercial successes of premium foods using high pressure processing found elsewhere now includes salsa, tapas, pre-chopped onions, various juices, flavourful fruit smoothies and apple sauce. The list is growing.

SEAFOOD

There has been considerable interest in the use of high pressure to both process and preserve fish products over the past decade. The processing of fish and seafood products has traditionally been labour intensive, expensive and fraught with injury due to the use of sharp knives particularly such as in the manual process of shucking or prising open of oysters. Shellfish are conventionally sold from chilled storage and are susceptible to viral, enzymatic and microbial spoilage. Certain shellfish products such as oysters, which are required to remain alive and fresh after shucking, have a shelf life limitation and hence limited distribution implication.

The use of high pressure has in a number of notable cases transformed the seafood industry. Using pressures in the range of 250 to 500 MPa, for 1 to 3 minutes, shellfish and other crustaceans are also now being processed and sold. Unlike conventional thermal processing which results in protein dehydration and a loss in weight, high pressure processing results in protein hydration. A number of companies in the US, in particular, have been successfully using high pressure to shuck shellfish meat from their shells. Following this success, there is more recently an increasing interest in using high pressure in the crab and lobster industries with benefits in raising the yield and quality of meat recovery. The range of products is now been extended to include mussels, scallops and clams.

The primary use of high pressure has been for the inactivation of harmful microorganisms and extension of shelf life. Yet, while a considerable number of microbial studies have been undertaken, concerns remain regarding the effectiveness of high pressure processing to eradicate harmful microorganisms either entirely or to acceptably low levels.

Many species of microorganism as well as viruses are known to contaminate oysters and other molluscan shellfish including *Vibrios, Salmonella, Listeria* and *Staphylococcus aureus.* These microorganisms are known to be sensitive to effects of high pressure (Styles et al., 1991; Murchie et al., 2005; Kingsley et al., 2007). The main difficulty with the use of pressure, however, is that many spores including those of *Clostridium botulinum* are relatively insensitive or resistant to pressure. Additional processing and preserving techniques are therefore required to ensure the complete eradication and absence of such pathogens. It is worth noting that those microorganisms which are susceptible to heat are not necessarily susceptible to pressure to the same extent. The bacterium *Escherichia coli* 0157, by way of example, though relatively heat labile is relatively pressure insensitive.

Like the oyster (*Crassostrea gigas*), the King scallop *Pecten maximus*, is a high-added value product which can be consumed raw, lightly cooked as in stir-frying, or baked. The microbial contaminants can lead to acute food poisoning leaving consumers cautious in purchase. Scallops are noted for their low fat content and high nutritional value through their content of polyunsaturated, notably eicosepentaenoic (20:5n-3) and docosahexaenoic (22:6n-3) fatty acids and also zinc (250 $mg.l^{-1}$), vitamins D and B12. For consumers, it is the distinctive flavour and texture as well as the healthy image that are the primary factors giving appeal to these premium products.

In addition to the inactivation of pathogens bacteria such as *Vibrio vulnificus*, which has been principally responsible for the highest fatality rate among foodborne pathogens in the US (Cook, 2003), the application of high pressure processing has, in a number of notable cases, transformed this industry in which it has been shown to be responsible for aiding the clean meat separation from lobsters, oysters, clams, and other fresh crustaceans and shellfish. This is attributed to the denaturing effects of high pressure on the specific protein that holds the meat to the shell. Facilitating the ease of release thereby allows maximum product yield without causing any mechanical damage to the product regardless of size. By manipulating the processing conditions, beneficial texture changes can also be created by improving the moisture retention ability of proteins, thus resulting is less water loss during storage or cooking as well as adding to improvements in texture including mouth feel.

Studies on the high pressure treatment effects on cod (*Gadus morhua*) muscle has shown that the oxidative stability lipids markedly decreases and thought to be due to the release of metal ions from complexes (Angsupanich and Ledward, 1998). Lipids in fish are more susceptible to oxidation by pressure than those in most meat producing animals since they have a high concentration of polyunsaturated fats.

High pressure has the additional effect of modifying the structure of connective proteins. It is generally accepted that the connective tissue, collagen, is relatively unaffected by pressure since it is generally stabilised by pressure insensitive hydrogen bonds. The effects of pressure on myofibrillar proteins is, however, more complex.

Myosin and actomyosin, which are the major component of skeletal muscle and plays a significant role in gel formation, in both fish and meat is denatured by high pressure to form a gel-like texture (Cheftel and Culioli, 1997, Angsupanich et al., 1999). High pressure can cause the depolymerisation of actin and actomysin and solubilisation of myofibrillar proteins (Gilleland et al., 1997). Protease active in muscle may also become more or less active due to the application of pressure.

Meats and Meat Products

Meats and meat products are composed largely of fats, lipids, proteins and water with other components such as haem and salts. Fresh meats are particularly susceptible to microbial contamination. With the effective reduction of harmful and food spoilage microorganisms, high pressure processing of meat and meat products enables quality in terms of texture, colour and nutritional content to be retained as well as extension of shelf-life (Cofrades et al., 2002). There is therefore a reduction on the dependence on the use of preservatives.

Used as a post-lethality intervention step for ready-to-eat meats, high pressure processing enables the shelf life of foods to be extended. This is particularly important for meat processors involved in preparing sliced delicatessen meats, where the risk of re-contamination with harmful pathogens, and in particular *Listeria monocytogenes*, and *Escherichia coli* may be the greatest. Eradication of harmful microorganisms through high pressure treatment can therefore extend the shelf life of meat products. With proper preparation, adequate flexible packaging and chilled distribution, pressure treated products can retain quality for up to 60 days. Using pressures of 600 MPa, the process cycle time is reduced substantially allowing increased production times.

Tenderness and textural quality are important aspects of meat acceptability for both meat and fish. The level of tenderness is dependent on numerous biological and technological factors including age, sex, species and muscle type as well as storage conditions since rigor toughness is resolved by calpain degradation of structural proteins (Koohmaraie et al., 1996). Tenderness can be enhanced using electrical stimulation, acid marination, injection of Ca^{2+} and ultrasound (Pommier 1992; Ertbjberg et al., 1999; Wheeler et al., 1992; Lyng et al, 1998).

There is much literature on the effects of pressure on the microbial quality of meat and how high pressure modifies the eating and mouth feel properties and qualities. Various studies have reported high pressure induced changes in texture and reduction in meat toughness at pressures of between 100 to 300 MPa exposed for up to 3 hours and temperatures of between 0°C to 10°C (MacFarlane et al. 1980, Suzuki et al., 1992). High pressure treatment of meat is responsible for causing three kinds of change in meat:

enzymatic, protein modification and structural proteins (Cheftel and Culioli 1997). Protease activity has been found to be enhanced through pressurisation as well as degradation of myofibrillar proteins along with pressure-induced modifications to muscle structure. Relating tenderness to the state of contraction of sarcomere and fibre diameter, high pressure treatment has been found to induce changes in textural properties in post-rigor beef (Jung et al., 2000).

As pressure and temperature affect the weak linkages maintaining the structural proteins in muscle in different ways, pressure treated meat and fish have different textures to their thermally treated counterparts. As pressure induced structures are to some extent stabilised by thermally labile hydrogen bonds, subsequent heat treatments may partially melt the pressure treated gel to give a texture and appearance not entirely dissimilar to that of the product treated by heat alone.

While commercial high pressure treatment of meat and fish projects is typically in excess of 250 MPa, lipids in both red and white meats and fish become susceptible to oxidation, due to the release of transition metal catalysts such as iron and copper from non-haem complexes present in the tissue. This has implications for both flavour and nutritional quality where products are processed and subsequently stored in air. The flesh of white meats such as turkey and pork take on a cooked appearance at pressures brought on about by the denaturation of myosin. This occurs above 400 MPa. In red meats, such as beef, myoglobin is also irreversibly denatured at pressures in excess of 400 MPa and a pigment produced that is spectrally similar to that found in cooked meats.

The application of high pressure induces a small amount of denaturation in meat proteins. After high pressure treatment, the texture of meats and colour are marginally altered. This effect can be used to provide a more tender meat product with less drip loss. However, Gelatinisation and structuring of meats through high pressure can also lead to new meat products with improved water binding capacities (Cofrades et al., 2002). Further, the re-growth of lactic acid bacteria which cause acidification and consequent organoleptic and textural modifications are reduced.

In addition to affecting the structural and haemoproteins in meat, several of the enzymic systems present in meat are modified by pressure, including those involved in post-mortem textural and colour changes. Using pressures on only 50 MPa, colour stability can be maintained by destroying the system responsible for catalysing the formation of the brown metmyoglobin. The desirable bright red colour due to oxymyoglobin can therefore be maintained for longer during retail display.

DIARY PRODUCTS

Diary products are generally defined as high-energy-yielding foods produced from cow's or domestic buffalo's milk. In general, raw milk for processing is mainly derived from cows but can also be sourced from other mammals including sheep, goats, yaks and horses. There are numerous diary food products and processes including various forms of milk including homogenised and pasteurised, skimmed milk, milk powder, evaporated and condensed milk amongst others as well as butter, cheeses and fermented products such as yoghurts produced using various cultures of *Lactobacillus* bacteria.

There is a growing body of literature concerning the effects of high pressure on diary products (Kolakowski et al., 1996, Molina et al., 2000, Okpala et al., 2010). The pressure-induced changes in structure and functionality of proteins up to 1000 MPa has been the subject of a number of early reviews (Balny *et al.*, 1989; Silva and Weber, 1993, Sawyer et al., 2002). Some work has focussed on the effects of milk, its constituents and its products including cheeses including the globular bovine milk protein β-lactoglobulin, which makes up approximately half of whey protein. The functional and structural properties of milk proteins are not only particularly sensitive to temperature through pasteurisation, they are also sensitive to high isostatic pressure. Structural changes can be induced in milk proteins above 200 MPa while above 500 MPa this causes non-reversible effects including unfolding of monomeric proteins, aggregation and formation of gel structures (Iametti *et al.*, 1997). Hayakawa *et al.* (1996) monitored the effects of high pressure on the α-helix content of β-lactoglobulin. When processed at 1000 MPa for 10 minutes, the content is reported to be up to 90% destroyed. Pressure-induced denaturation is reported as being more severe than temperature-induced denaturation. The denaturation of proteins by high pressure, however, is not identical to the temperature-induced process which is often irreversible because of the breakage of covalent bonds and/or aggregation of unfolded protein (Mozhaev *et al.*, 1996). The exact unfolding mechanism by which the pressure-induced protein is denatured remains elusive although it has been suggested that pressure causes changes in the structure of the protein molecules due to the cleavage of weak hydrogen bonds and van der Waals forces while covalent bonds remain unaffected. This consequently ensures the retention of essential vitamins and nutrients and thereby 'improving' or conversely not compromising the quality of product obtained.

In the production of cheese using high pressure, cheese yield and product shelf life has been found to increase due to the effects of rennet coagulation, microbial inactivation as well as effects on the physicochemical and sensory effects. Rennet coagulation involves a two-stage enzymatic process (O' Reilly et al., 2001), in which the Ca^{2+} ion level plays a crucial role in milk. (López-Fandiño, 2006). High pressure can also lead to changes in moisture distribution within cheese as well as varying colour effects (Johnston and Darcy, 2000; Capellas et al., 2001; Sheehan et al., 2005). Textural properties and sensory characteristics have also been obtained on high pressure treated fresh cheeses using pressures of range from 50 to 500 MPa. (Sheehan, et al., 2005; San Martín-González et al. 2007; Rynne at al., 2008).

Microbial inactivation by high pressure on fresh cheeses has been the focus of a number of studies (Gallot-Lavallée, 1998). In these, target microorganisms are typically inoculated in the cheese or milk, after which high pressure treatment is then applied. The main microorganisms used in these studies are Gram positive microorganisms such as *Listeria monocytogenes*, *Staphylococcus* spp., *Bacillus* spp. and Gram negative microorganisms such as *Escherichia coli* (Gao et al., 2006). A wide range of log reductions on microorganisms are reported resulting in variations in shelf-life results. High pressure treatment has been used to obtain high count reductions of *L. monocytogenes* in sliced cheese. Applying 400 MPa for 10 minutes at 2°C to salted vacuum-packaged curds has resulted in cheeses with a very low level of contaminant flora and little modifications of rennet or plasmin activities, enabling the assessment of enzymatic activities that may have occurred due to proteolysis (Trujillo et al., 2000).

The pH of the milk is an important factor in the stability of the milk proteins under high pressure. The pH dependence at lower pH values is due to electrostatic attraction which tends to favour the folded state. Cations (Ca^{2+}) present in milk therefore offer protection to microorganisms against inactivation by high pressure. Combining high pressure treatment with mild heat or nisin on the indigenous as well as inoculated microorganisms present in fresh cheese can improve inactivation rates especially for the Gram-negative bacteria (Hauben et al. 1998). Nisin when used on cheese has been shown to delay spoilage for approximately 1 month, while high pressure treatment approached 3 months depending on duration of treatment (Trujillo et al., 2000). The most effective treatment has been found by combining both nisin and high pressure treatment. Nisin is said to influence mainly the sporulated population, while the fractions resistant to both treatments were inactivated by nisin in combination with high pressure treatment.

Modification to the molecular structure of proteins have been shown to take place in the high pressure treatment of fresh cheese. Denaturation of protein by high pressure, however, cannot be likened to that of heat treatment because the latter brings about an irreversible reaction since the covalent bonds of the unfolded protein either break or aggregate (Mozhaev et al., 1996).

Moisture content is reported to increase in high pressure milk cheese by using three 1-minute cycles with pressures above 500 MPa on raw or pasteurized milk cheeses (Drake et al., 1997). High pressure processing of milk before cheese manufacture is reported to affect cheese yield (San Martin-González et al., 2007) although pasteurised or raw milk cheeses showed no differences in moisture content (Trujillo et al., 2002).

A fresh soft cheese has recently been reported in Australia with an inherent refrigerated shelf-life of around three weeks limited to spoilage yeasts has been extended to up to 8 weeks following high pressure treatment of pressures up to 600 MPa for 5 min at ambient temperature ($\leq$22 °C). Viable count reductions obtained a maximum of 7 log reduction (Daryaei et al., 2008).

Conclusion

The development and commercialisation of high pressure processing of foods worldwide over the past twenty years has gained momentum. Based on the principles of biochemistry and chemical technology, the application of high pressure is proving to be a clean and energy efficient way to process foods. While the many technical problems that have previously hampered high-pressure research and development have now been overcome, there is still much of the fundamental science of high pressure to be understood, particularly at the molecular level. Even so, the use of high pressure is an interesting technology with evident product advantages. Many new foods are now in the stages of development. It may be not too long to wait before high pressure becomes routinely used for the processing of foods that we take for granted and eat each day.

The cost of high-pressure processed food production remains a significant factor in the widespread use of this technology. Product and retails costs are is currently higher than those of conventionally heat processed foods due to the high cost of the equipment needed and short production speeds confining products to the largely niche markets or products with a significant added value.

High-pressure food processing is proving itself to be an increasingly effective way of eliminating microbial hazards when combined with adequate hygienic practices, such as the hygiene of personnel and sterilisation of equipment. Maximum internal volumes of high-pressure vessels are limited to the materials of construction that are able withstand the high internal pressures and this in term limits production rates.

The sales of high-pressure processed foods, however, continue to rise suggesting a continuing market demand for such foods. While the processing of many foods by high pressure is becoming increasingly popular as a minimalist approach to food processing, the principal technological constraints remain the high capital equipment costs. For a particular food protein there exists a minimum threshold pressure and thus process cost. Precise thresholds require detailed examination. Shorter production speeds are also currently being developed and, combined with extended product shelf life, the higher costs may be offset. The future may well eventually lead to high pressure processed foods being more commercially competitive, affordable and more widely available.

REFERENCES

Abramovič, H. and Klofutar, C. (1998) The temperature dependence of dynamic viscosity for some vegetable oils. *Acta Chim. Slov* 45(1):69-77.

Angsupanich, K. and Ledward, D., (1998) High pressure treatment effects on cod (*Gadus morhua*) muscle. *Food Chemistry* 63(1):39-50.

Angsupanich, K., Edde M. and Ledward, D. (1999) Effects of high pressure on the myofibrillar proteins of cod and turkey muscle. 47(1):92-99.

Balny, C., Masson, P. and Travers, F. (1989) Some recent aspects of the use of high-pressure for protein investigation in solution. *High Pressure Research* 2:1-28.

Butz, P., Funtenberger. S., Haberditzl, T. and Tauscher, B. (1996) High Pressure Inactivation of Byssochlamys nivea Ascospores and Other Heat Resistant Moulds *Lebensmittel-Wissenschaft und-Technologie* 29(5-6):404-410

Capellas, M., Mor-Mur, M., Trujillo, A.J., Sendra, E. and Guamis, B. (1997) Microstructure of high pressure treated cheese studied by confocal scanning light microscopy. In K.Heremans (Ed.), *High Pressure Research in the Biosciences and Biotechnology* (pp 391-394). Leuven, Belgium: Leuven University Press.

Capellas, M., Mor-Mur, M., Gervilla, R., Yutse, J. and Guamis, B. (2000) Effect of high pressure combined with mild heat or nisin on inoculated bacteria and mesophiles of goat's milk fresh cheese. *Food Microbiology* 17:633-641.

Capellas, M., Mor-Mur, M., Sendra, E. and Guamis, B. (2001) Effect of high-pressure processing on physico-chemical characteristics of fresh goat's milk cheese (Mató). *International Dairy Journal* 11:165 –173.

Cheah, P.B. and Ledward, D.A. (1995) High pressure effects on lipid oxidation. *JOACS* 1059-1063.

Cheftel, J.C. (1991) Applications des hautes pressions en technologie alimentaire. *Actualités des Industries Alimentaire et Agro-Alimentaire* 108 :141-153.

Cheftel, J.C. and Culioli, J. (1997) Effects of high pressure on meat: a review. *Meat Science* 46:211-236.

Chilton, P., Isaacs, N.S., Mackey, B., and Stenning, R. (1997) The effects of hydrostatic pressure on bacteria. In: Heremans, K. (Ed.), *High Pressure Research in the Biosciences and Biotechnology*.Leuven Univ. Press, Leuven, pp. 225– 228.

Cofrades, S., Carballo, J., Fernandez-Martin F. and Jimenez-Colmenero, F. (2002) High Pressure/thermal treatments effects on functionality of comminuted muscle from different meat species. *High Pressure Research* 22:721-723.

Cook, D.W. (2003) Sensitivity of *Vibrio* species in phosphate-buffered saline and in oysters to high hydrostatic pressure processing. *J. Food Protection* 66:2277-2282.

Daryaei, H., Coventry, M.J., Versteeg, C., and Sherkat, F. (2008). Effect of high pressure treatment on starter bacteria and spoilage yeasts in fresh lactic curd cheese of bovine milk. *Innovative Food Science and Emerging Technologies* 9:201-205.

Denys, S., Schülter, O., Hendrick, M., and Knorr, D. (2001) Impact of high hydrostatic pressure on water-ice transitions in foods. In. Hendricks M and Knorr D., (Eds) Ultra high pressure treatment of foods. Aspen, Gaitherburg, 2002.

Doolittle, A.K., Simon, I., Cornish, R.M. and Doolittle, D.B. (1960) Compression of liquids *AICheE J.,* 6:150-162.

Drake, M.A., Harrison, S.L., Asplund, M., Barbosa-Canovas, G. and Swanson, B.G. (1997). High pressure treatment of milk and effects on microbiological and sensory quality of Cheddar cheese. *Journal of Food Science* 62:843-845.

Earnshaw, R. (1996) High pressure food processing. *Nutrition and Food Science* 2:8-11.

Ertbjberg, P., Mielche, M.M., Larsen, L.M. and Moller, A.J. (1999) Relationship between proteolytic changes and tenderness in pre-rigor lactic acid marinated beef *J Science of Food and Agriculture* 79:970-978.

Funtenberg, S., Dumay, E.M. and Cheftel, J.C. (1995) Pressure-induced aggregation of β-lactoglobulin in pH 7.0 buffers *Lebens-Wiss. u.-Technol. 28:*:410-418.

Galazka, V.B. and Ledward, D.A. (1995) Developments in high pressure food processing. *Food Technology International Europe*, pp.123-125.

Gallot-Lavallée, T. (1998) Efficacité du traitement par les hautes pression sur la destruction de *Listeria monocytogenes* dans des fromages de chèvre au lait cru. *Sciences des Aliments,* 18:647-655.

Gao, Y.-L., Ju, X.-R. and Jiang, H.-H. (2006) Use of response surface methodology to investigate the effect of food constituents on *Staphylococcus aureus* inactivation by high pressure and mild heat. *Process Biochemistry* 41:362-369.

Gilleland, G.M., Lanier, T.C. and Hamann ,D.D. (1997) Covalent bonding in pressure-induced fish protein gels, *J. Food Science* 62:713-716.

Hauben, K.J.A., Bernaerts, K. and Michiels, C.W. (1998) Protective effect of calcium on inactivation on *Escherichia coli* by high hydrostatic pressure. *Journal of Applied Microbiology* 85:678–684.

Hayahsi, R. (1992) Utilization of pressure in addition to temperature in food science and technology, *High Pressure and Biotechnology* 224:185-193.

Hayakawa, I., Linko, Y. Y. and Linko, P. (1996) Mechanism of high pressure denaturation of proteins, *Lebens-Wiss. u.-Technol.* 29:756-762.

Hoŭska, M., Kubásek, M., Stromhalm, J., Landfeld, A., & Kamarád, J. (2004) Warming of olive oil processed by high isostatic pressure. *High Pressure Research* 24(2):303-308.

Iametti, S., Transidico, P., Bonomi, F., Vecchio, G, Pittia, P., Rovere, P and Dall'Aglio, G. (1997) Molecular modifications of β-lactoglobulin upon exposure to high pressure. *J. Agric Food Chem.* 45:23-29.

Irving, J. B., and Barlow, A. J. (1970) An automatic high pressure viscometer. *J. Phys. E: Sci. Instrum.* 4:232-236.

Johnston, D.E., and Darcy, P.C. (2000) The Effects of High Pressure Treatment on Immature Mozzarella Cheese. *Milchwissenschaft* 55:617-620.

Jung, S. Lamballerie-Anton, M. and Ghoul M. (2000) Modifications of ultrastructure beef and myofibrilar proteins of post-rigor beef treated by high pressure, *Lebensm-Wiss u-Technol.* 33:313-319.

Kelly, S.M. and Price, N.C. (1997) The application of circular dichroism to studies of protein folding and unfolding, *Biochemica et Biophsica Acta* 1338:161-185.

Kingsley, H.D, Hollimann D.R., Calci, K.R., Chen, H. and Flick, G.J. (2007) Inactivation of a Norovirus by high pressure processing. *Applied Environmental Microbiology* 73:551-558.

Knorr, D. (1993) Effects of high-hydrostatic-pressure processes on food safety and quality. *Food Technology* 47(6):156-161.

Knorr, D. (2002) High pressure processing for preservation modification and transformation of foods *High Pressure Research* 22:595-599.

Kolakowski, P., Reps, A. and Babuchowski, A. (1998) Characteristics of Pressure Ripened Cheese. *Polish Journal of Food & Nutrition Sciences*, 7(48):473-483.

Koohmaraie, M., Doumit, M.E. and Wheeler, T.L. (1996) Meat toughening does not occur when shortening is prevented *J. Animal Science* 74:2935-2942.

López-Fandiño, R. (2006) High pressure-induced changes in milk proteins and possible applications in the dairy technology. *International Dairy Journal* 16:1119-1131.

López-Fandiño, R., Carrascosa, A.V. and Olano, A. (1996) The effects of high pressure on whey protein denaturation and cheese-making properties of raw milk. *Journal of Dairy Science* 79:929-936.

Lyng, J.G., Allen, P. and McKenna, B.M. (1998) The effect on aspects of beef tenderness of pre- and post-rigor exposure to a high intensity ultrasound probe. *J Science of the Food and Agriculture* 78:309-314

MacFarlane, J.J., McKenzie, I.J. and Turner, R.H. (1980) Pressure treatment of meat: effects on thermal transitions and shear values *Meat Science* 5:307-317.

McKenna, B.M. and Lyng, J.G. (2001) Rheological measurements of foods. In *"Instrumentation and Sensors for the Food Industry"*, E. Kress-Rogers and C. J.B. Brimelow (eds), Second edition, CRC Press, Boca Raton, Florida, pp. 425-452.

Mertens, B. and Knorr, D. (1992) Developments of nonthermal processes for food preservation. *Food Technology* 46:124-133

Mertens, B. and Deplace, G. (1993) Engineering aspects of high pressure technology in the food industry, *Food Tech.* 47(6): 164-169.

Molina, E., Álvarez, M.D., Ramos, M., Olano, A. and López-Fandiño, R. (2000) Use of high-pressure treated milk for the production of reduced-fat cheese. *International Dairy Journal* 10:467-475.

Mozhaev, V.V., Heremans, K., Frank, J., Masson, P. and Balny, C. (1996) High pressure effects on protein structure and function. *Proteins: Structures Function and Genetics* 24:81-91.

Murchie, L.W., Cruz-Romero, M., Kerry, J., Linton, J., Patterson M., Smiddy, M. and Kelly A. (2005) High pressure processing of shellfish: a review of microbiological and other quality aspects. *Innovative Food Science and Emerging Technologies* 6(3):257-270.

Okpala, C.O.R., Piggott, J.R. and Schaschke, C.J. (2010) Influence of high-pressure processing (HPP) on the physicochemical properties of fresh cheese. *Innovative Food Science and Emerging Technologies* 11:61-67

O' Reilly, C.E., Kelly, A.L., Murphy, P.M. and Beresford, T.P. (2001) High pressure treatment: applications in cheese manufacture and ripening. *Trends in Food Science and Technology* 12:51-59.

Pittia, P., Wilde, P.J., Husband, F.A. and Clark, D.C. (1996) Functional and structural properties of β-lactoglobulin as affected by high pressure treatment, *Journal of Food Science* 61(6):1123-1128.

Pommier, S.A. (1992) Vitamin A, electrical stimulation, and chilling rate effects on lysosomal enzyme activity in aging bovine muscle *J Food Science* 57:30-35.

Price, N. C. (1995) Circular dichroism in protein analysis: A comprehensive desk review. In *Molecular Biology and Biotechnology.* ed. R. A. Meyers, pp179-185. VCH Publishers Inc.

Provencher, S.W. and Glöckner, J. (1981) Estimations of globular protein secondary structure from circular dichroism, *Biochemistry* 20:33-37.

Rao, M.A. (1977). Rheology of liquid foods – a review. *J. Texture Studies*. 8, 135-168.

Rasanayagam, V., Balasubramaniam V.M., Ting , E., Sizer, C., Anderson, C. and Bush, C. (2003) Compression heating of selected fatty food substances during high pressure processing *J Food Science* 68(1):254-259.

Rynne, N.M., Beresford, T.P., Guinee, T.P, Sheehan, E., Delahunty, C.M. and Kelly, A.L. (2008) Effect of high-pressure treatment of 1 day-old full-fat Cheedar cheese on subsequent quality and ripening. *Innovative Food Science and Emerging Technologies* 9:429-440.

Sawyer, L., Barlow, P.N., Boland, M.J., Creamer, L.K., Denton, H., Edwards, P.J.B., Holt, C., Jameson, G.B., Kontopidis, G., Norris, G.E., Uhrinova, S. and Wu S-Y. (2002) Milk protein structure – what can it tell the diary industry? *International Dairy Journal.*, 12:299-310.

Severini, C., Baiano, A., Rovere, P., Dall'aglio, G. and Massini. R. (2003) Effect of high pressure on olive oil oxidation and the Malliard reaction in model and food systems *Italian food and beverage technology* 31:8-12.

Santos, J.C.O., Santos, I.M.G. and Souza, A.G. (2005). Effect of heating and cooling on rheological parameters of edible vegetable oils. *J. Food Eng.*, 67, 401-405.

San Martín-González, M.F., Rodríguez, J.J., Gurram, S., Clark, S., Swanson, B.G. and Barbosa-Cánovas, G.V. (2007) Yield, composition and rheological characteristics of cheddar cheese made with high pressure processed milk. *LWT Food Science and Technology* 40:697-705.

Schaschke CJ, Holmberg. E Allio S (2006) Viscosity Measurement of Vegetable Oil at High Pressure Trans IChemE Part C 84(C3) 173-178

Sheehan, J.J., Huppertz, T., Hayes, M.G., Kelly, A.L., Beresford, T.P. and Guinee, T.P. (2005) High pressure treatment of reduced-fat Mozzarella cheese: Effects on functional and rheological properties. *Innovative Food Science and Emerging Technologies* 6:73-81.

Silva, J. and Weber, G. (1993) Pressure stability of proteins, *Annu. Rev. Phys. Chem.,* 44:89-113.

Styles, M.F., Hoover, D.G. and Farkas, D.F. (1991) Response to *Listeria monocytogenes* and *Vibrio parahaemolyticus* to high hydrostatic pressure *J. Food Science* 56:1404-1407.

Suzuki, A., Kim, K., Homma, N., Ikeuchi, Y. and Saito, M. (1992) Acceleration of meat conditioning by high pressure treatment. In Balny C., Hayashi R., Heremans K and Masson P., (Eds) High Pressure and Biotechnology Montrouge, France: Colloque INSERM pp.219-227.

Tedford, L.-A. Kelly, S.M. Price, N.C., Schaschke, C.J. (1998) Combined Effects of Thermal and Pressure Processing on Food Protein Structure Institution of Chemical Engineers Trans C, 76: 80-86

Toledo, R.T. (1999) Fundamentals of Food Process Engineering. Publ. Springer. pp332-335.

Trujillo, A.J., Capellas, M., Buffa, M., Royo, C., Gervilla, R., Felipe, X., Sendra, E., Saldo, J., Ferragut, V. and Guamis, B. (2000) Application of high pressure treatment for cheese production. *Food Research International* 33:311-316.

Trujillo, A.J., Buffa, M.N., Casals, I., Fernández, P. and Guamis, B. (2002). Proteolysis in goat cheese made from raw, pasteurized or high-pressure treated milk. *Innovative Food Science and Emerging Technologies* 3:309-319.

Vant, S.C., Glen, G.F., Kontopodis, G., Sawyer, L. and Schaschke, C.J. (2002) Volume changes to the molecular structure of β-lactoglobin processed at high pressure, *High Temperature-High Pressure.* 34:705-712.

Yuste, J., Capellas, M., Pla, R., Fung, D.Y.C. and Mor-Mur, M. (2001) High pressure processing for food safety and preservation: A review. *Journal of Rapid Methods and Automation in Microbiology* 9:1-10.

Wheeler, T.L., Crouse, J.D. and Koohmaraie, M. (1992) The effect of post mortem time of injection and freezing on the effectiveness of calcium chloride for improving beef tenderness *J Animal Science* 70 :308-314.

In: Focus on Food Engineering
Editor: Robert J. Shreck
ISBN: 978-1-61209-598-1

Chapter 4

Spray Drying of Açai (*Euterpe Oleracea* Mart.) Juice: Effect Of Process Variables and Type of Carrier Agent on Product's Quality and Stability

Renata V. Tonon[1], Catherine Brabet[2] and Míriam D. Hubinger[1]
[1]Faculty of Food Engineering – State University of Campinas,
P.O. Box 6121, 13083-862, Campinas, SP, Brazil
[2]Centre de Coopération Internationale en Recherche Agronomique pour le Développement (CIRAD) – Department PERSYST – UMR QualiSud,
Montpellier, France

Abstract

This chapter describes and discusses some results obtained through the study of the microencapsulation of açai juice by spray drying using different carrier agents. Initially, the influence of process conditions on the moisture content, process yield and anthocyanin retention was evaluated using a central composite design. From the conditions selected in this first section (inlet air temperature of 140°C, feed flow rate of 15 g/min and 6% of carrier agent), particles were produced using four types of carrier agents: maltodextrin 10DE, maltodextrin 20DE, gum Arabic and tapioca starch. These particles were then characterized with respect to water activity, bulk and absolute density, porosity, particle size distribution and morphology. The samples produced with maltodextrin 20DE and with gum Arabic exhibited the highest water activity and the lowest particles size, while those produced with tapioca starch were the less porous, with the lowest bulk density and highest mean diameter. Then, physical stability of particles, when exposed to different relative humidities, was evaluated through the construction of sorption isotherms and determination of the glass transition temperature. The samples produced with maltodextrin 10DE exhibited the highest critical water activity, being considered as the most stable powder. Glass transition temperature decreased with increasing moisture content, confirming the plasticizant effect of water on this property. Finally, anthocyanin stability of powders stored at different temperatures and relative humidities was evaluated. The increase of both these parameters resulted in higher anthocyanin degradation. Maltodextrin 10DE was the carrier agent that showed the best pigment protection, for all the conditions studied.

1. Introduction

Açai (*Euterpe oleracea* Mart.) is a typical palm berry from Amazonia, with great occurrence and economical importance in the Brazilian state of Pará. It consists of a round fruit, with a proportionally big seed (approximately 85%), little pulp content and diameter varying between 1.0 and 1.5 cm (Figure 1). Besides having high energetic content, açai is also rich in fibers, vitamin E, proteins, minerals and unsaturated fatty acids such as Omega-6 and Omega-9. When processed into juice, açai has a large market in the Amazonian region, with a consumption of 100,000-180,000 liters per day, being used to produce energetic beverages, ice cream, jelly and liquor.

Figure 1. Açai fruit before harvest.

In the last years, açai has been recognized for its functional properties for use in food and nutraceutical products, due to its high antioxidant activity, which is related to its high anthocyanin and phenolic content (Coisson et al., 2005; Schauss et al., 2006). The presence of anthocyanins has increased the attractiveness of açai, not only for its functional properties, which are encouraging many industries to produce açai capsules, but also because this fruit can be considered to be an important source of natural pigments, which have no toxic effects and can contribute to the reduction of synthetic pigments in foods. This explains the significant growth of açai production in the state of Pará, which increased from 257,000 tons in 2003 to 581,000 tons in 2008 (SAGRI, 2010), being commercialized within various Brazilian states, as well as being exported to other countries around the world.

However, due to its high perishability, açai has a short shelf life, even under refrigeration. Moreover, anthocyanins are very unstable in processing and storage. Thus, the food industry is constantly looking for processes that increase its shelf life and improve pigment stability.

The production of açai powder represents an alternative, aiming at improving the product's preservation. Powdered fruit juices have low water activity, which makes difficult or even hinders microorganism growing and deterioration reactions, thus increasing its shelf-life. In addition, the production of powdered fruit juices also has advantages, like the ease of transport, storage and handling of the final product, even for direct consumption or as an ingredient in the elaboration of other food products.

The process generally used in the production of powdered fruit juices is spray drying (Quek et al., 2007; Cano-Chauca et al., 2005; Righetto & Netto, 2005; Dib Taxi et al., 2003). It consists of the atomization of the liquid in a chamber that receives a hot air flow in such a way that the fast water evaporation allows for maintaining a low temperature inside the particles. The physicochemical properties of powders produced by spray drying depend on some process variables, such as the characteristics of the liquid feed (viscosity, particles size, flow rate) and of the drying air (temperature, pressure), as well as the type of atomizer. Therefore, it is important to optimize the drying process in order to obtain products with better sensory and nutritional characteristics and better process yield.

Despite all the advantages related to the spray drying process, the powders resulting from drying of fruit juices usually exhibit problems like stickiness and high hygroscopicity, due to the presence of low molecular weight sugars and acids, which have low glass transition temperature. Some of these problems can be solved by the addition of carrier agents to the product before atomization, such as carbohydrates (maltodextrins, starch products, gums, and cellulose), proteins and some lipids. Such agents, besides increasing powder T_g, are very useful for microencapsulation purposes.

Microencapsulation has been used in order to protect sensitive foods or ingredients against adverse conditions, increase their stability and promote controlled release. In the case of açai juice, microencapsulation can represent a promising technique, which can protect anthocyanins against factors like oxygen, light and heat (increasing its shelf life), besides resulting in better handling properties, providing lower water adsorption and making the product less hygroscopic.

Taking into account the potential of powdered açai juice as a natural colorant and its high nutritional value, this chapter describes and discusses some results obtained in the study of spray drying of açai juice, as well as the effect of different carrier agents on powder physical properties and anthocyanin stability upon storage.

2. Microencapsulation by Spray Drying

2.1. General Aspects

Microencapsulation is defined as a technique by which liquid droplets, solid particles or gas compounds are entrapped into thin films of a food grade carrier agent (Gharsallaoui et al., 2007).

According to Ré (1998), the main reasons for using microencapsulation in food products are: to protect food against adverse environmental conditions (light, moisture, oxygen, UV radiation) and nutritional losses, to incorporate controlled release mechanisms to the formulations, to mask or preserve flavors and aromas and, finally, to make the product more attractive, by promoting a better flexibility and control in the development of more tasty and nutritive products, in order to satisfy consumers' expectation.

There are many methods used for microencapsulation of food ingredients, which are divided in: physical (spray drying, freeze drying, spray cooling, spray chilling, spray coating, extrusion, fluidized bed drying, co-crystallization), chemical (interfacial polymerization, *in situ* polymerization) and physicochemical (simple and complex coacervation, solvent evaporation, molecular inclusion). The choice of a method depends on the economic factors, the core sensitivity, the desired microcapsule size, the physicochemical properties of the core and the wall material, as well as the mechanism of release (Jackson & Lee, 1991).

The most common process used for microencapsulation in the food industry is spray drying (or atomization), which is a well established and straightforward technology that results in particles with good quality (Gouin, 2004). Equipment is readily available and production costs are lower than most other methods. This process can be useful for drying heat-sensitive foods, pharmaceutical and other substances, since it promotes rapid water evaporation from the droplets, keeping low the temperature inside the particles.

Atomization results from the application of an energy which acts on the liquid until its disruption and disintegration, creating a spray of droplets. This spray is put in contact with the hot air and an immediate drying occurs, resulting in the collect of powder. According to Goula et al. (2004), in the early stages of drying, the droplet has a free liquid surface and evaporation from this surface is fast. The water removal cause the solute to be more concentrated at the surface, which depends on the evaporation speed and the rate at which the liquid can be replenished from the interior of the droplet. Due to this increased concentration, solids form a crust around a hollow particle, whose thickness depends on the drying rate (high initial drying rates lead to larger particles with thin shells, whereas, low initial drying rates lead to smaller particles with thicker shells).

2.2. Spray Drying of Fruit Juices

Spray drying, when performed at optimized conditions, has proved to be an efficient method for obtaining several types of food products. Drying of sugar-rich products such as fruit juices has great economical potential, since it results in products with very much reduced volume and longer shelf life, besides making their transport and storage easier.

However, fruit juice powders obtained by spray drying have some drawbacks in their handling properties, such as stickiness, higroscopicity and solubility, making their packaging and utilization substantially difficult. According to Bhandari et al. (1997), the sticky behavior of sugar and acid-rich materials is attributed to low molecular weight sugars such as fructose, glucose and sucrose, and organic acids such as citric, malic and tartaric acid, which constitute more than 90% of the solids in fruit juices and purees. According to the authors, the fast water removal that occurs during atomization results in a completely amorphous product, or even in a product with some micro-crystalline regions that disperses in an amorphous mass. The sugars and acids present in fruit juices have low glass transition temperature and are very hygroscopic in their amorphous state and lose free flowing nature at high moisture content (Roos & Karel, 1991).

Bhandari et al. (1997) described the main stages of spray drying, as well as the changes occurring in the sugar-rich products during drying. According to the authors, the droplets are initially dispersed individually in a large volume of the dryer, avoiding agglomeration of particles. Towards the lower part of the dryer, when the particles are already solid, agglomeration should neither occur. However, due to the presence of high sugar content with low T_g, the product can remain as syrup, even at low moisture content level. Depending on the product composition and the spray drying conditions, the surface of droplets may remain plastic, resulting in the stickiness to the drying chamber wall or even between the particles. Thus, the amorphous product obtained at the end of the drying process could be either syrup or a sticky powder, or even a relatively free-flowing powder (Bhandari & Howes, 1999).

In general, the glass transition temperature of sugar-rich foods is so low, that drying of these pure products is not economically viable. In this context, the addition of carrier agents with high molecular weight to the product before atomization is necessary, in order to increase its glass transition temperature and reduce powder stickiness.

2.3. Carrier Agents

The choice of the carrier agent depends on the physicochemical properties of the material to be dried, the process used to form the particle and its desired final properties. The ideal wall material must be non-reactive with the core material, exhibit low viscosity at high concentrations, be able to disperse or emulsify the active material and to stabilize the emulsion formed, entrap and keep the core material inside its structure during processing and storage. Moreover, it must completely release the core material (release can be controlled or not), provide the maximum protection against environmental conditions, be soluble in solvents acceptable in the food industry, show high availability and low cost (Gharsallaoui *et al.*, 2007; Desai & Park, 2005).

Typical carrier agents include maltodextrins, hydrophobically modified starches, gum Arabic, milk and soy proteins and mixtures of these materials. Several works can be found in the literature showing maltodextrin and gum Arabic as carrier agents used in the spray drying of fruit juices, such as acerola (Righetto & Netto, 2005), mango (Cano-Chauca *et al.*, 2005), cactus pear (Rodríguez-Hernández *et al.*, 2005) and camu-camu (Dib Taxi *et al.*, 2003).

Maltodextrins are products of starch hydrolysis, consisting of α-D-glucose units linked mainly by (1→4) glycosidic bonds and described by their dextrose equivalency (DE), which varies with the hydrolysis degree, determines their reducing capacity and is inversely related to their average molecular weight (BeMiller & Whistler, 1996). They are very useful for spray drying of food materials, mainly due to their low cost. According to Reiniccius (2001), maltodextrins are mainly used in materials that are difficult to dry, like fruit juices, flavorings, enzymes or sweeteners.

Gum Arabic is a natural plant exudate of Acacia trees, comprising a complex hetero-polysaccharide with a highly ramified structure, with a main chain comprising D-galactopyranose units joined by β-D-(1→3) glycosidic bonds. Side chains with different chemical structures are linked to the main chain by β-(1→6) bonds (BeMiller & Whistler, 1996). Gum Arabic is the only gum used in food products that exhibits high solubility and low viscosity in aqueous solution, which facilitates the spray drying process.

Tapioca starch is a fine flour refined from cassava root (*Manihot esculenta* Crantz) by natural fermentation, which has been used by some Brazilian companies as a carrier agent in the production of powdered fruit juices, in order to ensure that no genetically modified material is being used in the powder production – a condition required by some importing countries. Moreover, the use of such an agent has other advantages, such as absence of flavor and taste, high availability and low cost in Brazil.

3. Spray Drying of Açai Juice

3.1. Effect of Process Variables

In our study about microencapsulation of açai juice, we firstly evaluated the influence of some spray drying conditions (inlet air temperature, feed flow rate and maltodextrin concentration) on the powder physicochemical properties, according to a central composite design. For this, the açai pulp was filtered through a filter paper, in order to eliminate solids in suspension, facilitating the product's passage through the nozzle atomizer. Filtration increased the anthocyanin content per gram of dried mass and reduced the fat content (Table 1), making the product less susceptible to lipid oxidation. Then, maltodextrin 10DE was added to the filtered pulp under magnetic agitation, until complete dissolution.

Table 1. Composition of pure and filtered açai pulp (*Euterpe oleraceae* Mart.)

Analyzed item	Pure pulp	Filtered pulp	Analysis method
Moisture (% wet basis)	85.96 ± 0.11	96.94 ± 0.01	A.O.A.C. (1990)
Proteins (%)	1.43 ± 0.04	0.50 ± 0.05	A.O.A.C. (1990)
Lipids (%)	6.53 ± 0.03	0.21 ± 0.01	Bligh and Dyer (1959)
Fibers (%)	4.52 ± 0.22	n.d.*	A.O.A.C. (1990)
Total sugars (%)	0.48 ± 0.05	1.69 ± 0.12	A.O.A.C. (1990)
Ash (%)	0.44 ± 0.01	0.39 ± 0.01	A.O.A.C. (1990)
Acidity (% citric acid)	0.34 ± 0.02	0.32 ± 0.02	A.O.A.C. (1990)
Anthocyanins (mg/100g dried matter)	1265.14 ± 16.27	3920.48 ± 33.74	Francis (1982)

*n.d. = not detected

Spray drying process was performed in a laboratory scale spray dryer LabPlant SD-05 (Huddersfield, England), with a 1.5 mm diameter nozzle and main spray chamber of 500 mm × 215 mm. The mixture was fed into the main chamber through a peristaltic pump and the feed flow rate was controlled by the pump rotation speed. Drying air flow rate was 73 m^3/h and compressor air pressure was 0.06 MPa.

Table 2 shows the codified independent variables of the central composite design.

Table 2. Codified independent variables of the central composite design

Variables	- 1.68	- 1	0	+ 1	+ 1.68
Inlet air temperature (°C)	138	150	170	190	202
Feed flow rate (g/min)	5	9	15	21	25
Maltodextrin concentration (%)	10	14	20	26	30

Moisture content, process yield and anthocyanin retention were analyzed as responses. Process yield was calculated as the relationship between total solids content in the resulting powder and total solids content in the feed mixture. moisture contents were determined gravimetrically by drying in a vacuum oven at 70°C until constant weight (A.O.A.C., 1990)

Total anthocyanin content was determined by the spectrophotometric method, according to Francis (1982). Data were fitted to the following polynomial:

$$y = \beta_0 + \beta_1 x_1 + \beta_2 x_2 + \beta_3 x_3 + \beta_{11} x_1^2 + \beta_{22} x_2^2 + \beta_{33} x_3^2 + \beta_{12} x_1 x_2 + \beta_{13} x_1 x_3 + \beta_{23} x_2 x_3 \quad (1)$$

where β_n are constant regression coefficients; y is the response (moisture content, process yield and anthocyanin retention), and x_1, x_2 and x_3 are the coded independent variables (inlet air temperature, feed flow rate and maltodextrin concentration, respectively).

Table 3 shows the regression coefficients for the coded second order polynomial equation, the p values and the determination coefficients (R^2). Some non-significant terms were eliminated and the resulting equations were tested for adequacy and fitness by the analysis of variance (ANOVA). The fitted models were suitable, showing significant regression, low residual values and no lack of fit.

Table 3. Coded second-order regression coefficients for process yield, moisture content and anthocyanin retention

Coefficients	Process yield (%)	Moisture content (%)	Anthocyanin retention (%)
β_0	48.49	1.48	82.05
β_1	2.31	–0.61	–2.36
β_2	–3.74	0.31	NS
β_3	–3.21	NS	NS
β_{11}	NS	NS	NS
β_{22}	–1.64	NS	NS
β_{33}	–1.12	NS	NS
β_{12}	NS	–0.19	NS
β_{13}	NS	NS	NS
β_{23}	–1.27	NS	NS
R^2	0.926	0.907	0.703
p-value	<0.0001	<0.0001	<0.0001

NS: Non-significant (p>0.05)

Figures 2 to 4 show the response surfaces generated from the models obtained by statistic analyses of the results of process yield, moisture content and anthocyanin retention.

Moisture content was significantly influenced by inlet air temperature and feed flow rate, varying from 0.64 to 2.89%, which is in agreement with the values reported by Papadakis et al. (2006) for spray-dried raisin juice. Inlet air temperature was the variable that showed the greatest influence on powders moisture content. Higher inlet air temperatures imply in higher temperature gradient between the atomized liquid and the drying air, resulting in a greater driving force for water evaporation and thus in products with lower moisture content. A similar behavior was observed by Quek et al. (2007), Rattes & Oliveira (2007) and Grabowski et al. (2006), studying the spray drying of watermelon juice, sodium diclofenac and sweet potato puree, respectively. The feed flow rate negatively affected powders moisture content. The higher the flow rate, the shorter is the contact time between the liquid and the drying air, which makes the heat transfer less efficient and results in lower water evaporation. Hong & Choi (2007) also observed a more pronounced effect of the drying temperature than of the pump rate on the moisture content of spray-dried protein-bound polysaccharide from *Agaricus blazei* Murill. As illustrated in Figure 2(b), maltodextrin concentration did not influence powders moisture content.

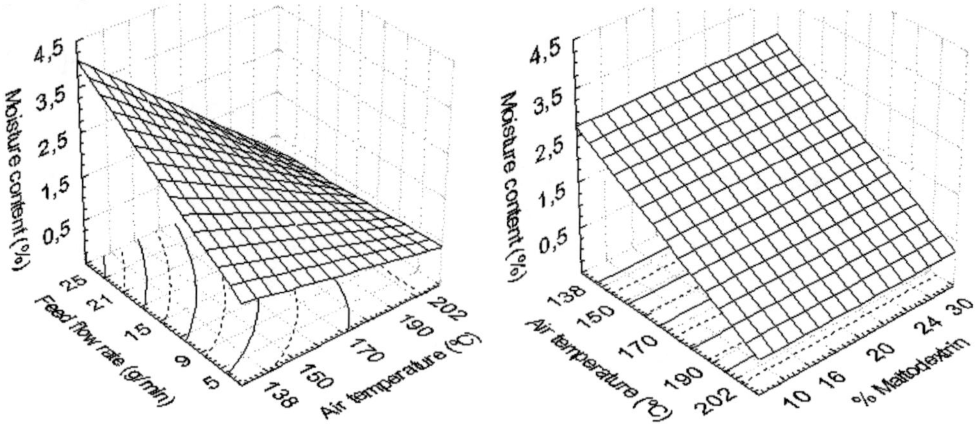

(A) (B)

From Tonon et al., 2008.

Figure 2. Response surface for moisture content, for (a) 20% maltodextrin, (b) feed flow rate of 15g/min.

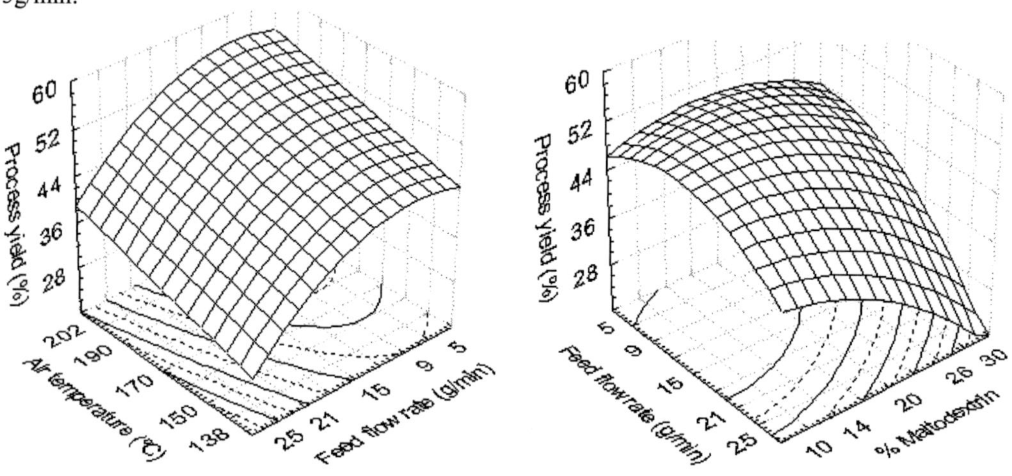

(A) (B)

From Tonon et al., 2008.

Figure 3. Response surface for process yield, for (a) 20% maltodextrin, (b) inlet air temperature of 170°C.

The highest process yields (50 - 55%) were also obtained when higher drying temperatures and lower feed flow rates were used, due to the more efficient heat and mass transfer's occurrence when higher inlet air temperatures are used. As stated before, increasing inlet air temperatures and decreasing flow rates also result in powders with lower moisture content, which are less susceptible to stickiness on the dryer chamber wall. Moreover, higher temperatures lead to the fast formation of a hard crust, which makes particles less "plastic" and thus susceptible to caking and stickiness, when compared to those produced at lower temperatures, which may be more pliable and collapsed. Maltodextrin concentration also showed a negative effect on process yield, probably due to the increase on mixture viscosity, which can lead more solids to paste in the main chamber wall, thus reducing process yield (Cai & Corke, 2000).

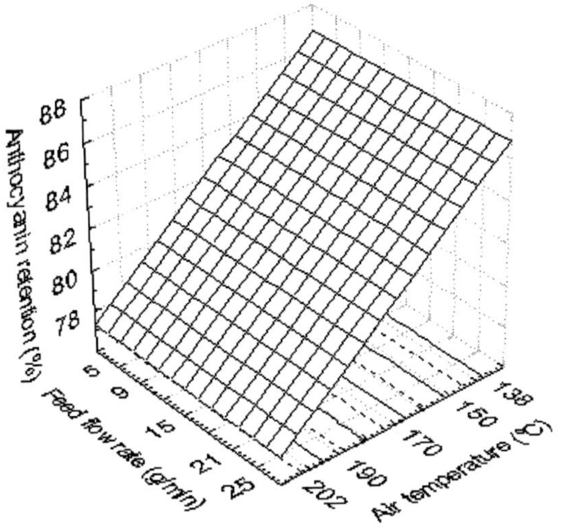

From Tonon et al., 2008.

Figure 4. Response surface for anthocyanin retention, for 20% maltodextrin.

Inlet air temperature was the only variable affecting anthocyanin retention, mainly due to the high sensitivity of these pigments to high temperatures. Moreover, powders produced at lower temperatures have a tendency to agglomerate, because of their higher moisture content. This agglomeration reduces the powder exposition to oxygen, protecting the pigments against oxidation (Quek et al., 2007).

3.2. Selection of the Best Drying Conditions

As color (as well as its associated functional properties) is the main attractive in powdered açai juice, the main criterion used for selection of the best process conditions was the anthocyanin retention. As discussed before, the only variable that influenced this response was the inlet air temperature (lower air temperatures resulted in better retention). Thus, the drying temperature chosen for further tests was 140°C.

Considering that the richest particles in anthocyanins were those produced with lower maltodextrin concentration (since there is lower "dilution effect" by maltodextrin) and that this variable did not influence pigment retention, the maltodextrin concentration selected was 10%.

Regarding to the feed flow rate, in the tests performed with feed rates higher than 15 g/min, there was a dripping inside the main chamber, that is, part of the mixture passed straight to the chamber and was not atomized, resulting in lower process yield and product wasting. On the other hand, for feed rates of 15 g/ml or lower, process yield was considered high. Thus, as lower feed rates represents higher energy spent, the feed flow rate selected as the best was of 15 g/min.

Therefore, the best process conditions selected from the central composite design were: inlet air temperature of 140ºC, feed flow rate of 15 g/min and maltodextrin concentration of 10%.

3.3. Additional Tests

According to the results obtained from the central composite design, anthocyanin retention was not influenced by maltodextrin concentration. This suggests that the use of lower concentrations could result in products with higher anthocyanin content (less "diluted" by maltodextrin). Thus, some additional tests were performed with lower maltodextrin concentrations, in the selected conditions of inlet air temperature and feed flow rate (140ºC and 15 g/min, respectively).

Tests were performed using 6% and 8% of maltodextrin (concentrations lower than 6% led to excessive powder stickiness on the chamber wall and insignificant process yield). Each sample was processed in triplicate and the results were statistically analyzed by the Duncan's test. They are presented in Table 4.

Table 4. Results of spray drying of açai juice using 10%, 8% and 6% maltodextrin (MD), with inlet air temperature of 140ºC and feed flow rate of 15 g/min

Tests	Moisture content (%)	Process yield (%)	Anthocyanin retention (%)
10% MD	2.54 ± 0.14^a	49.35 ± 3.52^a	85.84 ± 2.26^a
8% MD	2.46 ± 0.12^a	48.60 ± 0.56^a	84.78 ± 3.52^a
6% MD	2.32 ± 0.05^a	46.88 ± 1.48^a	85.11 ± 2.48^a

Different letters indicate significant difference between samples produced with different maltodextrin concentrations ($p \leq 0.05$).

According to Table 4, samples produced with varying maltodextrin concentrations did not show significant differences between each other with respect to moisture content, process yield and anthocyanin retention. Thus, taking into account that the powders produced with a lower amount of maltodextrin have higher anthocyanin content, the selected concentration for spray drying of açai juice was 6%.

From the conditions selected in this first part of our study (inlet air temperature of 140ºC, feed flow rate of 15 g/min and 6% of carrier agent), particles were produced using different types of carrier agents (maltodextrin 10DE, maltodextrin 20DE, gum Arabic and tapioca starch) and were characterized with respect to some physical and morphological properties, physical stability at different relative humidities, as well as anthocyanin stability during storage. The results are presented in the next topics.

4. Physical Characterization of Powdered Açai Juice Produced with Different Carrier Agents

The use of different carrier agents for powder production can result in different physicochemical properties, depending on the structure and the characteristics of each agent. According to Barbosa-Cánovas & Juliano (2005), knowledge of food properties is essential for optimizing processes and functionalities, as well as for cost reduction, especially in the case of powders produced or used in pharmaceutical and food industries.

Properties such as water activity are essential for powder stability and storage, since it represents the free water available for deterioration reactions. Density is an important property to be studied in dried mixtures. Knowledge of density is essential in industrial processes, for determination of storage, processing, packaging and distribution conditions. The products obtained by milling or drying are, in general, characterized by their bulk density, which considers the volume of the solid material and of all the porous closed or open to the atmosphere (Barbosa-Cánovas & Juliano, 2005). On the other hand, the absolute density only considers the volume of the solid material, excluding the pores' volume. Particle size distribution is also important for food powders in many aspects, such as processing, handling and shelf life, since it can influence the product taste, colour, texture and flavour (O'Hagan et al., 2005), which are the characteristics most considered by consumers. Microstructure, in turn, relates to powders' functionality, stability and flowability (Jackson & Lee, 1991; Shahidi & Han, 1993).

In this second part of the present chapter, we evaluated the physical properties of spray-dried açai juice produced with four different types of carrier agents: maltodextrin 10DE, maltodextrin 20DE, gum Arabic and tapioca starch. Powders were produced using the process conditions selected in the first part of our study and were characterized with respect to water activity, bulk and absolute density, intergranular porosity, particle size distribution and morphology.

Water activity was measured in an Aqualab 3TE (Decagon, Pullman, WA, USA) hygrometer, at 25°C.

Bulk density (ρ_{bulk}) was measured by weighing 2 g of sample and placing into a 50 ml graduated cylinder. Absolute density (ρ_{abs}) was determined by helium pycnometry, in an AccuPyc 1330 Automatic Gas Pycnometer (Micromeritics, Norcross, USA).

Intergranular porosity (ε) was calculated according to Equation (2):

$$\varepsilon = 1 - \frac{\rho_{bulk}}{\rho_{abs}} \qquad (2)$$

Particle size distribution was measured in a laser light diffraction instrument, Mastersizer S (Malvern Instruments, Malvern, UK), using isopropanol as the dispersion medium and particle morphology was evaluated by scanning electron microscopy (SEM), performed at 5 kV, in a LEO440i scanning electron microscope (LEICA Electron Microscopy Ltd, Cambridge, UK).

Results (Table 5) were obtained in triplicate and statistically analyzed by analysis of variance, using the software STATISTICA 5.0 (StatSoft, Tulsa, USA). Mean analysis was performed using Duncan's procedure at $p \leq 0.05$.

Table 5. Water activity, bulk density, absolute density and intergranular porosity of powders produced with different carrier agents

Carrier agent	Water activity	Bulk density (g/cm^3)	Absolute density (g/cm^3)	Porosity (%)
Maltodextrin 10DE	0.229 ± 0.006^a	0.390 ± 0.015^a	1.531 ± 0.004^a	74.50 ± 1.01^a
Maltodextrin 20DE	0.245 ± 0.002^b	0.370 ± 0.016^a	1.511 ± 0.004^b	75.49 ± 1.07^a
Gum Arabic	0.244 ± 0.002^b	0.377 ± 0.011^a	1.491 ± 0.008^c	74.70 ± 0.72^a
Tapioca starch	0.189 ± 0.002^c	0.480 ± 0.005^b	1.514 ± 0.001^b	68.33 ± 0.30^b

Different letters indicate significant difference ($p \leq 0.05$).

4.1. Water Activity

All the samples showed water activity values (a_w) below 0.3, which is very positive for powder stability, once it represents less free water available for microorganism growing and biochemical reactions and hence, longer shelf life (Fennema, 1996). A similar range of a_w values were obtained by Quek et al. (2007) for spray dried watermelon powders.

The particles produced with tapioca starch showed the lowest water activity, followed by those produced with maltodextrin 10DE. Particles produced with gum Arabic and with maltodextrin 20DE showed the highest water activity values and had no significant differences between them. Such results can be attributed to the chemical structure of gum Arabic and maltodextrin 20DE, which have a high number of ramifications with hydrophilic groups, and thus can easily bind to water molecules from the ambient air during powder handling after spray drying.

4.2. Bulk Density, Absolute Density and Intergranular Porosity

The powder produced with tapioca starch exhibited the highest bulk density, whereas the others did not significantly differ between each other. This highest bulk density can be explained by the highest molecular weight of tapioca starch. Starches are composed basically by two polymers, amylose and amylopectin, the latter showing higher molecular weight. Most of the starches contain about 20-30% of amylose and 70-80% amylopectin and these values vary according to the botanical source. Tapioca starch has higher amylopectin content when compared to corn starch (83% and 72%, respectively), which explains its higher molecular weight. The heavier the material, more easily it accommodates into the spaces between the particles, thus occupying less space and resulting in a higher bulk density.

Regarding to the absolute density, all the samples exhibited similar values. According to Table 5, the particles produced with maltodextrin 10DE showed a slightly higher density, followed by those produced with maltodextrin 20DE and tapioca starch, which did not differ between each other. The sample produced with gum Arabic showed the lowest absolute density.

Absolute density corresponds to the real solid density and does not consider the spaces between particles, in contrast to the bulk density, which takes into account all these spaces. Thus, the results indicate that the samples produced with maltodextrin 10DE, maltodextrin 20DE and gum Arabic have a higher number of interparticle spaces than the sample produced with tapioca starch. These results are expressed in the calculation of intergranular porosity, which measures exactly this quantity of spaces. Porosity is an important property in the case of microcapsules where the encapsulated material is susceptible to oxidation. The larger number of spaces between particles implies in more oxygen available to degradation reactions, leading to a faster loss of the compound being protected.

4.3. Particle Size Distribution

Figure 5 shows the particle size distribution for the powders produced with the different carrier agents. Particles exhibited a large range of sizes, with diameters varying from 0.1 to 41.0 μm, and showed a bimodal distribution, i.e. a distribution with two distinct peaks, each one representing a predominant size. This is particularly interesting in the case of powders, since the "population" of smaller particles can penetrate into the spaces between the larger ones, thus occupying less space. The presence of larger particles may be attributed to an incipient agglomeration process, where the formation of irreversible link bridges leads to the production of particles of greater size.

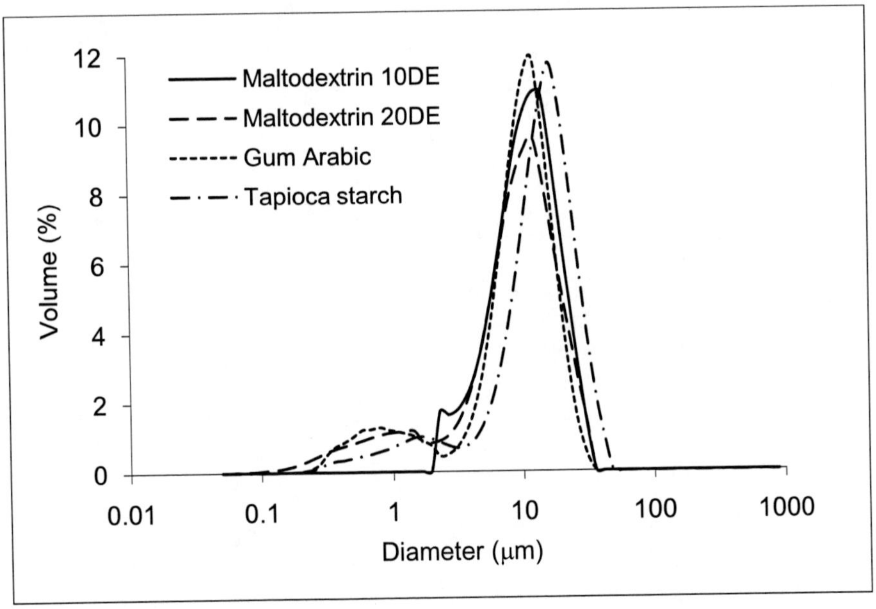

Figure 5. Particle size distribution of powders produced with different carrier agents.

The particle mean diameter varied from 9 to 14 μm, approximately, and the difference between the samples produced with the different carrier agents was small. The particles produced with gum Arabic exhibited the lowest mean diameter (9.33 μm), very similar to the particles produced with maltodextrin 20DE (9.41 μm). The sample produced with maltodextrin 10DE exhibited a mean diameter higher than both (10.94 μm), while the one produced with tapioca starch was even bigger (13.67 μm). This increased particle size is related to the molecular size of each carrier agent. The higher the maltodextrin DE, the higher is the degree of hydrolysis, and therefore the shorter are its chains. This explains the smaller particle size of the particles produced with maltodextrin 20DE, when compared to those produced with maltodextrin 10DE and with tapioca starch.

4.3. Particle Morphology

The SEM microphotographs of the powders produced with different carrier agents are shown in Figure 6. The resulting powders had particles of various sizes, for all the carrier agents, which agree with the results obtained for particle size distribution.

The particles produced with maltodextrin 10DE, maltodextrin 20DE and gum Arabic were very similar (Figures 6a, 6b and 6c), exhibiting a predominantly spherical shape, which is typical of materials produced by spray drying. Most of the particles exhibited a shriveled surface, due to the low inlet air temperature used (140°C), which leads to slower heat transfer and results in particles with more pliable and collapsed crust (Allamilla-Beltrán et al., 2005).

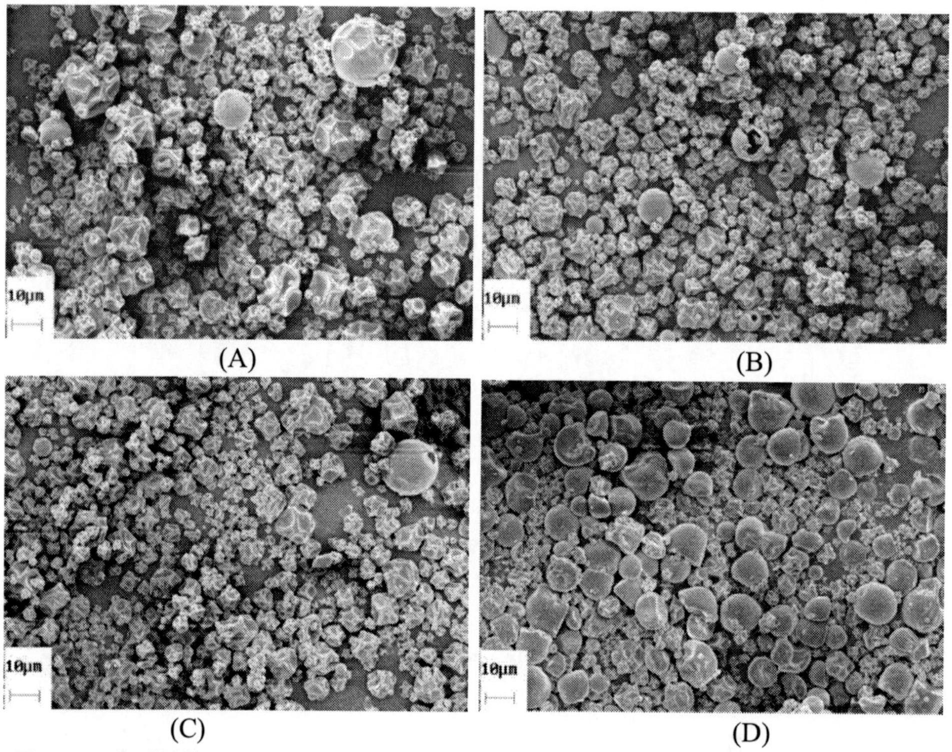

From Tonon et al., 2009b.

Figure 6. Micrographs of microcapsules produced with: (a) maltodextrin 10DE, (b) maltodextrin 20DE, (c) gum Arabic and (d) tapioca starch.

Most of the particles produced with tapioca starch exhibited a rounded shape and smooth surface. This type of morphology was also observed by Loksuwan (2007), when encapsulating β-carotene with native tapioca starch. Leonel (2007) evaluated the morphology of tapioca starch and observed a very similar structure to that observed here. Thus, as tapioca starch solubility is low, the rounded and smooth particles observed in Figure 6d are probably the granules of this agent which were not dissolved and did not form a "matrix" with the juice, while the shriveled ones are the particles of the dried juice.

5. Stability of Powdered Açai Juice

For many years, water activity (a_w) has been considered more important than total amount of water, concerning to quality and stability of foodstuff. In this context, water sorption isotherms are important thermodynamic tools for predicting the interactions between water and food components. They describe the relationship between water activity and the equilibrium moisture content of a food product and provide useful information for food processing operations such as drying, packaging and storage (Lomauro et al., 1985).

Roos (1995) reported that the plasticization of biosolids is a result of combined effects of water and temperature. According to the author, the prediction of food stability based only on sorption isotherms data is not enough, since certain physicochemical and structural processes such as stickiness, crispness, collapse, amorphous-to-crystalline transformations and the rates of non-enzymatic browning are not related to a monolayer value and they are better correlated to the glass transition temperature through plasticization by water or temperature.

The glass transition temperature (T_g) is defined as the temperature at which an amorphous system changes from the glassy to the rubbery state. Theoretically, in the glassy state, the high viscosity of the matrix (about 10^{12} Pa · s) does not allow the occurrence of diffusion-controlled reactions (Slade & Levine, 1991). However, as the temperature increases above T_g, various changes such as increase of free volume and specific heat, as well as decrease of viscosity, are noticed (Rahman, 2006). These factors control various time-dependent structural transformations, such as stickiness, collapse and crystallization, during food processing and storage.

Thus, the use of state diagrams that indicate the material's physical state, combined with the sorption isotherms, helps in the prediction of food stability, regarding to its physical characteristics. Several authors have coupled the concepts related to water activity with those of glass transition temperature in order to evaluate food stability, thus providing an integrated approach to the role of water in food (Slade and Levine, 1991; Sablani et al., 2004; Shrestha et al., 2007; Symaladevi et al., 2009).

The critical water content/water activity is the value at which the glass transition temperature of a product is equal to the room temperature. Above the room temperature, the amorphous powders are susceptible to deteriorative changes like collapse, stickiness and caking, resulting in overall quality loss. Thus, the critical water activity (a_{wc}) indicates the maximum relative humidity to what the product can be exposed, at a given temperature, without showing structural deteriorations and diffusion-controlled reactions.

In the case of açai powder, anthocyanin degradation is one of the most important reactions affecting product's quality during storage, since these pigments are sensitive to factors like temperature, light, pH, oxygen and others. Several works are found regarding to anthocyanin stability as affected by these factors, in products like spray dried black carrot extracts (Ersus & Yurdagel, 2007), crude extracts of *Ranunculus asiaticus* flowers (Amr & Al-Tamimi, 2007), black carrot juice (Kirca et al., 2007) and others. Anthocyanin degradation usually follows first-order kinetics, i.e., anthocyanin content exponentially decreases with time.

Thus, in this third part of the present chapter, we evaluate the stability of spray-dried açai juice produced with the four different types of carrier agents mentioned before (maltodextrin 10DE, maltodextrin 20DE, gum Arabic and tapioca starch). From the critical storage conditions determined by coupling sorption isotherms and glass transition temperature, powders were stored at different temperatures and relative humidity for 120 days, when anthocyanin stability was evaluated.

5.1. Physical Stability

The physical stability of powders produced with different carrier agents were evaluated through the construction of sorption isotherms and determination of the glass transition temperature.

Sorption isotherms were determined by the gravimetric static method, using eight saturated salt solutions with water activies varying from 0.113 to 0.843. Experimental data were fitted to the modified BET model with three parameters (Equation 3), which is able to accurately predict the equilibrium moisture content over a larger range of water activities than the two-parameters BET model (Brunauer et al., 1938).

$$X_e = \frac{X_m C_{BET} a_w [1-(n+1)(a_w)^n + n(a_w)^{n+1}]}{(1-a_w)[1+(C_{BET}-1)a_w - C_{BET}(a_w)^{n+1}]} \quad (3)$$

where X_e is the equilibrium moisture content, X_m is the monolayer moisture content, n is the number of adsorbed layers and C_{BET} is the model constant.

The glass transition temperature was determined by differential scanning calorimetry (DSC), in a TA-MDSC-2920 (TA Instruments, New Castle, USA) equipped with a mechanical refrigeration system (RCS – refrigerated cooling accessory). Açai powder samples (about 5 mg) were placed into aluminum pans and equilibrated over saturated salt solutions in desiccators at 25°C. After equilibrium was reached, samples were hermetically sealed, weighed and taken for DSC analysis, being heated at 10°C/min from -70 to 120°C, two times (since the second scanning reduces the enthalpy relaxation of the amorphous powder that appears in the first scan). Equipment calibration was performed with indium ($T_{melting}$ = 156.6°C) and verification with azobenzol ($T_{melting}$ = 68.0°C). Dry helium, 25 ml/min, was used as the purge gas. All analyses were done in triplicate and data were treated by the software Universal Analysis 2.6 (TA Instruments, New Castle, USA). The glass transition temperature was taken as the mid point of the second-order transition that produces a step change in the heat flow at the temperature of phase transition.

5.1.1. Sorption Isotherms

The sorption isotherms fitted to the three-parameter BET model are presented in Figure 7 and they were of type III according to Brunauer's classification (Brunauer et al., 1938). This type of curve was also observed by Wang et al. (2008) for freeze-dried Chinese gooseberry and by Gabas et al. (2007) for vacuum dried pineapple added of maltodextrin and gum Arabic. The estimated parameters for the three-parameters BET model are presented in Table 6. The model showed a good fit to experimental data, with high R^2 values and satisfactory mean relative deviation modulus (E).

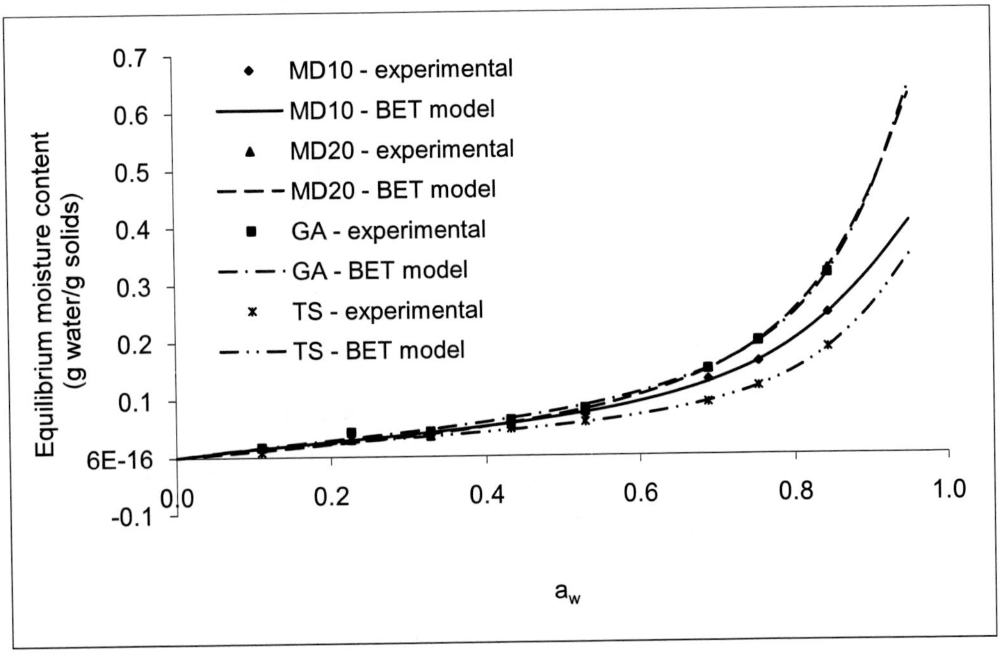

From Tonon et al., 2009a.

Figure 7. Sorption isotherms of spray dried açai juice produced with different carrier agents (MD10 = maltodextrin 10DE; MD20 = maltodextrin 20DE, GA = gum Arabic; TS = tapioca starch).

Table 6. Estimated BET parameters for açai juice powder produced with different carrier agents

Parameters	Carrier agents			
	Maltodextrin 10DE	Maltodextrin 20DE	Gum Arabic	Tapioca starch
X_m	0.045	0.058	0.054	0.031
C_{BET}	3.45	1.67	2.96	6.33
N	21.47	27.45	31.41	28.03
R^2	0.997	0.999	0.998	0.995
E (%)	6.58	8.60	6.08	15.46

The BET model is based on the monolayer moisture concept and provide the value of the monolayer moisture content of the material (X_m), considered as the safe moisture for dried foods during preservation, while most other models lack this parameter. The monolayer moisture content (X_m) indicates the amount of water that is strongly adsorbed to specific sites at the food surface and is considered an important value to assure food stability. The X_m values obtained for spray dried açai juice varied from 3.1% to 5.8% according to the BET model, which are in agreement with the values obtained by Righetto & Netto (2005) and Moraga et al. (2006), for spray-dried acerola and freeze-dried kiwi, respectively. Pérez-Alonso et al. (2006) determined the sorption isotherms for maltodextrin 10DE and they also obtained lower monolayer moisture contents for maltodextrin 10DE (6.96-7.35%) than for gum Arabic (8.11-11.0%) at temperatures of 25, 35 and 40°C. The authors attributed such results to a combination of factors, such as the conformation and topology of molecule and the hydrophilic/hydrophobic sites adsorbed at the interface.

According to Figure 7, the powders produced with tapioca starch showed the lowest water adsorption, followed by that produced with maltodextrin 10DE, while the samples produced with maltodextrin 20DE and gum Arabic were the most hygroscopic ones. Such differences in water adsorption can be explained by the chemical structure of each agent. Maltodextrin 20DE and gum Arabic have a great number of ramifications with hydrophilic groups and therefore, can easily adsorb moisture from the ambient air. Maltodextrin 10DE is less hydrolyzed, showing less hydrophilic groups and thus adsorbing less water. Tapioca starch is a native starch (not hydrolyzed), which explains its lower hygroscopicity. Cai and Corke (2000) and Ersus and Yurdagel (2007) also verified an increase of hygroscopicity with increasing maltodextrin's DE, working with microencapsulation of betacyanins and anthocyanins, respectively. The authors attributed such increase to the lower molecular weight of the maltodextrins with higher DE, which have shorter chains and, therefore, more hydrophilic groups.

Some physical changes could be observed on the powders stored at different relative humidities. When stored at relative humidities of 43% or lower, particles remained as a free-flowing powder, for all the carrier agents used. At 53%, particles showed a beginning of agglomeration, and the powder could not flow so easily. When stored at higher relative humidities, physical transformations were more evident. At a_w's above 0.69, the particles produced with maltodextrins and gum Arabic showed the formation of hard and dark blocks, resulting from the compaction, an advanced stage in caking associated with a pronounced loss of system integrity as a result of thickening of interparticle bridges owing to flow, reduction of interparticle spaces and deformation of particle clumps under pressure (Aguilera et al., 1995). The samples produced with maltodextrin 20DE and with gum Arabic, when stored at the highest relative humidity (84%), had the appearance of a highly sticky liquid. According to Aguilera et al. (1995), in this stage of caking the interparticles bridges disappear as a result of sample liquefaction and low molecular weight fractions are solubilized. The particles produced with tapioca starch, even when stored at higher water activities, were agglomerated but did not show the formation of blocks or liquefaction, which is probably related to its lower water adsorption as compared to the powders produced with the other agents.

5.1.2. Glass Transition Temperature

The glass transition temperature decreased with increasing moisture content due to the plasticizing effect of water. The same trend was observed by Goula et al. (2008), Wang et al. (2008), Moraga et al. (2004, 2006) and Telis & Sobral (2001), working with tomato, gooseberry, kiwi, strawberry and pineapple, respectively. The plasticizing effect of water was described by the Gordon-Taylor model (Equation 4), where T_{gw} was taken as –135°C (Johari et al., 1987).

$$T_g = \frac{w_s T_{gs} + k w_w T_{gw}}{w_s + k w_w} \quad (4)$$

where k is the constant; w is the weight fractions (g/g total); and the subscripts s and w represent solids and water.

Experimental data of T_g were well fitted to the Gordon-Taylor model, showing satisfactory values of R^2 and E. The fitted curves are shown in Figure 8. The estimated parameters are presented in Table 7.

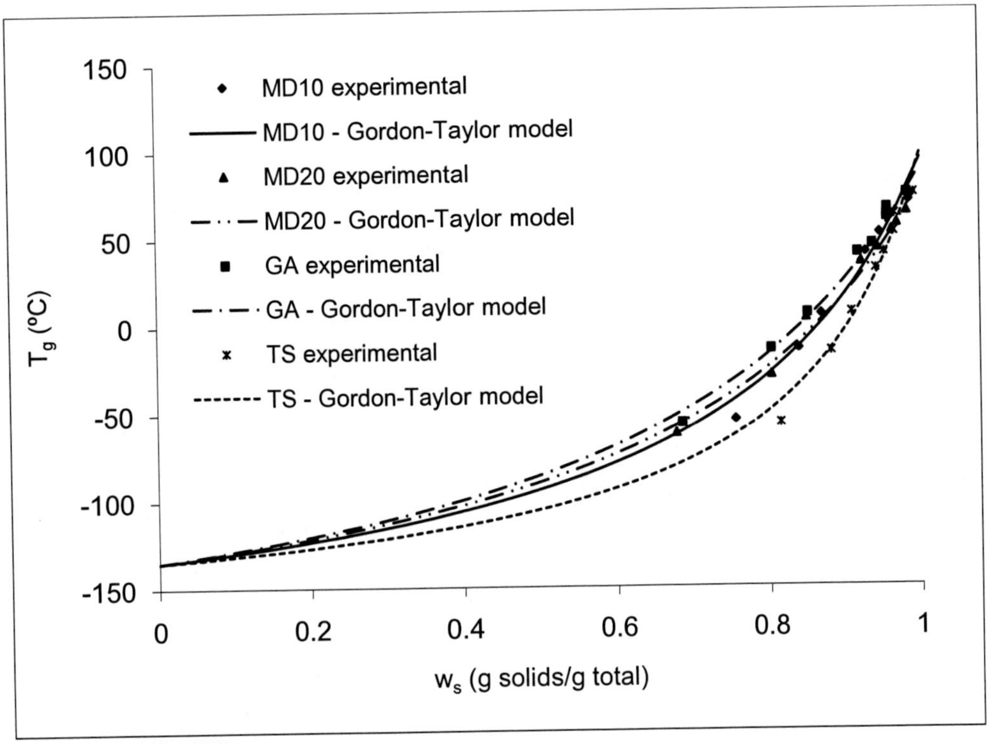

From Tonon et al., 2009a.

Figure 8. Glass transition temperature as a function of solids content for spray dried açai juice produced with different carrier agents (MD10 = maltodextrin 10DE; MD20 = maltodextrin 20DE; GA = gum Arabic; TS = tapioca starch).

Table 7. Estimated Gordon-Taylor parameters for açai juice powder produced with different carrier agents

Parameters	Carrier agents			
	Maltodextrin 10DE	Maltodextrin 20DE	Gum Arabic	Tapioca starch
T_{gs} (°C)	93.99	79.12	88.29	96.72
k	4.60	3.75	3.56	6.87
R^2	0.978	0.987	0.993	0.974
E (%)	1.83	1.49	1.32	1.89

According to Table 7, T_{gs} values varied from 79 to 97°C. The T_{gs} values obtained can explain the difficulty on drying the pure açai juice (without adding of a carrier agent). It has been shown that the addition of carrier agents such as maltodextrins and gum Arabic leads to a considerable increase on T_{gs}. Silva et al. (2006) verified an increase in the T_{gs} of freeze-dried camu-camu from 74.59 (pure pulp) to 125.45°C, when 30% of maltodextrin was added. Kurozawa et al. (2009) obtained a T_{gs} of 44.43°C for the pure chicken meat hydrolysate protein powder, while the addition of 10% of maltodextrin or gum Arabic led to T_{gs} values of 91.90 and 94.70, respectively. Thus, the spray dried pure açai juice might have a glass transition temperature much lower than the values obtained for the powders produced with additives. According to Truong et al. (2005), the sticky-point temperature is normally about 10-23°C higher than the glass transition temperature and, in spray drying, particles which are above this temperature stick to the dryer wall and degrade, and/or clump together, adversely affecting the free-flowing property. In the case of pure açai juice, considering its sugars and acids level, the sticky-temperature is much lower than 78°C (the outlet air temperature observed when the inlet air temperature was 140°C) and that would result in a high degree of stickiness and thus in an insignificant powder yield. With respect to the parameter k, the values obtained by Gordon-Taylor model were between 3.56 and 6.87, similar to those obtained for tomato, camu–camu, kiwi, garlic powder and some berries (Goula et al., 2008; Silva et al., 2006; Moraga et al., 2006; Rahman et al., 2005; Khalloufi et al., 2000). This parameter controls the degree of curvature of T_g dependence on water content (in a binary system) and can be related to the strength of the interaction between the system components (Gordon and Taylor, 1952).

The powder produced with maltodextrin 10DE showed higher glass transition temperatures as compared to that produced with maltodextrin 20DE, which is related to the decrease in the molecular weight, which decreases the T_g (Roos et al., 1996). The powder produced with tapioca starch was expected to have higher T_g than those produced with maltodextrins, since it is a native starch with higher molecular weight. However, the values obtained for such agent were similar or even lower than the obtained for the other powders. This can be attributed to the low solubility of tapioca starch at room temperature. When this agent was added to açai juice, it did not reach complete dissolution and a little quantity was precipitated inside the pipe of the spray dryer, during the process. Thus, the content of tapioca starch in the final powder was not the same that for maltodextrins and gum Arabic, which were totally soluble. As the amount of carrier agent was lower, the amount of solids content of juice was higher, which can explain the obtained T_g values, lower than the expected.

5.1.3. Critical Storage Conditions

The critical storage conditions for açai juice powder were determined by plotting the sorption isotherms and T_g data as a function of a_w, considering a room temperature of 25°C (Figure 9). The water content and T_g values were predicted by the BET and Gordon-Taylor models, respectively.

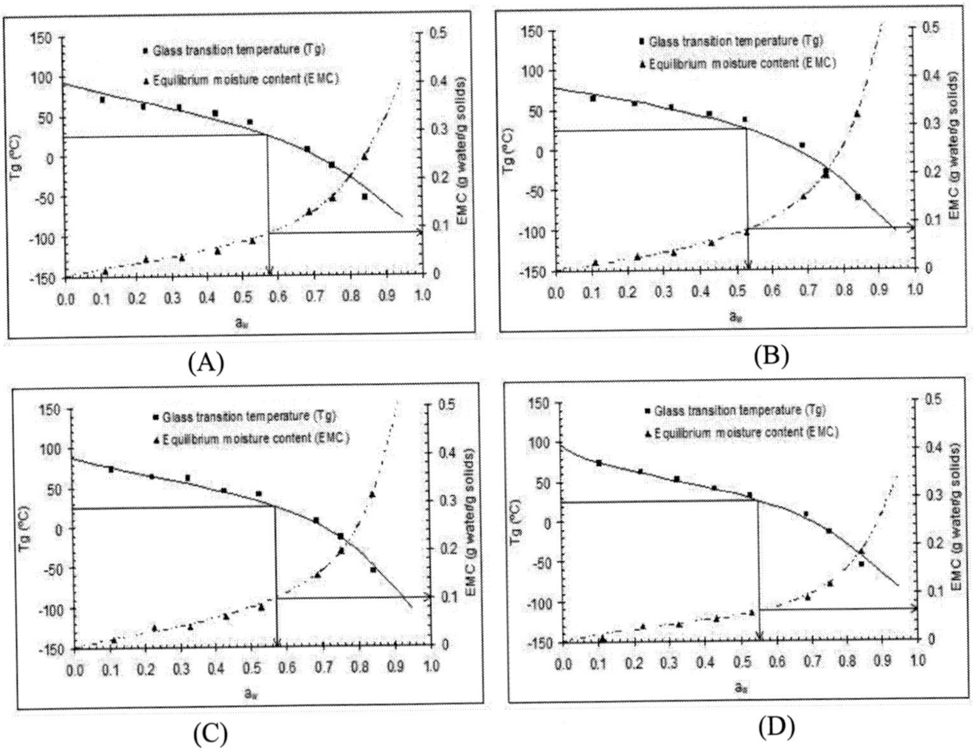

From Tonon et al., 2009a.

Figure 9. Variation of glass transition temperature (solid line) and equilibrium moisture content (dashed line) with water activity for spray dried açai juice produced with: (a) maltodextrin 10DE, (b) maltodextrin 20DE, (c) gum Arabic and (d) tapioca starch.

Table 8 shows the critical a_w and moisture content for the powders produced with the different carrier agents.

Table 8. Critical values for water activity (a_{wc}) and moisture content (X_c) of spray dried açai juice

Carrier agent	a_{wc}	X_c (g/g dry matter)
Maltodextrin 10DE	0.574	0.086
Maltodextrin 20DE	0.535	0.083
Gum Arabic	0.571	0.100
Tapioca starch	0.554	0.061

The critical water activities were similar for all the carrier agents, varying between 0.535 and 0.574. The powder produced with maltodextrin 10DE can be considered as the most stable, since it showed the highest a_{wc}, equal to 0.574. This means that when the powder is stored at 25°C, the maximum relative humidity to which it can be exposed is 57.4% and its moisture content is 8.6%. However, when stored at a relative humidity higher than 57.4% (at 25°C), or at a higher temperature (at a_w = 0.574), the powder will suffer physical transformations such as collapse, stickiness and caking.

Moraga et al. (2004, 2006) obtained very lower values of critical water activity and moisture content for freeze-dried kiwi (0.034% and 1.4%, respectively) and strawberry (0.110% and 2.0–4.0%) at 30ºC, which is probably related to the higher sugar and acid content present in these fruits, as compared to açai. Moreover, the authors did not use any additive in the powder production, resulting in lower T_g values.

5.2. Anthocyanin Stability

For the study of anthocyanin stability upon storage, powders were put in 50 × 10 mm Petri dishes, in such a way that a large surface area was exposed during storage. The dishes were stored in airtight plastic containers filled with $MgCl_2$ and $Mg(NO_3)_2$ saturated solutions, in order to provide relative humidity values of 32.8 and 52.3%, respectively. These relative humidities were chosen based on the previously determined critical water activities of powdered açai juice. The containers were stored at two different temperatures: 25ºC, representing the ambient temperature, and 35ºC, which is one of the temperatures recommended by Labuza and Schmidl (1985) for accelerated shelf-life studies.

Samples were analyzed each 15 days, during 120 days, with respect to total anthocyanin, by to the spectrophotometric method. The first-order reaction rate constants (k) and half-lives ($t_{1/2}$) were calculated according to the following equations:

$$ln\left(\frac{C_t}{C_0}\right) = -kt \qquad (5)$$

$$t_{1/2} = \frac{\ln 2}{k} \qquad (6)$$

where C_0 is the initial anthocyanin content and C_t is the anthocyanin content at the reaction time t.

Figure 10 shows the anthocyanin degradation of powdered açai juice produced with different carrier agents, during 120 days of storage. Anthocyanin degradation exhibited two first-order kinetics: the first one, with higher reaction rate constant, up to 45-60 days of storage (here designated t_2), and the second one, after t_2, with lower degradation rate. Similar behavior was observed by Matioli & Rodriguez-Amaya (2002) and by Desobry et al. (1997), in microencapsulated lycopene and β-carotene, respectively. According to the last authors, the period with higher reaction rate corresponds to the degradation of the superficial β-carotene (non-encapsulated) or to the internal β-carotene in contact with the oxygen present in pores or entrapped in bubbles, which leads to oxidation. Moreover, at times greater than t_2, the matrix density and distance to the entrapped material would limit oxygen transfer, which explains the lower degradation rate at this period. In the same way, for açai juice powder, the higher degradation rate can be attributed to the non-encapsulated material, which shows greater contact with oxygen, or even to the material in contact with the oxygen present in the interior of pores. Moreover, the higher water adsorption at beginning of storage also can be responsible for the higher degradation rate, since higher water content implies in higher molecular mobility.

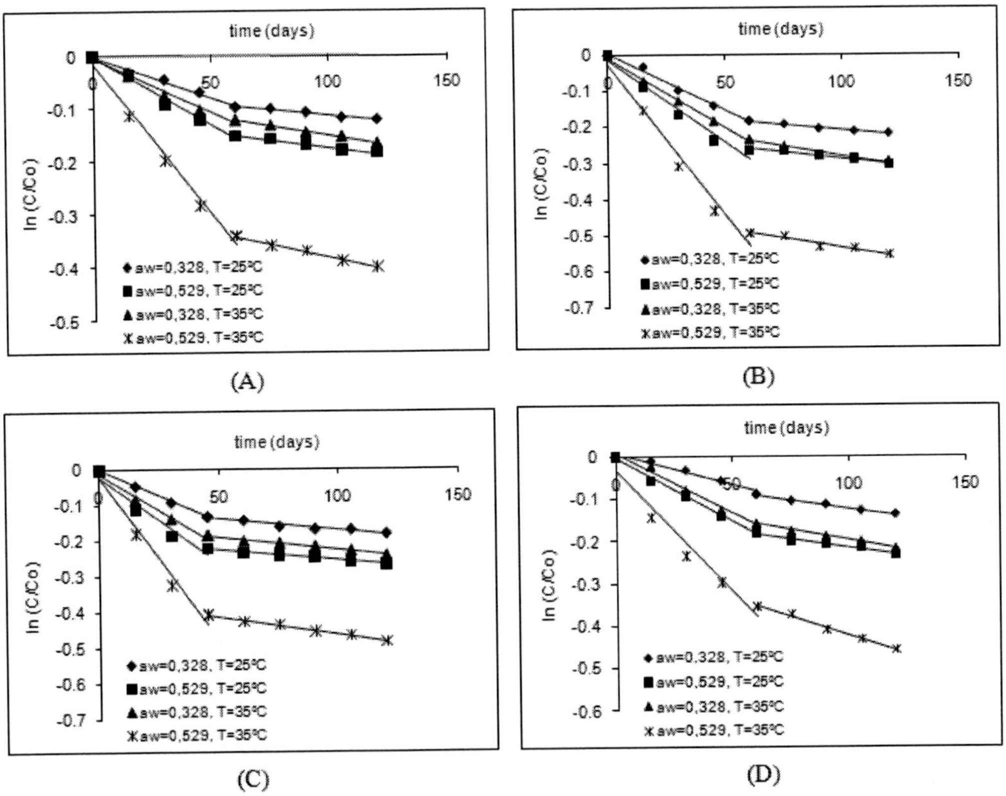

From Tonon et al., 2010.

Figure 10. Anthocyanin degradation kinetics of particles produced with: (a) maltodextrin 10DE, (b) maltodextrin 20DE, (c) gum Arabic and (d) tapioca starch.

As the particles exhibited two different first-order kinetics, two values of k and $t_{1/2}$ were calculated for each sample. However, actual half-life was determined as the time at which the anthocyanin content was reduced by 50% with respect to the zero time (Desobry et al., 1997). These values are presented in Table 9.

The increase of temperature led to a faster anthocyanin degradation, which was expected, since these pigments are highly thermo-sensitive. Pacheco-Palencia et al. (2007) evaluated the anthocyanin stability in the whole, semi-clarified and clarified açai pulp, and verified a degradation rate 3.5 times higher when samples were stored at 20°C than when they were stored at 4°C. Kirca et al. (2007) also observed a strong dependence of anthocyanin degradation on storage temperature, during heating (at 70, 80 and 90°C) and storage (at 20 and 37°C) of black carrot juice with different pH's and soluble solids.

The faster anthocyanin degradation at higher temperature may also be related to the presence of sugars, together with proteins, which can result in the Maillard reaction (non-enzymatic browning), which generally occurs during food processing at high temperatures or during food storage for long time. According to Von Elbe & Schwartz (1996), the presence of sugars or products resulting from their degradation can accelerate the anthocyanin degradation, since this reaction rate follows the rate of conversion of sugars to furfural.

Table 9. Kinetic parameters of anthocyanin degradation in the powdered açai juice produced by spray drying using maltodextrin 10DE, maltodextrin 20DE, gum Arabic and tapioca starch as carrier agents

Storage conditions		Storage time	k (days^{-1})	$t_{1/2}$ (days)	R^2	Actual $t_{1/2}$ (days)
T (ºC)	a_w					
			Maltodextrin 10 DE			
25	0.328	$t < t_2$	0.0014	495.11	0.987	1248.29
		$t > t_2$	0.0005	1386.29	0.970	
	0.529	$t < t_2$	0.0026	266.60	0.985	962.58
		$t > t_2$	0.0006	1155.25	0.996	
35	0.328	$t < t_2$	0.0021	330.07	0.985	880.35
		$t > t_2$	0.0007	990.21	0.996	
	0.529	$t < t_2$	0.0057	121.60	0.987	411.55
		$t > t_2$	0.0010	693.15	0.987	
			Maltodextrin 20 DE			
25	0.328	$t < t_2$	0.0032	216.61	0.991	909.58
		$t > t_2$	0.0006	1155.25	0.970	
	0.529	$t < t_2$	0.0045	154.03	0.971	677.64
		$t > t_2$	0.0007	990.21	0.943	
35	0.328	$t < t_2$	0.0039	177.73	0.995	516.15
		$t > t_2$	0.0011	693.15	0.946	
	0.529	$t < t_2$	0.0084	82.52	0.978	260.55
		$t > t_2$	0.0010	693.15	0.953	
			Gum Arabic			
25	0.328	$t < t_2$	0.0029	239.02	0.999	978.75
		$t > t_2$	0.0006	1155.25	0.943	
	0.529	$t < t_2$	0.0049	141.46	0.950	826.25
		$t > t_2$	0.0006	1155.25	0.976	
35	0.328	$t < t_2$	0.0041	169.06	0.985	767.07
		$t > t_2$	0.0007	990.21	0.986	
	0.529	$t < t_2$	0.0091	76.17	0.976	328.55
		$t > t_2$	0.0010	693.15	0.994	
			Tapioca starch			
25	0.328	$t < t_2$	0.0015	462.10	0.963	813.81
		$t > t_2$	0.0008	866.43	0.988	
	0.529	$t < t_2$	0.0030	231.05	0.996	695.18
		$t > t_2$	0.0008	866.43	0.962	
35	0.328	$t < t_2$	0.0028	247.55	0.985	655.27
		$t > t_2$	0.0009	770.16	0.997	
	0.529	$t < t_2$	0.0057	121.60	0.959	248.14
		$t > t_2$	0.0018	385.08	0.993	

Furfural, which is a derivative from aldopentoses, as well as hydroximetilfurfural, which is a derivative from keto-hexoses, are products resulting from the Maillard reaction that condense together with the anthocyanins, leading to the formation of compounds with brown coloration. This reaction is highly dependent on temperature, being accelerated by the presence of oxygen and occurring more frequently in fruit juices.

In a general way, the influence of temperature on the degradation rates was higher for the samples stored at the higher relative humidity, which indicates that water activity also plays an important role on anthocyanin degradation of powdered açai juice. The higher the water content, the higher the molecular mobility inside the food, which facilitates the physicochemical degradation reactions.

Garzón & Wrolstad (2001) and Cai et al. (1998) verified a considerable increase in the reaction rates and a decrease in half-lives of pelargonidin-based anthocyanins and Amaranthus betacyanin, respectively, when a_w was increased. Amr & Al-Tamimi (2007) observed a very pronounced negative effect of water activity on anthocyanin retention in crude extracts of *Ranunculus asiaticus* flowers. According to these authors, low a_w values resulted in lower conversion of anthocyanins to their hydrated carbinol base, which is less stable. This explains the greater pigment retention in the powders stored at this condition.

According to Slade & Levine, 1991, when the material is in the glassy state, the high viscosity of the matrix (about 10^{12} Pa · s) does not allow the occurrence of diffusion-controlled reactions. However, some authors have demonstrated that some diffusion-controlled reactions, such as non-enzimatic browning and lipid oxidation, may occur, even at the glassy state (Miao & Ross, 2004; Orlien et al. , 2000). This contradicts the theory of Slade and Levine (1991), according to which the storage at a temperature inferior to the glass transition temperature assures product stability. According to Schebor et al. (1999), factors like ageing of the glassy material, rotational mobility and diffusion throughout pre-existing holes (pores), due to defects and porosity in the structure, as well as the characteristic heterogeneity of food systems, can explain the occurrence of chemical reactions in foods, even in the glassy state. Thus, in the present work, although samples were stored at water activities lower than the critical aw, the diffusion of oxygen may have caused pigment degradation in the powdered açai juice particles studied.

Gradinaru et al. (2003) also verified anthocyanin degradation at temperatures below T_g, when evaluating the stability of freeze-dried anthocyanins extracted from *Hibiscus sabdariffa* flowers, stored at 40ºC at different water activities. The authors pointed out to the possibility of some sort of reactant mobility in the glassy state, indicating the insufficiency of using the calorimetrically determined T_g as an absolute threshold of stability. Moreover, according to the authors, macroscopic heterogeneities in the glassy matrix, non-homogeneous distribution of water and phase separation phenomena (demixing of reactants and inert matrix) are also likely to influence the apparent reaction rates and further explain why reactions do not cease below the measured T_g.

According to Figure 10, the increase in anthocyanin degradation was much more pronounced for the powders stored at the highest temperature (35ºC) and relative humidity (52.9%). As mentioned before, the water activity at which powders were exposed was chosen according to the critical a_w values observed for each sample, which were determined at 25ºC. However, at a higher storage temperature, the critical a_w is lower. Thus, is it is possible that the powders stored at 35ºC and 52.9% relative humidity were not in the glassy state, which can be the cause of the very higher degradation at these conditions, since molecular mobility is much greater, which can accelerate the oxidation reactions. Although there was no visual evidence that the material was in the "rubbery" state (except for samples produced with maltodextrin 20DE, which showed a beginning of agglomeration), it is important to consider that the glass transition occurs over a range of temperatures and is not defined as a specific value. Thus, it is possible that the samples stored at 35°C (at a_w of 0.529) were in this range.

Concerning to the different carrier agents used, the particles produced with maltodextrin 10DE had the highest half-life, in all the conditions studied, followed by those produced with gum Arabic. The particles produced with maltodextrin 20DE and with tapioca starch showed higher degradation rates and, consequently, lower half-lives, with respect to the others.

Rodríguez-Hernandez et al. (2005) observed better vitamin C retention in spray-dried cactus pear juice produced with maltodextrin 10DE than in the produced with maltodextrin 20DE, after drying. The authors attributed this greater retention to the better binding properties of maltodextrin 10DE, which have higher polymerization degree. Cai & Corke (2000) also observed that betacyanin retention in spray-dried *Amaranthus* pigments decreased with increasing 386 maltodextrin DE, after 16 weeks of storage. This was attributed to the higher hygroscopicity of maltodextrins with higher DE, which adsorb more water and thus are more susceptible to degradation reactions.

According to the evaluation of the powder's physical stability, the particles produced with maltodextrin 10DE had the highest critical water activity, being considered as the most stable, which was reflected in the lower anthocyanin degradation shown by them. The higher anthocyanin retention in this sample can also be related to its particle size distribution. According to Figure 5, all particles of the powder produced with maltodextrin 10DE showed diameters superior to 2 μm, while the other samples showed a little "population" of particles with mean diameter smaller than 1 μm. The smaller the particles, the larger the exposed surface area and, consequently, the faster the degradation of compounds susceptible to deterioration. In addition, the sample produced with maltodextrin 10DE had the highest absolute density, which implies in slower oxygen diffusion (Desobry, et al., 1997), thus retarding anthocyanin oxidation.

For the particles produced with gum Arabic, the second first-order kinetics started at 45 days of storage, differently from the others, in which it happened after 60 days. This can indicate a lower number of non encapsulated juice particles, or even be a consequence of the higher degradation rate observed at the first kinetics, which lead to faster degradation of the "non-protected" anthocyanins.

Although the particles produced with tapioca starch have exhibited one of the lowest half-lives, the reaction rate constant values of this sample, in the first degradation kinetics, were low and close to those of maltodextrin 10DE. This slow anthocyanin degradation can be related to the low hygroscopicity of these particles with respect to the others, which applies in lower water adsorption and thus, lower molecular mobility, making more difficult the occurrence of oxidation reactions. Moreover, the particles produced with tapioca starch had a lower porosity with respect to all the other samples, which represents lower quantity of porous containing oxygen available for pigment degradation. In the second kinetics, however, k values for this sample were in general higher than the obtained for the others. As previously discussed, the formation of the matrix occurs when the juice containing the dissolved carrier agent is exposed to the high temperature of the dryer chamber. Nevertheless, tapioca is a highly insoluble material and thus, it is not possible to affirm that a matrix system has been formed and that microencapsulation has occurred. Therefore, it is possible that tapioca starch has been used only as an aid for drying and that is probably why the protection provided by this agent along of storage was inferior to the other agents.

6. Conclusion

Spray drying was proved to be an efficient method aiming at obtaining an anthocyanin-rich product and extending açai juice's shelf life. Inlet air temperature showed significant effect on all the responses studied. Increasing temperature led to higher process yield and to lower moisture content and anthocyanin retention. Feed flow rate negatively influenced process yield and positively influenced moisture content. The increase on maltodextrin concentration also caused a reduction on process yield, probably due to the increase on feed viscosity. The use of different types of carrier agents resulted in powders with different characteristics and different behavior during storage. The critical conditions for storage at 25°C were determined based on the sorption isotherms and the glass transition temperature. The T_g of powders decreased with increasing moisture content, confirming the strong plasticizing effect of water on this property. With respect to anthocyanin stability, pigment degradation in the spray-dried açai juice produced with different carrier agents exhibited two first-order kinetics: the first one with higher reaction rate, up to 45–60 days of storage, and the second one with lower reaction rate, after this period. Temperature negatively influenced anthocyanin stability, due to the high sensitivity of these pigments to heat, and the increase of water activity also resulted in higher degradation, due to the higher molecular mobility, which allows easier oxygen diffusion. Maltodextrin 10DE was the material that produced particles with the highest critical water activity and highest half-lives, being considered as the most adequate carrier agent for producing microencapsulated açai juice.

References

Aguilera, M.J.; Del Valle, J.M.; Karel, M. (1995). Caking phenomena in amorphous food powder. *Trends in Food Science and Technology*, 6, 149-155.

Allamilla-Beltrán, L.; Chanona-Pérez, J.J.; Jiménez-Aparicio, A.R.; Gutiérrez-López, G.F. (2005). Description of morphological changes of particles along spray drying. *Journal of Food Engineering*, 67, 179-184.

Amr, A.; Al-Tamimi, E. (2007). Stability of the crude extracts of *Ranunculus asiaticus* anthocyanins and their use as food colourants. *International Journal of Food Science and Technology*, 42, 985-991.

Barbosa-Cánovas, G.V.; Juliano, P. (2005). Physical and chemical properties of food powders. In: Onwulata, C. (Ed.) *Encapsulated and powdered foods* (1st edition, pp. 39-71). Boca Raton, USA: Taylor & Francis.

BeMiller, J.N.; Whistler, R.L. (1996). Carbohydrates. In: Fennema, O.R. (Ed.) *Food Chemistry* (3rd edition, pp. 157-223), New York, USA: Marcel Dekker.

Bhandari, B.R., Howes, T. (1999). Implication of glass transition for the drying and stability of dried foods. *Journal of Food Engineering*, 40, 71-79.

Bhandari, B.R.; Data, N.; Howes, T. (1997). Problems associated with spray drying of sugar-rich foods. *Drying Technology*, 15, 671-684.

Brunauer, S.; Emmett, P.H.; Teller, E. (1938). Adsorption of gases in multimolecular layers. *Journal of the American Chemists'Society*, 60, 309-319.

Cai, Y.Z.; Corke, H. (2000). Production and properties of spray-dried *Amaranthus* Betacyanin Pigments. *Journal of Food Science*, 65, 1248-1252.

Cai, Y.; Sun, M.; Corke, H. (1998). Colorant properties and stability of Amaranthus betacyanin pigments. *Journal of Agricultural and Food Chemistry*, 46, 4491-4495.

Cano-Chauca, M.; Stringheta, P.C.; Ramos, A.M.; Cal-Vidal, J. (2005). Effect of the carriers on the microstructure of mango powder obtained by spray drying and its functional characterization. *Innovative Food Science and Emerging Technologies*, 5, 420-428.

Coisson, J.D.; Travaglia, F.; Piana, G.; Capasso, M.; Arlorio, M. (2005). *Euterpe oleracea* juice as a functional pigment for yogurt. *Food Research International*, 38, 893-897.

Desai, K.G.H.; Park, H.J. (2005). Recent developments in microencapsulation of food ingredients. *Drying Technology*, 23, 1361-1394.

Desobry, S.A.; Netto, F.M.; Labuza, T.P. (1997). Comparison of spray-drying, drum-drying and freeze-drying for β-carotene encapsulation and preservation. *Journal of Food Science*, 62, 1158-1162.

Dib Taxi, C.M.; Menezes, H.C.; Santos, A.B.; Grosso, C.R. (2003). Study of the microencapsulation of camu-camu (Myrciaria dubia) juice. *Journal of Microencapsulation*, 20, 443-448.

Ersus, S.; Yurdagel, U. (2007). Microencapsulation of anthocyanin pigments of black carrot (*Daucuscarota* L.) by spray dryer. *Journal of Food Engineering*, 80, 805-812.

Fennema, O.R. (1996). Water and ice. In: Fennema, O.R. (Ed.) *Food Chemistry* (3rd edition, pp. 17-93), New York, USA: Marcel Dekker.

Gabas, A.L.; Telis, V.R.N.; Sobral, P.J.A.; Telis-Romero, J. (2007). Effect of maltodextrin and arabic gum in water vapor sorption thermodynamic properties of vacuum dried pineapple pulp powder. *Journal of Food Engineering*, 82, 246-252.

Garzón, G.A.; Wrolstad, R.E. (2004). The stability of pelargonidin-based anthocyanins at varying water activity. *Food Chemistry*, 75, 185-196.

Gharsallaoui, A.; Roudaut, G.; Chambin, O.; Voilley, A.; Saurel, R. (2007). Applications of spray drying in microencapsulation of food ingredients: An overview. *Food Research International*, 40, 1107-1121.

Gordon, M.; Taylor, J.S. (1952). Ideal copolymers and the second-order transitions of syntetic rubbers. I. Non-crystalline copolymers. *Journal of Applied Chemistry*, 2, 493-500.

Gouin, S. (2004). Microencapsulation: Industrial appraisal of existing technologies and trends. *Trends in Food Science and Technology*, 15, 330-347.

Goula, A.M.; Adamopoulos, K.G.; Kazakis, N.A. (2004). Influence of spray drying conditions on tomato powder properties. *Drying Technology*, 22, 1129-1151.

Goula, A.M.; Karapantsios, T.D.; Achilias, D.S.; Adamopoulos, K.G. (2008). Water sorption isotherms and glass transition temperature of spray dried tomato pulp. *Journal of Food Engineering*, 85, 73-83.

Grabowski, J.A.; Truong, V.D.; Daubert, C.R. (2006). Spray-drying of amylase hydrolyzed sweetpotato puree and physicochemical properties of powder. *Journal of Food Science*, 71, E209-E217.

Gradinaru, G.; Biliaderis, C.G.; Kallithraka, S.; Kefalas, P.; Garcia-Viguera, C. (2003). Thermal stability of *Hibiscus sabdariffa* L. anthocyanins in solution and in solid state: effects of copigmentation and glass transition. *Food Chemistry*, 83, 423-436.

Hong, J.H.; Choi, Y.H. (2007). Physico-chemical properties of protein-bound polysaccharide from *Agaricus blazei* Murill prepared by ultrafiltration and spray drying process. *International Journal of Food Science and Technology*, 42, 1-8.

Jackson, L.S.; Lee, K. (1991). Microencapsulation and the food industry. *Lebensmittel-Wissenschaft und Technologie*, 24, 289-297.

Johari, G.P.; Hallbrucker, A.; Mayer, E. (1987). The glass-liquid transition of hyperquenched water. *Nature*, 330, 552-553.

Khalloufi, S.; El-Maslouhi, Y.; Ratti, C. (2000). Mathematical model for prediction of glass transition temperature of fruit powders. *Journal of Food Science*, 65, 842-848.

Kirca, A.; Özkan, M.; Cemeroglu, B. (2007). Effects of temperature, solid content and pH on the stability of black carrot anthocyanins. *Food Chemistry*, 101, 212-218.

Kurozawa, L.E.; Park, K.J.; Hubinger, M.D. (2009). Effect of maltodextrin and gum arabic on water sorption and glass transition temperature of spray dried chicken meat hydrolysate protein. *Journal of Food Engineering*, 91, 287-296.

Leonel, M. (2007). Analysis of the shape and size of starch grains from different botanical species. *Ciência e Tecnologia de Alimentos*, 23, 579-588.

Loksuwan, J. (2007). Characteristics of microencapsulated β-carotene formed by spray drying with modified tapioca starch, native tapioca starch and maltodextrin. *Food Hydrocolloids*, 21, 928-935.

Lomauro, C.J.; Bakshi, A.S.; Labuza, T.P. (1985). Evaluation of food moisture isotherm equations. 1: Fruit, vegetable and meat products. Lebensmittel-Wissenschaft und Technologie, 18, 111-117.

Matioli, G.; Rodriguez-Amaya, D.B. (2002). Lycopene encapsulated with gum Arabic and maltodextrin: Stability study. *Brazilian Journal of Food Technology*, 5, 197-203.

Miao, S.; Roos, Y.H. (2004). Comparison of nonenzymatic browning kinetics in spray-dried and freeze-dried carbohydrate-based food model systems. *Journal of Food Science*, 69, E222-E331.

Moraga, G.; Martínez-Navarrete, N.; Chiralt, A. (2004). Water sorption isotherms and glass transition in strawberry. *Journal of Food Engineering*, 62, 315-321.

Moraga, G.; Martínez-Navarrete, N.; Chiralt, A. (2006). Water sorption isotherms and phase transitions in kiwifruit. *Journal of Food Engineering*, 72, 147-156.

O'Hagan, P.; Hasapidis, K.; Coder, A.; Helsing, H.; Pokraja, G. (2005). Particle size analysis of food powders. In: Onwulata, C. (Ed.) *Encapsulated and powdered foods* (1st edition, pp. 215-245). Boca Raton, USA: Taylor & Francis.

Orlien, V.; Andersen, A.B.; Sinkko, T.; Skibsted, L.H. (2000). Hydroperoxide formation in rapeseed oil encapsulated in a glassy food model as influenced by hydrophilic and lipophilic radicals. *Food Chemistry*, 68, 191-199.

Pacheco-Palencia, L.A.; Hawken, P.; Talcott, S.T. (2007). Phytochemical, antioxidant and pigment stability of açaí (*Euterpe oleracea* Mart.) as affected by clarification, ascorbic acid fortification and storage. *Food Research International*, 40, 620-628.

Papadakis, S.E.; Gardeli, C.; Tzia, C. (2006). Spray drying of raisin juice concentrate. *Drying Technology*, 24, 173-180.

Quek, S.Y.; Chok, N.K.; Swedlund, P. (2007). The physicochemical properties of spray-dried watermelon powder. *Chemical Engineering and Processing*, 46, 386-392.

Rahman, M.S.; Sablani, S.S.; Al-Habsi, N.; Al-Maskri, S.; Al-Belushi, R. (2005). State diagram of freeze-dried garlic powder by differential scanning calorimetry and cooling curve methods. *Journal of Food Science*, 70, E135-E141.

Rahman, M.S. (2006). State diagram of foods: Its potential use in food processing and product stability. *Trends in Food Science and Technology*, 17, 129-141.

Rattes, A.L.R.; Oliveira, W.P. (2007). Spray drying conditions and encapsulating composition effects on formation and properties of sodium diclofenac microparticles. *Powder Technology*, 171, 7-14.

Ré, M. I. (1998). Microencapsulation by spray drying. *Drying Technology*, 16, 1195-1236.

Reiniccius, G.A. (2001). Multiple-core encapsulation - The spray drying of food ingredients. In: Vilstrup, P. (Ed.). *Microencapsulation of food ingredients* (1st edition, pp.151-185). Surrey, UK: Leatherhead Publishing.

Righetto, A.M.; Netto, F.M. (2005). Effect of encapsulating materials on water sorption, glass transition and stability of juice from immature acerola. *International Journal of Food Properties*, 8, 337-346.

Rodríguez-Hernández, G.R.; González-García, R.; Grajales-Lagunes, A.; Ruiz-Cabrera, M.A. (2005). Spray-drying of cactus pear juice (*Opuntia streptacantha*): Effect on the physicochemical properties of powder and reconstituted product. *Drying Technology*, 23, 955-973.

Roos, Y.; Karel, M. (1991). Water and molecular weight effects on glass transitions on amorphous carbohydrates and carbohydrate solutions. *Journal of Food Science*, 56, 1676-1681.

Roos, Y.H. *Phase transitions in foods*. San Diego, USA: Academic Press; 1995.

Roos, Y.H.; Karel, M.; Kokini, J.L. (1996). Glass transitions in low moisture and frozen foods: Effects on shelf life and quality. Scientific status summary. *Food Technology*, 50, 95-108.

Sablani, S.S.; Kasapis, S.; Rahman, M.S.; Al-Jabri, A.; Al-Habsi, N. (2004). Sorption isotherm and the state diagram for evaluating stability criteria of abalone. *Food Research International*, 37, 915-924.

SAGRI. – Secretaria Executiva de Agricultura do Pará. *Estatística.* Available from: http://www.sagri.pa.gov.br. Acess in 26th january 2010.

Schauss, A.G.; Wu, X.; Prior, R.L.; Ou, B.; Huang, D.; Owens, J.; Agarwal, A.; Jensen, G.S.; Hart, A.N.; Shanbrom, E. (2006). Antioxidant capacity ond other bioactivities of the freeze-dried amazoniam palm berry, *Euterpe oleracea* Mart. (Acai). *Journal of Agricultural and Food Chemistry*, 54, 8604-8610.

Schebor, C.; Buera, M.P.; Karel, M.; Chirife, J. (1999). Color formation due to non-enzymatic browning in amorphous, glassy, anhydrous, model systems. *Food Chemistry*, 65, 427-432.

Shahidi, F.; Han, X.D. (1993). Encapsulation of food ingredients. *Critical Reviews in Food Science and Nutrition*, 33, 501-547.

Shrestha, A.K.; Howes, T.; Adhikari, B.P.; Bhandari, B.R. (2007). Water sorption and glass transition properties of spray dried lactose hydrolysed skim milk powder. *LWT – Food Science and Technology*, 40, 1593-1600.

Silva, M.A.; Sobral, P.J.A.; Kieckbusch, T.G. (2006). State diagrams of freeze-dried camu-camu (*Myrciaria dubia* (HBK) Mc Vaugh) pulp with and without maltodextrin addition. *Journal of Food Engineering*, 77, 426-432.

Slade, L.; Levine, H. (1991). Beyond water activity: Recent advances based on an alternative approach to the assessment of food quality and safety. *Critical Reviews in Food Science and Nutrition*, 30, 115-360.

Symaladevi, R.M.; Sablani, S.S.; Tang, J.; Powers, J.; Swanson, B.G. (2009). State diagram and water adsorption isotherm of raspberry (*Rubus idaeus*). *Journal of Food Engineering*, 91, 460-467.

Telis, V.R.N.; Sobral, P.J. (2001). Glass transitions and state diagram for freeze-dried pineapple. *Lebensmittel-Wissenschaft und Technologie*, 34, 199-205.

Tonon, R.V.; Brabet, C.; Hubinger, M.D. (2008). Influence of process conditions on the physicochemical properties of açai (*Euterpe oleraceae* Mart.) powder produced by spray drying. *Journal of Food Engineering*, 88, 411-418.

Tonon, R.V.; Baroni, A.F.; Brabet, C.; Gibert, O.; Pallet, D.; Hubinger, M.D. (2009a). Water sorption and glass transition temperature of spray dried açai (*Euterpe oleracea* Mart.) juice. *Journal of Food Engineering*, 94, 215-221.

Tonon, R.V.; Brabet, C.; Pallet, D.; Brat, P.; Hubinger, M.D. (2009b). Physicochemical and morphological characterisation of açai (Euterpe oleraceae Mart.) powder produced with different carrier agents. *International Journal of Food Science and Technology*, 44, 1950-1958.

Tonon, R.V.; Brabet, C.; Hubinger, M.D. (2010). Anthocyanin stability and antioxidant activity of spray-dried açai (*Euterpe oleracea* Mart.) juice produced with different carrier agents. *Food Research International*, doi:10.1016/j.foodres.2009.12.013.

Truong, V.; Bhandari, B.R.; Howes, T. (2005). Optimization of co-current spray drying process of sugar-rich foods. Part II – Optimization of spray drying process based on glass transition temperature. *Journal of Food Engineering*, 71, 66-72.

Von Elbe, J.H.; Schwartz, S.J. Colorants. In: Fennema, O.R. (Ed.) *Food Chemistry* (3rd edition, pp. 651-722), New York, USA: Marcel Dekker.

Wang, H.; Zhang, S.; Chen, G. (2008). Glass transition and state diagram for fresh and freeze-dried Chinese gooseberry. *Journal of Food Engineering*, 84, 307-312.

In: Focus on Food Engineering
Editor: Robert J. Shreck

ISBN: 978-1-61209-598-1

Chapter 5

COMPUTED TOMOGRAPHY IN FOOD SCIENCE

Elena Fulladosa, Núria Garcia-Gil, Eva Santos-Garcés, Maria Font i Furnols, Israel Muñoz and Pere Gou
IRTA. Finca Camps i Armet, E-17121 Monells,
Girona, Spain

ABSTRACT

Computed tomography (CT) is one of the emerging technologies of interest to food science as it permits a non-destructive characterization of food products and their control throughout processing. This work describes the history and physical basis of this technology as well as the working principles of CT. It focuses on the latest research findings related to the application of this technology to different food products; especially dry-cured ham production as well as other issues like pig carcass classification. A revision of other X-ray technologies applied to food science is also included. In dry-cured ham production, CT helps the study of the factors which affect the salting/curing processes. These processes can be monitored because salt can easily be detected due to the differences in densities of meat and salt. Using experimental models, salt and water contents can be non-destructively determined at any moment during the process thus enabling the establishment of safety and quantity criteria in order to avoid either sensory defects or the microbiological hazards common in dry-cured ham. For carcass classification purposes, CT can be used to obtain the lean content of carcasses which is of interest to the food industry as it defines the commercial value of the pig. The estimation of the lean content is usually calculated from the physical measurements of subcutaneous fat depths and muscle thicknesses in specific locations. Devices for this task need to be calibrated and therefore, the dissection is the reference method most commonly used, but this method is difficult and time consuming. CT is an excellent tool for this task as it easily distinguishes the differences between lean, fat and bone.

1. INTRODUCTION

1.1. History

In the early 1900s, the Italian radiologist Alessandro Vallebona devised a method to represent a single slice of the body on radiographic film. This method was known as tomography, which was one of the pillars of radiologic diagnostics until the late 1970s. Later, it was gradually supplanted for the modality of Computed Tomography (CT). Godfrey Hounsfield conceived this idea in 1967 and it was publicly announced in 1972 (Hounsfield, 1973). Allan McLeod Cormack independently invented a similar process, and both Hounsfield and Cormack shared the Nobel Prize for Medicine in 1979 (Hounsfield, 1980).

Over the years, this first prototype has been modified and improved in order to obtain images more rapidly and with better resolution. In 1971, matrixes of values were composed of only 80 x 80 pixels whereas present matrixes are of 512 x 512 pixels. Since its introduction in the 1970s, CT has become an important tool in medical diagnostics.

1.2. CT Equipment

CT mainly consists of a gantry where X-rays are generated and where detectors are placed, an electric generator and a workstation where images are reconstructed (Figure 1). The X-ray source emits X-rays which pass partially through the studied object and reach the detectors. The X-ray source and detectors rotate together around the object; one exposure comprises 360 ° rotations. Many data scans are progressively taken as the object is scanned.

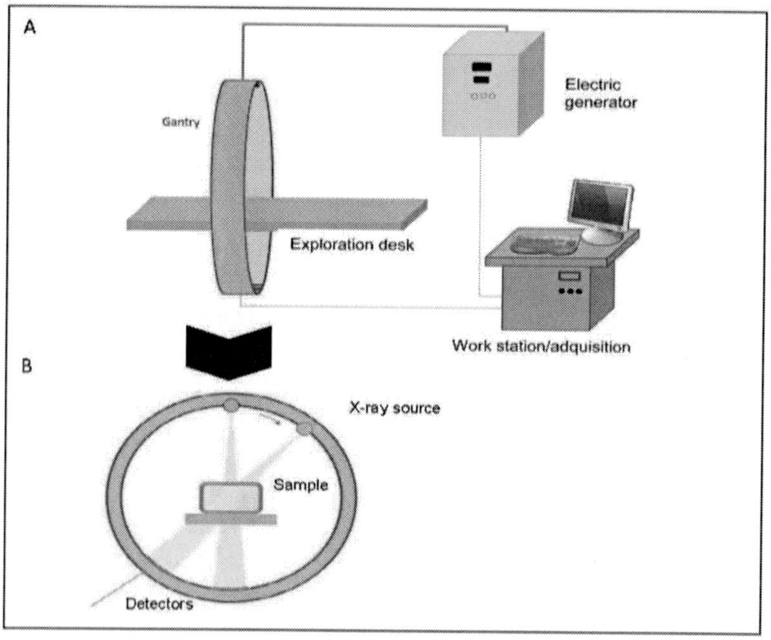

Figure 1. Schematic drawing of the components of CT equipment (A) and source-detector system of a 3rd generation CT scanner (B).

A CT examination starts with a digital projection radiograph ("scout view") of the region to be analyzed (Figure 2). The scout view is obtained by moving the exploration desk bearing object or sample to be analyzed through the X-ray beam without rotating the source or detectors. The scout view is used for the selection of slice locations. Once the location is selected, X-ray source and detectors rotate around the sample to obtain a high resolution image. The improvement of this technology has led to spiral CT, also known as helical CT, in which the gantry rotates continuously around the object to be scanned, while the object is simultaneously moved longitudinally. This allows a reduction of scanning time and improves accuracy when compared to the traditional CT.

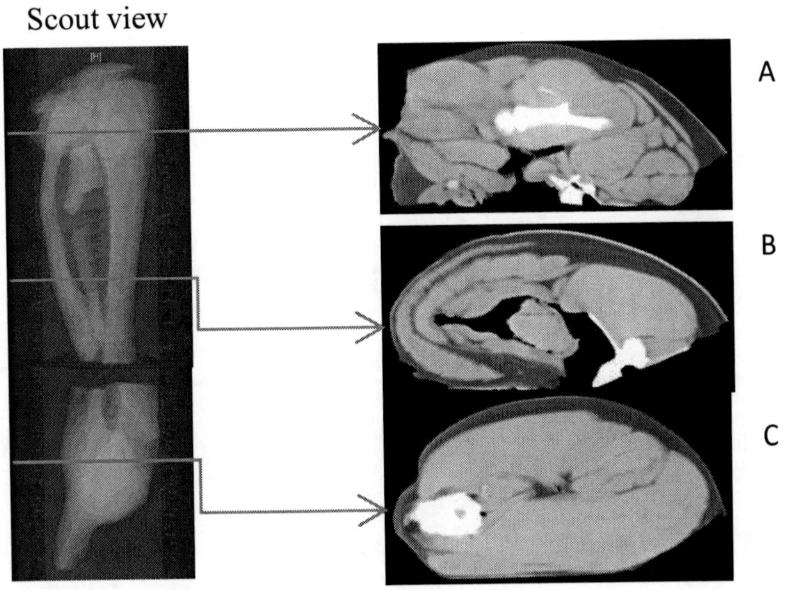

Figure 2. Scout view of a half pig carcass and cross-sectional section (tomogram) at the indicated position; the shoulder (A), loin and belly (B) and ham (C).

1.3. Physical Principles

On their way through tissues, emitted X-rays are attenuated, partly due to absorption of energy and partly due to the scattering. The attenuation may be expressed by the following equation:

$I = I_0 e^{-\mu d}$

where I is the intensity of the transmitted radiation (i.e., the radiation exiting from the tissue), I_0 is the intensity of the incident radiation (entering to the tissue), μ is the so-called total linear attenuation coefficient of the tissue, and d is the travelled distance of the radiation through the tissue (tissue thickness). The attenuation coefficient μ is determined by the atomic number and electron density of the tissue; the higher the atomic number and electron density, the higher the attenuation coefficient. Atomic number and electron density are thus the two parameters determining the X-ray attenuating properties of a tissue. All CT applications in medicine or in food technology are based upon the fact that different tissues provide different degrees of X-ray attenuation.

In addition, attenuation coefficient (μ) is also dependent on the incident X-ray energy. This means that for diagnostic purposes, in medicine or other applications in industry or the food technology field, the use of different incident energies may be useful.

Attenuation of a sample is calculated from intensity (I) collected by the detectors, located along the gantry, obtaining a matrix of attenuation values, also called CT values, which are expressed as Hounsfield units (HU). CT values are defined as the attenuation (μ) difference of a given matter (x) relative to water (w) (Kalender, 2005):

$$\text{CT value} = \frac{\mu_x - \mu_w}{\mu_w} \times 1000$$

This matrix of CT values is used to create an image in various grey tones called tomograms by using image reconstruction algorithms (Seeram, 2009). HU of different tissue varies according to the density of the tissue and the presence of other components such as salt. HU ranges between +1000 HU for bone which has a high density and -1000 HU for air which has a very low density (Figure 3). In these tomograms, brighter tones symbolize higher X-ray attenuations.

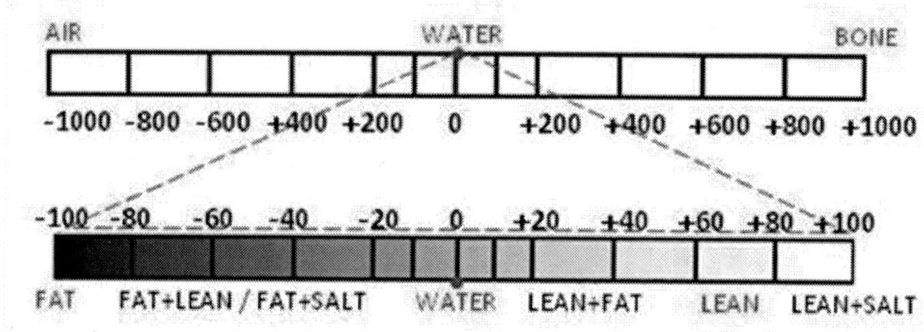

Figure 3. A representation of Hounsfield's scale in the range for food. Values are approximate since they depend on the exact composition of the food.

1.4. General Applications of CT

CT has been extensively used in medicine for diagnostic purposes, to quantify adipose tissue or determine bone density (Casas et al., 2004; Pedraza & Méndez-Méndez, 2004). CT can also be used for research, anatomical and veterinary purposes, for instance in dogs, exotic animals such as newly born rhinoceros, bats, turtles, lions, lynxs, etc (Donkó et al., 2009) (Figures 4 and 5). CT also has a place in industrial applications since can provide non-invasive volumetric information about an object, which in turn can be used for inspection, measurements, population studies or other analyses.

It is also a suitable tool in archaeology and palaeontology for the study of antiquities such as ancient mummies (Harwood-Nash, 1979), or geology for several petrophysical applications such as the characterization of rock materials (Wellington & Vinegar, 1987). By means of CT, the inner properties of wood such as density can be evaluated (Taylor et al., 1984; Funt &

Bryan, 1987; Lindgren, 1991). Fromm et al. (2001) used CT to analyze the general features of the conductive systems of a tree both the wood density and water state of conductive xylem, in two different species, spruce (*Picea abies*) and oak (*Quercus robur*).

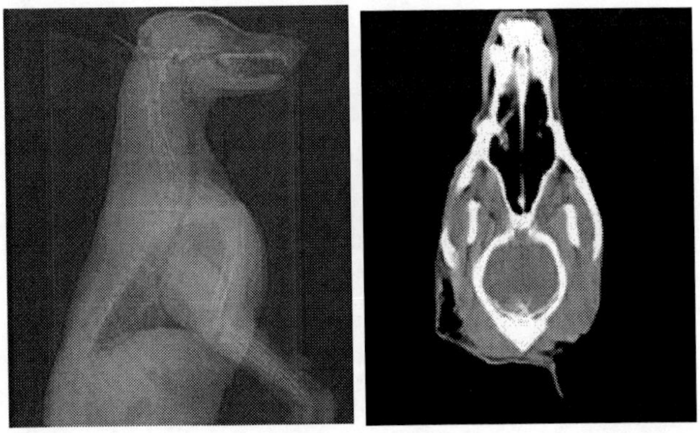

Figure 4. Scout view of a live dog and a head section.

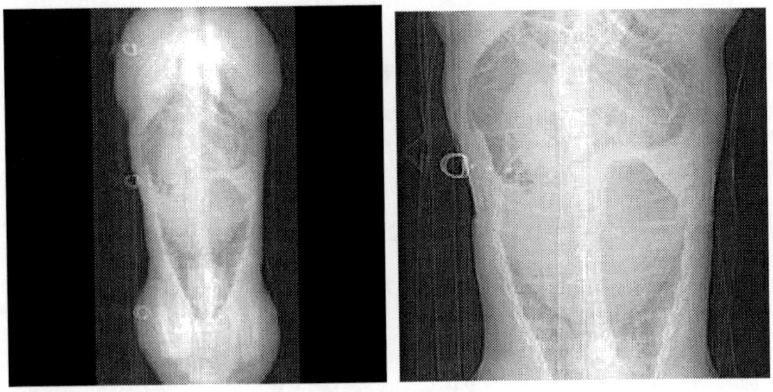

Figure 5. Scout view of a live pig.

In meat science, the intramuscular fat in meat is an important parameter which affects sensory properties such as juiciness and flavour (Wood et al., 2003). It also affects the visual acceptability of the meat. CT can be used to determine intramuscular fat or even the composition of some fatty acids in live animals, and this parameter could be useful in breeding programmes, or in carcass evaluation (Karamichou et al., 2006; Navajas et al., 2009). Adapting the scanning parameters and the reconstruction algorithm, it is also possible to visually determine marbling in meat (Font i Furnols et al., 2009a).

The knowledge of the body composition of live animals and the evolution and deposition of different tissues in different anatomical areas during growth is important in order to obtain a product which meets the requirements and demands of the market, taking into account animal type and genetics, production systems, the type of meat industry and the consumers. This information can also be used in breeding programs to improve precision in selection and the understanding of the animals' characteristics. Norsvin International AS in Norway uses CT for these purposes in pigs (Aass et al., 2009; www.norsvin.no), the Scottish Agricultural College in Scotland in sheep (Bünger et al., 2009) and Kaposvar University in Hungary in pigs, turkeys and rabbits (Donkó et al., 2009).

Applications to food technology processing are limited but they have increased in the last years because of the non-destructive character of this technology. CT has been found to be a powerful tool for studing salting processes in meat (Vestergaard et al., 2004, 2005; Håseth et al., 2008; Fulladosa et al., 2010) and is extensively explained in the following sections.

2. CT Applications in Food Processing: Dry-Cured Ham Elaboration Process

Some quality evaluations are still performed manually in food industry by trained inspectors. Manual evaluations are tedious, laborious, costly and inherently unreliable due to their subjective nature. Increasing demands for objectivity, consistency and efficiency have led to the development and use of non-destructive technologies which can predict specific parameters of foods objectively. In order to understand, improve and to better control food processes and quality, non-destructive techniques such as CT can be used.

CT has been found to be especially useful for studying curing processes. Salting and drying, together with smoking, are considered to be the oldest and main methods for the preservation of meat. Due to its high density, salt uptake produces a marked increase of CT attenuation values in meat and fish and can be easily followed during the process. During recent years, curing processes have been studied using CT.

2.1. Dry-Cured Ham Elaboration Process

The dry-cured ham elaboration process starts with a salting step, which is the essential operation in the production of dry-cured ham (Figure 6). Apart from the meat, salt is strictly the only ingredient needed when preparing these products. In addition to its contribution to flavour, salt is important in reducing the water activity (a_W) of the product. This is particularly important in order to control microbial spoilage in the early phase of production. In order to avoid microbiological hazards especially in the internal part of the product, after salting, hams are kept at a low temperature until the salt has been homogenized throughout the entire ham. Following this post-salting step, the drying process starts, where the characteristic texture and aromas of dry-cured ham are developed (Arnau et al., 1987).

Nevertheless, in order to avoid the production of dry-cured ham with sensory defects due to excessive or defective salt content, factors affecting the salting/post-salting processes need further study. Although dry-cured ham is elaborated by a traditional process, optimization of industrial operations to avoid variations in salt content, to ensure product safety and to obtain maximum sensory qualities are needed.

Furthermore, the finished product is characterized by its high salt content and represents a significant salt uptake for consumers. Because the reduction of dietary salt intake is perceived by consumers as having a positive effect on human health, the dry-curing industry is increasingly interested in producing a product with a lower salt content. Besides, the World Health Organization (WHO) also recommends reducing salt intake in order to prevent diseases such as hypertension (AESAN, 2005). Nevertheless, salt acts as a preservative and its reduction is not straightforward.

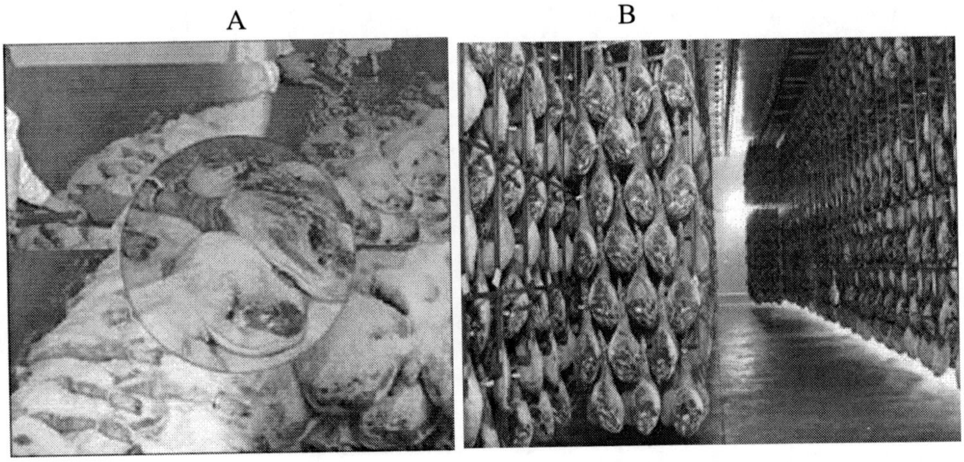

Figure 6. Traditional dry-cured ham process: A) salting and B) post-salting.

The simple NaCl reduction increases proteolytic activity during the traditional process of dry-cured ham and consequently increases the frequency of areas with anomalous texture and flavour within the ham (Arnau et al., 2007). Additionally, safety hazards may occur because of salt deficiency in critical areas. Therefore, the reduction of NaCl should be done in combination with innovative strategies which could both solve the current problems and contribute redesigning new elaboration processes.

2.2. Adaptation of CT to Dry-Cured Ham

Sørheim & Berg (1987) and Frøystein et al. (1989) were the first to quantify and describe the salt distribution in dry-cured hams by means of CT. Several authors have more recently demonstrated the linearity and the strong correlation between CT signals and salt concentrations in meat. Quantification of salt concentrations in cured pork loin was achieved by Vestergaard et al. (2004), who demonstrated that it is possible to measure salt penetration in a model system such as pork loin using CT. Subsequently, Vestergaard et al. (2005) studied salt distribution in dry-cured hams by using CT and image analysis. Håseth et al. (2007) were the first to provide a mathematical calibration model between CT values and salt content in dry-cured ham, improving it in a later study (Håseth et al., 2008). These studies showed that the combination of various energies (scans taken at 110 and 130 kV) improves accuracy of the models. Furthermore, it has been observed that fat and water content highly influence the precision of the models (Håseth et al., 2008; Fulladosa et al., 2010). Recently, Fulladosa et al. (2010) developed the first models for water prediction in dry-cured ham using combination of two or three energies.

To adapt or calibrate CT equipment in order to optimize the dry-cured ham elaboration process, CT attenuation values at 80, 120 and 140 kV (HU_{80}, HU_{120} and HU_{140}, respectively) were used to fit the analytically measured salt and water contents (calibration models). The predictability of such models can be given by the coefficient of determination (R^2) and the Root Mean Square Error of Prediction (RMSEC) value (Næs et al., 2002). Prediction models for salt and water content in dry-cured ham, to be applied during salting and post-salting steps in lean areas has an error of prediction of 0.3% and 1.5%, respectively (Fulladosa et al., 2010).

In Figure 7, measured salt (A) and water (B) content versus the predicted values using the developed models are shown. Correlation in the case of salt is high for all the sampled areas of the ham (SM, BF and ST). In the case of water, an important effect on the sampled area, *Semimembranosus* muscle (SM), *Biceps femoris* (BF) or *Semitendinosus* muscle (ST) is observed. This fact is mainly related to the fat content of the samples since the major deviations of the predictions are found in ST muscle, which is the muscle with the highest fat content.

A

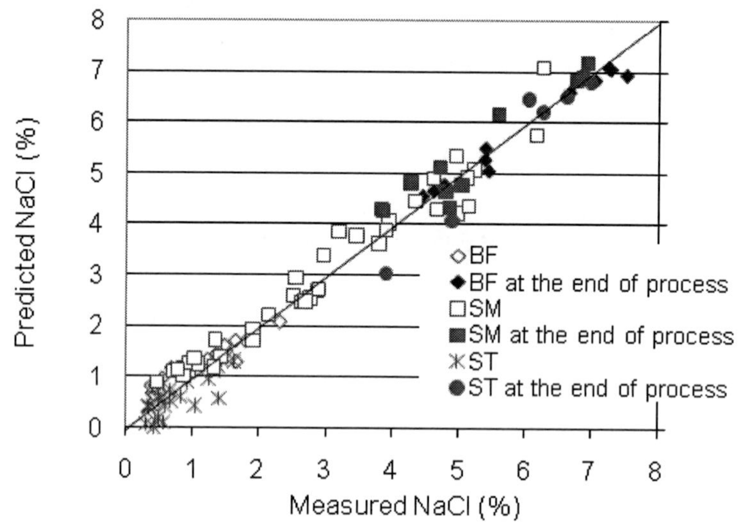

B

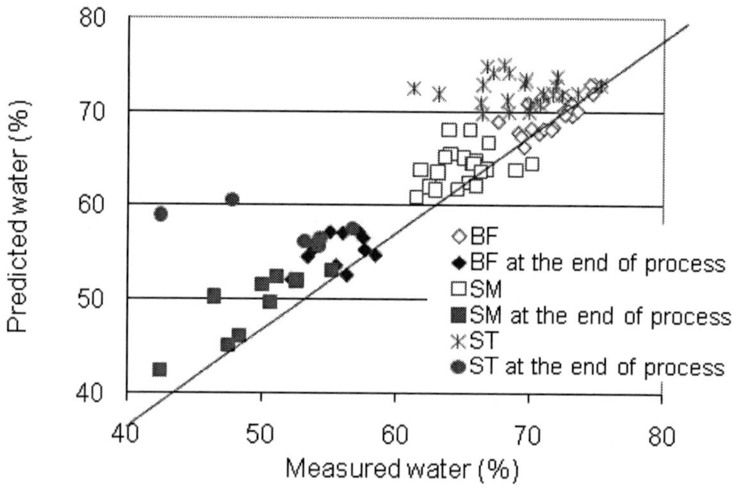

Figure 7. Predicted versus analytical salt and water contents in dry-cured ham during the post-salting and drying process. Different colours represent samples obtained from *Biceps femoris* muscle (BF), *Semimembranosus* muscle (SM) and *Semitendinosus* muscle (ST).

Since the drying level of the sample has a great influence on the prediction, specific models to be applied during drying process were also developed (Santos-Garcés et al., 2010). Matrixes of values for salt and water contents may also be useful when studying elaboration processes. Distribution diagrams of salt or water are 2-dimensional graphic representations which allow the distribution of salt or water content to be distinguished visually in the scanned slice as each colour represents a given salt or water content (Figure 8).

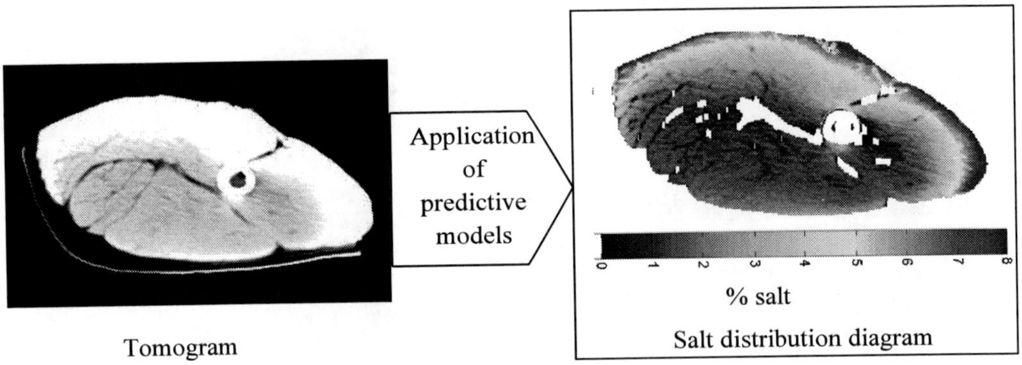

Figure 8. Salt distribution diagram obtained after the application of the prediction models on matrixes of attenuation values obtained at different energies.

2.3. CT Used to Study the Factors Affecting the Salting Processes

Traditionally, each dry-cured ham producer applies a process based on experience and accepts a percentage of defective dry-cured hams (non commercial). Any modification of the process (for instance, a change of raw material properties or the amount of added NaCl) needs a specific time to adjust to the conditions of the new product, which up to now has been done by trial and error methods. Nowadays, CT technology offers the chance to design safe processes for specific raw properties (pH, weight, fatness, fresh/frozen origin) and processing conditions (trimming, shaping and salting conditions: time, temperature, salt characteristics, etc.) in a more accurate, faster and cheaper manner.

Concerning raw material, it is known that pH is a relevant parameter which affects absorption, diffusion and distribution of salt in dry-cured hams. CT shows that hams with high pH_{SM24} ($pH \geq 6.2$ measured at 24 hours post-mortem in the *Semimembranosus* muscle) absorb less salt in comparison to those with a low pH_{SM24} ($pH \leq 5.55$).

Concerning the shape of the hams, brine remains longer on the lean surface of flat hams during salting procedure (Arnau, 2007). A flat shape can be a characteristic of the ham itself or caused by applying pressure perpendicularly to the skin at the cushion distal part from the aitch bone, which is a useful application in industry. CT is useful for viewing how salt is being distributed in the pressed and non-pressed product. In Figure 9, the evolution of two hams from the same carcass, one of them pressed; throughout of the elaboration process using CT is depicted, in which brighter tones represent salt. Pressed hams absorb a higher amount of salt than non-pressed hams (Figure 9B vs. 9F) but dry faster (Figure 9D vs. 9H). Differences in salt content due to pressing would explain in part the variability in salt uptake among hams salted at different levels within the salt pile.

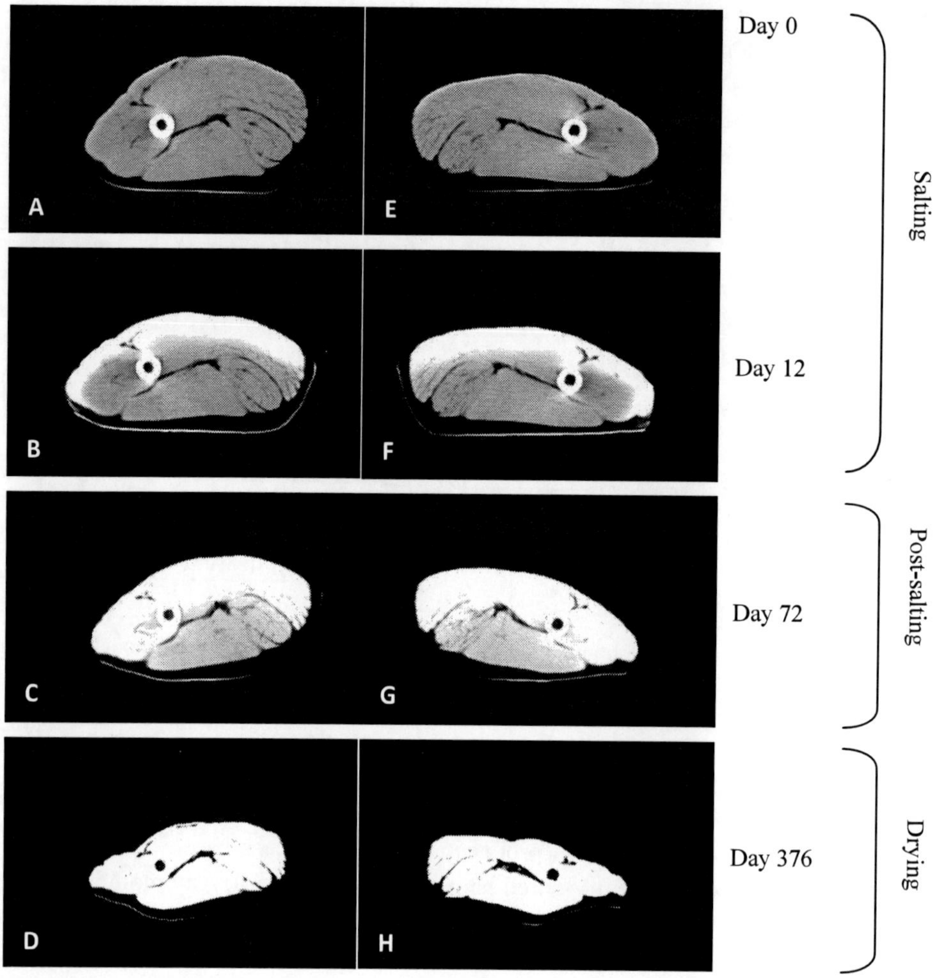

Figure 9. CT cross sections taken throughout the elaboration process (salting, post-salting and drying) from one non-pressed dry-cured ham (A-D) and from its pressed pair (E-H). Note that the pressed ham shows a higher salt content after salting (Day 12) and is reduced in size at the end of the elaboration process (Day 376).

The use of frozen/thawed hams as raw material is a common practice in dry-cured ham production because it allows a time reduction of 61% when compared with the traditional process (Barat et al., 2005). CT technology demonstrates that both salt uptake and salt distribution in frozen/thawed hams are higher in comparison with fresh hams (Figure 10) because frozen/thawed material shows a lower water holding capacity. These facts permit a time reduction during the different stages (salting, post-salting and drying) of the elaboration process without affecting the quality of the final product. The composition of hams also has a great influence on salt-uptake and salt-distribution. Fatty hams absorb less amounts of salt because fat acts as a barrier for salt penetration. On the other hand, the diffusion of salt in the *Biceps femoris* muscle in these kinds of hams is also slower compared to lean hams (Figure 10). Therefore, CT can be used to adjust processing conditions for each class of ham.

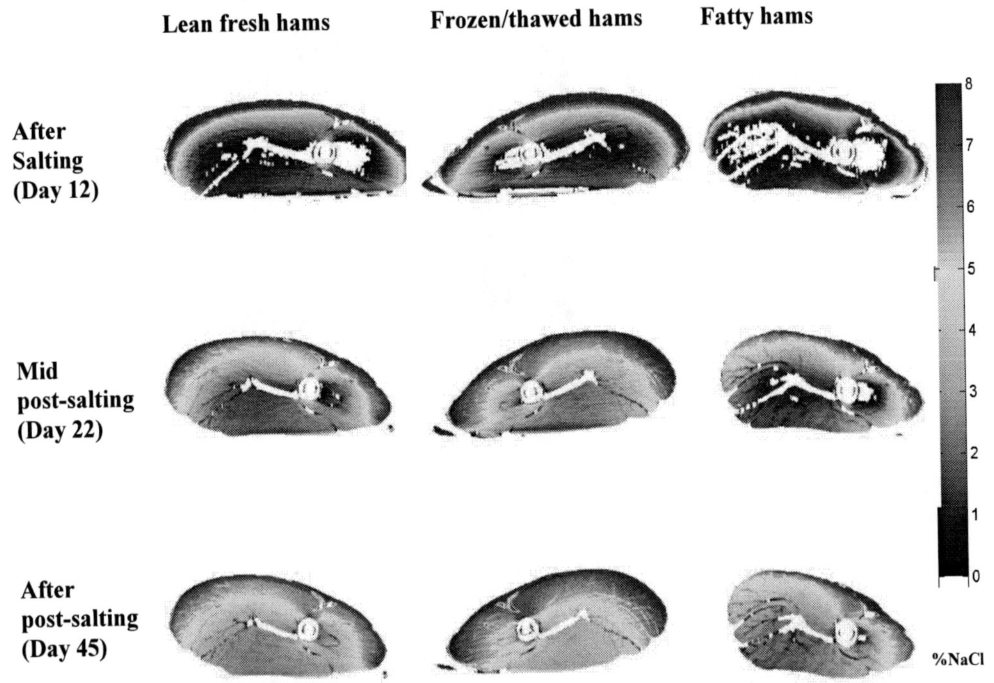

Figure 10. Distribution diagrams for salt as a function of time (in days) during post-salting periods in different types of raw ham: lean fresh hams, frozen/thawed hams and fatty hams.

Other operations before salting such as skin trimming affect salt-uptake and salt-distribution. Hams with the skin trimmed in a "V" shape absorb a higher amount of salt compared to non-trimmed hams (Figure 11). Using CT, Frøystein et al. (1989) showed that salt absorption in dry-cured ham takes place mainly through the lean tissue but that variability in the thickness of the subcutaneous fat led to variability in salt uptake. Using the prediction models previously developed by Fulladosa et al. (2010) to quantify salt and water contents in dry-cured hams, the amount of these components at different time points of the elaboration process can be estimated. This analysis showed how the salt reached the most critical internal parts in the V-shaped hams faster. The presence of salt in critical areas helps to reduce the a_w (Comaposada et al., 2000) and contributes to increase stability with respect to the temperature increase during the process.

2.4. CT to Ensure Food Safety

Monitoring salt and water contents by means of CT permits decisions to be taken at the critical points of the elaboration process when either raw material or processing parameters mentioned in the section above (such as skin trimming, pressing of hams, etc.) are changed. One of the critical points of the dry-cured ham elaboration process is the moment when the temperature is increased after post-salting period. At this point, salt content in the most internal and critical areas must be sufficient to ensure the safety of the product.

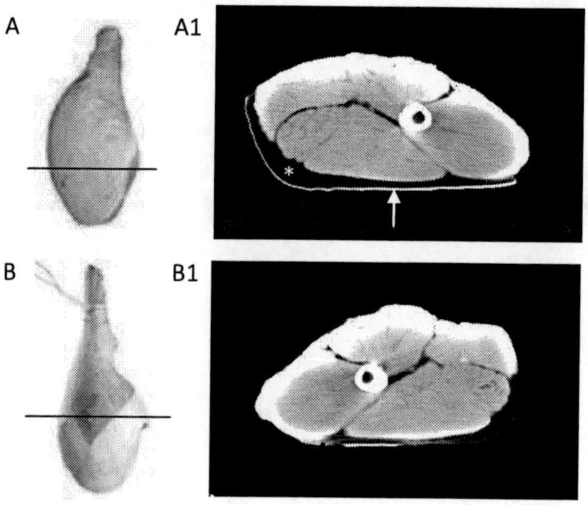

Figure 11. Raw hams from the same carcass; The skin was kept on the ham (A) and the skin of the second ham was trimmed in a "V" shape (B). The position of each CT scan of the hams is shown in pictures A1 and B1. CT cross sections from one dry-cured ham with skin and from its "V" shape trimmed pair obtained after salting are shown in the diagrams. Note that in A1 scan the skin (→) and subcutaneous fat (*) are natural barriers for salt penetration. The thin subcutaneous fat layer and the little presence of skin in B1 allowed salt penetration (Brighter areas represent salt).

CT can be used to design the correct processes for new products such as dry-cured hams with reduced salt content. In this case, hams are kept at a low temperature (below 5 ºC) during post-salting, until the salt content of the most critical area of the hams is equal to the content shown by hams traditionally salted (non-reduced salt content). Extending the post-salting period for dry-cured hams with reduced salt content is a strategy which guarantees microbiological stability during processing of this type of product (Figure 12) as the salt content in the most critical area at the end of post-salting is the same as that in standard salted dry-cured hams. The length of the post-salting period depends on the type of hams and the processing conditions.

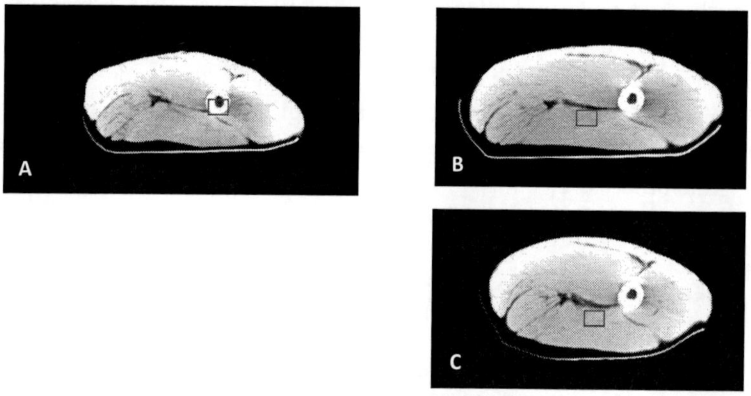

Figure 12. Dry-cured ham with standard salting showing a salt content of 1.20 % in the most critical area (indicated by a red square) after 45 days of post-salting (A). Dry-cured ham with reduced salt content after 45 days of post-salting. The salt content in the most critical area at this time is 1.09 % (B). Therefore, to achieve the same salt content as hams with standard salting the post-salting period needs to be extended (C).

2.5. Sensory Defects: Crust and Hollow Defects

Sensory defects in dry-cured ham can also be detected, studied and evaluated using CT. On the one hand, since development of a crust on dry-cured ham surface is related to hams with low intramuscular fat and a great amount of intermuscular fat, CT may be useful to identify green hams with these characteristics and therefore used when selecting the most appropriate genetic line, thus avoiding this defect. On the other hand, evaluation of crust level in meat industry is normally done by pressing the ham surface with the fingers but this operation is sometimes subjective and must be done by a trained inspector. In the finished and sliced product, the assessment for crustiness can be done by using visual scales. In order to develop such scale, CT can help to select those samples which have a different intensity of this technological defect. It can also be used to instrumentally evaluate different levels of crust by using colour images of water distribution. Water distribution images of a ham without crust and a ham with moderate crust on the surface are shown in Figure 13.

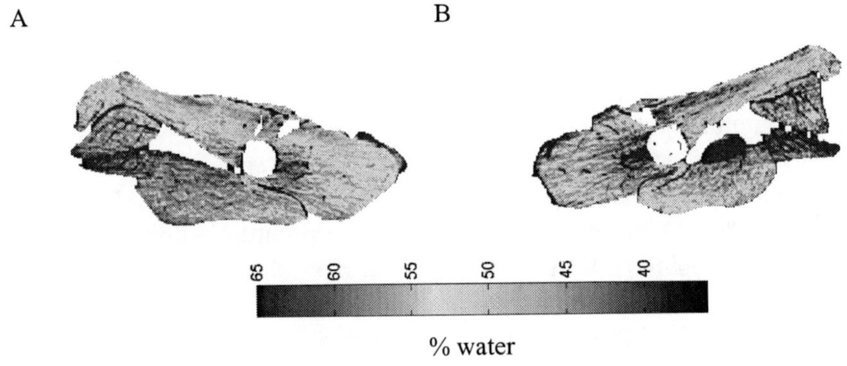

Figure 13. Images of water distribution representing two different levels of crustiness on the ham surface. No crust (A), moderate crust (B).

Hollow defect in dry-cured ham is mainly due to a rapid retraction of the ham during drying which may induce formation of cavities around the coxofemoral joint (Figure 14). These cavities house appropriate conditions for microbial development. CT can be used to detect and determine the degree of damage by measuring the extent of the cavities when using different elaboration conditions. It may help to optimize industrial processes in order to avoid this problem.

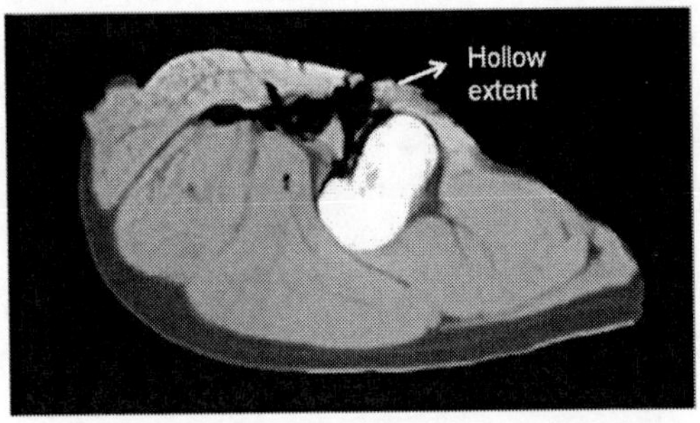

Figure 14. Tomogram of a dry-cured ham slice with a hollow defect.

3. CT Applications in Pig Carcass Classification

The lean yield of a carcass (or a cut) is an important quality parameter of interest to the pig industry and its definition varies between countries, as reviewed in Pomar et al. (2009). It is defined as the proportion of tissues of interest in a carcass and measured following a reference method. The commercial value of the carcasses is usually measured by means of weight and lean yield. In the European Union (EU) the classification of pig carcasses uses the lean meat percentage (LMP), which is the proportion of dissected lean in the carcass, although its definition has changed over the years. The main objective of the classification of pig is improving market transparency and is essential for the purpose of price recording and application of intervention arrangements (Council Regulation (EC) 1234/2007).

The estimation of the lean content is usually carried out from the physical measurements of subcutaneous fat depths and muscle thicknesses in specific locations (Engel et al., 2003), which is measured on line with different types of probes, or devices with different degrees of automation, and accuracies (Font i Furnols & Gispert, 2009). Moreover, these devices can be based on reflectance (Fat-O-Meat'er and Classification Center of Carometec A/S, Herlev, Denmark; Hennessy Grading Probe of Hennessy Grading System Ltd, Auckland, New Zealand; Capteur Grass-Maigre of Sydel Corporation, Lorient Cedex, France; etc.), ultrasounds (Autofom and Ultrafom of Carometec A/S, Herlev, Denmark; Ultrameater of CSB-System International, Geilenkirchen, Germany), vision (VCS200-0 of e+V Technology GmbH, Oranienburg, Germany; Vision of Rovi-Tech S.A., Presles, Belgium), electromagnetism (TOBEC of Meat Quality Inc., Springfield, Illinois, USA) or lineal measurements in the midline of the carcass (Zwei-Punkt-Messverfahren, ZP method). These devices need to be calibrated and therefore, the dissection is the reference method most commonly used. In the EU there is a specific legislation (Council Regulation (EC) 1234/2007 and Commission Regulation (EC) 1249/2008) which lays out the guidelines to be followed when carrying out the calibration: cutting and dissection of a minimum of 120 carcasses, representative of the pig population, and an error of prediction (RMSEP) lower than 2.5%. Until now, the dissection has had to be done manually by trained butchers following the EU reference method (Walstra & Merkus, 1995). However it is hard, difficult and time consuming (6-10 h to dissect half a carcass, depending on the type of dissection: simplified or full), expensive and it has a mean error between butchers of 1.96% (Nissen et al., 2006) which averaged between butchers is 0.98% (Judas, 2008).

An European project on pig carcass classification was carried out (GR6RD-CT99-0127 – EUPIGCLASS, www.eupigclass.net), in which one objective was to develop indirect methods for predicting the LMP of a pig carcass which are highly correlated with the reference dissection method resulting in a recommendation. One of the indirect methods studied was CT, and after the good results obtained in the project (Allen, 2003), together with other results obtained from previous works (Vangen, 1984; Horn, 1995; Jopson et al., 1995), CT was included in the EU legislation for concerning carcass classification (Commission Regulation (EC) 1249/2008) as equipment that could replace the dissection of carcasses, if "satisfactory comparative dissection results are provided". In this section of the chapter there is review of the investigations carried out to adapt CT for pig carcass dissection for use as a suitable reference.

3.1. CT Scanning of the Carcasses

Dissection trials following the European legislation are usually carried out on the left half carcass only, and for this reason, normally only the left half pig carcasses are fully scanned (Figure 15). The characteristics of CT scanning differ slightly between different investigations. The most common potential is between 137-140 kV, the intensity is between 145-180 mA and the thickness of the slices 10 mm (Romvári et al., 2003; Judas et al., 2007; Vester-Christensen et al., 2009; Font i Furnols et al., 2009b). An average of 140-145 images were taken for each carcass. The LMP (Font i Furnols et al., 2009b) or the lean meat weight (LMW) (Judas et al., 2007; Lyckegaard et al., 2006) are estimated in different investigations.

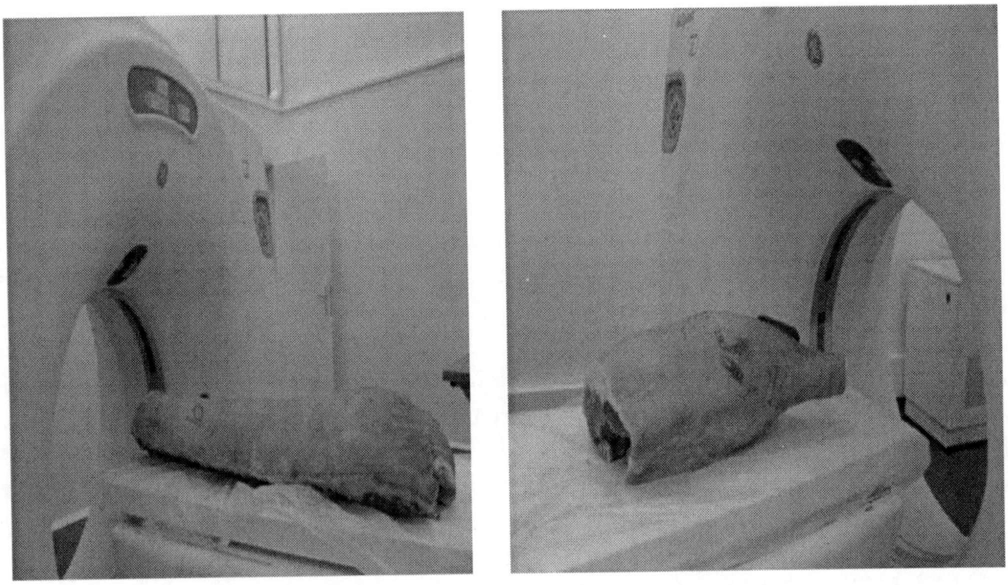

Figure 15. Scanning of a half carcass (ham and rest of the carcass) by means of CT.

3.2. Image Processing

Three different approaches of the analysis of the data/images obtained by CT are explained.

3.2.1. Partial Least Square (PLS) Regression and Other Statistical Estimations

From all the images of one carcass, the frequency of voxels or the volume associated with each attenuation Hounsfield value can be obtained. The comparison of these frequencies for carcasses with similar weights allows a differentiation between carcasses which depends on their leanness or fatness. Figure 16 shows this frequency for 3 different types of carcasses depending on their lean content which is assessed by the fat thickness measured with Fat-O-Meat'er at 6 cm of the midline and between the 3rd and 4th last ribs, being Lean (fat thickness $\leq$ 12 mm), Medium (fat thickness between 12 and 17 mm) and Fat (fat thickness > 17 mm).

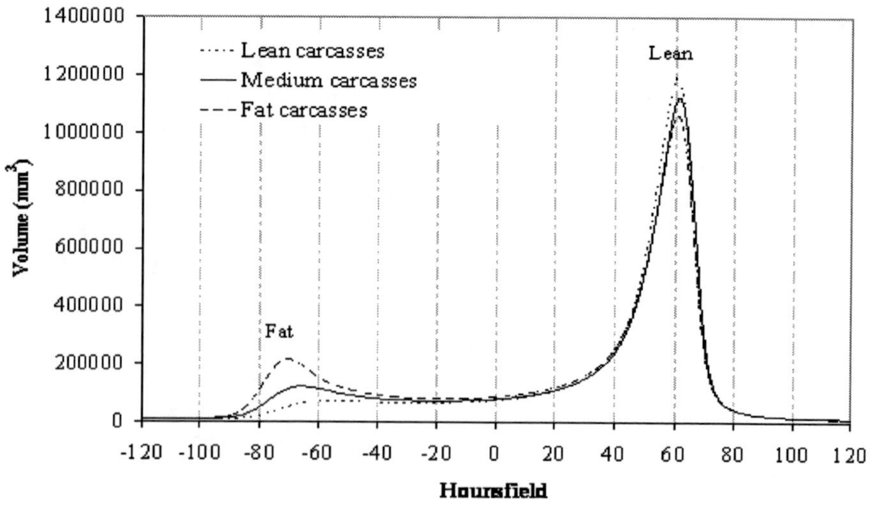

Figure 16. Volume associated with each attenuation Hounsfield value for lean, medium and fat carcasses (Font i Furnols et al., 2009b).

One of the ways to analyse the frequency of attenuation values from CT data is the Partial Least Square (PLS) regression. This is a very useful statistical technique when prediction variables are highly correlated as it extracts a low number of latent factors (linear combinations of the prediction variables) that explain the maximum covariance between prediction and response variables. PLS is superior to ordinary linear regression (OLR) for this type of data because it copes better with the variations of intramuscular fat content: PLS uses different densities to calculate the coefficients of the regression equation while OLR uses an average density of muscle tissues, which can depending on the intramuscular fat content (Judas et al., 2007). Therefore, PLS allows the avoidance of the Partial Volume Effect (PVE) problem, *i.e.* voxels which have more than one class of tissue, because it is not necessary to classify voxels into fat, lean or bone (Font i Furnols et al., 2009b).

Dobrowolski et al. (2003) were the first to use PLS to CT data from 60 scanned and dissected carcasses with an error of 232 g, which corresponds to 1% of the total LMW.

It is important to define the range of HU values that should be included for the estimation of the lean content because depending on the range used the error of prediction can widely differ (Font i Furnols et al., 2009b). Different ranges have been used in different investigations. Judas et al. (2007) applied this technique to obtain the LMW with a prediction error of 274 g analysing the spectra of 136 carcasses from 10 to 110 HU. Christensen et al. (2006) applied PLS to the spectral range between -500 and +1500 HU of 57 carcasses and obtained and error of prediction of the LMW of 339 g. Font i Furnols et al. (2009b) also applied PLS to estimate the lean meat percentage with a prediction error of 0.82% studying spectra form -100 to +120 HU. In the latter investigation, the regression coefficients for the different HU attenuation values, multiplied by the volume associated with each variable, were plotted (Figure 17) together with the volume associated with each HU value for the average number of carcasses scanned (depicted by the line on the graph). It can be seen that coefficients corresponding to HU values lower than 20 were negative, indicating a negative effect on the determination of the LMP. The variables placed at the lean area were the major positive variables in the prediction of the LMP.

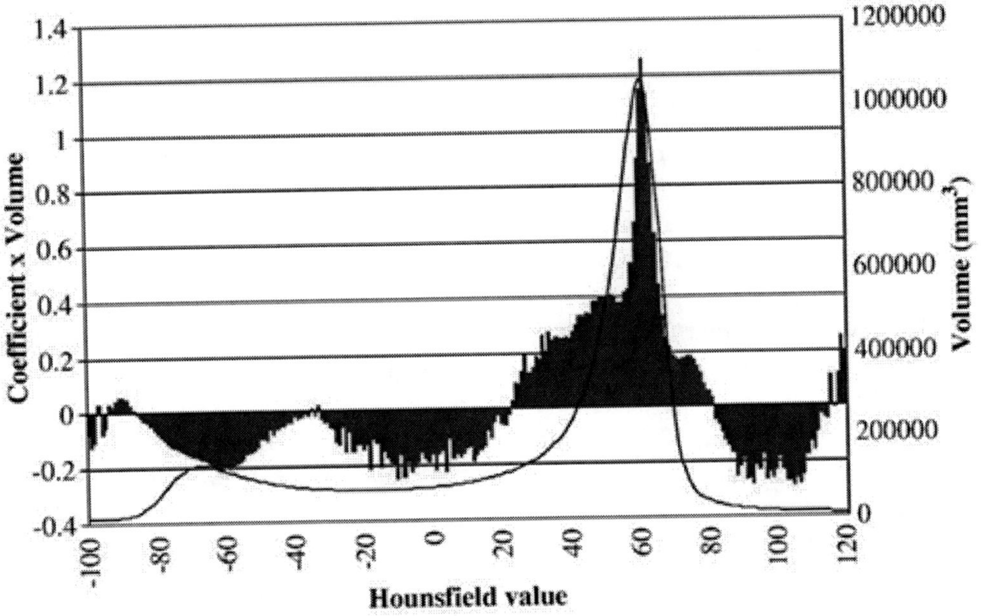

Figure 17. Coefficients of the regression equation multiplied by the volume associated with each variable (attenuation Hounsfield value) (intercept = 61.34). The line is the volume associated with each Hounsfield value for the average carcass scanned (Font i Furnols et al., 2009b).

The relationship between the estimated lean meat percentages with CT using PLS and the dissected lean meat percentage is shown in Figure 18.

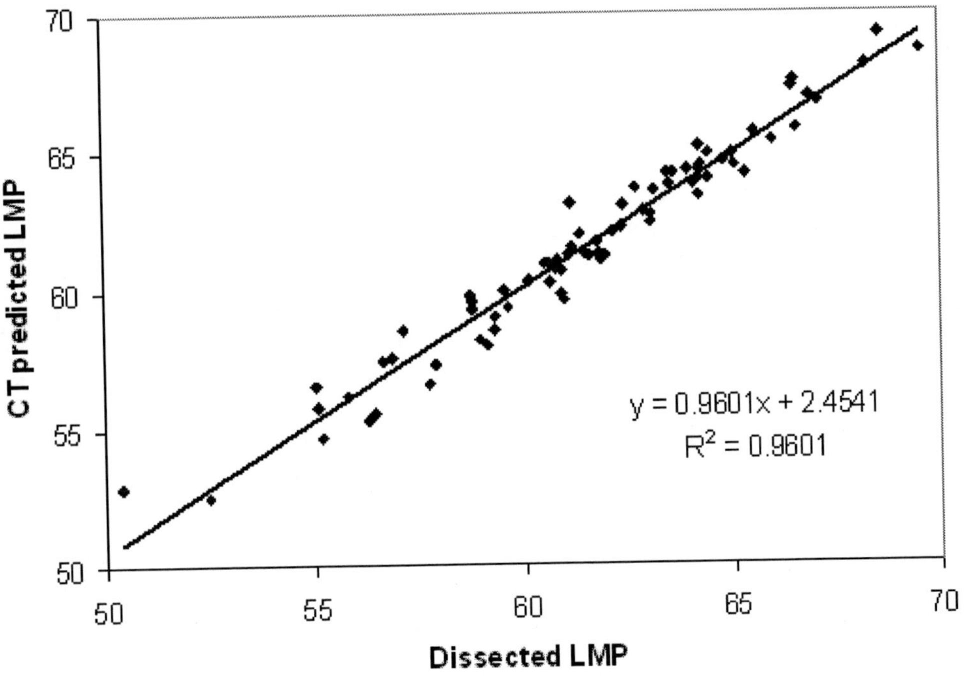

Figure 18. Lean meat percentage predicted with CT and dissected.

Collewet et al. (2005) evaluating pig carcasses with magnetic resonance imaging (MRI), used PLS to predict the LMW, obtaining a prediction error of 465 g which can be reduced to 326 g if the side effect is removed using the error propagation law.

Studying lamb carcass composition, Johansen et al. (2007) used PLS to analyse the two-dimensional histograms and Parallel Factor Analysis (PARAFAC) (Bro, 1997) and multi-way PLS (NPLS) (Bro, 1996) to analyse the three-dimensional histograms, The Hounsfield range was between -1024 and +1256. Errors of prediction of the LMW were 805 g, 907 g and 772 g using PLS, PARAFAC and NPLs, respectively.

3.2.2. Segmentation Estimation

Image segmentation divides an image into regions (fat, lean and bone) that can be considered homogenous with respect to a given criterion. Therefore the different voxels of the CT images are classified into fat, lean and bone. For some voxels the classification is clear, but for those voxels with PVE it is less clear. For this purpose different segmentation techniques can be used.

Lyckegaard et al. (2006) applied the Owen-Hjort-Mohn algorithm contextual Bayesian classification scheme (Larsen, 2000) and post-processing analysis using mathematical morphology for the bone marrow. Following this they applied different tissue densities to estimate the LMW with a prediction error of 584 g. Furthermore, Vester-Christensen et al. (2009) also applied the maximum-a-posteriori (MAP) probability to classify voxels into lean, fat and bone with post-processing analysis for the bone marrow and the skin. Moreover, they evaluated another estimation which including the PVE in the model using the value of the posterior probability of each voxel pertaining to either meat, fat or bone class. The incorporation of the PVE in the model did not significantly decrease the prediction error with respect to the MAP model (79.0 *vs* 83.6 g).

Monziols et al. (2006) analysed pig primal cuts (ham, loin, shoulder and belly) and obtained images with MRI using an adaptation of the Markovian segmentation algorithm (Shattuck et al., 2001) which took into account the PVE. For the segmentation which labels each pixel as fat, muscle or fat/muscle partial volume considering the signal and the neighbourhood of each pixel. The fat/muscle pixels were then split into fat and muscle following a partial volume detection method developed by Monziols et al. (2005). The errors of prediction of the ham, loin, shoulder and belly muscle were 82, 122, 129 and 81 g, respectively and the error of prediction of the carcass LMW from the 4 cuts was 284 g.

Teran (2009) used the excess entropy image segmentation that was first introduced by Crutchfield & Packard (1983) and improved it to be later used to find the optimal thresholds of a 2D or 3D image automatically, Bardera et al. (2009) (Figure 19). Post-processing analysis for skin using mathematical morphology and for intermuscular fat and lean affected by bone reflection using partial volume detection method were introduced. The final error of prediction of the LMP was 1.16%.

Johansen et al. (2007) analysing lamb carcasses and Collewet et al. (2005) analysing pig carcasses measured with MRI, used Otsu thresholding technique to separate fat, muscle and bone. When all the pixels were classified, they were added according to the type of tissue to estimate the fat and muscle content. In lamb, Johansen et al. (2007) found an error of 463 g for fat and 657 g for muscle. Collewet et al. (2005) found and error of prediction of the LMW of 657-665 g and of the LMP of 1.28-1.29%.

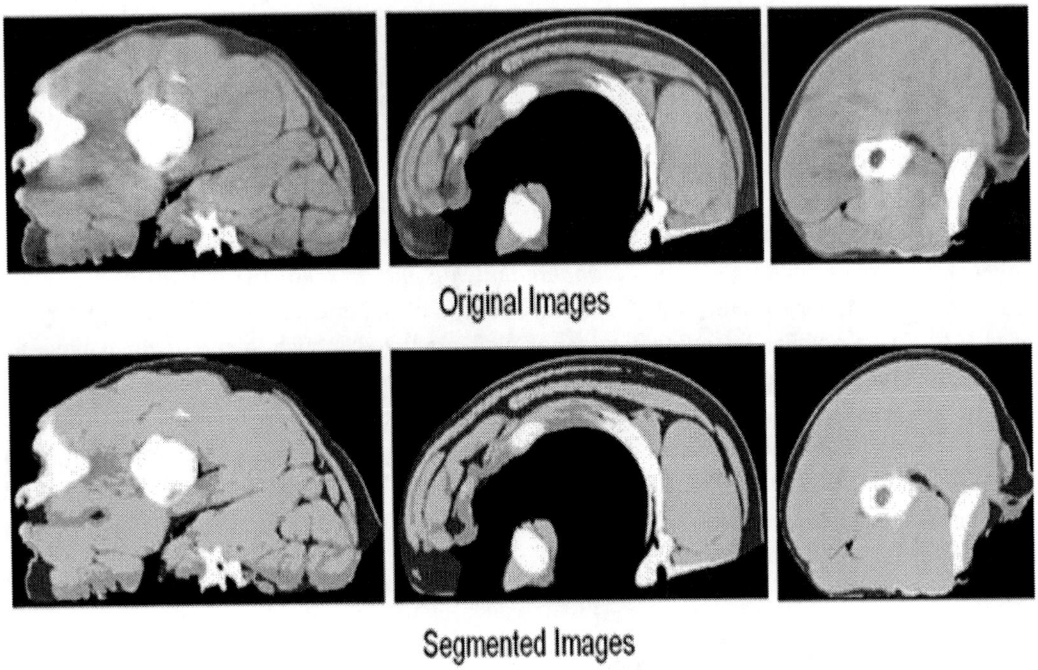

Figure 19. Original and segmented images with excess entropy segmentation technique from the shoulder (left), loin and belly (medium) and ham (right).

Navajas et al. (in press) evaluated beef cuts with CT. They classified the different voxels by means of the STAR 4.8 software (Mann et al., 2008) using the tissue thresholds estimated in Navajas et al. (2009), which were between -254 and 29 for fat and between 30 and 133 for muscle. Following this, tissue densities were applied using a fixed value for bone (Jopson, 1993) and a regression equation for fat and muscle (Fullerton, 1980). The determination coefficient of the regression of LMW by dissection and CT was 0.97.

3.2.3. Density Estimation

After the segmentation, the density estimation of every type of tissue in each voxel is usually applied to obtain the weight of the tissues of interest. Voxels are classified in fat, lean and bone, and with the voxel volume, the volume of each tissue can be estimated by means of the densities of the different tissues. For example, LMW = muscle density x voxel volume of lean. As has been explained above, the classification of the voxels into the different tissues can be done by different type of segmentation techniques.

However, in some cases, the LMW can be estimated by means of densities but without a previous classification of the voxels into the different tissues. Picouet et al. (in press) estimated the LMW based on the area of the histogram of the whole half carcass in a range of -250 to +800 HU, using a density of the tissues which varied linearly with the Hounsfield attenuation value. No post-processing of the data was carried out. The error obtained after correcting the bias was 290 g. Moreover, these authors estimated the LMP based on the ratio between the area of the lean peak in the calculated histograms and the area of the histogram of the whole half carcass. In this case the prediction error was 1.48%.

4. Other Applications in Food

4.1. Fish Industry

Several studies have obtained successful results for the prediction of fat contents in different fatty fish species, such as Atlantic salmon (*Salmo salar*) or halibut (Rye, 1991; Kolstad et al., 2004; Folkestad et al., 2008), Cyprinid species (*Cyprinus carpio L., Ctenopharyngodon idella Val.* and *Hypophthalmichtis molitrix Val.*) (Romvári et al., 2002; Hancz et al., 2003) or Atlantic cod (*Gadus morhua*) (Kolstad et al., 2008). However, Romvári et al. (2002) found that fat content was not well predicted in fish species which have an extremely low fat composition, as in the case of Pike-perch (*Stizostedion lucioperca L.*). Recent studies have demonstrated that CT can also be useful to control and optimize manufacturing processes in the fish industry. For instance, to study salt content distribution in Atlantic salmon (Segtnan et al., 2009) or in Atlantic cod (Håseth et al., 2009. The study of protein levels in fish species however, presented more difficulties (Gjerde, 1987; Rye, 1991; Romvári et al., 2002; Håseth et al., 2009).

4.2. Freezing and Thawing Processes

Freezing and thawing processes of food have become common practice in industry. Because the density of water in solid phase is lower than in liquid phase, freezing/thawing processes can be followed easily using CT. Figure 20 shows tomograms obtained during the freezing/ thawing of a green ham (A) and pork loin (B). Fresh hams (A1) show higher attenuation values than hams after 21 h freezing (A2) or hams which are completely frozen (A3). During thawing process the inverse phenomenon can be observed (A4-A5). In the case of pork loin, the time for freezing/thawing process is reduced, achieving a complete freezing after 12 hours (B3). The study of different freezing/thawing conditions using CT would provide additional information to enable a thorough study of the consequences of this process.

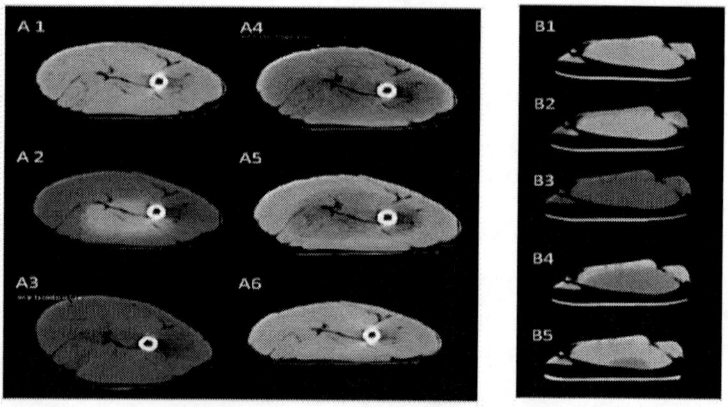

Figure 20. Tomograms obtained during the freezing/thawing process of a green ham (A) and pork loin (B). Dry cured ham: fresh (A1), 21 h freezing (A2), 87 h freezing (A3), 24 h thawing (A4), 78 h thawing (A5) and 196 h thawing (A6). Pork loin; fresh (B1), 4 hours freezing (B2), 12 h freezing (B3), 6 h thawing (B4) and 12 h thawing (B5).

4.3. Fruit and Cheese

CT has also been proved to be an efficient method for evaluating food quality and has been used for many food products, including fish, meat, fruits and vegetables. As an example, CT was used to monitor the internal quality changes in peaches during ripening. Results showed that X-ray CT number was directly related with density, moisture content and titratable acidity whereas the soluble solids and pH are inversely related with CT number. Thus, CT can be used as an effective tool in the evaluation of the internal quality of peach (Barcelon et al., 1999). Different structures of fruits (pear, orange and banana) (Figure 21) and degree of ripening in pineapple (Figure 22) can also be studied using tomograms.

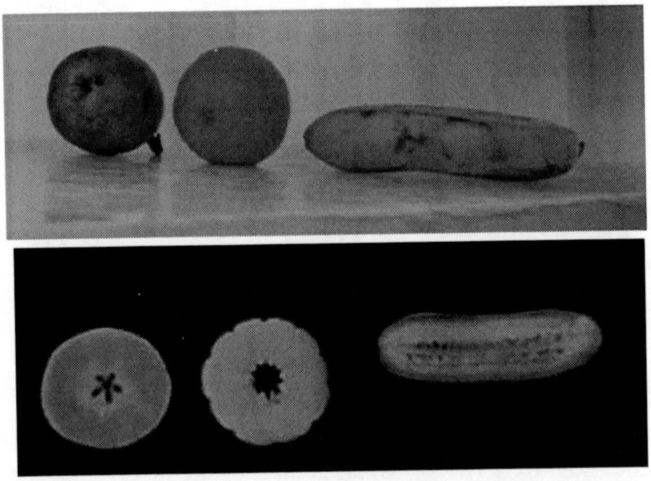

Figure 21. Photograph and tomograms of a pear, an orange and a banana.

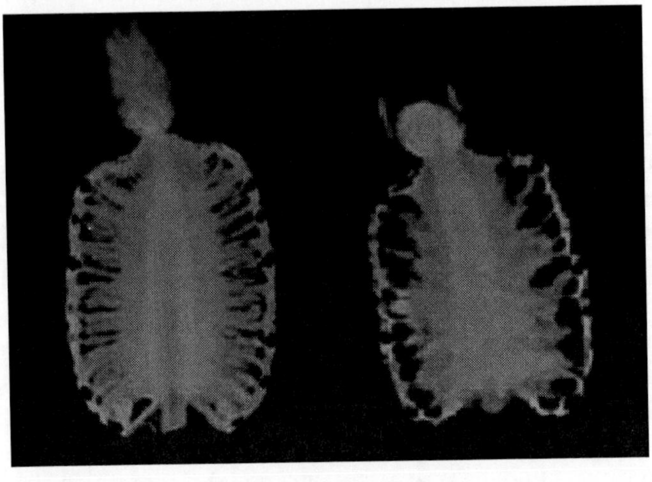

Figure 22. Degree of ripening of a pineapple: Unripe (A), ripe (B).

Cheese researchers have always wanted to study the factors which affect the product without damaging it. Thus, in the case of Jarlsberg cheese, CT was proved to be a very useful tool in successfully showing eye formation as the gas was produced in cheese without cutting or puncturing the product (Strand, 1985, Abrahamsen et al., 2006; Kraggerud et al., 2009,). In

Figure 23, a photograph and a tomogram of a section of Emmental and Parmesan cheeses are presented.

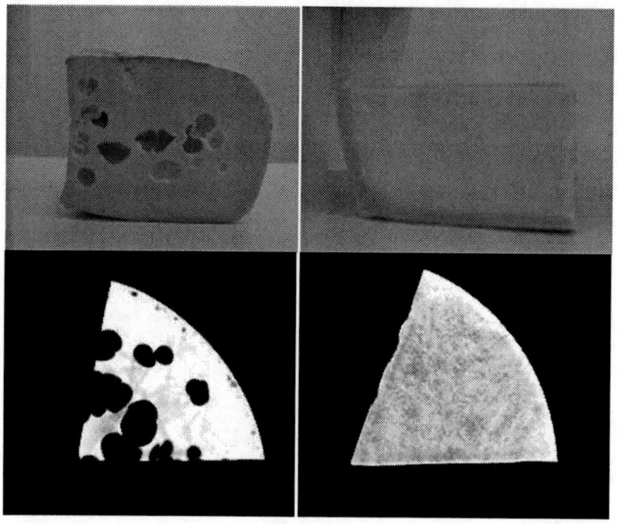

Figure 23. Photograph and tomograms of Emmental and Parmesan cheeses.

5. Other X-Ray Technologies

Other X-ray based equipment can also be used for a wide variety of studies. One of them is Micro-computed tomography (μCT) which is used to create cross-sections of a 3D-object without destroying the original model but the pixel sizes of the cross-sections are in the micrometer range. This technology, which is a combination of X-ray microscopy and tomographical algorithms, is the high spatial resolution version of the computed axial tomography and provides three-dimensional images of the interior of a material with a resolution of 10 μm.

These scanners have been typically used for small animals, biomedical samples, foodstuffs, microfossils, and other studies for which minute detail is necessary. Modern food processing is increasingly concerned with the production of products with complex aerated microstructures which determine, to a large extent, the mechanical and sensory properties of these products. The development and analysis of such new materials require non-invasive techniques for visualization and measurement of the internal microstructure. Lim & Barigou (2004) studied the cellular microstructure of a number of food products (aerated chocolate, mousse, marshmallows and muffins) using μCT. μCT has also been successfully used in food science for determining the inner structure of bread and for quantifying breadcrumb microstructure by image analyze (Falcone et al., 2004, 2005). In more recent studies, μCT has also been used in the bread industry to analyzed the microstructure of cereal products (Babin et al., 2005), the bubble growth and foam setting during breadmaking (Babin et al., 2006), and to study the bubble size distribution in wheat flour dough (Bellido et al., 2006). In a study where the effect of freezing conditions was analyzed, Mousavi et al. (2005) demonstrated the capability of the technique to characterize ice crystal microstructure of mycoprotein products of paste, dough and steamed dough. Other authors have used the technique to study air bubbles in dairy products, herbs in fat and pores in rice kernels (Van Dalen et al., 2003), the bubble formation and dispersion characteristics in chocolate (Haedelt et al., 2005), the relationship between texture, mechanical properties and structure of cornflakes (Chaunier et al., 2007) and extruded starches (Babin et al., 2007).

In addition, it has been observed that µCT has potential applications in plant science research. For instance, in the case of 'Conference' pears (*Pyrus communis* cv. Conference), Lammertyn et al. (2002) succeeded in measuring non-destructively, the spatial distribution of core breakdown symptoms. Subsequently, Lammertyn et al. (2003) also studied the time course of core breakdown disorder symptoms in 'Conference' pears. In another study, the three-dimensional network of gas-filled intercellular spaces in the sarcocarp of Cucumber fruit (*Cucumis sativus L.*, cv. Mugen) was studied by µCT (Kuroki et al., 2004). Other results showed that this technology supplies a high resolution quantitative analysis of the three-dimensional topology of the pore space in apple tissue (Mendoza et al., 2007). Likewise, Léonard et al. (2008) determined the total pore volume and size distribution of dried banana slices by means of µCT.

In the field of meat sciences, Frisullo et al. (2009) demonstrated that µCT was able to provide an accurate percentage of fat volume plus detailed information on the structure of the fat present in salami samples. µCT also enables a rapid estimation of intramuscular fat in meat, providing a more accurate description of the fat microstructure and meat quality (Frisullo et al., 2010).

X-ray inspection technology offers exceptional contamination detection, such as that found in glass, metal, stone and high-density plastics, and for products packaged in foil or metalized film. In addition, X-ray systems can simultaneously perform a wide-range of on-line quality checks. These include measuring mass, counting components, identifying missing or broken products, monitoring fill levels, inspecting seal soundness and checking for damaged packaging. It may also be useful in the cheese industry. Kraggerud et al (2009) developed a method using a low resolution online X-ray instrument which was found promising for quality control as it facilitates a non-destructive monitoring of eye formation of cheese throughout the ripening period. It may also be useful to visualize the product and the amount in eyes before cutting the cheese into portions of desired weight. Use of this more simple technology is also on going in meat science. Dual Energy X-ray Absorptiometry (DXA) has additionally been used to determine the chemical composition of live pigs and carcasses and the composition of dissected tissues in pig and sheep carcasses (Mitchell et al., 1996, 1998; Pomar & Rivest, 1996; Marcoux et al., 2003; Mercier et al., 2006).

6. Conclusion

CT has extensive potential in food science because it allows the observation of the inner part of food and enables the differentiation not only between different tissues but also within the same tissue, identifying different properties such as salt content, degree of ripening, water content, etc. Dry-cured ham elaboration processes may be optimized using this technology in order to obtain safer products of better quality. CT is also a useful and powerful technology for determining the composition of live animals and carcasses and consequently to be used as reference in pig carcass classification.

Acknowledgments

This work was partially supported by the TRUEFOOD European commission Integrated Project within the Sixth RTD Framework Program (Contract no. FOOD-CT-2006-016264). The information in this document reflects only the authors' views and the Community is not liable for any use that may be made of the information contained therein. It was also partially financed by INIA-Plan Específico de Investigación de Teruel (contract no.PET 2007-08-C11-08).

References

Aass, L., Hallenstvedt, E., Dalen, K., Kongsro, J., & Vangen, O.. Datatomography (CT) as a method to measure meat- and fat quality in live pigs? II Workshop on the use of Computed Tomography (CT) in pig carcass classification. Other CT applications: live animals and meat technology, Monells, Spain, 16-17 Abril 2009

Abrahamsen, R. K., Byre, O., Steinsholt, K., & Strand, A. H. (2006). Jarlsberg cheese: Jarlsberg cheese history and development (1st edition). Poland: BZGraf SA.

AESAN, Agencia Española de Seguridad Alimentaria y Nutrición (2005). Estrategia NAOS. Estrategia para la Nutrición, Actividad Física y Prevención de la obesidad. Ministerio de Sanidad y Consumo.
<http://www.naos.aesan.msc.es/naos/ficheros/investigacion/publicacion1estrategianaos.pdf>.

Allen, P. (2003). WP3 Summary. EUPIGCLASS Final Workshop, 6-7 October 20013. Roskilde, Denmark. http://www.eupigclass.net/Work3/04_Paul_WP%203Summary-filer/frame.htm

Arnau, J., Hugas, M., & Monfort, J.M. (1987). Jamón curado: Aspectos técnicos (1st edition). Barcelona: Institut de Recerca i Tecnologia Agroalimentàries.

Arnau, J. (2007). Factores que afectan a la salazón del jamón curado. *EUROCARNE*, 160: 59-76.

Arnau, J., Serra, X., Comaposada, J., Gou, P., & Garriga, M. (2007). Technologies to shorten the drying period of dry- cured meat products. *Meat science*, 77 (1), 81-89.

Babin, P., Della Valle, G., Chiron, H., Cloetens, P., Hoszowska, J., Pernot, P., Réguerre, A.L., Salvo, L., & Dendievel, R. (2006). Fast x-ray tomography analysis of bubble growth and foam setting during breadmaking. *Journal of cereal science*, 43 (3), 393-397.

Babin, P., Della Valle, G., Dendievel, R., Lassoued, N., & Salvo, L. (2005). Mechanical properties of breadcrumbs from tomography based finite element simulations. *Journal of materials science*, 40 (22), 5867-5873.

Babin, P., Della Valle, G., Dendievel, R., Lourdin, D., & Salvo, L. (2007). X-ray tomography study of the cellular structure of extruded starches and its relations with expansion phenomenon and foam mechanical properties. *Carbohydrate polymers*, 68 (2), 329-340.

Barat, J. M., Grau, R., Ibáñez, J. B., & Fito, P. (2005). Post-salting studies in Spanish cured ham manufacturing. Time reduction by using brine thawing–salting. *Meat science*, 69 (2), 201–208.

Barcelon, E.G., Tojo, S., & Watanabe, K. (1999). X-ray computed tomography for internal quality evaluation of peaches. *Journal of agricultural engineering research*, 73 (4), 323-330.

Bardera, A., Boada, I., Feixas, M., & Sbert, M. (2009). Image segmentation using excess entropy. *Journal of signal processing systems*, 54 (1-3), 205-214.

Bellido, G. G., Scanlon, M. G., Page, J. H., & Hallgrimsson, B. (2006). The bubble size distribution in wheat flour dough. *Food research international*, 39 (10), 1058-1066.

Bro, R. (1996). Multiway calibration. Multilinear PLS. *Journal of chemometrics*, 10 (1), 47-61.

Bro, R. (1997). PARAFAC. Tutorial and applications. *Chemometrics and intelligent laboratory systems*, 38 (2), 149-171.

Bünger, L., Navajas, E. A., Lambe, N. R., Macfarlane, J., McLean, K. A., Glasbeyu, C. A., & Simm, G.(2009) Use of computer tomography in UK meat sheep breeding: research and practice. II Workshop on the use of computed tomography (CT) in pig carcass classification. Other CT applications: live animals and meat technology, Monells, Spain, 16-17 Abril 2009.

Casas, J. D., Díaz, R., Valderas, G., Mariscal, A., & Cuadras, P. (2004). Prognostic value of CT in the early assessment of patients with acute pancreatitis. *American journal of roentgenology*, 182 (3), 569-574.

Chaunier, L., Valle, G. D., & Lourdin, D. (2007). Relationships between texture, mechanical properties and structure of cornflakes. *Food research international*, 40 (4), 493-503.

Christensen, L. B., Lickegaard, A., Borggaard, C., Romvári, R., Olsen, E. V., Branscheid, W., & Judas, M. (2006). Contextual volume grading vs. spectral calibration. Proceedings of the international congress of meat science and technology, 13–18 August 2006, Dublin, Ireland.

Collewet, G., Bogner, P., Allen, P., Busk, H., Dobrowolski, A., Olsen, E., & Davenel, A. (2005). Determination of the lean meat percentage of pig carcasses using magnetic resonance imaging. *Meat science*, 70 (4), 563-572.

Comaposada J., Gou P., & Arnau J. (2000). The effect of sodium chloride content and temperature on pork meat isotherms. *Meat science*, 55 (3), 291-295.

Commission Regulation (EC) 1249/2008 of 10 December 2008 laying down detailed rules on the implementation of the Community scales for the classification of beef, pig and sheep carcases and the reporting of prices thereof.

Council Regulation (EC) No 1234/2007 of 22 October 2007 establishing a common organisation of agricultural markets and on specific provisions for certain agricultural products (Single CMO Regulation).

Crutchfield, J. P., & Packard, N. H. (1983). Symbolic dynamics of noisy chaos. *Physica D: Nonlinear Phenomena*, 7 (1-3), 201-223.

Dobrowolski, A., Romvári, R., Davenel, A., Marty-Mahe, P., & Allen, P. (2003). Using PLS on data provided by x-ray CT, MRI and VIA. EUPIGCLASS Final Workshop, 6-7 October. Roskilde, Dinamarca.

Donkó, T., Petrási, Zs., Holló, G., Petneházy, Ö., Bogner, P., & Repa, I. (2009). Application of CT cross sectional imaging in animal research at Kaposvár University. II Workshop on the use of Computed Tomography (CT) in pig carcass classification. Other CT applications: live animals and meat technology, Monells, Spain, 16-17 Abril 2009.

Engel, B., Builst, W.G., Walstra, P., Olsen E., & Daumas, G. (2003). Accuracy of prediction of percentage lean meat and authorization of carcass measurements instruments: adverse effects of incorrect sampling of carcasses in pig classification. *Animal science*, 76 (2), 199-209.

EUPIGCLASS. European Project: GR6RD-CT99-0127. www.eupigclass.net

Falcone, P. M., Baiano, A., Zanini, F., Mancini, L., Tromba, G., Montanari, F., & Del Nobile, M. A. (2004). A novel approach to the study of bread porous structure: phase-contrast x-ray microtomography. *Journal of food science*, 69 (5), 38-43.

Falcone, P. M., Baiano, A., Zanini, F., Mancini, L., Tromba, G., Dreossi, D., Montanari, F., Scuor, N., & Del Nobile, M.A. (2005). Three-dimensional quantitative analysis of bread crumb by x-ray microtomography. *Journal of food science*, 70 (3), 265-272.

Folkestad, A., Wold, J. P., Rørvik, K-A., Tschudi, J., & Haugholt, K. H. (2008). Rapid and non-invasive measurements of fat and pigment concentrations in live and slaughtered Atlantic salmon (Salmo salar L.). *Aquaculture*, 280 (1-4), 129-135.

Font i Furnols, & M., Gispert, M. (2009). Comparison of different devices for predicting the lean meat percentage of pig carcasses. *Meat science*, 83 (3), 443-446.

Font i Furnols, M., Teran, F., & Gispert, M. (2009b) Estimation of lean meat content in pig carcasses using x-ray computed tomography images and PLS regression. *Chemometrics and intelligent laboratory systems*, 98 (1), 31-37.

Font i Furnols, M., Teran, M. F., & Gispert, M. (2009a). Determination of intramuscular fat of the loin with CT. Preliminary results. II Workshop on the use of computed tomography (CT) in pig carcass classification. Other CT applications: live animals and meat technology, Monells, Spain, 16-17 Abril 2009.

Frisullo, P., Laverse, J., Marino, R., & Del Nobile, M. A. (2009). X-ray computed tomography to study processed meat microstructure. *Journal of food engineering*, 94 (3-4), 283-289.

Frisullo, P., Marino, R., Laverse, J., Albenzio, M., & Del Novile, M. A. (2010). Assessment of intramuscular fat level and distribution in beef using x-ray microcomputed tomography. *Meat science*, 85 (2), 250-255.

Fromm, J. H, Sautter, I., Matthies, D., Kremer, J., Schumacher, P., & Ganter, C. (2001). Xylem water content and wood density in spruce and oak trees detected by high-resolution computed tomography. *Plant physiology*, 127, 416-425.

Frøystein, T., Sørheim, O., Berg, S. A., & Dalen, K. (1989). Salt distribution in cured hams, studied by computer x-ray tomography. *Fleischwirtch*, 69 (2), 220-222.

Fulladosa, E., Santos-Garcés, E., Picouet, P.,. & Gou, P. (2010). Salt and water content prediction by computed tomography in dry-cured hams. *Journal of food engineering*, 96 (1), 80-85.

Fullerton, G. D. (1980). Tissue imaging and characterisation. In Medical Physics of CT and ultrasound (eds. G.D. Fullerton and J.A. Zagzebski), pp. 125-162. Medical Physics Monograph, American Institue of Physics.

Funt, B. V., & Bryan, E. C. (1987). Detection of internal log defects by automatic interpretation of computer tomography images. *Forest products journal*, 37 (1), 56-62.

Gjerde, B. (1987). Predicting carcass composition of rainbow trout by computerized tomography. *Journal of animal breeding and genetics (Zeitschrift Fur Tierzuchtung Und Zuchtungsbiologie),* 104 (1-2), 121-36.

Haedelt, J., Pyle, D. L., Beckett, S.T., & Niranjan, K. (2005). Vacuum-induced bubble formation in liquid-tempered chocolate. *Journal of food science*, 70 (2), 159-164.

Hancz, C., Romvári, R., Szabó, A., Molnár, T., Magyary, I., & Horn, P. (2003). Measurement of total body composition changes of common carp by computer tomography. *Aquaculture research*, 34 (12), 991-997.

Harwood-Nash, D. C. F. (1979). Computed tomography of ancient Egyptian mummies. *Journal of computer assisted tomography*, 3 (6), 769-773.

Håseth, T., Egelandsdal, B., Bjerke, F.,. & Sørheim, O. (2007). Computed tomography for quantitative determination of sodium chloride in ground pork and dry-cured hams. *Journal of food science*, 72 (8), 420-427.

Håseth, T., Høy, M., Egelandsdal, B, & Sørheim, O. (2009). Nondestructive analysis of salt, water, and protein in dried salted cod using computed tomography. *Journal of food science*, 74 (3), 147-153.

Håseth, T., Høy, M., Kongsro, J., Kohler, A., Sørheim, O.,. & Egelandsdal, B. (2008). Determination of sodium chloride in pork meat by computed tomography at different voltages. *Journal of food science*, 73 (7), 333-339.

Horn, P. (1995).Using x-ray computed tomography to predict carcass leanness in pigs. National swine improvement federation conference and annual meeting, Clive, Iowa.

Hounsfield, G. N. (1973). Computerised transverse axial scanning (tomography) Part 1. Description of system. *British journal of radiology*, 46, 1016-1022.

Hounsfield, G.N. (1980). Computed medical imaging: Nobel lecture, December 8, 1979. *Journal of Computed Assisted Tomography*, 4 (5), 665–674.

Johansen, J., Egelandsdal, B., Røe, M., Kvaal, K., & Aastveit, A. H. (2007). Calibration models for lamb carcass composition analysis using computerized tomography (CT) imaging. *Chemometrics and intelligent laboratory systems*, 87 (2), 303–311.

Jopson, N. B. (1993). Physiological adaptations in two seasonal cervids. PhD Thesis. University of New England, New South Wales, Australia.

Jopson, N. B., Kolstad, K., Sehested, E., & Vangen, O. (1995). Computed tomography as an accurate and cost effective alternative to carcass dissection. In Proceedings of the Australian association of animal breeding and genetics, 11, 635-639.

Judas, M. (2008). Rules for the implementation of CT as a reference dissection method. II CT Workshop on the use of computed tomography (CT) in pig carcass classification. Other CT applications: live animals and meat technology, 16–17 April 2009, Monells, Girona, Spain.

Judas, M., Höreth, R., & Branscheid, W. (2007). Computed tomography as a method to analyse the tissue composition of pig carcases. *Fleischwirtschaft international*, 1, 56-59.

Kalender, W. A. (2005). Computed tomography: fundamentals, system technology, image quality, applications (2nd edition, pp. 304). Erlangen, Germany.

Karamichou, E., Richardson, R. I., Nute, G. R., McLean, K. A., & Bishop, S. C. (2006). Genetic analyses of carcass composition, as assessed by x-ray computer tomography and meat quality traits in Scottish Blackface sheep. *Animal science*, 82 (2), 151-162.

Kolstad, K., Mørkøre, T., & Thomassen, M. S. (2008). Quantification of dry matter % and liquid leakage in Atlantic cod (Gadus morhua) using computerised X-ray tomography (CT). *Aquaculture*, 275 (1–4), 209-16.

Kolstad, K., Vegusdal, A., Baeverfjord, G., & Einen, O. (2004). Quantification of fat deposits and fat distribution in Atlantic halibut (Hippoglossus hippoglossus L.) using computerized x-ray tomography (CT). *Aquaculture*, 229 (1-4), 255-64.

Kraggerud, H., Wold, J.P., Høy, M., & Abrahamsen, K. (2009). X-ray images for the control of eye formation in cheese. *International journal of diary technology*, 62 (2) 147-153.

Kukori, S., Oshita, S., Sotome, I., Kawagoe, Y., & Seo, Y. (2004). Visualization of 3-D network of gas-filled intercellular spaces in cucumber fruit after harvest. *Postharvest biology and technology*, 33 (3), 255-262.

Lammertyn, J., Dressalaers, T., Van Hecke, P., Jancsók, P., Wevers, M., & Nicolaï, B. M. (2002). MRI and x-ray CT study of spatial distribution of core breakdown in 'Conference' pears. *Magnetic resonance imaging*, 21 (7), 805-815.

Lammertyn, J., Dresselaers, T., Van Hecke, P., Jancsónk, P., Wevers, M., & Nicolaï, B. M. (2003). Analysis of the time course of core breakdown in 'Conference' pears by means of MRI and x-ray CT. *Postharvest biology and technology*, 29 (1), 19-28.

Larsen, R. (2000). 3-D contextual Bayesian classifiers. *IEEE Transactions on Image Processing*, 9 (3), 518-524.

Léonard, A., Blancher, S., Nimmol, C., & Devahastin, S. (2008). Effect of far-infrared radiation assisted drying on microstructure of banana slices: An illustrative use of x-ray microtomography in microstructural evaluation of a food product. *Journal of food engineering*, 85 (1), 154-162.

Lim, K. S., & Barigou, M. (2004). X-ray micro-computed tomography of cellular food products. *Food research international*, 37 (10), 1001-1012.

Lindgren, L. O. (1991). Medical CAT-scanning: x-ray absorption coefficients, CT-numbers and their relation to wood density. *Wood science and technology*, 25 (5), 341-349.

Lyckegaard, A., Larsen, R., Christensen, L. B., Vester-Christensen, M., & Olsen, E. V. (2006). Contextual analysis of CT scanned pig carcasses. Proceedings of the international congress of meat science and technology, 13–18 August 2006, Dublin, Ireland.

Mann, A. D., Glasbey, C. A., Navajas, E. A., McLean, K. A., & Bünger, L. (2008). STAR: Sheep tomogram analysis routines (Version 4.9). BioSS software documentation.

Marcoux, M., Bernier, J. F., & Pomar, C. (2003). Estimation of Canadian and European lean yields and composition of pig carcasses by dual-energy x-ray absorptiometry. *Meat science,* 63 (3), 359-356.

Mendoza, F., Verboven, P., Mebatsion, H. M., Kerckhofs, G., Wevers, M. & Nicolaï, B. (2007). Three-dimensional pore space quantification of apple tissue using x-ray computed microtomography. *Planta*, 226 (3), 559-570.

Mercier, J., Pomar, C., Thériault, M., Goulet, F., Marcoux, M., & Castonguay, F. (2006). The use of dual-energy x-ray absorptiometry to estimate the dissected composition of lamb carcasses. *Meat science*, 73 (2), 249-257.

Mitchell, A. D., Conway, J. M., & Pott, W. J. E. (1996). Body composition analysis of pigs by dual-energy x-ray absorptiometry. *Journal of animal science*, 74 (11), 2663-2671.

Mitchell, A. D., Scholz, A. M., Pursel, V. G., & Evock-Clover, C. M. (1998). Composition analysis of pork carcasses by dual-energy x-ray absorptiometry. *Journal of animal science*, 76 (8), 2104-2114.

Monziols, M., Collewet, G., Bonneau, M., Mariette, F., Davenel, A., & Kouba, M. (2006). Quantification of muscle, subcutaneous fat and intermuscular fat in pig carcasses and cuts by magnetic resonance imaging. *Meat science*, 72 (1), 146-154.

Monziols, M., Collewet, G., Mariette, F., Kouba, M., & Davenel, A. (2005). Muscle and fat quantification in MRI gradient echo images using a partial volume detection method. Application to the characterization of pig belly tissue. *Magnetic resonance imaging*, 23 (6), 745-755.

Mousavi, R., Miri, T., Cox, P. W., & Fryer, P. J. (2005). A novel technique for ice crystal visualization in frozen solids using x-ray micro-computed tomography. *Journal of food science*, 70 (7), 437-442.

Næs, T., Isaksson, T., Fearn, T. & Davies, T. (2002). A user-friendly guide to Multivariate Calibration and Classification. NIR Publications, Chichester, UK.

Navajas, E. A., Glasbey, C. A., Fisher, A. V., Ross, D. W., Hyslop, J. J., Richardson, R. I., Simm, G., & Roehe, R. (2009). Assessing beef carcass tissue weights using computed tomography spirals of primal cuts. *Meat science*, 84 (1), 30-38.

Navajas, E. A., Richardson, R. I., Fisher, A. V., Hyslop, J. J., Ross, D. W., Prieto, N., Simm, G., & Roehe, R. Predicting beef carcass composition using tissue weight of a primal cut assessed by computed tomography. *Animal*. doi:10.1017/S1751731110001096 (in press).

Nissen, P. M., Busk, H., Oksama, M., Seynaeve, M., Gispert, M., Walstra, P., Hansson, I., & Olsen, E. (2006). The estimated accuracy of the EU reference dissection method for pig carcass classification. *Meat science* 73 (1), 22-28.

Norsvin International AS. Norway. www.norsvin.no

Pedraza, S. & Méndez-Méndez, J. (2004). Valor pronóstico de la tomografía computarizada en la hemorragia subaracnoidea aneurismática aguda. *Revista de neurología*, 39 (4), 359-363.

Picouet, P., Teran, F., Gispert, M., & Font i Furnols, M. Lean content prediction in pig carcasses, loin and ham by computerized tomography (CT) using a density model. *Meat science*, doi: 10.1016/j.meatsci.2010.04.039 (in press).

Pomar, C., Marcoux, M., Gispert, M., Font i Furnols, M., & Daumas, G. (2009). Determining the lean content of pork carcasses. In: Improving the sensory and nutritional quality of fresh meat. Ed. Joseph P. Kerry & D. Ledward. Woodhead Publishing, CRC Press. England.

Pomar, C., & Rivest, J. (1996). Évaluation d'une méthode à balayage aux rayons-x pour prédire la composition corporelle des porcs vivants. In Proc. Journées de recherche en zootechnie, C. P. A. Q., 30-31 mai.

Romvári, R., Dobrowolski, A., Horn, P., & Allen, P. (2003). CT examination of pig carcasses. EUPIGCLASS Final Workshop, Roskilde, Denmark, 6-7 October 2003, http://www.eupigclass.net/Work3/05_Robert_CT.htm

Romvári, R., Hancz, C., Petrási, Z., Molnár, T., & Horn, P. (2002). Non-invasive measurement of fillet composition of four freshwater fish species by computer tomography. *Aquaculture international*, 10 (3), 231-40.

Rye, M. (1991). Prediction of carcass composition in Atlantic salmon by computerized-tomography. *Aquaculture*, 99 (1-2), 35-49.

Santos-Garcés, E., Gou, P., Garcia-Gil, N., Arnau, J., & Fulladosa, E. (2010). Non-destructive analysis of aw, salt and water in dry-cured hams during drying process by means of computed tomography. *Journal of food engineering*, 101 (2), 187-192.

Seeram, E. (2009). Computed tomography: physical principles, clinical applications, and quality control. Philadelphia, P.A.: W.B. Saunders Co.

Segtnan, V., Høy, M., Sørheim, O., Kohler, A., Lundby, F., Wold, J., & Ofstad, R. (2009). Noncontact salt and fat distributional analysis in salted and smoked salmon fillets using x-ray computed tomography and NIR interactance imaging. *Journal of agricultural and food chemistry*, 57 (5), 1705-1710.

Shattuck, D. W., Sandor-Leahy, S. R., Schaper, K. A., Rottenberg, A., & Leahy, R. M. (2001). Magnetic resonance image tissue classification using a partial volume. *Neuroimage*, 13 (5), 856-876.

Sørheim, O., & Berg, S. A. (1987). Computed x-ray tomography (CT) as a non destructive method to study salt distribution in meat. In rapid analysis in food processing and food control. Loen, Norway. (p. 87)

Strand, A. H. (1985). Skanning av Jarlsbergost på datatomograf. Meieriteknikk, 39 (1), 2-7.

Taylor, F. W., Wagner, F. G., McMillin, C.W., Morgan, I. L., & Hopkins, F. F. (1984). Locating knots by industrial tomography. A feasibility study. *Forest products journal*, 34 (5), 42-46.

Teran, F. (2009) Utilización de la tomografía computerizada para la determinación de la distribución de la grasa y el magro de la canal y su cuantificación. Universitat de Girona. July, 2009.

Van Dalen, G., Blonk, H., Van Aalst, H., Hendriks, C. L. (2003). 3D imaging of foods using x-ray microthomography. *G. I. T. Imaging & microscopy*, 3, 18-21.

Vangen, O. (1984). Evaluation of carcass composition of live pigs based on compouted tomography. Proceedings of the 35th annual meeting of the EAAP, August 6-9, 1984. The Hague, The Netherlands.

Vester-Christensen, M., Erbou, S. G. H., Hansen, M. F., Olsen, E. V., Christensen, L. B., Hviid, M., Ersbøl, B. K., & Larsen, R. (2009). Virtual dissection of pig carcasses. *Meat science*, 81 (4), 699-704.

Vestergaard, C., Erbou, S. G., Thauland, T., Adler-Nissen, & J., Berg, P. (2005). Salt distribution in dry-cured ham measured by computed tomography and image analysis. *Meat science*, 69 (1), 9-15.

Vestergaard, C., Risum, J., & Adler-Nissen, J. (2004). Quantification of salt concentrations in cured pork by computed tomography. *Meat science*, 68 (1), 107-113.

Walstra, P., & Merkus, G. S. M. (1995). Procedure for the assessment of lean meat percentage as a consequence of the new EU reference dissection method in pig carcass classification. DLO-Research institute of animal science and health (ID-DLO) Zeist, The Netherlands.

Wellington, S. L., & Vinegar, H. J. (1987). X-ray computerized tomography. *Journal of petroleum technology*, 39 (8), 885-898.

Wood, J. D., Richardson, R. I., Nute, G. R., Fisher, A. V., Campo, M. M., Kasapidou, E., Sherard, P. R., & Enser, M. (2003). Effect of fatty acids on meat quality: a review. *Meat science*, 66 (1), 21-32.

In: Focus on Food Engineering
Editor: Robert J. Shreck

ISBN: 978-1-61209-598-1

Chapter 6

High Pressure Processing of Meat, Meat Products and Seafood

Marco Campus*
Porto Conte Ricerche Srl, 07041,
Loc. Tramariglio, Alghero (SS), Italy

Abstract

High Pressure Processing (HPP) allows decontamination of foods with minimal impact on their nutritional and sensory features. The use of HPP to reduce microbial loads has shown great potential in the muscle-derived food industry. HPP has proven to be a promising technology and industrial applications have grown rapidly, especially in the stabilization of ready-to-eat meats and dry-cured products, satisfying the demands of regulatory agencies such as the United States Department of Agriculture-Food Safety and Inspection Services (USDA-FSIS). Applications also extend to seafood products and HPP has been used in a wide range of operations, from non-thermal decontamination of acid foods to combined pressure-heating treatments to inactivate pathogenic bacteria, pressure supported freezing and thawing, texturization, and removal of meat from shellfish and crustaceans. Research has also been conducted on the impact of the technology on quality features. Processing-dependent changes in muscle foods include changes in colour, texture and water-holding capacity, with endogenous enzymes playing a major role in the phenomena. This review summarizes the current approaches to the use of high hydrostatic pressure processing, focusing mainly on meat, meat products and seafood. Recent findings on the microbiological, chemical and molecular aspects, along with commercial and research applications, are described.

Introduction

Mild preservation technologies aim at energy saving and being environmentally friendly, mild for the food but destructive for pathogenic and spoilage microorganisms. In this way,

---

* Tel.: +39 079 998 400; fax: +39 079 998 577.
E-mail address: campus@portocontericerche.it (M. Campus)

their use preserves the natural features of the product to a high extent. The implementation of new technologies in the food industry, such as high pressure processing (HPP), oscillating magnetic fields (ohmic heating, dielectric heating, microwaves), controlled instantaneous decompression (CID), intense light pulses (ILP), X-rays and electron beams, has prompted research on different approaches to their use in the food industry during the last decade (Hugas, Garriga and Monfort, 2002; Aymerich, Picouet and Monfort, 2008). However, not all mild technologies can be regarded as totally safe. In this respect, HHP offers promising possibilities for the processing and preservation of muscle-derived foods. From the pioneering experiments carried out at the end of the 19th century (Hite, 1899) on the inactivation of microorganisms in milk, high pressure processing has been used in a wide variety of applications. Addressing the growing demand for minimally processed foods that are safe and with superior sensory and nutritional features, the food industry has employed HPP to develop products that have the quality of fresh foods but an extended shelf life, without the use of preservatives. Areas of experimentation for industrial applications in the meat sector include: optimization of HPP conditions to inactivate target microorganisms for each product and commercial presentation, new packaging systems and combination with natural antimicrobial substances to enhance the shelf life extension, development of new meat products based on cold gelification of starches, HPP-thermal coagulation of proteins, selective enzymatic inactivation, meat separation. In recent years, HPP has satisfied the requirements of regulatory agencies such as the United States Department of Agriculture-Food Safety and Inspection Services (USDA-FSIS), which issued a letter-of-no-objection (LNO) in 2003 for the use of HPP as an effective post-packaging intervention method in controlling *Listeria monocytogenes* in ready-to-eat (RTE) meat and poultry products for US companies. Similar approval for the control of *L. monocytogenes* has been granted by other agencies. For example, Health Canada recently issued a similar LNO for the control of *L. monocytogenes* in cured and uncured RTE pork products. HPP is now commonly used by many US and Canadian processors to meet the FSIS requirements. In the European community, HPP foods are classified as "novel foods". Nevertheless, if a novel food can be shown to be substantially equivalent to a traditional food already on the market, it can be treated at a national regulation level without the need to adhere to the novel food regulation (Garriga et al., 2004). HPP has shown great potential, spreading throughout the world almost exponentially since 2000 (Fig. 1 A), especially in the vegetable and meat industries (Fig. 1 B). The most popular applications to meat-based products concern ready-to-eat, cooked and dry-cured meat products and seafood. This review summarizes research findings and practical concepts for the use of HPP as an effective technology to improve the safety of meat and seafood products while maintaining high quality and an extended shelf life.

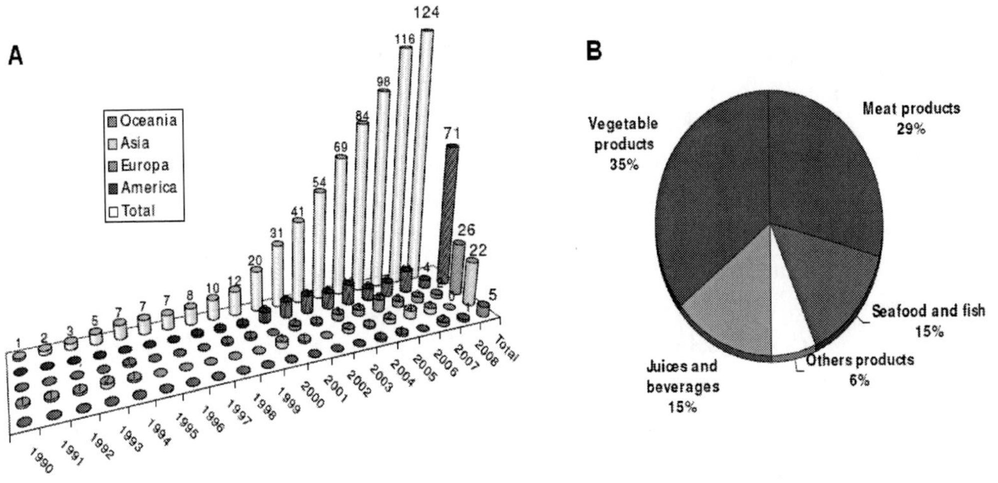

Figure 1 A. Evolution of HPP industrial machines installed on continents B. Industrial HPP machines number versus food industries. Courtesy of Mark de Boevere, NC Hyperbaric.

Governing Principles of High Pressure Processing

High Pressure Processing (HPP) is also referred to as High Hydrostatic Pressure (HHP) or Ultra-High Pressure Processing (UHP). The packaged food, usually under vacuum in a flexible package, is placed in a pressure vessel containing a pressure-transmitting liquid (water or aqueous solutions) and submitted to pressures ranging from 100 to 900 MPa, although the pressure level most often used is from 100 to 600 MPa depending on the product. (Jiménez-Colmenero and Borderias, 2003). The pressure is produced by a hydraulic pump (indirect system) or by a piston (direct system) and is isostatically transmitted inside the pressure vessel to the food product instantaneously and uniformly. In contrast to conventional processes such as thermal treatments, the process is independent of the product and equipment size and geometry because the pressure transmission is not mass/time-dependent, thus minimizing the treatment time. Chemical reactions and physical phenomena (breaking and formation of molecular interactions, ionization, phase transitions, reaction kinetics, etc.) are affected by HPP according to Le Chatelier's principle, which predicts that application of pressure shifts equilibrium to the state that occupies the smallest volume. Hence pressure favours reactions accompanied by a volume decrease, and vice versa (Heremans, 1982, Gross and Jaenicke, 1994). For example, pressure opposes to reactions such as transition from water to ice, resulting in the lowering of the freezing point with increasing pressure (Knorr et al, 1998). Hydrogen bond formation is stabilized by HPP (Heremans, 2002), along with the breaking of ions, as this leads to a decrease in volume, while covalent bonds are not affected (Norton and Sun, 2008). As a consequence, HPP modifies macromolecules such as proteins, inducing changes in their secondary, tertiary and quaternary structure, disrupting cell structures to some extent, affecting membrane proteins and lipid conformation (Kato et al., 2002), and inactivating enzymes (Butz and Tauscher, 2002). Most small molecules, such as vitamins and flavour compounds, are not affected, allowing preservation of the nutritional value and sensory appeal (Linton and Patterson, 2000). This is a major advantage of HPP with respect to conventional heat treatments and is highly appreciated by the food industry (McClements et al., 2001; Hoover et al.,1989; Smelt, 1998; Téllez et al., 2001). HPP also causes a temperature rise due to compressive work against intermolecular forces, known as adiabatic heating. The amount of the temperature increase in the treated food and in the pressure-transmitting medium depends on the food composition, pressurization rate (pressure ramp employed) and the geometry of the processing equipment (Hartmann and Delgado, 2002; Otero et al., 2007). In pressure-assisted thermal sterilization (PATS) the rise in temperature from adiabatic heating can be advantageous. The successful use of compression heating can result in reduction of processing time and, as a consequence, higher product quality and lower energy consumption. Use of compression heating could also be made to increase inactivation of microorganisms in food where an initial preheating top high temperatures was already achieved (Wilson et al., 2008).

Process Optimization

High pressure processing is applied to foodstuffs mainly to control microbial loads and/or enzymatic activity. Although the pressure applied to a food can be assumed uniform (Delgado, 2003), the technique cannot avoid temperature gradients inside the high pressure vessel. Different temperature-time profiles during the process in different locations of the pressurized food may result in non-uniform effect, which can be more or less pronounced depending on the pressure-temperature degradation kinetics of the examined component (Denys et al., 2000, Van der Plancken *et al.* 2008).

During processing, heat transfer takes place between components with different compression heating. When the compression heat of the high pressure vessel wall, of the pressure transmitting medium, of the packaging and of the product differ, based on the laws of heat transfer, temperature gradients may rise in different locations and at different times in the processed material (Grawet et al., 2010). Density differences within the pressurizing medium lead to a downward draft of fluid near the wall (if the walls are colder than the interior) and rising flow in the middle (free convection phenomenon) (Rauh *et al.* 2009, Khurana and Karwe 2009). Moreover, temperature gradients has been reported to rise as an effect of pressure medium addition in injection pressurizing systems, if the pressure medium heats up due to compression at the moment when additional pressure medium is being injected (forced convection phenomenon) (Abdul Ghani and Farid 2007, Khurana and Karwe 2009). Pressure is the predominant process parameter for the inactivation kinetics of vegetative cells in high pressure pasteurization applications; therefore non-uniformity of the process field is limited. However, studies on the inactivation of spores under high pressure – high temperature conditions, showed that temperature become a determining variable (Ju et al. 2008, Zhu et al. 2008, Barbosa-Canovas and Juliano 2008, Juliano et al. 2009). For process optimization purposes, the detection of the point of the lowest and higher impact is necessary to verify the effects of the treatment on safety and quality features. In HPP, the point of lowest impact only coincides with the minimum of the temperature field in the case of a synergistic effect between temperature and pressure. Different methods have been developed or are under development to monitor non-uniformities in HPP: 1) direct monitoring of temperature profiles inside the vessel 2) the use of enzymatic pressure-temperature-time indicators (pTTIs) 3) numerical simulation of the temperature distributions. In direct monitoring, thermocouples must be positioned across the whole volume of the pression vessel to demonstrate the whole temperature fields. Up to date, only wired systems are available, and this requires special attention especially to the points of sealing of the thermocouples' passage through the vessel wall. Moreover, sensors mustn't affect the free movements of the flow inside the vessel. Therefore, direct monitoring of the temperature of the whole volume inside the vessel becomes technically too complex on an industrial scale.

The ideal sensor to monitor non-uniformities during processing should show a pressure-temperature-time dependence, should, preferably, be easily and accurately measurable and do not disturb the actual process. Moreover, in order to be used for process impact evaluation on a specific target attribute, the pressure-temperature sensitivity of the indicator should match as closely as possible the pressure-temperature sensitivity of the target (Van der Plancken *et al.* 2008). In this respect, enzymatic pressure-temperature-time indicators (pTTIs) have been successfully applied to demonstrate the non-uniformity in a HP vessel (Denys et al., 2000, Grauwet et al., 2009). Every pTTI is characterized by its application window, which is defined as the pressure-temperature-time range in which the indicator characteristics show a clear pressure-temperature-time dependency (Grauwet et al., 2010). Grauwet et al., (2009) studied the suitability of *Bacillus subtilis* α-amylase to show non-uniformity by positioning the sensor at different axial and radial positions in a vertical single vessel system, demonstrating that indicators located at the bottom of the vessel and more closely to the vessel wall were less affected, and attributing higher residual activity to lower temperatures at specific positions. Since the pressure-temperature stability of enzymes is solvent dependent, solvent engineering, that is the change of the solvent in order to obtain the targeted sensitivity to the treatment, has been successfully applied (Grauwet et al., 2010), with the purpose to shift the range of the enzymes inactivation in different HP treatments range (intense HP treatment, mild HP treatment). Once a pTTI has been identified and tested for its p-T sensitivity and kinetic data are acquired, it must be validated under real process conditions and implemented in several applications. For a review on pTTIs see Van der Plancken *et al.* (2008). Numerical simulation allow to calculate the complete temperature and velocity fields within the vessel and food product the pressure and temperature dependent impact distribution of the high pressure process can be accurately obtained. Simulations are based on the conservation equations of mass, momentum and energy and the transport equation of chemical substances (Delgado et al., 2008). Based on these balance equations, the simulation compute the temperature changes due to added work of compression and conductive and convective heat transfer processes. This is coupled to the calculation of thermo-fluid-dynamic phenomena such as the resulting free and forced convection. The transport of the fluid through regions of different temperatures results in a treatment history of the fluid (e.g. the food) during the process. The numerical simulations need experimental information about the pressure and temperature dependent thermo-physical properties (i.e. thermal conductivity, viscosity, density, thermal capacity) of the treatment media (e.g. pressure transmitting medium, food product). Numerical models will be an essential tool to properly design uniform high pressure processes in terms of temperature control. The compression heating behaviour of foods has been studied recently and polynomial functions have been proposed to model the under-pressure fluid dynamics of pressure-transmitting media, liquid foods, fatty foods and oils, with the goal of maximum heating in a high pressure process or determination of the initial temperature required to reach a target temperature under high pressure conditions independently of sample size (Rasanayagam et al., 2003; Otero et al., 2007; Buzrul et al., 2008, Knoerzer et al., 2010).

The relative magnitude of heat transfer mechanisms (conduction and free convection) and overall scale of the system influence the uniformity of the treatment (Otero et al., 2007). Hartmann and Delgado (2002) used computational fluid dynamics (CFD) and dimensional analyses to determine the timescales of convection, conduction and bacterial inactivation, and their respective contribution to the efficiency and uniformity of conditions during HPP. In a pilot scale system, they showed that when processed fluids exhibit larger convection than inactivation timescale, intensive fluid motion and convective heat transfer result in more homogeneous bacterial inactivation, while non-uniformities in the inactivation process were dominant when the convection timescale were significantly smaller and the conduction timescale were significantly larger than inactivation timescales. For a comprehensive review on modelling and simulation of high pressure treatments see Delgado et al., (2008).

Filling ratio of the HP vessel influences the process uniformity. Convection heating is the predominant heat transfer mechanism when the filling ratio of the vessel is low, while heat transfer slows, and efficiency decreases, when large samples, with a high filling ratio inside the vessel, are processed. Otero et al. (2007) found that convective currents have least effect on heat transfer when this ratio is large. As a consequence, when the filling ratio is reduced, thermal re-equilibrium is reached sooner. The thermal properties of the pressure vessel boundaries, which are in contact with the pressure-transmitting medium, affect the uniformity of the process. Insulated materials with compression heating properties (then that the temperature of the insulation increases as the product is pressurized) prevent heat transfer from the product being treated to the surrounding medium and to the cooler pressure vessel wall, with a substantial increase in efficiency (Hartmann et al., 2004). Moreover, industrial-scale systems result in greater efficacy of bacterial inactivation than pilot-scale ones because compression heating persists for a longer time (Hartmann and Delgado, 2002; Otero et al., 2007). A strong coupling also exists between spatial concentrations of surviving microorganisms and low-temperature zones of packaging materials. In fact, low thermal conductive packages improve the uniformity of treatment, avoiding heat exchanges from the food to the pressure fluid, with up to a 2 Log cfu decrease per tenfold reduction of package thermal conductivity. For a comprehensive review of applied engineering aspects, see Norton and Sun, 2008.

HPP EQUIPMENT AND PROCESSES

HPP is primarily practiced as a batch process where pre-packaged food products are treated in a chamber surrounded by water or another pressure-transmitting fluid. Semi-continuous systems have been developed for pumpable foods where the product is compressed without a container and subsequently packaged "clean" or aseptically. The primary components of an HPP system include a pressure vessel; closure(s) for sealing the vessel; a device for holding the closure(s) in place while the vessel is under pressure (e.g., yoke); high-pressure intensifier pump(s); a system for controlling and monitoring the pressure and (optionally) temperature a product-handling system for transferring product to and from the pressure vessel. Normally, perforated baskets are used to insert and remove pre-packaged food products from the pressure vessels. Systems also have provisions for filtering and reusing the compression fluid (usually water or a food-grade solution) (Balasubramaniam, et al., 2008). For most applications, products are held for 3–5 min at 600 MPa. Approximately 5–6 cycles/hr are possible, allowing time for compression, holding, decompression, loading, and unloading. Slightly higher cycle rates may be possible using fully automated loading and unloading systems. After pressure treatment, the processed product is removed from the vessel and stored/distributed in a conventional manner. Liquid foods can be processed in a batch or semi-continuous mode. In the batch mode, the liquid product is pre-packaged and pressure-treated as described above for packaged foods. Semi-continuous operation requires two or more pressure vessels, each equipped with a free-floating piston that allows each vessel to be divided into two chambers. One chamber is used for the liquid food; the other for the pressure-transmitting fluid. The basic operation involves filling one chamber with the liquid food to be treated. The fill valve is closed and then pressure-transmitting fluid is pumped into the second chamber of the vessel on the opposite side of the floating piston. Pressurization of the fluid in this second chamber results in compression of the liquid food in the first. After an appropriate holding time, the pressure is released from the second chamber. The product discharge valve is opened to discharge the contents of the first chamber, and a low-pressure pump injects pressure-transmitting fluid into the second chamber, which pushes on the piston and expels the contents of the product chamber through the discharge valve. The treated liquid food is directed to a sterile tank from which sterile containers can be filled aseptically (Farkas and Hoover, 2000).

Avure Technologies (22408 66th Avenue South Kent, WA 98032, USA), NC Hyperbaric (C/ Condado de Treviño 59.Polígono de Villalonquéjar, 09001 Burgos – Spain), and Uhde (Friedrich-Uhde-Strasse 1544141 Dortmund, Germany) are major suppliers of commercials scale pressure equipment. Both horizontal and vertical pressure vessel configurations are available (commercial size from 30 to 600-liter capacity) for batch HPP equipment. Avure Technologies also make semi-continuous systems for the processing of liquid beverages such as juices. While commercial pressure vessels have the pressure limit of 700 MPa, research machines can go up to 1400MPa. A commercial scale high pressure vessel costs approximately between $500,000 to $2.5 million dollars depending upon the equipment capacity, and the extent of automation. Currently, HPP treatment costs are quoted as ranging from 4–10 cents/lb, including operating cost and depreciation, and are not "orders of magnitude" higher than thermal processing, as is often thought (Sàiz et al., 2008).With two 215-litre HPP units operating under typical food processing conditions, a throughput of approximately 9 thousand tons per year is achievable. High throughput is accomplished by using multiple pressure vessels. Factory production rates beyond 18 thousand tons per year are now in operation. As demand for HPP equipment grows, capital cost and operating cost will continue to decrease. Consumers benefit from the increased shelf life, quality, and availability of value-added and new types of foods, which are otherwise not possible to make using thermal processing methods (http://grad.fst.ohio-state.edu/hpp/faq.html).

EFFECT OF HPP ON MICROORGANISMS

Mechanism of Cell Inactivation

The inactivation of microorganisms by HPP is the result of a combination of factors (Simpson and Gilmour, 1997) including changes in the cell membranes, cell wall, proteins, and enzyme-mediated cellular functions. Cell membranes are the primary sites of pressure-induced damage, with consequent alterations of cell permeability, transport systems, loss of osmotic responsiveness, organelle disruption and inability to maintain intracellular pH. In a model system of protein and lipid membrane, Kato et al. (2002) observed a decrease in the lipid bilayer fluidity and a reversible conformational change of transmembrane proteins at pressures of 100 MPa or lower, leading to functional disorder of membrane-bound enzymes. At pressures of 100-220 MPa, there was a reversible phase transition in parts of the lipid bilayer, which passed from the liquid crystalline to gel phase; there was also dissociation and/or conformational changes of the protein subunits, which could cause separation of protein subunits and gaps between protein and lipid bilayer, creating transmembrane tunnels. A pressure of 220 MPa or higher irreversibly destroyed and fragmented the gross membrane structure due to protein unfolding and interface separation, which was amplified by the increased pressure. The presence of a cell wall does not mean that pressure resistance is enhanced; indeed, Ludwig et al. (2002) suggested that pressure may induce mechanical stresses on the microbial cell wall which, in turn, may interact with inactivation mechanisms. Bud scars, nodes to the cell wall and separation of the cell wall from the membrane were observed by Ritz et al. (2001) and Park et al. (2001) with electronic microscopy. Moreover, models proposed to define the mechanical behaviour of cells under pressure predicted heterogeneous mechanical stresses under high hydrostatic pressure (Hartmann and Delgado, 2004; Hartmann et al., 2006). Protein denaturation and changes in the active centres have also been observed, together with changes in enzyme-mediated genetic mechanisms such as replication and transcription, although DNA itself is highly stable due to the fact that α-helical structures are supported by hydrogen bonds. The inactivation by HPP depends on the type of microorganism and its growth phase, the pressure applied, the processing time, the composition of the food, temperature, pH and water activity (Tewari, Jayas, and Holley, 1999). In general, it is assumed that Gram-negatives and cells in the growth phase are more sensitive than Gram-positives and cells in the stationary phase, respectively. Nevertheless investigations have shown that cell disruption is highly specific to the geometry of the bacteria rather than to the Gram type. Ludwig and Schreck (1997) reported morphological changes for the rod-shaped *Escherichia coli* and *Pseudomonas aeruginosa,* whereas *Staphylococcus aureus* (cocci) was more resistant to pressure. On the other hand, Schreck et al. (1999), working with *Mycoplasma pneumoniae,* found no correlation between Gram type and pressure sensitivity, while there was a correlation between cell shape and pressure sensitivity.

Hurdles technology

Application of the hurdles technology concept has been proposed as an approach to increase the microbicidal effect of the process at lower pressures. Hurdles technology relies on the synergistic combination of moderate doses of two or more microbe-inactivating and/or growth-retarding factors. The use of antimicrobials, such as bacteriocins and lysozyme (Hauben et al., 1996; Kalchayanand et al., 1998; Masschalck et al., 2001; Garriga et al., 2002), has been shown to have a synergic effect on bacterial inactivation. For example, Gram-negative bacteria such as *E. coli* or *Salmonella,* which are normally insensitive to bacteriocins of lactic acid bacteria as they lack specific receptors, can be sensitized to nisin or other bacteriocins when pressurized (Kalchayanand et al., 1994). Mechanisms of transient and persistent sensitization of bacteria to antimicrobial compounds by high pressure in buffer systems have also been described (Masschalk et al., 2001).

Effect of Processing Conditions and Food Composition on Microbial Inactivation and Survival Following HPP Treatment

As a general rule, cell death rate increases with increasing pressure but it does not follow a first order kinetics and a tail of inactivation is sometimes present (Garriga, et al., 2002; Kalchayanand, er al., 1998). Moreover, temperature plays an important role in microbial inactivation by HPP. At optimal growth temperatures, inactivation is less than at higher or lower temperatures of growth because membrane fluidity can be more easily disrupted at no optimal growth temperatures (Smelt, 1998). The nature of growth media can also affect the pressure resistance of the microorganisms (García-Graells, Masschalck, and Michiels, 1999). Therefore, inactivation experiments conducted in buffers or synthetic media cannot always be extrapolated and applied to real situations. Archer (1996), reported that in real food situations the microbial safety and stability are determined by the effect of food composition both during and after the HPP treatment. In fact, bacterial survival after HPP can be greatly increased when treated in nutritionally rich media, e.g., meat, containing substances like carbohydrates, proteins, and fat (Simpson and Gilmour, 1997) that showed a protective effect. Patterson et al., (1995) reported the different sensitivity of *L. Monocytogenes* and *E. coli* O157:H7 when treated in poultry meat and buffer systems. The same treatment reduces *E. coli* O157:H7 in 6 log CFU in buffer while only 2.5 log in poultry meat. A low water activity (a_W) protects microorganisms against pressure and even at the same a_W the solute is important; in glycerol they are more sensitive than in mono-o-disaccharide while trehalose has a protective effect (Smelt, 1998). Resistant or sub lethally injured cells could be able to grow during storage (Chen and Hoover, 2003; Garriga, et al., 2002; Patterson et al., 1995). In this respect, tests in real food matrices followed during the shelf life of the product should be recommended to assure the safety of the product.

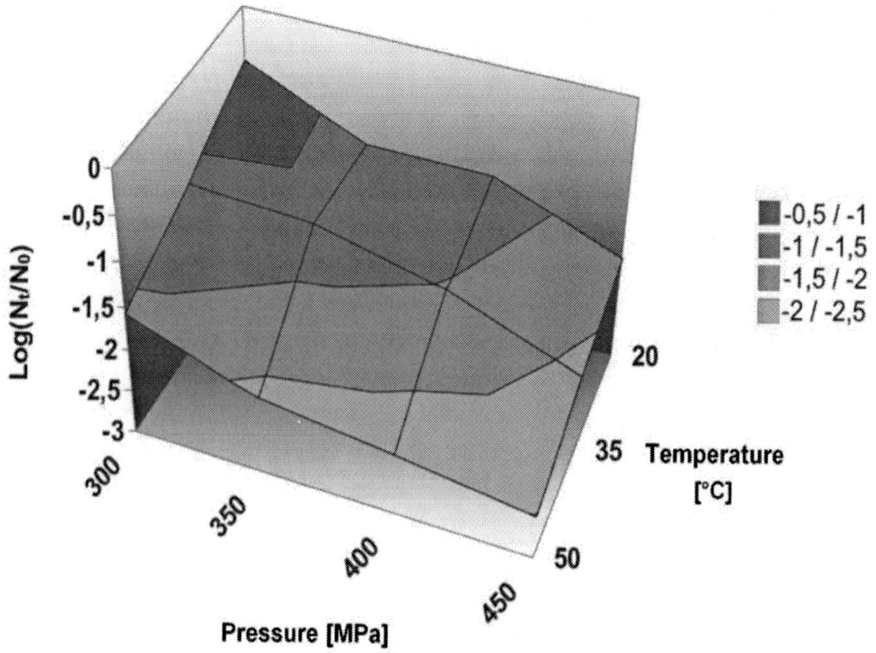

Figure 2 Combined HP-thermal treatment of sea urchin gonads. Logarithmic reduction of Total Aerobic Count following High Pressure treatment.(N_t=cfu/g after treatment; N_0=cfu/g before treatment).

HPP Pasteurization and Sterilization

Several studies have reported the antimicrobial effect of HPP in meat products and the results are summarized in Figure 3. For pasteurization, treatment is in the range of 300-600 MPa for a short period of time, which inactivates the vegetative pathogenic and spoilage microorganisms (>4 Log units). Nevertheless, response of pathogenic bacteria to HP treatments is variable, and depends on the temperature applied. In fact, it has been observed that bacteria exhibit the biggest pressure resistance at temperatures between 20 and 30 °C (Fig. 2). For example, studies on the inactivation of *E. coli* O157:H7 in poultry meat showed a 1 log decimal reduction, when the product is treated at 400 MPa and 20°C for 15 minutes. The same results as for 50°C heat treatment alone. When treatments at 400 MPa are combined with a temperature of 50°C, a 6 log reduction was achieved (Patterson and Kilpatric, 1998). The greatest challenge in the use of high pressure is the inactivation of bacterial spores. Differences in response to pressure between different species, and between strains of the same species, are frequent (Heinz and Knorr, 2002). For examples, spores of *Clostridum sporogenes* in fresh chicken breast required a pressure of 680 MPa to 1 hour to achieve a relevant (5 log) inactivation (Crawford et al., 1996), while other workers found that a 1500 MPa treatment of *C. Sporogenes* in liquid media led only to a 1.5 log reduction (Maggi, et al., 1996). Spores of *Bacillus subtilis,* a food borne pathogen associated mainly with meat or vegetables in pastry, cooked meat or poultry products, are thought to be susceptible to pressure induced germination by use of pressure between 100-600 MPa (Wuytack et al, 1998). Another inactivation treatment (heat shock, pressure cycling and the application of germinating agents, etc.) must be used to obtain significant inactivation of spores (Furukawa, Nakahara, and Hayakawa, 2000; Kalchayanand et al., 2004) and hurdles (low pH, low a_W, low temperature, antimicrobial substances) must be placed to prevent the outgrowth of surviving spores (Smelt, 1998; Stewart et al., 2000). Pressure induced germination may enable an inactivation of spores by mild heat or pressure treatment. However this concept cannot reliably be adopted commercially due to the distribution in the variability of the effects of high pressure on spore germination.

Among food borne pathogens associated to meat and poultry consumption, *C. perfringens* type A is of major concern, ranking as the third most common food-borne illness (McClane, 2007). *C. perfringens* showed to be pressure resistance, and pressures of 100-200 MPa have a negligible effect on spores germination. Recently, a strategy has been successfully employed to induce germination of spores and subsequently inactivate the bacteria by High Pressure. The strategy has been developed on poultry meat (Akhatar, et al., 2009) and consisted of: 1) a primary heat treatment (80°C, 10 min.) to pasteurize and denature meat proteins and to activate *C. Perfringens* spores for germination, 2) cooling of the product to 55°C in about 20 minutes and further incubation at 55°C for 15 minutes for spore germination 3) inactivation of germinated spores by pressure assisted thermal processing (586MPa at 73°C for 10 minutes). The efficiency of the strategy requires the bioavailability of L-asparagine and KCl necessary for rapid spore germination (Paredes-Sabja, et al., 2007)

The practical feasibility of HPP to inactivate unconventional pathogens such as prions and agents of Transmissible Spongiform Encephalopathies (TSE), such as scrapie and BSE, has also been investigated. Prions are highly resistant to pressure and 700-1000 MPa at high temperature are needed to reduce infectivity. Pressure treatments at 60°C for 2 h (Fernández-García et al., 2004) are needed to reduce the survival rate over the infected meat product by 47%. At 60–80°C, an efficient pressure inactivation of infectious scrapie prions PrPres was observed during short pressure treatments at 800 MPa (3 × 5 minute cycles) (Heindl et al., 2008). Discrepancies between *in vivo* infectivity counts and the results of enzyme immunoassays revealed that the infectivity was inactivated faster and much more efficiently than PrPres was degraded, and researchers concluded that pressure affected a highly infectious subpopulation of scrapie prions. In mechanically recovered meat products (hot dogs) artificially infected with BSE prions, Brown et al. (2003) found that infectivity levels (assayed by western blots of PrPres) were significantly reduced compared with untreated controls: from $\approx 10^3$ LD_{50} per g at 690 MPa to $\approx 10^6$ LD_{50} per g at 1200 MPa, with a running temperature of 121-137°C. They concluded that application of commercially practical conditions of temperature and pressure could ensure the safety of processed meats against bovine spongiform encephalopathy contamination. The same authors (Cardone et al., 2006) reported a reduction of the level of infectivity of prions from 10^3 to 10^6 mean lethal doses (LD50) per gram of tissue after a combined pressure-temperature treatment, whereas autoclave, alkali and bleach treatments had been ineffective.

Target microrganism	Product	Initial counts (cfu/g)	Cell load reduction (Log_{10} cfu/g)	Treatment	Reference
L. monocytogenes	Cooked ham	2.6	1.9, after 42 days at 6°C	400 MPa,10 min.	Aymerich, et al. (2005)
Total Viable Count	Bovine meat (*Biceps femoris*)	4.0	2.5, after treatment. Counts reached 4.0 Log after 4 weeks at 4°C	520 MPa, 4.2 min. 10°C	Jung, et al. (2003)
L. monocytogenes	Dry Cured Ham	4.65	Total Inactivation (<1 cfu/g)	600 MPa, 9 min.	Tanzi, et al. (2004)
S. aureus	Marinated beef Cooked ham Dry Cured ham	3.62 3.70 2.74	2.67, after treatment. 1.12, after treatment. 0,55 after treatment.	600 MPa, 6 min., 31°C	Hugas, et al. (2002)
Salmonella spp.	Cooked ham	3.8	Total Inactivation (<1 cfu/g), after 120 days, 4°C	600 MPa, 6 min., 31 °C	Garriga, et al. (2002)
E. coli O157:H7	Raw minced meat	5.9	5, after treatment	700 MPa, 1 min., 15 °C	Gola, et al. (2000)
L. monocytogenes	Sliced beef cured ham	4.0	2, after 210 days , 6°C	500 MPa, 5 min, 18 °C	Rubio, et al. (2007)
C. freundii *P. fluorescens* *L. innocua*	Minced beef muscle	7	5, after treatment, 20°C	300 MPa 10 min, 20 °C 200 MPa, 20 min, 20 °C 400 MPa, 20 min, 20 °C	Carlez, et al. (1993)
L. monocytogenes	Iberian ham	6.3	3.6, after 60 days, 8°C	450 MPa, 10 min, 12 °C	Morales, et al. (2006)
Lactic Acid Bacteria	Marinated beef Cooked ham Dry Cured ham	4.94 5.63 4.23	3,99, after treatment 4,57, after treatment 1,58, after treatment	600 MPa, 6 min., 31°C	Hugas, et al. (2002)
Campylobacter jejuni	Pork slurry	6-7	6, after treatment	400 MPa, 25min., 10°C	Shigehisa, et al. (1991)
Toxoplasma gondii	Ground pork meat	Viable tissue cysts	Inactivation (cysts not viables)	300 MPa	Lindsay, et al. (2006)

Figure 3. Microbial inactivation by HPP in meat products.

EFFECT OF HPP ON COLOR

The colour of meat depends on the amount and type of myoglobin, although other proteins such as haemoglobin and cytochrome C may also play a role in beef, lamb, pork, and poultry; in addition, the optical (scattering) properties of the meat surface can influence colour evaluation. Myogloblin contains a prosthetic group, the heme ring, with a centrally located iron atom. The ligand present (oxygen, carbon dioxide, carbon monoxide) and the valence of iron dictate muscle colour (Mancini and Hunt, 2005). Moreover, the colour of dry-cured products is mainly due to the presence nitrosylmyoglobin (meat products with added $NaNO_2$ and KNO_3), and metmyoglobin. Defaye et al. (1995) showed that high pressure treatment of myoglobin caused partial denaturation, but the process was reversible. It is also known that the effect of high pressure treatment on myoglobin solutions depends on the temperature at which the pressure treatment occurs. Carlez et al. (1995) reported that meat discolouration in pressure processed meat was due to: (1) a whitening effect due to myoglobin denaturation and/or to heme displacement or release, and (2) oxidation of the ferrous myoglobin to ferric myoglobin above 400 MPa. Cheah and Ledward (1997) improved the colour stability with the application of pressure of 80-100 MPa for 20 minutes, as measured by the rate of metmyoglobin formation in beef muscles post-slaughter. However, pressure treatment of these muscles at 7 to 20 days post-slaughter did not improve their colour stability. Cheftel and Culioli (1997) suggested that pressure processing of fresh red meat causes drastic changes, especially in redness, and thus cannot be suitable of practical applications. In contrast, pressure processing of cured meat or white meat is unlikely to cause problems in this respect.

Studies on dry-cured ham reported an increase in lightness (measured as CIE L* parameter) and a decrease in redness (CIE a* parameter) when the ham was pressurized (Andres et al., 2004; Tanzi et al., 2004). However, Serra et al. (2007) only reported higher lightness in several muscles of dry-cured ham, while Marcos et al. (2007) found no effect on colour after pressure treatment at 400 MPa of low acid fermented sausages. Campus et al. (2008) reported changes during storage of pressurized dry-cured loin under vacuum at refrigerated temperatures. The a* and L* values showed a significant reduction after 2 days of storage in all treatments, except at 300 MPa, and the a* values were maintained during the storage time. However, the L* parameter showed a significant reduction in all treatments after 10 days of storage. In addition, the differences observed among treated and control samples (higher lightness and decreased redness in pressurized samples) were maintained during the vacuum storage. Andres et al. (2004) studied the changes in a modified atmosphere and reported an increase of lightness and a loss of red colour of pressurized and non-pressurized dry-cured ham during storage, which stabilized with time. In summary, studies indicate that HPP induces drastic changes in fresh meat, while the changes observed in dry-cured meat products are negligible in terms of acceptability.

Studies carried out on fresh fish indicate marked changes in appearance at higher pressure, to the point that fish no longer appears fresh (Fig. 4), with slight differences between species. Colour changes are usually marked by an increase in lightness (L*) and yellowness (b*) values along with a decrease of redness (a*), the former more marked in species with a higher proportion of red muscle (Oshima et al., 1993; Hurtado et al., 2000;

Amanatidou et al., 2000; Chevalier et al., 2001). On the other hand, Gòmez-Estaca et al. (2007) reported an increase of all CIE L*a*b* parameters in cold-smoked dolphinfish.

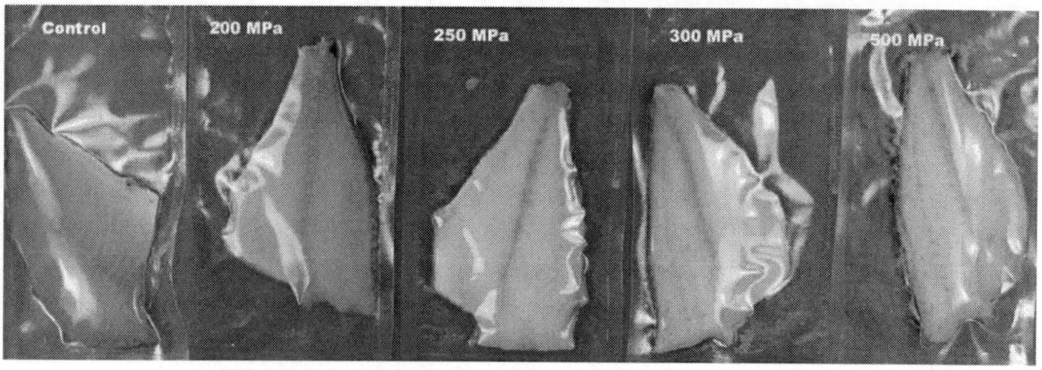

Figure 4. Color changes in Sea Bream (*Sparus aurata* L.) fillets following high pressure treatment.

EFFECT OF HPP TREATMENT ON ENDOGENOUS ENZYMES RELATED TO QUALITY

Severe biochemical changes take place during postmortem in muscle cells. During anaerobic glicolisys, glicogen reserves are depleted, ATP gradually hydrolyze as a consequence of pH fall due to ionic pumps failure, actin and myosin bind to form an irreversible acto-myosin complex, with the onset of rigor. Upon rigor onset, muscle elasticity decreases and at its completion, the tissue reaches its maximum toughness. Post-rigor tenderization occurs within hours, and differences can be observed depending on muscle fibers type, muscle type, individual animals and species, (Sentandreu, Coulis and Ouali, 2002), the main determinant of ultimate tenderness being the extent of proteolysis of key target proteins within muscle fibres (Koohmaraie and Geesink, 2006; Taylor, et al., 1995). Moreover, enzymes are involved in flavour and taste development in dry cured meat products (Toldrà and Flores, 1998).

Myofibrillar proteins, such as actin, myosin, tropomyosin, troponin T, nebulin and titin, along with cytoskeletal desmin and costamere vinculin, are subjected to cleavage by proteolytic enzymes post mortem. Little or no changes are reported in connective tissue as a consequence of pressure treatment (Suzuki, et al., 1993). Candidate systems responsible of muscle proteins degradation has been identified in different eso and endoproteases. The most studied are the calpain/calpastatin system and the lysosomal cathepsins. Also, caspases, a family of cysteine aspartate-specific proteases, along with the proteasome complex, have been involved in muscle tenderization, although their role is still controversial (Kemp, et al., 2010).

Calpains are the most extensively researched proteases with regard to meat science and it is widely accepted their contribute to meat tenderisation (Sentandreu et al., 2002; Koohmaraie and Geesink, 2006). Calpains are a large family of citoplasmatic cysteine Ca^{2+}- dependent proteases; in skeletal muscle, calpains (m, μ and p94) form a system with their specific endogenous inibitor, calpastatin. Calpains are able to degrade myofibrillar proteins including nebulin, titin, troponin-T and desmin (Huff-Lonergan et al., 1996.). Degradation of myofibrils by calpains have been correlated with post-mortem proteolysis and meat tenderisation by some authors (Geesink, et al., 2006). Data on the effect of HPP on calpains are lacking. Chéret et al., (2006) reported that m and μ calpain from sea bass subjected to high pressure treatment are readily inactivated at 300 MPa, probably due to structural modification and dissociation of calpain subunits. Evidence suggest that inibition of calpain activity by high pressure avoid degradation of cytoscheletal proteins, such as desmin, reducing the water holding capacity of muscle, namely, increasing the drip loss. The ability of muscle to retain water is strictly related to the post-mortem events such as pH decline, proteolysis and protein oxidation (Huff-Lonergan, et al., 1996), and it is very important in fish both from a quality, nutritional and, consequently, commercial point of view.

As rigor progresses, the space for water to be held in the myofibrils is reduced, and fluid can be forced into the extra-myofibrillar spaces where it is more easily lost as drip as a consequence of lateral shrinkage of the myofibrils occurring during rigor, which can be transmitted to the entire cell if proteins that link myofibrils together and myofibrils to the cell membrane (such as desmin) are not degraded (Kristensen and Purslow, 2001; Melody et al., 2004). Desmin is a known calpain substrate. HPP treatements at 300 and 400 MPa of sea bream muscle (Campus et al., 2010), were associated to reduced degradation of desmin, correlated to a decreased water holding capacity.

Similar results were obtained in enhanced pork loins (Davis et al., 2004), where reduced degradation of desmin has been correlated with a minor retention of fluids by muscle.

Cathepsins are a group of enzymes comprised of both exo and endo-peptidases. The cathepsins known to be expressed in muscle tissue include six cysteine peptidases (cathepsins B, L, H, S, F and K) and one aspartic peptidase, i.e. cathepsin D (Sentandreu et al., 2002). They are located in lysosomes and are mostly active at acidic pH. Being located inside lysosomes, their role in meat tenderization is on debate, although free cathepsin activity, especially that of cathepsins B, H and L, has been correlated with meat tenderness from early postmortem to the end of the ageing period (Calkins and Seidman, 1988; Johnson, et al., 1990). During post-mortem, lysosomal enzymes become accessible to muscle structural proteins. This is due to the progressive disruption of lysosomes through membrane breakdown, which occurs by the decrease in pH at high temperature post mortem (Moeller, et al., 1977) or by the failure of ionic pumps in lysosomal membranes during rigor development, consecutively to the depletion of ATP stores (Hopkins, 2000).

Ohmori et al. (1991) indicated that the application of high pressure in fresh meat produces an enhancement of the endogenous cathepsin proteolytic activity participating in meat conditioning, probably due to the release of proteases from lysosomes to the cytoplasm and by the denaturation of the tissue protein. This permeabilisation, or even disruption of the lysosomal membrane, has been observed in model systems (Kato, et al., 2002) or directly by microscopy (Jung, et al., 2000) and results in a higher proteolytic activity in pressurised samples. Kurth (1986) reported a retention of activity of cathepsin B (in solution) under pressures (150 MPa), and even an enhancement in some pressure-heat combinations. Homma et al. (1994) studied the effect of high pressure in bovine muscle (100–500 MPa, 5 min) and found an increase in activity of cathepsins B and L and inactivation of aminopeptidase B, also named as RAP, and cathepsin H, an aminoendopeptidase. These authors also measured the activity of these enzymes in crude extracts to determine the pressure effect on the enzyme itself and reported that all the measured enzymes lost activity as the applied pressure was increased; cathepsins B and L decreased gradually with increasing pressure. In dry cured meat products, were the breakdown of the lysosomal membrane has supposedly already taken place as a consequence of the pH decrease during post-mortem metabolism, other autors (Campus et al., 2007) observed a reduction of activity of catepsins B and L at pressure of 400 MPa, along with the loss of activity of other enzymes (aminopeptidases and dipeptidylpeptidases). Research findings indicate that the magnitude of proteolytic activity of lysosomal proteases following high pressure treatments is the balance of two contrasting phenomena: the release of cathepsins from the lysosomes due to pressure-induced membrane damage and the inactivation of the released enzymes by pressure.

Proteolytic enzymes,which are related to fish spoilage, are generally more susceptible to HPP than their mammalian counterparts, since fish are adapted to cold habitats and their enzymes tend to have a more flexible structure (Low and Somero, 1974). HPP treatment of Sheephead and Bluefish Cathepsin C resulted in a near total inactivation after treatment at 300 MPa for 30 minutes, were on bovine Cathepsin C treatment showed little or no effect.

In conclusion, the effect of high pressure on enzyme activity will depend on the enzyme itself, on the nature of the medium (substrate availability, pH, ionic strength etc.) and on the processing conditions (pressure, temperature, time), and this will affect texture and taste of the product.

HPP Impact on Texture

The effect of HPP on the texture of muscle foods has been investigated since 1973, when Macfarlane (1973) reported the potential use of HPP for pressure-induced tenderization of meat. Since then, many authors have reported that pre-rigor treatment for a few minutes at 100-200 MPa induces meat tenderization (Elgasim and Kennic, 1980; Ohmori et al., 1991). On the other hand, high pressure post-rigor treatments of beef muscle have no beneficial effects, whereas combined pressure-heating treatments (150 MPa, 55-60°C, 30 minutes) are effective in contrasting cold-shortening effects due to pre-rigor excision combined with exposure to low temperatures (Bouton et al., 1977; Macfarlane, 1985), even though they result in brown discolouration. Post-rigor meat tenderization without browning discolouration can be achieved using higher pressure (up to 300 MPa) for a few minutes without heat treatment.

Other authors have investigated textural changes induced by high pressure treatments in fish. Ashie and Simpson (1996) performed a puncture test and reported a decrease of ‘‘strength values” when blue fish was subjected to pressure above 200 MPa for over 10 minutes and also in fish treated at pressure above 300 MPa. The authors also reported a decrease of elasticity with increasing pressure just after treatment. Campus et al. (2010) reported a decrease in elasticity of sea bream muscle with increasing pressure just after treatment, but the elasticity was maintained during storage in samples treated at higher pressures. The phenomenon has been related to reduced degradation of cytoskeletal proteins (assayed by western blotting) due to blockade of proteolytic activity by HPP.

Pressure-induced texture modifications have been used to affect myofibrillar proteins and their gel-forming properties, raising the possibility of the development of processed muscle-based food. As previously stated, high pressure can affect molecular interactions (hydrogen bonds, hydrophobic interactions and electrostatic bonds) and protein conformation, leading to protein denaturation, aggregation or gelation (Messens, Van Camp, and Huyghebaert, 1997). Depending on the muscle source and other parameters such as protein concentration, pH and ionic strength, various changes occur in myofibrillar proteins depending on the pressure-temperature conditions (Acton and Dick, 1984; Hamm, 1981; Sano, 1988). Jiménez Colmenero (2002) reviewed pressure-assisted gelation of muscle protein systems. As an example, pressurization (100-500 MPa, 10 minutes, 0°C) has been found to favour the formation of structures with greater breaking strength in gels from fish meat mince, Alaska pollack surimi (above 200 MPa) and minced chum salmon meat (Okazaki et al., 1997).

Effect of HPP on Lipid Oxidation

HPP seems to promote lipid oxidation in meat products. Studies reported a generally more rapid increase in thiobarbituric acid reactive substances (TBARS) values in pressurized minced meat (Cheah and Ledward, 1996), deboned turkey meat (Tuboly, et al., 2003) and chicken breast muscle (Orlien, Hansen, and Skibsted, 2000), compared to control samples. Catalysis of lipid oxidation seems to take place during pressurization and has been related with the release of non heme iron and membrane damage. Cheah and Ledward (1996), Cheah and Ledward (1997) have reported that the effect of HPP on oxidative stability of lipids in pork meat depends on the applied pressure, with a value between 300 and 400 MPa constituting the critical pressure to induce catalysis. They also reported that denatured forms of proteins play an important role in catalyzing lipid oxidation. On the other hand, Orlien and Hansen (2000) reported that lipid oxidation at higher pressure was not related to the release of non-heme iron or catalytic activity of metmyoglobin, but could be linked to membrane damage. Cheftel and Culioli (1997), indicated that pressure-induced oxidation may limit the usefulness of this technology for meat-based products unless the use of oxygen free packaging or adding antioxidants. Removing oxygen or adding carbon dioxide prior to pressurization may be useful to prevent the pressure-induced lipid oxidation. Few studies have reported the effect of pressure treatment on TBARS in dry cured products. Andres et al., 2006 reported significantly higher values of TBARS in dry-cured ham treated at 400 MPa and stored for 39 days in a modified atmosphere with 5% residual oxygen, indicating a decrease in oxidative stability during storage. Marcos et al. (2007), found no differences in TBARS values in pressurized fermented sausages at 400MPa, indicating that most of the lipid oxidation had already occurred during ripening. This agrees with the results of Campus et al. (2007) where no differences in TBARS values were detected among samples of pressurized fry cured loin (400 MPa, 10', 20°C) after 45 d of vacuum storage. Markedly, oxidation of lipids is a key event for the development of aroma compounds in dry cured products (Toldrà and Flores, 1998). Cod muscle pressurized at 200 to 600 MPa for 15 to 30 minutes caused an increase in lipid oxidation measured as peroxides value. In mackerel lipid oxidation was even more marked (Ohshima et al., 1993). When a mix of sardine oil and defatted sardine meat was treated at 100MPa for 30-60 min, the peroxide value and TBARS value of samples in cold storage increased more rapidly with processing time than did the control samples (Tanaka et al., 1991). However, when sardine oil was treated alone up to 500 MPa, oxidation was minimal. It was concluded that fish oil oxidation was accelerated by pressure treatment in the presence of fish muscle (Wada, 1992). This can be related to the catalyzing power of metal ions present in fish meat.

COMMERCIAL APPLICATIONS OF HPP ON MEAT AND SEAFOOD

Minimally Processed and Cooked Meat Products

The application of HPP to fresh meat products is limited by the resulting discolouration, as previously stated, but it remains a powerful tool to control risks associated with *Salmonella* spp. and *Listeria monocytogenes* in minimally processed and dry-cured meat products. Murano et al. (1999) obtained a 10 Log_{10} reduction of the most resistant strain of *L. monocytogenes* in fresh pork sausage with a treatment of 400 MPa at 50°C for 6 min. The efficacy of treatment against spoilage microorganisms resulted in a shelf life extension of 23 days in storage at 4°C, with no substantial impact on the sensory qualities. Garriga, Aymerich and Hugas (2002) showed that HPP treatment could extend the shelf life of marinated beef loin by controlling the growth of spoilage and pathogenic bacteria. After vacuum skin-packaged sliced marinated beef loin was treated by HPP (600 MPa, 6 minutes, 31°C), the aerobic, psychrophilic and lactic acid bacteria counts showed at least a 4 Log_{10} reduction after treatment and they remained below the detection limit during the chilled storage of 120 days, helping to prevent off-flavours. In contrast, untreated samples reached 10^8 cfu/g after 30 days in the same conditions. Commercial applications of HPP to processed meat products include several ready-to-eat pork and poultry products, as summarized in Figure 5.

Country (year)	Product	Process	Packaging	Shelf-life	Achievements of high pressure and comments
Spain (1998)	Cooked Sliced ham and "tapas" (pork and poultry cuts)	400 MPa 10 min at 8 °C	Vaccum packed and MAP	2 months	Sanitization with minimum color and taste modifications
USA (2001)	Cooked Sliced ham, pork meat products and Parma ham	240 MPa, 90 s	Vacuum packed		Sanitization with minimum color and taste modifications. *Listeria* destruction
USA (2001)	Poultry ready-to-eat products		MAP		Sanitization with minimum color and taste modifications. *Listeria* destruction
USA (2002)	Spicy sliced precooked chicken and beef for fajitas		MAP	21 days	Sanitization with minimum color and taste modifications. *Listeria* destruction. Fajitas are made of HPP meat, onions, peppers and guacamole.
Spain (2002)	Sliced ham, chicken and turkey products. Cooked and Serrano ham, Chorizo.	500 MPa, 4to 10 minutes at 8 °C	No	2 months for cooked products	Sanitization with minimum color and taste modifications. *Listeria* destruction, Increase of shelf life and additives reduction.
Italy (2003)	Parma ham, salami, mortadella		Vaccum packed		Sanitization with minimum color and taste modifications. *Listeria* destruction. Products for USA and Japan export.
Japan (2005)	Cooked pork meat products, nitrite free: ham, sausages, becon		Vacuum packed	4 weeks	Shelf-life increase, sanitization.
Germany (2005)	Smoked german ham: whole, sliced and diced products		Vacuum packed		Sanitization. *Listeria* destruction. Products for USA export

Figure 5. HPP meat products avaiable in the market.

Dry Cured Meat

Microorganisms in commercial dry-cured products are mainly present on the surface and reach the sliced product during slicing and packaging operations. Moreover, the operations of boning, sectioning, slicing, involves the risk of contaminations by pathogens.

Tanzi et al. (2004) investigated sensory and microbiological properties of dry cured Parma hams treated with high pressure. High-pressure treatment (9 minutes at 600 MPa) allowed reduction of *Listeria monocytogenes* to negligible levels in samples. Treatments affected color (slight decrease in CIEL a* parameter, redness) and saltiness (enhanced perception), with changes inversely related to the age of the ham.

Sliced, skin vacuum packed dry cured Spanish ham samples, treated by HPP at 600 MPa for 6 min., showed a significant reduction of at least 2 Log_{10} cycles for spoilage associated microorganisms after treatment. The surviving microorganisms were kept at low levels during the storage period; contributing to the preservation of the organoleptic freshness during shelf life (120 days) and helping to prevent off-odours and off-flavours. *Listeria monocytogenes* was present (in 25 g) in one untreated sample, but absent in all HPP treated samples during the whole storage period (Garriga et al., 2002). The retention of quality characteristics of HPP treated dry cured products during chilled storage has been investigated by some authors (Rubio, et al. 2007; Serra et al., 2007) Deterioration of sensorial qualities in treated ham (500 MPa, 5 min) occurred during storage, limiting the shelf life to 90 days. HPP treated, dry cured products, sliced and packed under vacuum, are actually commercialized by industries, mainly for export purposes. Thus, HPP offers the possibility of implementation of commercial commodities and products portfolio of meat industries (Fig 5).

Seafood

HPP is successful in killing *Vibrio parahaemolyticus* and *V. vulnificus* in oysters without compromising their sensory attributes (Lopez-Caballero et al., 2000; He et al., 2002; Cook et al., 2003; Kural and Chen, 2008). In the United States, the presence of *Vibrio vulnificus* in molluscan shellfish causes the highest fatality rate among food-borne pathogens (Cook et al., 2003). Pressures of 205–275 MPa at temperatures of 10-30°C and treatment times of 1-3 minutes are typically used for raw oysters. For a 5-Log reduction of pressure-resistant strains of *V. parahaemolyticus* in live oysters, the pressure treatment needed to be ≥350 MPa for 2 minutes at temperatures between 1 and 35°C and ≥300 MPa for 2 minutes at 40°C (Kural et al., 2008). The product maintained the sensory characteristics of fresh oysters for an extended shelf life.

Recently the State of California recognized HPP as a valid process to reduce pathogenic *Vibrio* bacteria in fish products. Moreover, in a study of the hepatitis A virus and calicivirus, Kingsley et al. (2002) reported that HPP has the potential of making raw shellfish free of infectious viruses. HPP also denatures the oyster's adductor muscle and induces the shell to open spontaneously (Fig. 6). HPP shucking reduces the need for manual shucking and increases the quantity of meat removed from the shell (Murchie et al., 2005). A heat shrink plastic band is placed around each oyster shell prior to the HPP treatment to keep the shell closed during distribution and storage. However, changes in body colour and other descriptive characteristics were observed at higher pressures. The optimum pressure for oyster shucking (loosening a high percentage of adductor muscles but causing minimal changes) may vary with the oyster species and growing conditions, and also would need to be determined for individual processors (He et al., 2002).

Country (year)	Product	Process	Packaging	Shelf-life	Achievements of high pressure and comments
USA (1999)	Oyster Sauce for oyster dish	200 to 350 MPa 1 to 2 min	No packaging (Plastic band)	10 to 15 days	Opening of the shells (kept closed by a plastic band). Destruction of *Vibrio vulnificus*. Marketing of fresh and frozen opened oysters.
USA (2001)	Oysters	240 MPa, 90 s	No packaging (Plastic band)		Opening of the shells (kept closed by a plastic band). Destruction of *Vibrio*.
USA (2001)	Oysters				Opening of the shells (kept closed by a plastic band). Destruction of *Vibrio*.
USA (2001)	Oysters				Opening of the shells (kept closed by a plastic band). Destruction of *Vibrio*.
Canada (2004)	Seafood		No		Opening of the shells.
Spain (2004)	Ready to eat fishes: Salomon, Hake		Skin vacuum packed	2 months	Reconstituited sanitized sliced fish with minimum colour and taste modifications. *Listeria* destruction. Increase of shelf-life and additives reduction. Ready to eat 1,5 minutes in microwave oven.
Italy (2004)	Desalted cod	600 MPa	Vacuum packed		Shelf-life increase, sanitization.
S. Korea (2006)	Oysters				Opening of the shells (kept closed by a plastic band). Destruction of *Vibrio*.

Figure 6. HPP treated seafood avaiable in the market.

Recent advances in the use of HPP to improve the quality of cold-smoked salmon have been reviewed by Lakshmanan et al., 2003, altough a consistent amount of work is still needed to conclude the usefulness of HPP to improve the quality of cold smoked fish without affecting its sensory profile.

HPP is successfully employed to treat other types of seafood, such as lobsters (Figure 6), at pressure between 250 and 500 Mpa, improving microbiological quality and product yields.

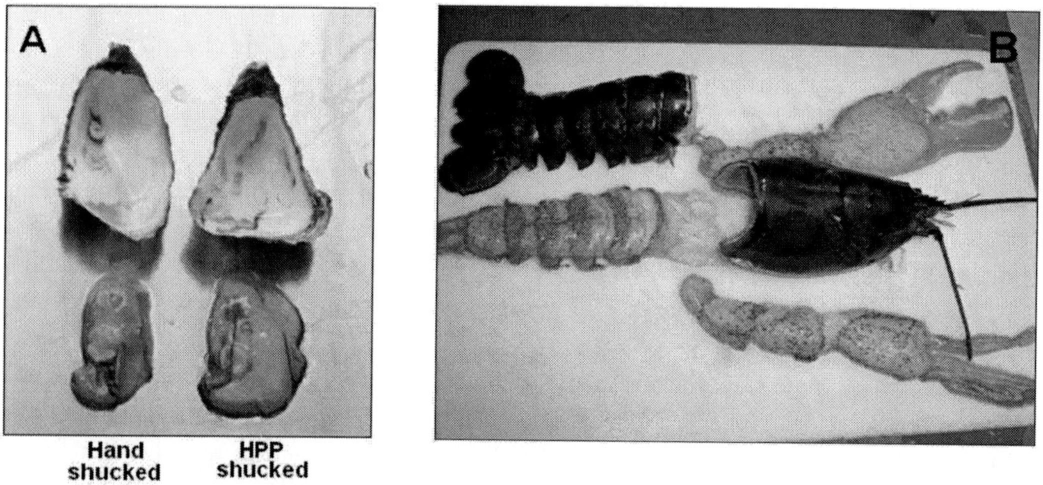

Figure 7. Meat removal from seafood by HPP. A. Increase of extraction yield in HPP treated oysters, compared to hand shucking. () B. Complete removal of meat from HPP treated lobster. A) Courtesy of Mark de Boevere, NC Hyperbaric. B) Courtesy of Alberto Vimercati, Avure Technologies.

Texturizing effects of HPP have been used to increased the gel strength of uncooked surimi by 2 to 3 fold by making protein substrates more accessible to transglutaminase which increases intermolecular cross-link formation and gel strength (Ashie and Lanier, 2007). These improvements in textural characteristics have created a high demand for HPP products from both the food service and retail sectors. Other potential uses of HPP technology which may be applicable to the seafood industry include pressure-assisted freezing (Kalichevsky, Knoor, and Lilliford, 1995) and HPP-thawing (Murakami et al., 1992; Rouillé et al., 2002; Schubring, et al., 2003).

Conclusion

In the last decade, HPP technology has proved to be a useful tool to improve meat and seafood safety and quality. Regulations recognize HPP as a post-packaging step in the control of food-borne pathogens (particularly *Listeria monocytogenes*) in meat products, and combined HP-thermal treatment is effective in sterilizing foods with limited impact on their nutritional and sensory qualities. Seafood processors are increasingly using HPP to inactivate bacterial pathogens and viruses in shellfish and to increase the extraction yield. Processors of crustaceans are using HPP to shuck lobsters and crabs, completely recovering meat from the shell, thereby increasing the processing efficiency and product yield and creating new markets. Texturizing effects over proteins has also been used to enhance the characteristics of already existing products and the development of new formulates. The development of high-efficiency HPP machines has reduced processing costs to acceptable levels. Last but not least, HPP as a low-temperature treatment is an environmentally friendly and waste-free technology.

References

Abdul Ghani, A.G. & Farid, M.M. (2007). Numerical simulation of solid-liquid food mixture in a high pressure processing unit using computational fluid dynamics. *Journal of Food Engineering.* 80, 1031-1042.

Aktar, S., Paredes-Sabja, D., Torres J.A., Sarker, M., R. (2009). Strategy to inactivate *Clostridium perfringens* spores in meat products. *Food microbiology*, 26, 272-277.

Acton, J. C., Dick, R. L. (1984). Protein-protein interaction in processed meats. *Reciprocal Meat Conference Proceedings*, 37, 36-42.

Amanatidou, A., Schlüter, O., Lemkau, K., Gorris, L. G. M., Smid, E. J., Knorr, D. (2000). Effect of combined high pressure treatment and modified atmospheres on the shelf-life of fresh Atlantic salmon. *Innovative Food Science and Emerging Technologies*, 1, 87-98.

Andres, A. I., Adamsen, C. E., Møller, J. K. S., Skibsted, L. H. (2004). High pressure treatment of dry-cured Iberian ham. Effect on radical formation, lipid oxidation and colour. *European Food Research and Technology*, 219, 205-210.

Archer, D. L. (1996). Preservation microbiology and safety: evidence that stress enhances virulence and triggers adaptive mutations. *Trends in Food Science and Technology*, 7, 91-95.

Ashie, I. N. A., Simpson, B. K. (1996). Application of high hydrostatic pressure to control enzyme related seafood texture deterioration. *Food Research International,* 29, 5-6.

Ashie, I. N., Lanier, T. C. (1999). High pressure effects on gelation of surimi and turkey breast muscle enhanced by microbial transglutaminase. *Journal of Food Science,* 64, 704-708.

Aymerich, T., Picouet, P.A., Monfort. J.M. (2008). Decontamination technologies for meat products. *Meat Science,* 78, 114-129.

Aymerich, M. T., Jofré, A., Garriga, M., Hugas, M. (2005). Inhibition of *Listeria monocytogenes* and *Salmonella* by natural antimicrobials and high hydrostatic pressure in sliced cooked ham. *Journal of Food Protection*, 68, 173-177.

Barbosa-Canovas, G.V. & Juliano, P. (2008). Food sterilization by combining high pressure and thermal energy. In: *Food Engineering: Integrated Approaches*. New York: Springer; . 9-46.

Bouton, P. E., Ford, A. L., Harris, P. V., Macfarlane, J. J., O'Shea, J. M. (1977). Pressure-heat treatment of post rigor muscle: Effects on tenderness. *Journal of Food Science*, 42, 132-135.

Brown, P., Meyer, R., Cardone, F., Pocchiari, M. (2003). Ultra-high-pressure inactivation of prion infectivity in processed meat *Proceedings of the National Academy of Sciences of the United States of America,* 100, 6093-6097.

Butz, P., Tauscher, B. (2002). Emerging technologies: chemical aspects. *Food Research International*, 35, 279-284.

Buzrul, S., Alpas, H., Largeteau, A., Bozoglu, F., Demazeau, G. (2008). Compression heating of selected pressure transmitting fluids and liquid foods during high hydrostatic pressure treatment. *Journal of Food Engineering,* 85, 466-472.

Calkins, C. R., Seidman, S. C. (1988). Relationship among calcium-dependent protease, cathepsin B and H, meat tenderness and the response of muscle to aging. *Journal of Animal Science*, 66, 1186-1193.

Campus, M., Addis, M. F., Cappuccinelli, R., Porcu, M. C., Pretti, L., Tedde, V., Secchi, N., Stara, G., Roggio, T. (2010). Stress relaxation behaviour and structural changes of muscle tissues from Gilthead Sea Bream (*Sparus aurata* L.) following High Pressure Treatment. *Journal of Food Engineering,* 96, 192-198.

Campus, M., Flores, M., Martinez, A., Toldrà, F. (2008). Effect of high pressure treatment on colour, microbial and chemical characteristics of dry cured loin. *Meat Science,* 80, 1174-1181.

Cardone, F., Brown, P., Meyer, R., Pocchiari, M. (2006). Inactivation of transmissible spongiform encephalopathy agents in food products by ultra high pressure–temperature treatment. *Biochimica et Biophysica Acta*, 1764, 558-562.

Carlez, A., Rosec, J. P., Richard, N., Cheftel, J. C. (1993). High pressure inactivation of *Citrobacter freundii*, *Pseudomonas fluorescens* and *Listeria innocua* in inoculated minced beef muscle. *Lebensmittel-Wissenschaft und-Technologie*, 26, 357-363.

Carlez, A., Veciana-Nogues, T., Cheftel, J. C. (1995). Changes in colour and myoglobin of minced beef meat due to high pressure processing. *Lebensmittel-Wissenschaft und-Technologie*, 28, 528-538.

Cheah, P. B., Ledward, D. A. (1996). High pressure effects on lipid oxidation in minced pork. *Meat Science*, 43, 123-134.

Cheah, P. B., Ledward, D. A. (1997). Catalytic mechanism of lipid oxidation following high pressure treatment in pork fat and meat. *Journal of Food Science*, 62, 1135-1139.

Cheftel, J. C. (1995). High-pressure, microbial inactivation and food preservation. *Food Science and Technology International*, 1, 75–90.

Cheftel, J. C., Culioli, J. (1997). Effect of high pressure on meat: a review. *Meat Science*, 46, 211-234.

Chen, H., Hoover, D. G. (2003). Bacteriocins and their food applications. *Comprehensive Reviews in Food Science and Food Safety*, 2, 81-100.

Chéret, R., Aránzazu, H.A., Delbarre-Ladrat, C., de Lambarrie, M., Verrez-Bagnis, V., (2006). Proteins and proteolytic activity changes during refrigerated storage in sea bass (Dicentrarchus labrax L.) muscle after high-pressure treatment. *European Food Research and Technology* 222, 527–535.

Chevalier, D., Le Bail, A., Goul, M. (2001). Effect of high pressure treatment (100-200 MPa) at low temperature on turbot (*Scophthalmus maximus*) muscle. *Food Research International*, 34, 425-429.

Cook, D. W. (2003). Sensitivity of *Vibrio* species in phosphate-buffered saline and in oysters to high hydrostatic pressure processing. *Journal of Food Protection*, 66, 2277-2282.

Crawford, Y.J., Murano, E.A., Olson, D.G., and Shenoy, K. (1996). Use of high hydrostatic pressure and irradiation to eliminate *Clostridium sporogenes* spores in chicken breast. *Journal of Food protection*, 59, 711-715.

Davis, K.J., Sebranek, J.G., Huff-Lonergan, E., Lonergan, S.M. (2004). The effects of aging on moisture-enhanced pork loins. *Meat Science,* 66, 519-524.

Defaye, A., Ledward, D. A., MacDougall, D. A., Tester, R. F. (1995). Renaturation of metmyoglobin subjected to high isostatic pressure. *Food Chemistry,* 52, 19-22.

Delgado, A.H.C. (2003). Pressure treatment of food: Instantaneous but not homogeneous effect. Advances in High Pressure Bioscience and Biotechnology II, 459-464.

Delgado, A., Rauh, C., Kowalczyk, W. & Baars, A. (2008). Review of modelling and simulation of high pressure treatment of materials of biological origin. *Trends in Food Science and Technology* 19: 329-336

Denys, S., Ludikhuyze, L. R., Van Loey, A. M., Hendrickx, M. E. (2000). Modeling conductive heat transfer and process uniformity during batch high-pressure processing of foods. *Biotechnology Progress,* 16, 92-101.

Elgasim, E. A., Kennick, W. H. (1980). Effect of pressurization of pre-rigor beef muscles on protein quality. *Journal of Food Science*, 45, 1122-1124.

Farkas, D. and Hoover, D. (2000). High pressure processing: Kinetics of microbial inactivation for alternative food processing technologies. Journal of Food Science (Supplement): 47-64.

Fernández-García, A., Heindl, P., Voigt, H., Bütter, M., Wienhold, D., Butz, P., Starke, J., Tauscher, B., Pfaff, E. (2004). Reduced proteinase K resistance and infectivity of prions after pressure treatment at 60°C. *Journal of General Virology,* 85, 261-264.

Furukawa, S., Nakahara, A., Hayakawa, I. (2000). Effect of reciprocal pressurization on germination and killing of bacterial spores. *International Journal of Food Science and Technology*, 35, 529-532.

García-Graells, C., Masschalck, B., Michiels, C. (1999). Inactivation of *Escherichia coli* in milk by high hydrostatic pressure treatment in combination with antimicrobial peptides. *Journal of Food Protection,* 62,1248-1254.

Garriga, M., Aymerich, M. T., Costa, S., Monfort, J. M., Hugas, M. (2002). Bactericidal synergism through bacteriocins and high pressure in a meat model system during storage. *Food Microbiology*, 19, 509-518.

Garriga, M., Aymerich, M. T., Hugas, M. (2002). Effect of high pressure processing on the microbiology of skin-vacuum packaged sliced meat products: cooked pork ham, dry cured pork ham and marinated beef loin. *Profit Final Project Report FIT060000200066.*

Garriga, M., Grèbol, N., Aymerich, M.T., Monforta, J.M., Hugas M. (2004). Microbial inactivation after high-pressure processing at 600 MPa in commercial meat products over its shelf life. *Innovative Food Science and Emerging Technologies*, 5, 451-457.

Gòmez-Estaca, J., Gòmez-Guillén, M. C., Montero, P. (2007). High pressure effects on the quality and preservation of cold-smoked dolphinfish (*Coryphaena hippurus*) fillets. *Food Chemistry*, 102, 1250-1259.

Gola, S., Mutti, P., Manganelli, E., Squarcina, N., Rovere, P. (2000). Behaviour of *E. coli* O157:H7 strains in model system and in raw meat by HPP: microbial and technological aspects. *High Pressure Research*, 19, 481-487.

Grauwet, T., Van der Plancken, I., Vervoort, L., Hendrickx, M. & Van Loey, A. (2009). Investigating the potential of *Bacillus subtilis* α-amylase as a pressure-temperature-time indicator for high hydrostatic pressure pasteurization processes. *Biotechnology Progress* 4, 1184-1193.

Grauwet, T., Van der Plancken, I., Vervoort, L., Hendrickx, M. & Van Loey, A. (2010). Solvent engineering as a tool in enzymatic indicator development for mild high pressure pasteurization processing. *Journal of Food Engineering*, 97, 301-310.

Gross, M., Jaenicke, R. (1994). Proteins under pressure: influence of High Hydrostatic Pressure on structure, function and assembly of proteins and protein complexes. *European Journal of Biochemistry*, 221, 617-630.

Hamm, R. (1981). Post-mortem changes in muscle affecting the quality of comminuted meat products. In R. Lawrie (Ed.), Development in meat science (pp. 93-124). London: Elsevier Applied Science.

Hartmann, C., Delgado, A. (2002). Numerical simulation of convective and diffusive transport effects on a high pressure induced inactivation process. *Biotechnology and Bioengineering*, 79, 94-104.

Hartmann, C., Delgado, A. (2004). Numerical simulation of the mechanics of a yeast cell under high hydrostatic pressure. *Journal of Biomechanics*, 37, 977-987.

Hartmann, C., Schuhholz, J. P., Kitsubun, P., Chapleau, N., Le Bail, A., Delgado, A. (2004). Experimental and numerical analysis of the thermofluidynamics in a high-pressure autoclave. *Innovative Food Science and Emerging Technologies*, 5, 399-411.

Hartmann, C., Mathmann, K., Delgado, A. (2006). Mechanical stresses in cellular structures under high hydrostatic pressure. *Innovative Food Science and Emerging Technologies*, 7, 1-12.

Hauben, K., Wuytack, E., Soontjens, C., Michiels, C. (1996). High pressure transient sensitization of *Escherichia coli* to lysozyme and nisin by disruption of outer-membrane permeability. *Journal of Food Protection,* 59, 350-355.

He, H., Adams, R. M., Farkas, D. F., Morrissey, M. T. (2002). Use of high-pressure processing for oyster shucking and shelf-life extension. *Journal of Food Science,* 67, 640-645.

Heindl, P., Fernandez Garcia, A., Butz, P., Trierweiler, B., Voigt, H., Pfaff, E., Tauscher, B. (2008). High pressure/temperature treatments to inactivate highly infectious prion subpopulations. *Innovative Food Science and Emerging Technologies,* 9, 290-297.

Heinz, V, and Knorr, D. (2002). Effect of high pressure on spores. In M. E. G. Hendrickx & D. Knorr (Eds.), Ultra high pressure treatment of foods (pp. 77-114). New York: Academic/Plenum Publishers.

Heremans, K. (1982). High pressure effects on proteins and other biomolecules. Ann. Rev. Biophys. Bioeng, 11, 1-21.

Heremans, K. (2002). The effects of pressure on biomaterials. In M. E. G. Hendrickx & D. Knorr (Eds.), Ultra high pressure treatment of foods (pp. 23-51). New York: Academic/Plenum Publishers.

Hite, B. H. (1899). The effects of pressure in the preservation of milk. *Bulletin of the West Virginia University Agricultural Experimental Station Morgantown*, 58, 15-35.

Homma, N., Ikeuchi, Y., Suzuki, A. (1994). Effects of high pressure treatment on the proteolytic enzymes in meat. *Meat Science*, 38, 219-228.

Hoover, D. G. (1993). Pressure effects on biological systems. *Food Technology*, 47, 150-155.

Hoover, D. G., Metrick, C., Papineau, A. M., Farkas, D. F., Knorr, D. (1989). Biological effects of high hydrostatic pressure on food microorganisms. *Food Technology*, 43, 99-107.

Hopkins, D. L. (2000). The relationship between actomyosin, proteolysis and tenderisation examined using protease inhibitors. PhD thesis, University of New England, Australia.

Huff-Lonergan, E., Mitsuhashi, T., Beekman, D. D., Parrish, F. C., Olson, D. G., Robson, R. M. (1996). Proteolysis of specific muscle structural proteins by mu-calpain at low pH and temperature is similar to degradation in postmortem bovine muscle. *Journal of Animal Science,* 74, 993-1008.

Hugas, M., Garriga, M., Monfort, J. M. (2002). New mild technologies in meat processing: high pressure as a model technology. *Meat Science*, 62, 359-371.

Hurtado, J. L., Montero, P., Borderìas, J. (2000). Extension of shelf life of chilled hake (*Merluccius capaensis*) by high pressure. *Food Science and Technology International*, 6, 243-249.

Jiménez-Colmenero, F., Borderias, A. J. (2003). High-pressure processing of myosystems. Uncertainties in methodology and their consequences for evaluation of results. *European Food Research and Technology*, 217,461-465.

Jiménez Colmenero F. (2002). Muscle protein gelation by combined use of high pressure/temperature. *Trends in Food Science and Technology*, 13, 22-30.

Johnson, M. H., Calkins, C. R., Huffman, R. D., Johnson, D. D., Hargrove, D. D. (1990). Differences in cathepsin B+L and calcium-dependent protease activities among breed types and their relationship to beef tenderness. *Journal of Animal Science*, 68, 2371-2379.

Ju, X.R., Gao, Y.L., Yao, M.L. & Qian, Y. (2008). Response of Bacillus cereus spores to high hydrostatic pressure and moderate heat. *LWT - Food Science and Technology,* 41,2104-2112.

Juliano, P., Knoerzer, K., Fryer, P.J. & Versteeg, C. (2009). *C. botulinum* inactivation kinetics implemented in a computational model of a high-pressure sterilization process. *Biotechnoly Progress* 25, 163-175.

Jung, S., Ghoul, M., de Lamballerie-Anton, M. (2003). Influence of high pressure on the color and microbial quality of beef meat. *Lebensmittel-Wissenschaft und-Technologie*, 36, 625-631.

Jung, S., Anton, M. L., Taylor, R. G., Ghoul, M. (2000). High-pressure effects on lysosome integrity and lysosomal enzyme activity in bovine muscle. *Journal of Agricultural and Food Chemistry*, 48, 2467-2471.

Kalchayanand, N., Sikes, T., Dunne, C. P., Ray, B. (1994). Hydrostatic pressure and electroporation have increased bactericidal efficiency in combination with bacteriocins. *Applied and Environmental Microbiology*, 60, 4174-4177.

Kalchayanand, N., Dunne, C. P., Ray, B. (1998). Interaction of hydrostatic pressure, time and temperature of pressurization and pediocin AcH on inactivation of foodborne bacteria. *Journal of Food Protection*, 61, 425-431.

Kalchayanand, N., Dunne, C. P., Sikes, A., Ray, B. (2004). Germination induction and inactivation of *Clostridium* spores at medium-range hydrostatic pressure treatment. *Innovative Food Science and Emerging Technologies,* 5, 277-283.

Kalichevsky, M. T., Knorr, D., Lilliford, P. J. (1995). Potential food applications of high-pressure effects on ice-water transitions. *Trends in Food Science and Technology*, 6, 253- 259.

Kato, M., Hayashi, R., Tsuda, T., Taniguchi, K. (2002). High pressure-induced changes of biological membrane. Study on the membrane-bound Na+/K+-ATPase as a model system. *European Journal of Biochemistry*, 269, 110-118.

Kemp, C. M., Sensky, P. L., Bardsley, R. G., Buttery, P. J., Parr, T. (2010). Tenderness – an enzymatic view. *Meat Science,* 84, 248-256.

Kingsley, D. H., Hoover D. J., Papafragkou, E., Richards, G. P. (2002). Inactivation of Hepatitis A Virus and Calicivirus by high hydrostatic pressure. *Journal of Food Protection*, 65, 1605-1609.

Knoerzer, K., Buckow, R., Sanguansri, P., Versteeg, C. (2010). Adiabatic compression heating coefficients for high-pressure processing of water, propylene-glycol and mixtures – a combined experimental and numerical approach. *Journal of Food Engineering*, 96, 229-238.

Knorr, D., Schlueter, O., & Heinz, V. (1998). Impact of high hydrostatic pressure on phase transitions of foods. *Food technology*, 52, 42-45.

Koohmaraie, M., Geesink, G. H. (2006). Contribution of postmortem muscle biochemistry to the delivery of consistent meat quality with particular focus on the calpain system. *Meat Science*, 74, 34-43.

Kristensen, L., Purslow, P. P. (2001). The effect of ageing on the water-holding capacity of pork: role of cytoskeletal proteins. *Meat Scie*nce, 58, 17-23.

Kural, A. G., Chen, H. Q. (2008). Conditions for a 5-log reduction of *Vibrio vulnificus* in oysters through high hydrostatic pressure treatment. *International Journal of Food Microbiology,* 122, 180-187.

Kural, A. G., Shearer, A. E. H., Kingsley, D. H., Chen H. (2008). Conditions for high pressure inactivation of *Vibrio parahaemolyticus* in oysters. *International Journal of Food Microbiology*,127,1-5.

Khurana, M. & Karwe, M.V. (2009). Numerical prediction of temperature distribution and measurement of temperature in a high hydrostatic pressure food processor. *Food Bioprocess Technol.* 2: 279-290.

Kurth, L. B., (1986). Effect of pressure-heat treatment on cathepsin B1 activity. *Journal of Food Science,* 51, 663-664.

Lakshmanan, R., Piggott, J. R., Paterson, A. (2003). Potential applications of high pressure for improvement in salmon quality. *Trends in Food Science and Technology,* 14 354-363.

Leistner, L., Gorris, L. G. M. (1995). Food preservation by hurdle technology. *Trends in Food Science and Technology*, 6, 41-46.

Lindsay, D. S., Collins, M. V., Holliman, D., Flick, G. J., Dubey, J. P. (2006). Effects of high-pressure processing on *Toxoplasma gondii* tissue cysts in ground pork. *Journal of Parasitology*, 92, 195-196.

Linton, M., Patterson, M. F. (2000). High pressure processing of foods for microbiological safety and quality. *Acta Microbiologica et immunologica Hungarica*, 47, 175-182.

Lopez-Caballero, M. E., Perez-Mateos, M., Montero, P., Borderias, A. J. (2000). Oyster preservation by high-pressure treatment. *Journal of Food Protection,* 63, 196-201.

Low, P. S., Somero, G. N. (1974). Temperature adaptation of enzymes. A proposed molecular basis for the different catalytic efficiencies of enzymes from ectotherms and endotherms. *Comparative Biochemistry and Physiology,* 49, 307-12.

Ludwig H., and Schreck,Ch. (1997). The inactivation of vegetative bacteria by pressure. High Pressure Research in the Biosciences and Biotechnology. Leuven University Press. Ed. by K. Heremans. Leuven University Press, 221-224.

Ludwig, H., van Almsick, G., Schreck, C. (2002). The effect of high hydrostatic pressure on the survival of microorganisms. In Y. Taniguchi, H. E. Stanley, & H. Ludwig (Eds). *Biological systems under extreme conditions*, (pp. 239-256). Berlin; Springer.

McClane, B.A., 2007. Clostridium Perfringens. In: Doyle, M.P., Beuchat, L.R. (Eds.), Food Microbiology: Fundamentals and Frontiers, third ed. ASM Press, Washington, D.C., pp. 423-444.

Macfarlane, J.J. (1973). Pre-rigor pressurization of muscle: Effect on pH, shear value and taset panel assessment. *Journal of Food Science*, 38, 294-298.

Macfarlane, J. J. (1985). High pressure technology and meat quality. In *Developments in meat science 3,* ed. R. A. Lawrie, (pp. 155-184). London: Elsevier Applied Science.

Maggi, A., gola, S., Rovere, P., Miglioli, L., Dall'aglio, G., and Loenneborg, N.G. (1996). Effects of combined high pressure-temperature treatments on *Clostridium sporogenes* spores in liquid media. *Industria conserve,* 71, 8-14.

Mancini, R. A., Hunt M. C. (2005). Current research in meat color. *Meat Science,* 71, 100-121.

Marcos, B., Aymerich, T., Guardia, M. D., Garriga, M. (2007). Assessment of high hydrostatic pressure and starter culture on the quality properties of low-acid fermented sausages. *Meat Science*, 76, 46-53.

Masschalck, B., Van Houdt, R., Van Haver, E. G. R., Michiels, C. W. (2001). Inactivation of Gram-negative bacteria by lysozyme, denatured lysozyme and lysozyme-derived peptides under high hydrostatic pressure. *Applied and Environmental Microbiology*, 67, 339-344.

Master, A. M., Krebbers, B., Van den Berg, R. W., Bartels, P. V. (2004). Advantages of high pressure sterilisation on quality of food products. *Trends in Food Science and Technology*, 15, 79-85.

Mayer, R. S. (2000). Ultra high pressure, high temperature food preservation process. US Patent 6,177,115, B1 (Richard S. Mayer, Tacoma, WA, USA).

McClements, J. M., Patterson, M. F., Linton, M. (2001). The effect of growth stage and growth temperature on high hydrostatic pressure inactivation of some psychrotrophic bacteria in milk. *Journal of Food Protection*, 64, 514-522.

Melody, J. L., Lonergan, S. M., Rowe, L. J., Huiatt, T. W., Mayes, M. S., Huff-Lonergan, E. (2004). Early postmortem biochemical factors influence tenderness and water-holding capacity of three porcine muscles. *Journal of Animal Science,* 82, 1195-1205.

Messens, W., Van Camp, J., Huyghebaert, A. (1997). The use of high pressure to modify the functionality of food proteins. *Trends in Food Science and Technology*, 8, 107-112.

Moeller, P. W., Field, P. A., Dutson, T. R., Landmann, W. A., Carpenter, Z. L. (1977). High temperature effects on lysosomal enzymes distribution and fragmentation of bovine muscle. Journal *of Food Science*, 42, 510-512.

Morales, P., Calzada, J., Nuñez, M. (2006). Effect of high-pressure treatment on the survival of *Listeria monocytogenes* Scott A in sliced vacuum-packaged Iberian and Serrano cured hams. *Journal of Food Protection*, 69, 2539-2543.

Murakami, T., Kimura, I., Yamagishi, T., Yamashita, M., Sugimoto, M., Satake, M. (1992). Thawing of frozen fish by hydrostatic pressure. In C. Balny, R. Hayashi, K. Heremans & P. Masson, High pressure and biotechnology (pp. 329-331). London: Colloque INSERM/J. Libby Eurotext Ltd.

Murano, E. A., Murano, P. S., Brennan, R. E., Shenoy, K., Moreira, R. G. (1999). Application of high hydrostatic pressure to eliminate *Listeria monocytogenes* from fresh pork sausage. *Journal of Food Protection*, 62, 480-483.

Murchie, L. W., Cruz-Romero M., Kerry J., Linton, J., Patterson, M., Smiddy, M., Kelly, A. (2005). High pressure processing of shellfish: a review of microbiological and other quality aspects. *Innovative Food Science and Emerging Technologies,* 6, 257-270.

Norton, T., Sun, D. W. (2008). Recent advances in the use of high hydrostatic pressure as an effective processing technique in the food industry. *Food Bioprocess Technology,* 1, 2-34.

Ohmori, T., Shigehisa, T., Taji, S., Hayashi, R. (1991). Effect of high pressure on the protease activities in meat. *Agricultural and Biological Chemistry*, 55, 357-361.

Okazaki, E., Ueda, T., Kusaba, R., Kamimura, S., Fukuda, Y., & Arai, K. (1997). Effect of heating on pressure-induced gel of chum salmon meat. In K. Heremans (Ed.), High pressure research in the bioscience and biotechnology (pp. 371–374). Leuven, Belgium: Leuven University Press.

Oshima, T., Ushio, H., Koizumi, C. (1993). High pressure treatments of fish and fish products. *Trends in Food Science and Technology*, 4, 370-374.

Orlien, V., Hansen, E., Skibsted, L. H. (2000). Lipid oxidation in high-pressure processed chicken breast muscle during chill storage: critical working pressure in relation to oxidation mechanism. *European Food Research and Technology*, 211, 99-104.

Otero, L., Mólina, A., Ramos, A., Sanz, P. D. (2002). A model for a real thermal control in high-pressure treatment of foods. *Biotechnology Progress*, 18, 904-908.

Otero, L., Ramos, A. M., de Elvira, C., Sanz, P. D. (2007). A model to design high-pressure processes towards an uniform temperature distribution. *Journal of Food Engineering,* 78, 1463-1470.

Paredes-Sabja., D., Gonzales, M., Sarker., M.R., Torres, J.A. (2007). Combined effects of hydrostatic pressure, temperature and pH on the inactivation of spores of *Clostridium perfringens* type A and *Clostridium sporogenes* in buffer solutions. *Journal of Food Science,* 72, 202-207.

Park, S. W., Sohn, K. Y., Shin, J. H., Lee, H. J. (2001). High hydrostatic pressure inactivation of *Lactobacillus viridescens* and its effects on ultrastructure of cells. *International Journal of Food Science and Technology*, 36, 775-781.

Patterson, M. F., Quinn, M., Simpson, R., Gilmour, A. (1995). Sensitivity of vegetative pathogens to high hydrostatic-pressure treatment in phosphate-buffered saline and foods. *Journal of Food Protection*, 58, 524-529.

Patterson, M.F., Kilpatrick, D.J. (1998). The combined effect of high hydrostatic pressure and mild heat on inactivation of pathogens in milk and poultry. Journa of food protection, 61, 432-436.

Rauh, C., Baars, A. & Delgado, A. (2009). Uniformity of enzyme inactivation in a short-time high-pressure process. *Journal of Food Engineering* 91, 154-163.

Rasanayagam, V., Balasubramaniam, V. M., Ting, E., Sizer, C. E., Bush, C., Anderson, C. (2003). Compression heating of selected fatty food materials during high pressure processing. *Journal of Food Science,* 68, 254-259.

Ritz, M., Tholozan, J. L., Federighi, M., Pilet, M. F. (2001). Morphological and physiological characterization of *Listeria monocytogenes* subjected to high hydrostatic pressure. *Applied and Environmental Microbiology*, 67, 2240-2247.

Rouillé, J., Lebail, A., Ramaswamy, H. S., Leclerc, L. (2002). High pressure thawing of fish and shellfish. *Journal of Food Engineering*, 53, 83-88.

Rubio, B., Martínez, B., García-Gachán, M. D., Rovira, J., Jaime, I. (2007). Effect of high pressure preservation on the quality of dry cured beef ''Cecina de Leon''. *Innovative Food Science and Emerging Technologies*, 8, 102-110.

Sano, T., Ohno, T., Otsuka-Fuchino, H., Matsumoto, J. J., Tsuchiya, T. (1994). Carp natural actomyosin: thermal denaturation mechanism. *Journal of Food Science*, 59, 1002-1008.

Sáiz, A.H., Mingo, S.T., Balda F.P.,and Samson C.T. (2008). Advances in design for successful commercialhigh pressure food processing. *Food Australia*, 60, 154-156.

Schubring, R., Meyer, C., Schlüter, O., Boguslawski, S., Knorr, D. (2003). Impact of high pressure assisted thawing on the quality of fillets from various fish species. *Innovative Food Science and Emerging Technologies*, 4, 257-267.

Schreck, C., Layh-Schmidt, G., & Ludwig, H. (1999). Inactivation of. Mycoplasma pneumoniae by high hydrostatic pressure. Pharmaceutical Industry, 61(8), 759-762.

Sentandreu, M. A., Coulis, G., Ouali, A. (2002). Role of muscle endopeptidases and their inhibitors in meat tenderness. *Trends in Food Science and Technology*,13,398-419.

Serra, X., Sarraga, C., Grebol, N., Guardia, M. D., Guerrero, L., Gou, P., Masoliver, P., Gassiot, M., Monfort, J. M., Arnau, J. (2007). High pressure applied to frozen ham at different process stages 1. Effect on the final physicochemical parameters and on the antioxidant and proteolytic enzyme activities of dry-cured ham. *Meat Science*, 75, 12-20.

Serra, X., Grebol, N., Guardia, M. D., Guerrero, L., Gou, P., Masoliver, P., Gassiot, M., Sarraga, C., Monfort, J. M., and Arnau, J. (2007). High pressure applied to frozen ham at different process stages 2. Effect on the sensory attributes and on the color characteristics of dry-cured ham. *Meat Science*, 75, 21-28.

Shigehisa, T., Ohmori, T., Saito, A., Taji, S., Hayashi, R. (1991). Effects of high pressure on the characteristics of pork slurries and the inactivation of micro-organisms associated with meat and meat products. *International Journal of Food Microbiology,* 12, 207-216.

Simpson, R. K., Gilmour, A. (1997). The effect of high hydrostatic pressure on *Listeria monocytogenes* in phosphate-buffered saline and model food systems. *Journal of Applied Microbiology*, 83, 181-188.

Smelt, J. P. P. M. (1998). Recent advances in the microbiology of high pressure processing. *Trends in Food Science and Technology*, 9, 152-158.

Stewart, C., Dunne, C., Sikes, A., Hoover, D. (2000). Sensitivity of spores of *Bacillus subtilis* and *Clostridium sporogenes* PA 3679 to combinations of high hydrostatic pressure and other processing parameters. *Innovative Food Science and Emerging Technologies*, 1, 49-56.

Suzuki, A., Watanabe, M., Ikeuchi, Y., Saito, M., Takahashi, K. (1993). Effects of high pressure treatment on the ultrastructure and thermal behaviour of beef intramuscular collagen. *Meat Science,* 35, 17-25.

Tanaka, M., Xueyi, Z., Nagashima, Y., Taguchi T. (1991). Effect of high pressure on lipid oxidation in sardine meat. *Nippon Suisan Gakkaishi*, 57, 957-963.

Tanzi, E., Saccani, G., Barbuti, S., Grisenti, M. S., Lori, D., Bolzoni, S., Parolari, G. (2004). High pressure treatment of raw ham. Sanitation and impact on quality. *Industria Conserve*, 79, 37-50.

Taylor, R. G., Geesink, G. H., Thompson, V. F., Koohmaraie, M., Goll, D. E. (1995). Is Z-disk degradation responsible for postmortem tenderization? *Journal of Animal Science*, 73, 1351-1367.

Téllez, S. J., Ramírez, J. A., Pérez, C., Vázquez, M., Simal, J. (2001). Aplicación de la alta pressión hidrostática en la conservación de los alimentos. *Ciencia y Tecnología Alimentaria*, 3, 66-80.

Tewari, G., Jayas, D. S., Holley, R. A. (1999). High pressure processing of foods: an overview. *Sciences des Aliments*, 19, 619-661.

Toldrà, F., Flores, M. (1998). The role of muscle proteases and lipases in flavor development during the processing of dry-cured ham. *CRC Critical Review of Nutrition and Food Science*, 38, 331-352.

Tuboly, E., Lebovics, V. K., Gaál, O., Mészáros, L., Farkas, J. (2003). Microbiological and lipid oxidation studies on mechanically deboned turkey meat treated by high hydrostatic pressure. *Journal of Food Engineering*, 56, 241-244.

Van der Plancken, I., Grauwet, T., Oey, I., Van Loey, A. & Hendrickx, M. (2008). Impact evaluation of high pressure treatment on foods: considerations on the development of pressure-temperature-time integrators (pTTIs). *Trends Food Science and Technology,* 19, 337-348.

Wada, S. 1992. Quality and lipid change of sardine meat by high pressure treatment in. High Pressure and Biotechnology, C. Balny, R. Hayashi, K. Heremans, P. Masson) ed. Colloque INSERM/John Libbey Eurotext Ltd, Montrouge, France 235-238.

Wilson, D. R., Lukasz, D., Stringer, S., Moezelaar, R. And Brocklehurst, T.F. (2008). High pressure in combination with elevated temperature as a method for the sterilisation of food. *Trends in Food Science and Technology*. 19, 289-299.

Wuytack, E.Y., Boven, S., and Michelis, C.W. (1998). Comparative study of pressure induced germination of Bacillus subtilis spores at low and high pressure. Applied and Environmental Microbiology, 64, 3220-3224.

In: Focus on Food Engineering
Editor: Robert J. Shreck
ISBN: 978-1-61209-598-1

Chapter 7

ADVANCED MODELING OF FOOD CONVECTIVE DRYING: A COMPARATIVE STUDY AMONG FUNDAMENTAL, ARTIFICIAL NEURAL NETWORKS AND HYBRID APPROACHES

Stefano Curcio[1], Maria Aversa[2] and Alessandra Saraceno[3]
Department of Engineering Modeling,
University of Calabria, 87030, Rende (CS), Italy

Mathematical modeling represents an effective support to design and control industrial processes. Different approaches can be used to develop reliable models aimed at investigating how system responses may change, with time, under the influence of both external disturbances and manipulated variables. In the present chapter, it will be shown how various kinds of advanced models, based either on fundamental, or on artificial neural networks or on hybrid modeling, can be used to predict the behavior of a typical industrial process: the convective drying of food. The main aims of convective drying are a decrease of food water content and an increase of its temperature; both the above effects improve food preservation since microbial spoilage is favored by low temperature and high moisture content. Drying does actually preserve foods by decreasing water activity, thus stopping the micro-organisms growth. If food drying is performed in uncontrolled conditions, the level of water content could be not sufficiently low to stop the activity of micro-organisms, which continue proliferating. The starting point of any kind of automatic control is definitely represented by the availability of a predictive model of the process under study.

Fundamental or theoretical modeling is based on the formulation of transport models analyzing the simultaneous transfer of momentum, of heat and of water (both as vapor and as liquid), occurring in air as well as in food. An exhaustive analysis of all the complex transport phenomena involved in drying process, however, is rather onerous and time consuming for practical purposes, since the resulting system of non-linear partial differential equations can

[1] Email: stefano.curcio@unical.it
[2] Email: maria.aversa@unical.it
[3] Email: alessandra.saraceno@unical.it

only be solved by means of numerical methods. Moreover, some physical transformations involved in drying process are not yet completely understood and, therefore, are very difficult to interpret by proper mathematical relationships. For the above reasons, several simplified approaches have been proposed to describe food convective drying.

On the other hand, a model based on Artificial Neural Networks (ANNs) does not make use of any transport equation that could help to determine, on the basis of fundamental principles, the mutual relationships existing between the inputs and the outputs. ANNs, in fact, are a data-driven method capable to learn from examples and represent a particularly useful tool to describe tough phenomena since no *a priori* knowledge is required. ANNs are composed of interconnected computational elements, called neurons or nodes. Each neuron receives input signals from the related units, elaborates these stimuli by an activation or transfer function and, eventually, generates an output signal, which is transferred to other neurons. A neural model is, generally, rather complicated, since it requires a large number of connections and, therefore, a great number of parameters that are to be estimated. Moreover, it is worthwhile observing that since extrapolation based on ANN predictions is an unreliable procedure, it is often necessary to perform many different experiments in order to train the network in a range as wider as possible of practical situations.

A good trade-off between a theoretical and a "pure" neural network approach is represented by hybrid modeling, which allows predicting the behavior of complex systems, in a more efficient way. Hybrid model predictions are indeed given as a combination of both theoretical and "pure" neural network models, together concurring in the obtainment of system responses. The main advantage of a hybrid system regards the possibility to describe some well-assessed phenomena by means of a quite simple fundamental approach. Some others, that could be very difficult to interpret, are, instead, described by rather straightforward "cause-effect" models, based on ANN.

MODELS FORMULATION

Fundamental Modeling

When dry and warm air flows around a moist and cold food sample, a simultaneous transfer of both heat and water occurs. Heat is transferred from air to the material; water is transported from the food core, then to its surface and, eventually, to air. The rates of heat and mass transfer depend on both temperature and concentration differences and, also, on air velocity field. Convective drying is definitely the most common method for food preservation, so several different approaches were proposed to model this process. The models available in the open literature may be subdivided in two categories: simplified and complex approaches. The latter are based on the formulation of transport models analyzing the simultaneous transfer of momentum, of heat and of water (both as vapor and as liquid), occurring in air as well as in food; the former, instead, are quite simple and are based either on simplification hypotheses, not applicable in several real cases, or on the utilization of empirical correlations, necessary to estimate - by means of a set of transport coefficients – the heat and water fluxes at food-air interface(s). On developing fundamental models aimed at predicting the convective drying behavior, the solid foods are, generally, regarded as porous hygroscopic materials containing a great amount of physically bound water [1]. Hygroscopic materials are characterized by a limit moisture content below which internal vapor pressure, expressed in terms of both moisture content and temperature, is lower than that of liquid water at the same temperature [2]. Unbound water, on the other hand, exerts its full vapor pressure and is mostly held in the voids of the solid. Water removal from hygroscopic substances is a rather complex phenomenon, since both unbound and bound water are actually to be transported; in fact, once unbound water has been removed, a significant amount of bound water may still be present. Generally, bound water is removed by progressive vaporization within the solid matrix, followed by diffusion and pressure driven transport of water vapor through the solid [1]. An exhaustive analysis of all the complex transport phenomena involved in the drying process was regarded as being too onerous and time consuming for practical purposes [3]. For this reason, many simplified approaches were proposed to model either the moisture transport only [4-6] or the simultaneous transfer of heat and moisture occurring during food drying [7, 8], even accounting for the variation of physicochemical properties of the food material [9]. The simultaneous presence of both liquid water and vapor within the solid makes drying modeling even more difficult [10]. Datta and his coworkers have already presented detailed and general multiphase models describing different heat and mass transfer processes in foods (convective heating, baking, frying, microwave heating, etc.) for which internal evaporation may play a significant role [1, 11-15]. Attention in Datta's papers, however, was focused on the formulation of transport models analyzing only food behavior; the transport phenomena occurring at food-air interfaces were described in terms of heat and mass transfer coefficients, estimated from the available literature data. To a certain extent, this might limit the model accuracy since, as reported in the literature [16, 17], even small errors in the estimation of transfer coefficients could lead to large deviations between predicted and real values, thus determining inappropriate equipment design or severe processing problems. In general, it would be advisable to account also for the transport phenomena occurring in the drying air in order to estimate the actual transport rates at the food/air interfaces without resorting to any empirical correlation [18, 19]. This is particularly necessary when food shape is not regular or it even changes with time because of, for instance, shrinkage. In a previous paper [18], some of the authors of this contribution actually formulated a complete theoretical model describing convective food drying. The model accounted for the simultaneous transfer of momentum, heat and mass in air as well as of heat and mass transfer within the food; the attention, however, was focused on the analysis of those cases characterized by very weak inner evaporation, which, therefore, was neglected. Moisture transport inside the food was modeled referring to an effective diffusion coefficient that did not make any distinction between the transport of liquid water, usually expressed in terms of a capillary diffusion coefficient, and that of vapor, expressed by a molecular diffusion coefficient [1]. In a subsequent improvement [20], another model was developed so to simulate drying behavior also when inner evaporation could not be neglected and a unique mass balance equation, expressing the transport of total moisture in food, was actually inadequate to describe properly the process. In particular, a multiphase approach, based on the conservation of both liquid water and vapor, was adopted. Moreover, the turbulent momentum transfer of drying air, considered as a gas mixture owing to its relative humidity always different from zero, was described by the k-ω model [21], which – as compared to the k-ε model used in [18] – allowed improving the model predictions, especially in the boundary layer developing close to the food surfaces. The k-ω approach does, indeed, have several advantages that are related to its higher accuracy in boundary layer modeling, to the easier integration due to a proper definition of a viscous sub-layer and to the more accurate description of the transition occurring close to the solid boundaries [21]. Both the general transport models [18 and 20] led to a rather complicated system of unsteady partial differential equations whose solution, performed by the Finite Elements Method, allowed predicting the drying behavior of foods available in all shapes and over a wide range of process and fluid-dynamic conditions.

Both the mathematical and the numerical complications involved in the solution of a complete and general model, as those presented in [18] and in [20], may suggest developing, however, much simpler models.

To allow an appropriate comparison among the different approaches presented in the present contribution, a transport model describing the simultaneous bi-dimensional heat and moisture transfer occurring within the food sample only will be formulated so to analyze the influence of some of the most important operating variables on carrots drying rate [22]. In the following, the main assumptions and the mathematical equations used to model, by a simplified fundamental approach, the unsteady-state behavior of a convective dryer will be shown. It was supposed that drying air was continuously supplied to the dryer inlet section and flowed around the food in the axial direction (y), parallel to its main dimension (Figure 1). Heat and mass transfer resistance were assumed negligible across the net on which the vegetable was placed [23-24]. On the basis of the above hypothesis, the system under investigation could be considered as a classical symmetric one; a symmetry axis, in fact, might be identified and only half of the original domain was taken in consideration. Moreover, any possible variation occurring with reference to the spatial coordinate *z* was assumed not relevant for the present study. This allowed analyzing a 2D geometry so that food sample could be considered as a slab. Each dependent variable was expressed as a function of two spatial coordinates, *x* and *y*, and of time, *t*.

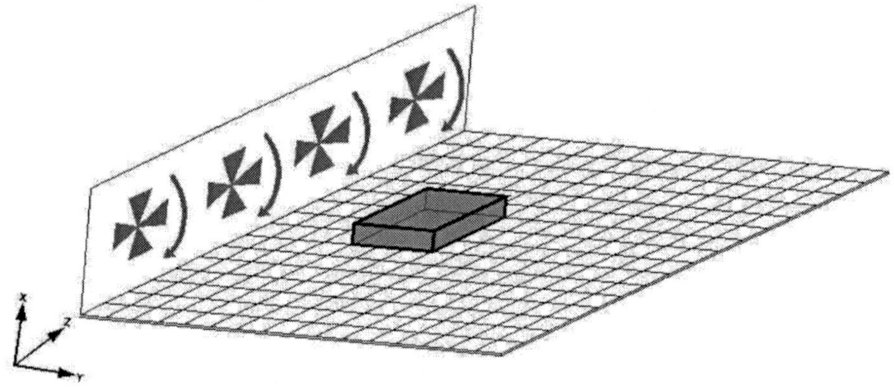

Figure 1. Schematic representation of the dryer under study.

The model accounted for the variation of air and food physical properties, defined in terms of the local values of temperature and of moisture content. As far as the food sample was concerned, a continuum approach was chosen; it was supposed, in fact, that the actual multiphase hygroscopic porous medium could be replaced by a fictitious continuum, any point of which had variables and parameters, which were continuous functions of point spatial coordinates and of time. Convective contributions in the transport equations written for the food were neglected, assuming weak inner evaporation. Therefore, heat and mass transfer in the product occurred only by conduction and diffusion, respectively. Shrinkage effects were assumed negligible in the range of average moisture content taken in consideration.

On the basis of the above discussion, water and heat transfers occurring during food drying process were modeled by the unsteady state mass and energy balances, respectively, whereas evaporation, occurring at air-food interfaces only, was considered by defining a proper set of boundary conditions expressed in terms of heat and mass transfer coefficients estimated by the semi-empirical correlations available in the literature [25].

The energy balance in the solid, based on the Fourier's law, led to

$$\rho C_p \, \partial T / \partial t = \underline{\nabla} \bullet \left(k_{eff} \underline{\nabla} T \right) \quad (1)$$

where ρ was the food density, C_p was its heat capacity, T was the temperature, t was the time and k_{eff} the effective thermal conductivity of food.

The mass balance, based on the Fick's law, led to:

$$\partial C / \partial t = \underline{\nabla} \bullet \left(D_{eff} \underline{\nabla} C \right) \quad (2)$$

where C was the water concentration in food and D_{eff} the effective diffusion coefficient of water in food. The subscript *eff* represented the condition that food transport properties were evaluated as effective indexes, thus accounting for a possible combination of different transport mechanisms occurring in the food.

Supposing that the above material parameters and transport properties (ρ, C_p, k_{eff}, D_{eff}), in the most general case, depended on the local values of food moisture content and of temperature, Eqs. 1-2 formed a system of unsteady, non-linear partial differential equations (PDEs). The expressions of ρ, C_p, k_{eff}, D_{eff}, as derived by Ruiz-López, Córdova, Rodríguez-Jimenes, García-Alvarado [26] in the case of carrot slices, were used (Tab. 1).

Table 1. Correlations used to estimate carrots properties [26]

$\rho = 440 + 90 \cdot X$	[Kg/m^3]
$C_p = 1750 + 2345 \cdot \left(X / 1 + X \right)$	[J/(Kg K)]
$k_{eff} = 0.49 - 0.443 \cdot \exp(-0.206 \cdot X)$	[W/(m K)]
$D_{eff} = 2.8527 \cdot 10^{-10} \exp(0.2283369 \cdot X)$	[m^2/s]

X (Kg water / Kg dry solid) is the food moisture content on a dry basis

The initial conditions were straightforward since it was assumed that before drying process actually began (t = 0), food moisture content and its temperature had definite values, i.e. C_0 and T_0, that were specified before performing each simulation.

The boundary conditions relative to Eq. 1, applied to each external surface of the food sample, where no accumulation occurred, expressed the physical condition that the heat transported by convection from air to food was partially used to raise sample temperature by conduction and partially to allow free water evaporation:

$$h(T_{db}-T_s) = -\underline{n}\bullet\left(-k_{eff}\underline{\nabla}T\right) - \lambda \cdot N_s \tag{3}$$

where λ was the latent heat of vaporization for water, N_s was the diffusive flux of water at the food surface, h was the heat transfer coefficient, T_{db} was air dry bulb temperature, T_s was the temperature at the food surface; $\underline{n}$ was a generic unity vector normal to the surface.

The boundary conditions relative to Eq. 2, applied to each external surface of food sample, where no accumulation occurred, expressed the balance between the diffusive flux of liquid water transported from the product core to the surface and the flux of vapor that left the food surface and was transferred to the drying air:

$$-\underline{n}\bullet\left(-D_{eff}\underline{\nabla}C\right) = k_c\left(C_i - C_{gb}\right) \tag{4}$$

where k_c was the mass transfer coefficient, C_{gb} was the bulk concentration of water in air, C_i was the water concentration evaluated in gaseous phase at the food/air interface.

An equilibrium relationship between the water concentration in the air and the water concentration on the food surfaces actually exposed to the drying air, was also formulated [27]:

$$\gamma_w x_w f_w = \hat{\phi}_w y_w p \tag{5}$$

The superscripts v and l (vapor and liquid, respectively) were omitted with the understanding that

γ_w, the activity coefficient of water and f_w , the fugacity of water, were referred to the

liquid phase; $\hat{\phi}_w$, the fugacity coefficient of water, was instead referred to the vapor phase. x_w and y_w were the molar fractions of water in food and in air, respectively, p was the pressure within the drying chamber. The fugacity of water was expressed by:

$$f_w = \phi_w^{sat} P_w^{sat} \exp\left[\frac{V_w\left(p - P_w^{sat}\right)}{RT}\right] \tag{6}$$

where P_w^{sat} was the vapor pressure of water, expressed in terms of the local values of temperature; the exponential term reported in eq. 6 is known as the Poynting factor.

Introducing the quantity Φ_w:

$$\Phi_w = \frac{\hat{\phi}_w}{\phi_w^{sat}} \exp\left[-\frac{V_w\left(p - P_w^{sat}\right)}{RT}\right] \quad (7)$$

The following relationship was obtained:

$$\gamma_w x_w P_w^{sat} = \Phi_w y_w p \quad (8)$$

At low pressures (up to at least 1 bar), vapor phase usually approximated ideal gases ($\hat{\phi}_w = \phi_w^{sat} = 1$) and the Poynting factor differed from unity by only a few parts per thousand; moreover, the values of $\hat{\phi}_w$ and ϕ_w^{sat} differed significantly less from each other than from unity and their influence in Eq. 7 tended to cancel. Thus, the assumption that $\Phi_w = 1$ introduced a little error for low-pressure vapor-liquid-equilibrium (VLE) and allowed simplifying Eq. 8 to:

$$\gamma_w x_w P_w^{sat} = y_w p \quad (9)$$

It is worthwhile observing that in hygroscopic materials, like most of the foods, the parameter γ_w accounts for the effects related to the amount of physically bound water so it is usually expressed as a function of both food moisture content and of its temperature [26]. Once the activity coefficient was known for the particular food under examination, Eq. 9 allowed calculating the molar fraction of water in the vapor phase and, therefore, the value of C_i appearing at the right-hand side of the above Eq. 4.

As it will be specified in the following, heat and mass transfer coefficients (Eq. 3 and 4) were estimated on the basis of the well-known semi-empirical correlations (Figure2, Table 2) expressing the dependence of Nusselt number upon Reynolds and Prandtl numbers and, by means of the Chilton-Colburn analogy, of Sherwood number on Reynolds and Schmidt numbers, respectively [25]. The above described fundamental model made use of three different correlations, one for each of the surfaces exposed to drying air, to evaluate the local values of heat transfer coefficients from the Nusselt number, *Nu*, expressed in terms of characteristic Reynolds, **Re**, Prandtl, **Pr**, and Grashof, ***Gr***, numbers [25, 28].

The Prandtl number was defined as $\mathrm{Pr} = {C_{pa} \cdot \eta_a}/{k_a}$. In the previous relationships, *x* and *y* defined the local position on each of the food surfaces; $h_{(x,y)|i}$, characterized by a numeric subscript (*i*) corresponding to each boundary, was the local value of heat transfer coefficient; u_0 was air velocity at the inlet section of the drier; *g* was the acceleration of gravity. All the physical and transport properties defining each dimensionless number were evaluated at film conditions (the subscript *f* was omitted). It is worthwhile observing that, as far as the rear face of food sample was concerned, the Nusselt number was actually expressed in terms of the

Grashof number since it was supposed that air circulation in proximity to boundary 3 was so limited that free convection prevailed over forced convection. Once the heat transfer coefficients was estimated, the Chilton-Colburn analogy was used to calculate the mass transfer coefficients referred to each corresponding boundary [25].

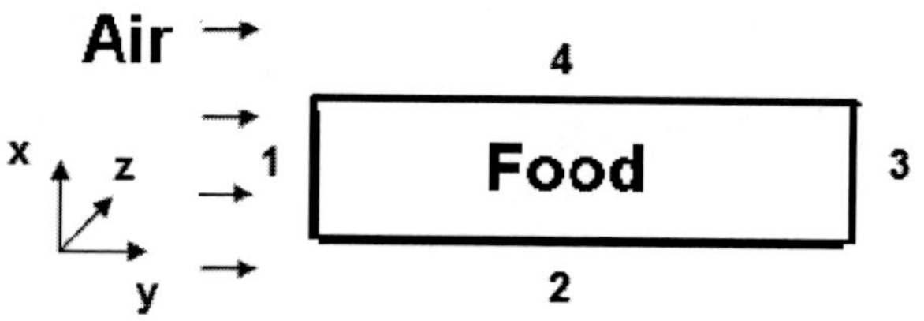

Figure 2. Schematic representation of the carrots sample.

Table 2. Semi-empirical correlations used to estimate heat transfer coefficients on each food surface

Boundary	Semi-empirical relationship	Definition of dimensionless numbers
Impact surface (identified as boundary 1)	$Nu_x = 0.25 \mathrm{Re}_x^{0.588} \mathrm{Pr}^{1/3}$	$Nu_x = h_{x\|1} \cdot x / k_a$; $\mathrm{Re}_x = \rho_a \cdot u_0 \cdot x / \eta_a$
Surface parallel to air flow (identified as boundaries 2 and 4)	$Nu_y = 0.648 \mathrm{Re}_y^{0.5} \mathrm{Pr}^{1/3}$	$Nu_y = h_{y\|2,4} \cdot y / k_a$; $\mathrm{Re}_y = \rho_a \cdot u_0 \cdot y / \eta_a$
Rear surface (identified as boundary 3)	$Nu_x = 0.59 (\mathrm{Pr} \cdot Gr_x)^{1/4}$	$Nu_x = h_{x\|3} \cdot x / k_a$; $Gr_x = x^3 \cdot \rho_a \cdot g \cdot \Delta\rho_a / \eta_a^2$

Thin-layer Model

Other kinds of simplified approaches were proposed in the literature to model the drying process of vegetables. Among these models, the so-called *thin-layer equation* [29-32] is actually the most common. The *thin-layer equation* is based on the assumption that moisture decrease is proportional to the instantaneous difference between material moisture content (assumed uniform within the food) and the moisture content in equilibrium with drying air. In addition, the model assumes a uniform temperature distribution within the food sample. The above assumptions led to a unique equation modeling the drying behavior:

$$\frac{dX}{dt} = k \cdot (X - X_e) \tag{10}$$

where X (Kg $_{water}$ / Kg $_{dry\ solid}$) was the food moisture content on a dry basis, X_e was the moisture content in equilibrium with drying air and t was the drying time. The parameter k was the so-called *drying constant* and represented a measure of drying process rate; it is a function of material moisture content, of the product size, of the uniform temperature distribution developing in the food and, finally, of air characteristics, i.e. its humidity, temperature and velocity. Several thin-layer models were proposed [32], differing from each other for the number of parameters to be estimated. In the case of Eq. 10, also known as the "Newton's model", only one parameter was considered. Integrating Eq. 10, the following relationship, resulting in an exponential equation according to a pseudo-first order reaction kinetics [29, 33], was obtained:

$$\frac{(X - X_e)}{(X_o - X_e)} = \exp(-k \cdot t) \tag{11}$$

where X_o (Kg $_{water}$ / Kg $_{dry\ solid}$) was the initial moisture content evaluated at t=0. The Newton's model is usually used by fitting the experimental data, expressed as the time evolution of food moisture content on a dry basis, thus estimating the k value. Because of its main features, the procedure can be applied only when a series of measurements has been actually obtained for a particular drying experiment, whereas it is meaningless when no experimental data exist.

Artificial Neural Network (ANN) Model

Artificial neural networks (ANNs) are a data-driven method capable to learn from examples capturing the functional relationships existing between the input(s) and the output(s). This feature makes ANNs a particularly useful tool to model phenomena difficult to be described by a model-based approach because no a priori knowledge is required. ANNs are composed of interconnected computational elements called neurons or nodes. Each neuron receives input signals from the related units, elaborates these stimuli by an activation or transfer function and generates an output signal that can be transferred to other neurons. Any kind of differentiable mathematical function could be used as neuron activation function, but the most common, due to their forecasting capability, are the sigmoid logistic function, the hyperbolic tangent function and the linear function. Even if the prediction of each single neuron could be imperfect and bias-affected, the outcome of the interconnection(s) among neurons is a computational tool capable to learn from examples and to provide accurate predictions even with examples never seen before [34]. Neurons are organized in a multi-layer structure which allows obtaining the output(s) signal starting from a definite set of the input(s). The interconnection between the nodes takes place by means of a weight, a constant that reflects the strength of the connection that is responsible for the signal propagation. Many different artificial neural network structures were proposed but the most common is the multi-layer perceptron (MLP). A MLP is composed by an input layer that receives the input information about the process, an output layer that produces the response(s) of the neural network and a certain number of hidden (intermediate) layers that are located between the input and the output layers [35].

To determine the network structure, it is necessary to specify the following information [35]: a) the number of both input and output variables; b) the number of layer (s); c) the number of neurons comprised in each layer; d) the activation function of each neuron. The number of input and output variables is problem-dependent since it is related, respectively, to the number of independent and dependent variables of the problem under consideration. On the contrary, the number of layers and the number of neurons in each layer is the result of an optimization process. Even if several methods were proposed to accomplish this task [36-38], a general procedure was not yet proposed and the network architecture is usually determined according to heuristic guidelines and trial and error procedures. After determining the network architecture, to complete the network definition, it is necessary to perform the so-called neural network training. During the training phase the network learns how to correlate the input to the output variables. More specifically, the network is submitted to a certain number of input and output data and, according to an error minimization algorithm, it changes the network weights values that could be considered as the key elements in which the network knowledge is stored [39]. Only a certain number of the available experimental points are used during the training phase; the remaining experimental points are used during a post-training analysis, called the test phase. During the test phase, the neural network is called to predict the output values corresponding to an input combination never exploited before and, therefore, a set of experimental points that do not belong to the training set. Even if the test phase is a post-training analysis only, it usually plays a fundamental role in the ANN architecture definition; in fact, the convergence of the above-mentioned trial-and-error procedure is usually considered achieved with reference to a performance index just based on test points. The test phase of the network is usually performed in the definition domain in which the training procedure was achieved. As a matter of fact, the forecasting capability of the neural networks outside this definition range cannot be guaranteed; due to the intrinsic black-box nature of neural networks models, the validity domain does indeed strictly depends on the range of data used in model definition [40]. Another kind of post-training analysis about neural network performance, it is the so-called validation phase. Similarly to the test phase, the network is called to predict the experimental points excluded from both the training and the test sets; unlike test phase, the evaluation of model prediction reliability does not influence the definition of neural network architecture.

Some authors developed ANNs models to describe the drying process of different vegetables: carrots [41], ginseng [42], tomato [43], cassava and mango [44]. In the present contribution the forecasting capability of neural networks was utilized so as to predict the time evolution of carrots drying. A preliminary theoretical analysis of the process was carried out with the aim of choosing the input and the output variables that resulted as the most representative of process dynamics. This step holds a fundamental position in neural network definition owing to the black-box nature of neural modeling. The output variables, in fact, should be sufficient to exhaustively describe the process dynamics as well as the input variables should be sufficient to properly predict the time evolution of the chosen output variables.

Drying processes are aimed at water removal from a matrix and, as a consequence, the variable chosen as the neural network output was the moisture content of the carrot sample;

actually, the time evolution of dimensionless moisture content, *M(t)*, was considered as more representative of process behavior and, therefore, of major interest as far as drying modeling was concerned:

$$M(t)=\frac{X(t)-X_e}{X_o-X_e} \qquad (12)$$

The choice to use a dimensionless form of carrots moisture content was taken so to allow an easy comparison between neural network and thin-layer model predictions. Moreover, by definition, at the beginning of each drying run *M (t=0)* had a value of 1, independently on the sample chosen to perform the experiment.

Hybrid neural model (HNM)

In the previous sections, different alternative modeling approaches of food convective drying were presented showing that the dynamics of the same process can indeed be described in a completely different way. The existence of so many modeling alternatives is due to the intrinsic complexity of the drying phenomenon that involves, from a physical standpoint, a simultaneous transfer of heat and of water (both as liquid and as vapor) taking place by convection, conduction and diffusion.

Theoretical models describe the process dynamics by means of a fundamental approach that usually results in coupled nonlinear partial differential equations and, as a consequence, in numerical simulations that are time consuming and difficult to be incorporated in an on-line control software.

On the other side, black-box models are able to describe only the input-output dynamics of the process without accounting for any physical relationship characteristic of the system under consideration. Empirical models can approximate the drying kinetics by several line segments, high order polynomials and neural networks: the common characteristic of these models is that they have a narrower validity range but require only a limited number of simple arithmetic operations [45].

A reasonable trade-off between theoretical and empirical approach is represented by hybrid modeling, leading to a so-called "grey-box" model capable of good performance in terms of data interpolation and extrapolation. Hybrid model predictions are given as a combination of both a theoretical and a "pure" neural network approach, together concurring to the obtainment of system responses. The main advantage of hybrid modeling regards the possibility of describing some well-assessed phenomena by means of a theoretical approach, leaving the analysis of other aspects, very difficult to interpret and describe in a fundamental way, to rather simple "cause-effect" models [40, 46-49]. Two kinds of HNMs can be generally defined depending on the interactions existing between the neural and the theoretical blocks. In a model based on a parallel architecture (Figure 3) the inaccuracy in the predicted value from the fundamental part is minimized by the addition of the residuals calculated by the neural network. In a model based on a series architecture (Figure 4) a process variable, which is difficult to measure, is estimated by a neural network and, then, fed to the theoretical block as an input to it. Finally, the output coming out from the fundamental part is checked with the experimental value for convergence.

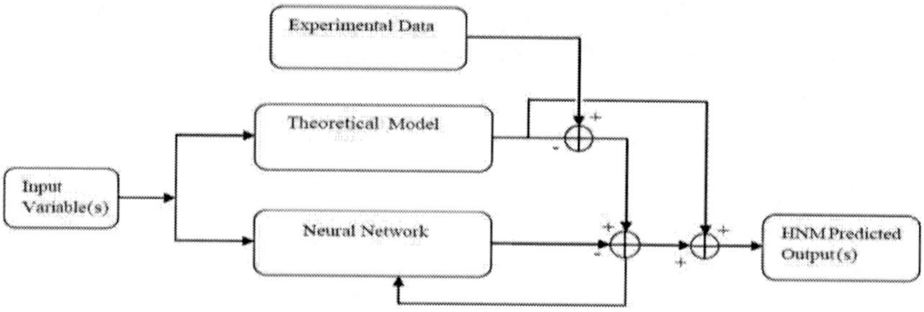

Figure 3. HNM structure based on a parallel architecture.

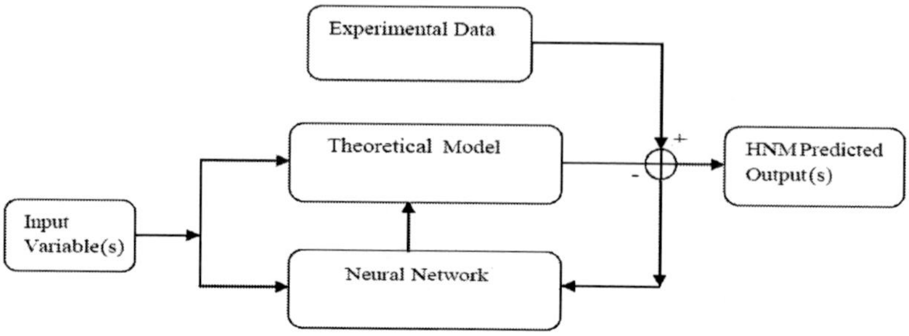

Figure 4. HNM structure based on a serial architecture.

Even if the hybrid neural approach is, in principle, characterized by a higher generalization capability than artificial neural networks, no specific work dealing with food drying modeling by HNM has been published in the literature.

MODELS IMPLEMENTATION

Theoretical Model Development

The above-described theoretical model leading to a system of unsteady non-linear partial differential equations (Eqs. 1-2) was solved by the Finite Elements Method (FEM) developed by the commercial package Comsol Multiphysics 3.4. The domain corresponding to food sample was discretized into 3550 triangular finite elements with a thicker mesh close to the boundaries actually exposed to the drying air. The total number of degrees of freedom was equal to about 15000. On a dual core personal computer running under Linux, a typical drying process duration of 5 hours was simulated, on average, in about 2 minutes, using the time-dependent nonlinear direct solver already implemented in the Comsol package. It is worthwhile remarking that the proposed model does not need any parameters adjustment or experimental data from drying experiments. Only the relationships expressing both physical and transport properties as a function of food temperature and moisture content and the specification of a set of input variables that can be varied within a definite range of physical significance are actually necessary to simulate food drying behavior at different operating conditions.

Experimental Runs

In order to develop the ANN, the hybrid and the thin-layer models, several experiments of carrots drying were performed. Carrot samples of different dimensions were dried by air in a lab-scale convective dryer (Memmert Universal Dryer model UFP 400). The carrots, bought in a local market, were cut in slab shapes. The slab side (L) had an initial value of 30 mm. Three different values of initial slab thickness (d), i.e. 5, 10 and 15mm, were chosen to perform the present experimental analysis thus leading to a total number of three kinds of food samples. For each sample the weight was monitored, with respect to time, by a precision balance (Mettler AE 160), with an accuracy of ± 0.0001 g. The lab-scale dryer allowed monitoring air temperature, by a Dostmann electronic Precision Measuring Instrument (P 655), its humidity (by a rh 071073 probe) and the inlet velocity, by a H 113828 probe. The convective flow of drying air was obtained by a line of fans placed along the edge of dryer internal tray. Two values of air velocities, i.e. 2.8 and 2.2 m/s, and three values of air dry bulb temperature, i.e. 50˚C, 70˚C, 85˚C, were chosen; air absolute humidity, was kept constant throughout all the experiments and equal to 10.32 g water/m^3 dry air. The food samples were placed on a wide-mesh perforated tray. The dryer characteristics allowed analyzing six samples at the same time. Food weight was periodically measured during each experiment; each test was repeated twice to ascertain its reproducibility. The difference between instantaneous food weight and dry solid weight allowed evaluating the total amount of water contained in food. The operating conditions chosen to perform the present experimental analysis are summarized in Table 3.

Table 3. Operating conditions of the experimental analysis

	d = 5mm	d = 10mm	d = 15mm
Air conditions (uo, Tdb)	Run N°	Run N°	Run N°
2.2m/s, 50°C	1	7	13
2.8m/s, 50°C	2	8	14
2.2m/s, 70°C	3	9	15
2.8m/s, 70°C	4	10	16
2.2m/s, 85°C	5	11	17
2.8m/s, 85°C	6	12	18

Thin Layer Model Development

To utilize the thin-layer model (Eq. 10) the experimental results corresponding to each of the performed runs (Table 3) were fitted so as to estimate the drying constant *k*. A commercial curve fitting software package was used to accomplish this task and to calculate all the parameters having a statistical significance. To calculate the moisture content in equilibrium with drying air, X_e, the psychrometric diagram was used. For the fixed value of air absolute humidity in which all the experiments were performed, the X_e values were equal to 0.02014, 0.0094, 0.00576 for the runs performed at an operating temperature of 50, 70 and 85° C, respectively.

ANN Model Development

With the aim to predict the time evolution of carrots moisture content, a set of significant input variables, i.e. the drying time (t), the dry bulb temperature (T_{db}), the air velocity (u_o), the characteristic sample size (d) and the relative humidity (U_r), was identified. From a preliminary sensitivity analysis it was showed that the above variables, among all the parameters that could affect the drying progress, exhibited the highest influence on the transport phenomena involved in the process and, therefore, on its performance. The input-output structure of the developed neural model is reported in Figure 5.

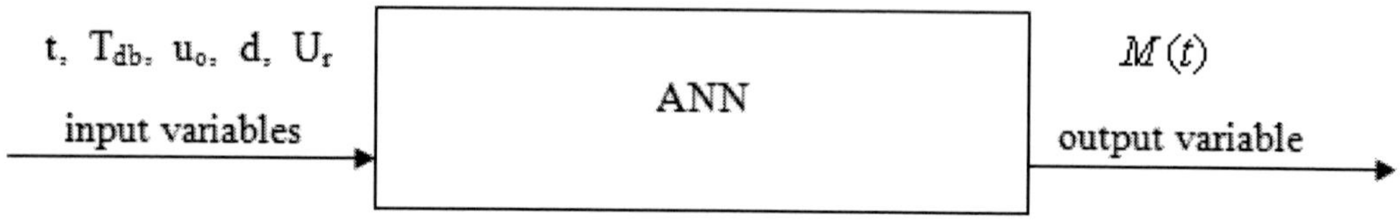

Figure 5. Neural model structure

After specifying the input-output structure of the model, the neural network architecture was defined and the training procedure was set-up: the experimental points relative to the 18 drying runs, were randomly split into three groups, reserving 2/3 of data (12 runs) to the training phase, 1/6 of the remaining data (3 runs) were used to test the predictions of the neural network during its development, whereas the residual 3 runs were used to validate the ANN predictions in three conditions never exploited neither during learning, nor during test. A multi-layer perceptron (MPL) feed-forward architecture with a pyramidal structure, having a decreasing number of neurons from input to output layer, was developed and implemented by Matlab Neural Network Toolbox, Ver. 4.0.1. To train each of the tested networks, thus estimating their weights and their biases, the Bayesian regularization was used [39]. The neuron transfer function was a hyperbolic tangent for both input and intermediate layers, whereas a linear transfer function was chosen for the output layer.

The choice of the "best" network architecture was realized by a trial-and-error procedure as it is suggested in the literature [50, 51]. The number of hidden levels and the number of neurons belonging to every single layer were determined through iterative cycles according to the block diagram shown in Figure 6.

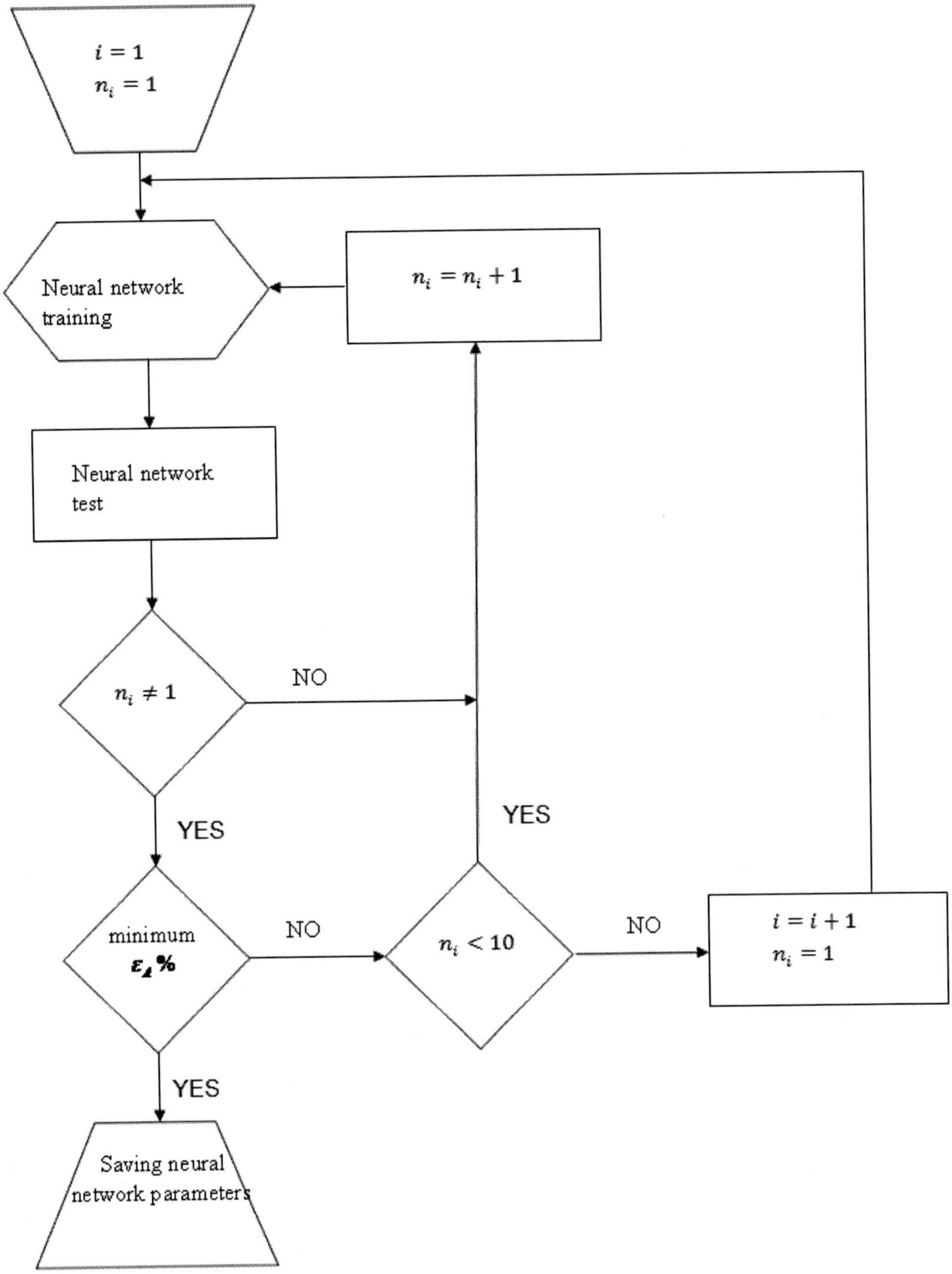

(i: generic hidden layer, n_i: number of neurons in the ith layer)

Figure 6. Algorithm for ANN development

It was supposed that the procedure convergence was achieved when, during the test phase, the percentage average error (calculated on the basis of the whole test dataset) between the predictions of the neural model and the corresponding experimental points,

$\left(\varepsilon_A\% = \frac{\left|M(t)_{\exp} - M(t)_{ANN1}\right|}{\min(M(t)_{\exp} - M(t)_{ANN1})} \cdot 100 \right)$ reached a minimum, set equal to 10%. On the basis of the above-described iterative procedure a neural model, ANN1, was eventually developed; it consisted of three different layers, according to a rather simple architecture (Figure7):

- An input layer with five neurons;
- A single hidden layer containing two neurons;
- A single neuron output layer.

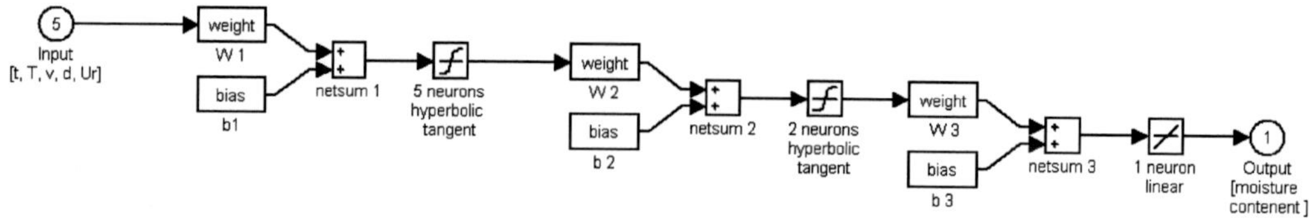

Figure 7. ANN1 structure

The training procedure reached convergence within 196 epochs, i.e. the number of iteration of the back-propagation algorithm; the sum of squared errors (SSE) registered during the training phase was equal to $4.9 \cdot 10^{-2}$.

Hybrid Neural Model Development

The choice to model the drying process by means of a hybrid neural model HNM could be considered as a consequence of a critical analysis of the other approaches (theoretical and empirical) presented in the previous sections to model convective drying of foods. This analysis showed the inherent limits of both theoretical and empirical models in describing vegetables drying. Theoretical models, in fact, implied the utilization of rather complicated numerical methods, whereas "cause-effect" models did not make use of any fundamental equation that, instead, might be helpful to achieve a more precise knowledge of the dynamic process under study. On developing the "theoretical" part of a HNM, in general, it should be advisable to adopt rather simple equation(s) so as to avoid the introduction of additional difficulties related to the solution of the equations. On the basis of the above consideration, the HNM was realized coupling a rather simple theoretical relationship, as represented by the solution of thin-layer model (Eq. 11) and a neural network aimed at determining, by the estimation of a model parameter (the drying constant *k*), the relationship between the process rate and the operating variables influencing drying behavior. Finally, the chosen model architecture was a classical serial model structure (Figure 8).

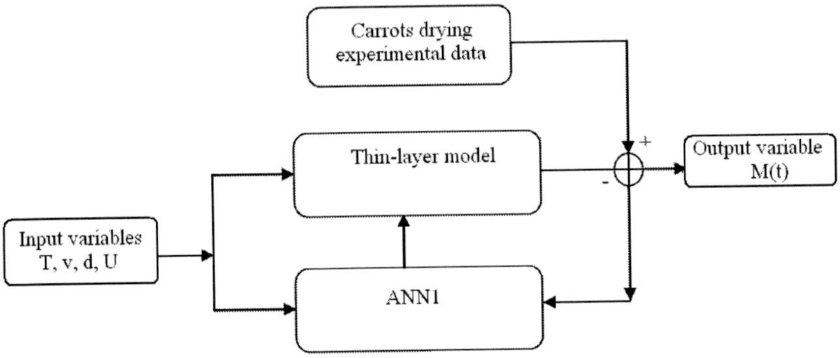

Figure 8. Hybrid neural model structure

Having determined the general architecture of the HNM, the inputs to the neural block were specified. It is worthwhile evidencing that, with respect to the pure neural model, it was not necessary to input the process time, since the thin-layer model is capable of calculating, once the *k* parameter has been estimated by the neural network, the time evolution of food moisture content. Actually, on developing a pure neural model referred to an intrinsic transient process, it was necessary to explicitly feed the network with an array of process times that permitted calculating how food moisture content (the network response) changed with time under the influence of the operating variables. On the contrary, in the HNM, the time-dependency was directly provided by the theoretical part of the model, i.e. by the solution (Eq. 11) of the ordinary differential equation represented by Eq. 10. On the basis of the above considerations, it is expected, therefore, that the neural model architecture contained in the HNM is much simpler than that of pure neural model, thus achieving a significant reduction of simulation time. Another interesting improvement related to the formulation of the present hybrid approach concerned the actual role played by the drying constant *k*, i.e. an ignorance parameter in which to include everything that might be difficult to express by proper mathematical relationships. Comparing the pure neural model and the hybrid neural model structures, it is worthwhile noting that the role played by the neural network is certainly less important in hybrid modeling, according to the initial statement that hybrid models are to provide a "less-dark" description of physical phenomena. Moreover, the theoretical part of the hybrid model plays a sort of filter function with respect to neural network model prediction; the estimated values of *k*, in fact, are used in the integration of Newton equation and, therefore, it is expected that the predictions errors due to the pure black-box nature of the model are limited.

After the HNM definition, it was necessary to define the neural part of the model, thus specifying the neural network architecture and set-up the training procedure. Similarly to the pure neural model, the experimental points relative to the 18 drying runs, were randomly split into three groups, reserving 2/3 of data (12 runs) to the training phase, 1/6 of the remaining data (3 runs) were used to test the predictions of the HNM during its development, whereas the residual 3 runs were used to validate the hybrid neural model predictions in three conditions never exploited neither during learning, nor during test.

In order to allow a proper comparison between different approaches, the runs chosen for both the test and the validation phases of the hybrid model were the same of those used to develop the pure neural model. A MLP feed-forward architecture was used and the same transfer functions and the training procedure described in the previous section to develop ANN1 were utilized. A second neural network, ANN2, was finally achieved; ANN2, which – as expected - is more slender than ANN1, had the following characteristics (Figure9):

- An input layer with two neurons and a hyperbolic tangent transfer function;
- A single neuron output layer with a linear transfer function.

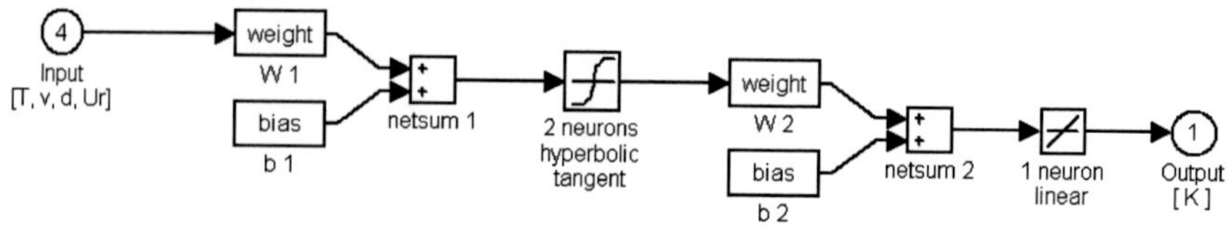

Figure 9. ANN2 structure

Comparison of Models Performance

All the developed models were tested in the experimental conditions summarized in Table 3 to verify the reliability of their predictions and to compare their performance. Before analyzing some of the most interesting results, it is worthwhile remarking the main differences existing among thin-layer, fundamental and ANN models. Actually, any fundamental model requires a deep knowledge of the transport phenomena occurring in the process, whereas both the thin–layer and ANN models require a great number of experimental data to achieve a higher accuracy. Both physical modeling and experimental tests are highly time-consuming; nevertheless, physical modeling is generally more complex than the experimental procedure. In most cases, also a fundamental approach requires a set of experimental data; in particular, both the physical and the transport properties of foods, necessary to solve the heat and mass balance equations, are very difficult to estimate and are available only for certain types of foods. A fundamental model is capable to describe the profiles of moisture content developing in food as a function of space coordinates, thus allowing determining, for instance, if the local moisture content is still so high to promote microbial spoilage. Thin-layer and ANN models are capable to predict the time-evolution of average moisture content, since the collected experimental results usually do not provide any information about temperature and moisture content distributions. A complete analysis of the transport phenomena occurring in food is very difficult to be modeled by a fundamental approach due to the structural and thermodynamic modifications, not yet fully understood, involved in food drying; for this reason, ANN and thin-layer models represent an attractive alternative since they require a series of accurate experimental tests that can be obtained much more easily than an accurate and comprehensive physical and mathematical modeling.

The results obtained during the fitting procedure, obtained using the thin-layer model, are summarized in Table 4, where, as well as the *k* values relative to each experimental run, the most meaningful statistical parameter are reported.

The very high value of r^2 coefficient indicates a very good agreement between the Newton's model and the experimental drying curves. For all the experiments an increase in drying constant value is observed, as both air temperature and its velocity increase and food characteristic dimension decreases. This last effect is to be ascribed to three main factors occurring at lower thickness: the shorter path through which water is transported within the solid matter; the increased exposed surface per unit volume; the decrease of both internal and external mass transfer resistances [52].

Figs. 10, 11 and 12 show, in some typical conditions, the drying curves obtained using three of the proposed models (the HNM performance will be analyzed more in detail afterwards). As far as ANN model is concerned, the predictions shown in Figs. 10, 11 and 12 are referred to some experiments belonging to the training, the test and the validation datasets, respectively. It is worthwhile noting that fundamental model exhibits a lower accuracy with respect to thin-layer approach that, instead, gives more reliable predictions.

As a general comment, it should be observed that the formulated theoretical model was a rather simple one that estimated the transfer rates at the food/air interface by a set of semi-empirical correlations, thus not accounting for the actual influence on drying rate of air velocity field developing in the drying chamber. Moreover, the proposed simplified fundamental model was based on a unique mass balance equation, expressing the transport of total moisture in food, thus not accounting for the actual transport of water both as liquid and as vapour. As far as the ANN model predictions are concerned, it can be observed that when the experimental data belong to the training set (Figure 10) ANN gives a very good representation of the actual time evolution of drying process. When, instead, the experimental data belong to the test set (Figure 11) and the neural network is called to predict a condition never exploited during the training process, but actually used to achieve the final network architecture, a lower accuracy of ANN model is observed with maximum relative errors between ANN predictions and experimental data of about 10%. Finally, when the experimental data belong to the validation set (Figure 12), i.e. to an input combination definitely unknown to the network, an even lower accuracy of ANN model is observed with relative errors between ANN predictions and experimental data reaching at most 40 %. Thin-layer model, as already reported in Table 4, gives always very good performance since it actually represents a best-fitting procedure (actually based on a simple theoretical model) of existing experimental data; however, it is worthwhile remarking that, due to its main features, thin layer model cannot be absolutely applied to any situation never exploited during the experimental tests.

From this preliminary analysis, it is possible to derive some interesting considerations. A) The theoretical model should be as accurate as possible to provide reliable predictions; even small errors in the estimation of transfer coefficients lead to huge deviations between estimated and real values of temperature and moisture content, thus determining an unfair design of drying equipments or severe problems during food processing. Since the developed theoretical model did not depend on any adjustable parameter, it was capable of estimating drying behavior over a wide range of operating conditions. However, the formulation of so many simplification hypotheses, necessary to reduce the numerical complications and to achieve the PDEs solution in a reasonable time so as to allow the incorporation of the model in an on-line control system, determined significant discrepancies between the theoretical predictions and the measured data. B) The developed neural network model gave very accurate predictions of actual system behavior when it was tested within the range used for training.

Table 4. Results of drying curves fitting procedure using Newton model

	d= 5mm				d = 10mm				d = 15mm			
Air conditions (uo, Tdb)	r^2	*k* [1/min]	*k* Standard Error	Test N°	r^2	*k* [1/min]	*k* Standard Error	Test N°	r^2	*k* [1/min]	*k* Standard Error	Test N°
2.8m/s, 85°C	0.987	3.84E-02	0.00193	6	0.992	1.84E-02	0.000467	12	0.993	1.33E-02	0.000248	18
2.8m/s, 70°C	0.994	2.85E-02	0.001121	4	0.999	1.20E-02	0.000151	10	0.99	8.30E-03	0.000192	16
2.8m/s, 50°C	0.978	1.45E-02	0.000827	2	0.995	7.90E-03	0.000149	8	0.988	5.88E-03	0.000124	14
2.2m/s, 85°C	0.991	2.93E-02	0.001076	5	0.995	1.34E-02	0.000222	11	0.999	9.17E-03	6.01E-05	17
2.2m/s, 70°C	0.995	2.06E-02	0.000623	3	0.996	1.02E-02	0.000194	9	0.995	7.17E-03	0.000109	15
2.2m/s, 50°C	0.988	1.00E-02	0.000366	1	0.984	6.00E-03	0.000171	7	0.997	4.36E-03	4.05E-05	13

However, ANN model exhibited worse performance when it was called to interpret the time evolution of drying process in conditions never exploited during the learning phase (test and validation), thus proving that extrapolation based on neural model predictions is definitely unreliable when it is performed outside the training range. C) The thin layer model offered very precise predictions, but represents an invalid and an inapplicable procedure in those cases in which no experimental data is actually available since it cannot allow the estimation of drying constant, *k*; however, due to its "semi-theoretical" nature, thin layer model may provide useful indications about the time dependency of food moisture content.

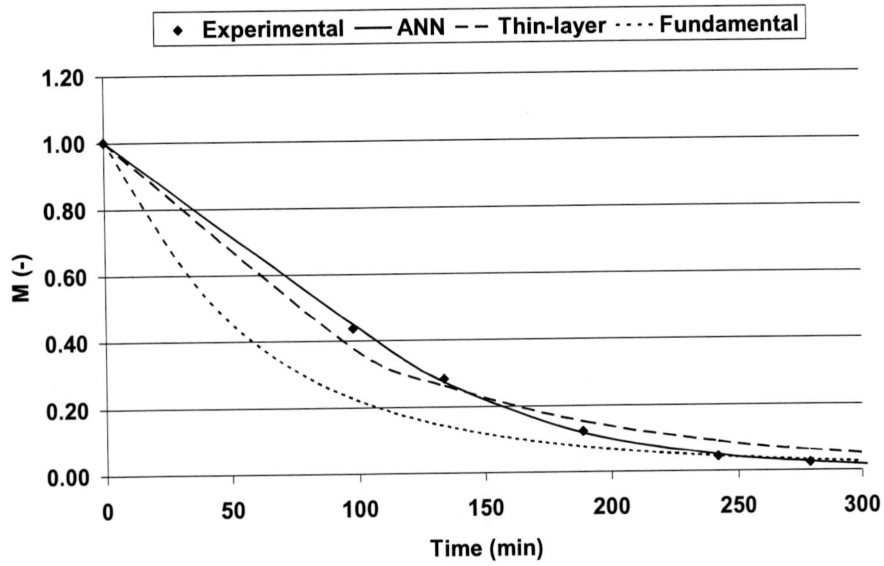

Figure 10. Simulation results relative to run N°1 (training phase). Operating conditions: Tdb=50; u_0=2,2; d=5 mm.

As described in the previous sections, a Hybrid Neural Model (HNM) may, in principle, combine the best features of the above-described models. For instance, it could allow extending the applicability of thin-layer model on the basis of the learning ability provided by an artificial neural network model, which operates as a parameter(s) estimator and allows evaluating, by means of drying constant, a series of complex phenomena difficult to be expressed by proper mathematical relationships.

Figure 13 shows a comparison between HNM predictions, considering only the points belonging to both test and validation datasets and the corresponding experimental results, expressed in terms of dimensionless moisture content M(t) (Eq. 12). It is worthwhile observing that the neural part of the hybrid model (ANN2) was used so as to predict the values of drying constant, *k*, in situations never exploited during the training phase. The choice to refer to test and validation datasets is due to the necessity of evaluating the forecasting capability of HNM model in those situations in which ANN1 showed rather unreliable predictions, with relative errors reaching about 40 %. A remarkable agreement between the model predictions and the experimental points can be observed since a straight line characterized by unitary slope and a nil intercept was used to fit the data with a very high correlation coefficient, $R^2 = 0.99$.

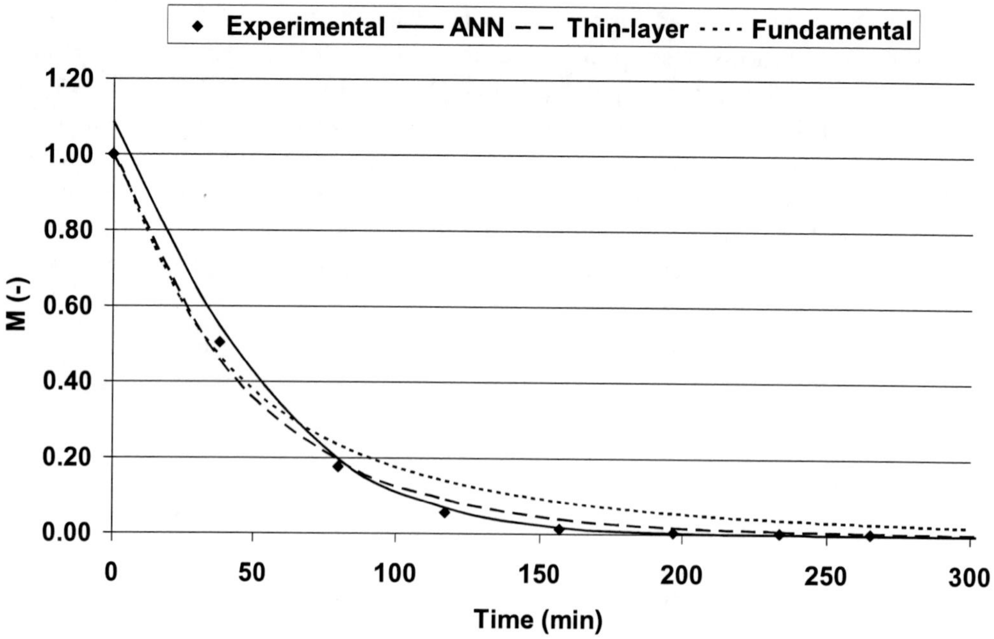

Figure 11. Simulation results relative to run N° 3 (test phase). Operating conditions: Tdb=70; u_0=2,2; d=10 mm.

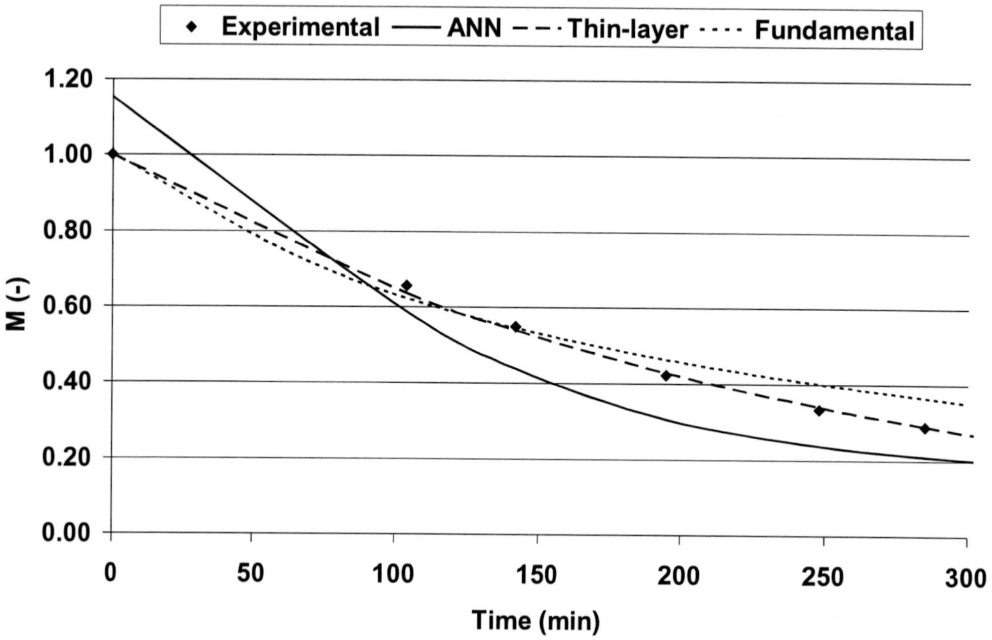

Figure 12. Simulation results relative to run N° 13 (validation phase). Operating conditions: Tdb=50; u_0=2,2; d=15 mm.

Figs. 14-16 show a comparison, expressed as the time evolution of dimensionless moisture content, between HNM and ANN1 model predictions during both test and validation phases. From Figure 14, obtained considering a set of experimental points belonging to the test dataset, it can be observed that both the models performance is comparable since HNM and ANN predictions are in good agreement with the experimental data. This behavior is to be ascribed to the trial and error procedure used to define ANN1 and ANN2 architectures and, in particular, to the control performed on percentage error even during the test phase.

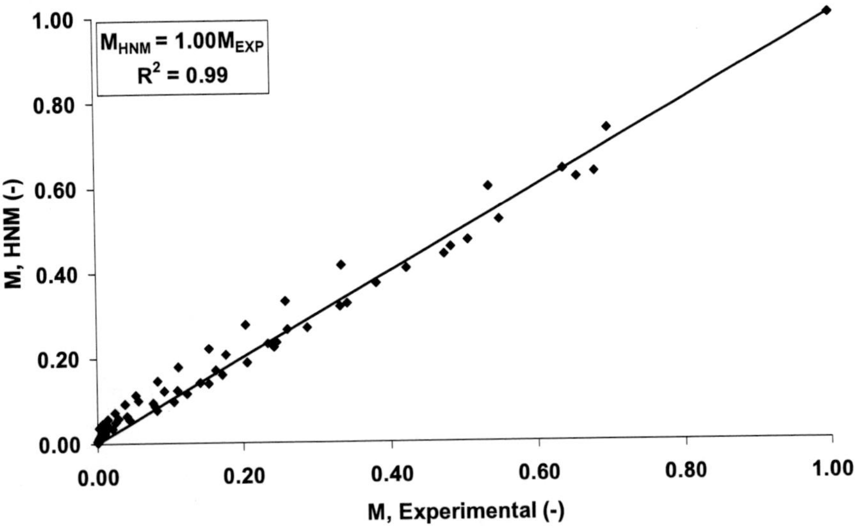

Figure 13. HNM performance

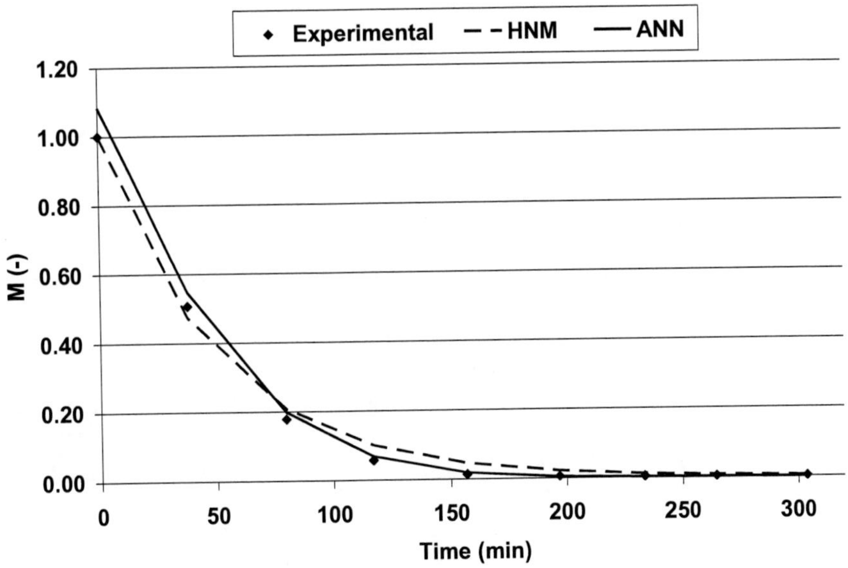

Figure 14. Comparison between HNM and ANN model (test phase). Run N°3: Tdb=70; u_0=2,2; d=5 mm.

However, HNM is more accurate in predicting the system behavior at the beginning of drying process due to its semi-theoretical nature that does indeed require the precise knowledge of an initial condition, which has to be fixed before each HNM simulation is performed.

Figs. 15-16 show a comparison between HNM and ANN model performance in two typical situations, belonging to the validation dataset and, therefore, definitely never exploited during the training phase by both ANN1 and ANN2. Actually, the behavior of the two models is rather different since HNM shows excellent prediction ability, comparable to that obtained during the test phase, whereas the pure ANN model is not capable to accurately reproduce the measured decrease of carrots moisture content.

As a matter of fact, the proposed HNM is characterized by a high level of reliability and, therefore, represents a powerful tool that, as compared to all the other models described in the present contribution, allows obtaining very precise predictions of the actual system behavior in all the tested conditions. The obtained results demonstrate that the combination of an even simple theoretical model with a straightforward neural model consisting of two neurons only is capable to fairly widen the applicability of pure neural models outside the training range, thus strengthening the model performance. The theoretical part of HNM does indeed play a role of filtering function with respect to ANN2 model predictions, thus limiting the errors introduction typical of a black-box model and determining a generalized improvement of model performance. On the contrary, the inherent data-based nature of pure artificial neural networks is responsible for a narrower validity of ANN models that, actually, makes any extrapolation outside the training range improper and uncertain.

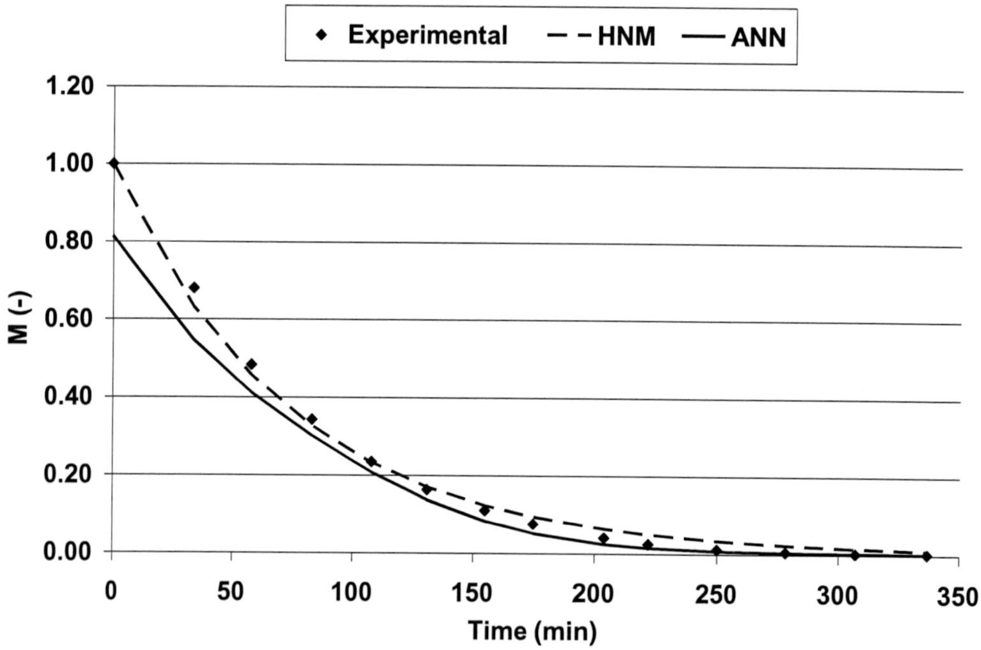

Figure 15. Comparison between HNM and ANN model (validation phase). Run N°11: Tdb=85; u_0=2,2; d=10 mm.

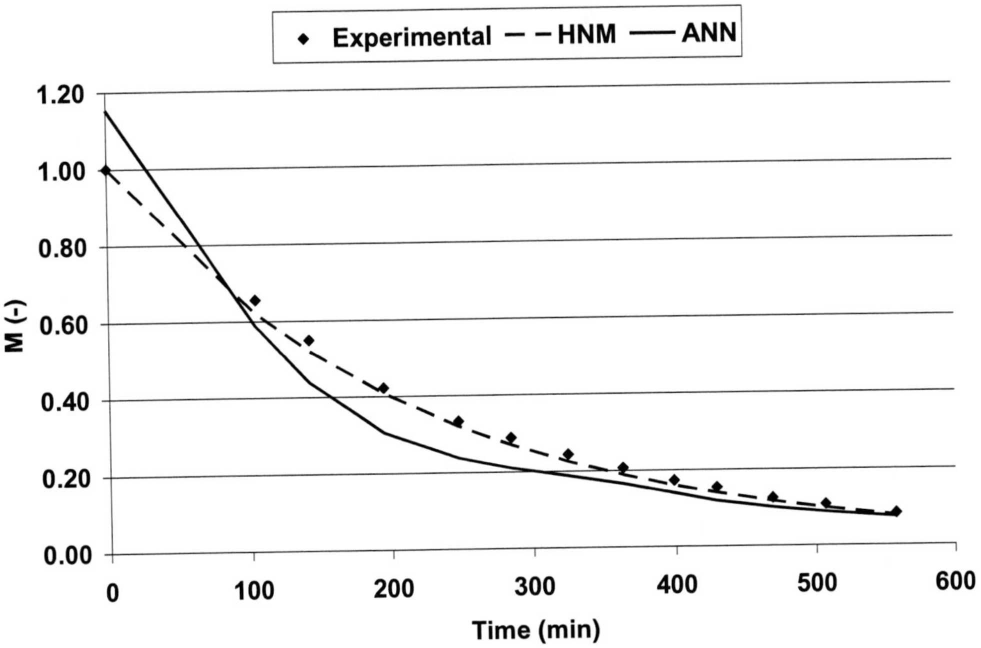

Figure 16. Comparison between HNM and ANN model (validation phase). Run N°13: Tdb=50; u_0=2,2; d=15 mm.

CONCLUDING REMARKS

In the present contribution different approaches of food convective drying modeling were presented and critically analyzed with the aim of describing the advantages and the drawbacks of each of them. Theoretical modelling, in principle, could predict the drying behavior of foods available in all shapes and over a wide range of process and fluid-dynamic conditions. However, when a fundamental model was based, as in the present case, on a great number of simplification hypotheses, significant discrepancies between the theoretical predictions and the measured data were generally observed. Among the straightforward theoretical models, it was showed that the thin-layer model provided useful and accurate indications about the time evolution of food moisture content, even though it did not allow any generalization being inapplicable when no experimental data were actually available.

On the other side, the developed neural network model reproduced very well the actual system behavior when the inputs combination belonged to the chosen training range, but it was characterized by unfair predictions when it was tested with a set of experimental points never exploited during network development (validation points). The hybrid paradigm, as proposed in the present contribution, was characterized by a high level of reliability and represented a powerful tool offering very precise predictions of the actual system behavior. The obtained results demonstrated that the proper combination of an even simple theoretical model with a straightforward neural model was capable to fairly widen the applicability of pure neural models outside the training range, thus allowing the utilization of HNM for process optimization purposes and for the implementation of efficient on-line control applications.

ACKNOWLEDGMENT

This work was supported by: Food Science & Engineering Interdepartmental Center of University of Calabria and L.I.P.A.C., Calabrian Laboratory of Food Process Engineering (Regione Calabria APQ-Ricerca Scientica e Innovazione Tecnologica- I atto Integrativo, Azione 2- Laboratori Pubblici di Ricerca "Mission oriented" Interfiliera).

REFERENCES

[1] Datta, A. K. Porous media approaches to studying simultaneous heat and mass transfer in food processes. I: Problem formulations. *Journal of Food Engineering*, 2007, 80, 80-95.

[2] McCabe W. L.; Smith J. C.; Harriot P. Unit operations of chemical engineering; McGraw-Hill Int.; Singapore, 1985.

[3] Wang, W.; Chen, G.; Mujumdar, A.S. Physical Interpretation of Solids Drying: An Overview on Mathematical Modeling Research. Drying Technology 2007, 25, 659–668.

[4] Chin, S. K.; Law, C. L.; Supramaniam, C. V.; Cheng, P. G. Thin-Layer Drying Characteristics and Quality Evaluation of Air-Dried Ganoderma Tsugae Murrill. *Drying Technology*, 2009, 27, 975–984.

[5] Bon, J.; Rossello, C.; Femenia, A.; Eim, V.; Simal, S. Mathematical Modeling of Drying Kinetics for Apricots: Influence of the External Resistance to Mass Transfer. *Drying Technology*, 2007, 25, 1829–1835.

[6] Arrieche L. S.; Sartori, D. J. M. Fluid Flow Effect and Mechanical Interactions during Drying of a Deformable Food Model. Drying Technology 2008, 26, 54–63.

[7] Zuniga, R.; Rouaud, O.; Boillereaux, L.; Havet, M. Conjugate Heat and Moisture Transfer During a Dynamic Thermal Treatment of Food. Drying Technology 2007, 25,1411–1419.

[8] Rahman, S. M. A.; Islam, M. R.; Mujumdar, A. S. A Study of Coupled Heat and Mass Transfer in Composite Food Products during Convective Drying. *Drying Technology*, 2007, 25, 1359–1368.

[9] Pakowski, Z.; Adamski, A. The Comparison of Two Models of Convective Drying of Shrinking Materials Using Apple Tissue as an Example. *Drying Technology*, 2007, 25, 1139–1147.

[10] [10] Chen, X. D. Moisture Diffusivity in Food and Biological Materials. *Drying Technology*, 2007, 25, 1203–1213.

[11] Datta, A. K. Porous media approaches to studying simultaneous heat and mass transfer in food processes. II: Property data and representative results. *Journal of Food Engineering*, 2007, 80, 96-110.

[12] Zhang J.; Datta A. K. Some considerations in modeling of moisture transport in heating of hygroscopic materials. Drying Technology 2004, 22 (8), 1983-2008.

[13] Ni H.; Datta A. K. Heat and moisture transfer in baking of potato slabs. *Drying Technology*, 1999, 17 (10), 2069-2092.

[14] Ni, H.; Datta, A. K.; Torrance, K. E. Moisture transport in intensive microwave heating of wet materials: a multiphase porous media model. *International Journal of Heat and Mass Transfer*, 1999, 42, 1501–1512.

[15] Zhang, J.; Datta, A. K.; Mukherjee, S. Transport processes and large deformation during baking of bread. AIChE Journal 2005, 51(9), 2569–2580.

[16] Kondjoyan, A.; Boisson, H.C. Comparison of Calculated and Experimental Heat transfer Coefficients at the Surface of Circular Cylinders Placed in a Turbulent Cross-flow of Air. *Journal of Food Engineering*, 1997, 34, 123-143.

[17] Verbdryer, P.; Nicolaï, B. M.; Scheerlinck, N.; De Baerdemaeker, J. The Local Surface Heat Transfer Coefficient in Thermal Food Process Calculations: A CFD Approach. *Journal of Food Engineering*, 1997, 33, 15-35.

[18] Curcio S.; Aversa M.; Calabro` V.; Iorio G. Simulation of food drying: FEM analysis and experimental validation, Journal of Food Engineering 2008, 87, 541–553

[19] Marra F.; De Bonis M. V.; Ruocco G. Combined microwaves and convection heating: a conjugate approach, Journal of Food Engineering 2009, doi: 10.1016/j.jfoodeng. 2009.09.012.

[20] Curcio S., A multiphase model to analyze transport phenomena in food drying processes, *Drying Technology*, 2010, in press.

[21] Wilcox, D.C. Turbulence Modeling for CFD, 2nd Ed.; DCW Industries Inc., 2004.

[22] Aversa, M.; Curcio, S.;Calabrò, V.; Iorio, G. An analysis of the transport phenomena occurring during food drying process. Journal of Food Engineering 2007, 78 (3), 922-932.

[23] Thorvaldsson, K.; Janestad, H. A Model for Simultaneous Heat, Water and Vapour Diffusion. *Journal of Food Engineering*, 1999, 40, 167-172.

[24] Viollaz, P.E.;Rovedo, C.O. A Drying Model for Three-Dimensional Shrinking Bodies. *Journal of Food Engineering*, 2002, 52, 149-153.

[25] [25] Perry, R.H.;Green, D. Perry's Chemical Engineers' Handbook. Mc Graw-Hill, New York.

[26] Ruiz-López, I.I.; Córdova, A.V.; Rodríguez-Jimenes, G.C.; García-Alvarado, M.A. Moisture and Temperature Evolution During Food Drying: Effect of Variable Properties. *Journal of Food Engineering*, 2004, 63, 117-124.

[27] Smith, J.M. Van Ness, H.C.; Abbott, M.M. Chemical Engineering Thermodynamics IV edition. McGraw-Hill, New York.

[28] Welty, J. R.; Wicks, C. E.; Wilson, R. E.; Rorrer, G. Fundamentals of Momentum, Heat and Mass Transfer (IV edition). John Wiley & Sons., New York, 2001.

[29] Ertekin, C.; Yaldiz, O. Drying Of Eggplant And Selection of a Suitable Thin Layer Drying Model. *Journal of Food Engineering*, 2004, 63, 349-359.

[30] Doymaz, İ.Convective Air Drying Characteristics of Thin Layer Carrots. *Journal of Food Engineering*, 2004, 61, 359-364.

[31] Krokida, M.K.; Karathanos, V.T.; Maroulis, Z.B.; Marinos-Kouris, D. Drying Kinetics of Burdens Vegetables. Journal of Food Engineering 2003, 59, 391-403.

[32] Akpinar, E. K.; Bicer, Y; Yildiz, C. Thin layer drying of red pepper. *Journal of Food Engineering*, 2003, 59, 99–104.

[33] Senadeera, W.; Bhandari, B.R.; Young, G.; Wijesinghe B. Influence of shapes of selected vegetable materials on drying kinetics during fluidized bed drying. *Journal of Food Engineering*, 2003, 58, 277-283.

[34] Reilly, D.L; Cooper, L.N. An overview of neural networks: early models to real world systems. In: Zornetzer, S.F.; Davis, J. L.; Lau, C. (Eds), An Introduction to neural and Electronic Networks. Academic Press, New York, 1990, 227-248.

[35] Zhang, G.; Patuwo, B.E.; Hu, M. J. Forecasting with Artificial Neural Network: the state of Art, *Int. J. of Forecasting*, 1998, 14, 35-62.

[36] Roy, A.; Kim, L.S.; Mukhopadhyay, S. A polynomial time algorithm for the construction and training of a class of multilayer perceptron. Neural Networks 1993, 6, 535-545.

[37] Wang, Z.; Massimo, C.D.; Tham, M.T.; Morris, A.J. A procedure for determining the topology of multilayer feed-forward neural networks. *Neural Network*, 1994, 7 (2), 291-300.

[38] Murata, N.; Yoshizawa, S.; Amari, S. Network information criterion determining the number of hidden units for an artificial neural network model. *IEEE transaction on Neural Networks*, 1994, 5 (6), 865-872.

[39] Demuth, H.; Beale, M. Neural Network Toolbox User's Guide. Natick, The MathWorks, 2000.

[40] van Can, H.J.L.; te brake, H.A.B.; Dubbelman, S.; Hellinga, C.; Luyben, K. C. A. M. Heijnen, J.. Understanding and Applying the Extrapolation Properties of Serial Grey-Box models. AIChE J. 1998, 44, 1071-1089.

[41] Erenturk, S.;Erenturk, K. Comparison of genetic algorithm and neural network approaches for the drying process of carrots. *Journal of Food Engineering*, 2007, 78, 905-912.

[42] Martynenko A.I.;Simon X. Yang.Biologically Inspired Neural Computation for Ginseng Drying Rate. *Biosystems Engineering*, 20069, 5 (3), 385–396.

[43] Movagharnejad, K.; Nikzad, M. Modeling of tomato drying using artificial neural network. *Computers and Electronics in Agriculture*, 2007, 59, 78–85.

[44] Hernàndez-Pèrez, J.A.; Garcia-Alvarado, M.A.; Trystram, G.; Heyd, B. Neural networks for the heat and mass transfer prediction during drying of cassava and mango, *Innovative Food Science and Emerging Technologies*, (2004), 5, 57–64.

[45] Hernàndez-Pèrez, J.A.; Garcia-Alvarado, M.A.; Trystram, G.; Heyd, B. Application of an Artificial Neural Network for Moisture transfer prediction considering Shrinnkage during drying of foodstuffs. In: Welti-Chanes, J.;Vélez-Ruiz, J.F.; Barbosa-Cànovas, G.V. (Eds.). *Transport Phenomena in Food Processing*, CRC press, 2003, 183-197.

[46] Agarwal, M. Combining neural and conventional paradigms for modelling, prediction and control. *Int. J. Syst. Sci.* 1997, 28, 65-81.

[47] Psichogios, D. D.; Ungar, L.H. A hybrid neural network-First principle approach to process modeling. AIChE J.1992, 38 (10), 1499-1511.

[48] Curcio, S., Calabrò, V., Iorio, G. Reduction and control of flux decline in cross-flow membrane processes modeled by artificial neural networks and hybrid systems. *Desalination*. 2009, 236 (1-3), 234-243

[49] Saraceno, A., Curcio, S., Calabrò, V., Iorio, G. A hybrid neural approach to model batch fermentation of "ricotta cheese whey" to ethanol. Computers and Chemical Engineering. 2009, in press.

[50] Curcio, S., Scilingo, G., Calabrò, V., Iorio, G. Ultrafiltration of BSA in pulsating conditions: An artificial neural networks approach. *Journal of Membrane Science*. 2005, 246 (2), 235-247

[51] Curcio, S., Calabrò, V., Iorio, G. Reduction and control of flux decline in cross-flow membrane processes modeled by artificial neural networks. *Journal of Membrane Science*. 2006, 286 (1-2), pp. 125-132

[52] Bird, R.B.; Stewart, W. E.; Lightfoot, E. N. Fenomeni di trasporto. Ambrosiana, Milano, 1979.

In: Focus on Food Engineering
Editor: Robert J. Shreck

ISBN: 978-1-61209-598-1

Chapter 8

RESPONSE TO STRESS CONDITIONS OF MICROORGANISMS TREATED WITH NATURAL ANTIMICROBIALS IN CHEESE WHEY

Mariana von Staszewski[1,2], Rosa J. Jagus[1*], Sandra L. Mugliaroli[1], Laura Hernaez[1] and Giselle Lehrke[1]

[1] Laboratorio de Microbiología Industrial, Departamento de Ingeniería Química, Facultad de Ingeniería, Universidad de Buenos Aires.

[2] Consejo Nacional de Investigaciones Científicas y Técnicas, Argentina. (CONICET)

ABSTRACT

Cheese whey is a by-product that recently has gained attention because it represents a source of food ingredients of high nutritive value. It is generally processed by ultra filtration and spray drying to produce whey powder and protein concentrates. However, the concentrated fluid obtained from the membrane process could be used directly in the formulation of foods if it maintains a proper microbial stability, avoiding the drying process which can be rather expensive and affects the nutritional and functional properties.

Food-processing methods have been developed to interfere with bacterial homeostasis, prevent growth, or kill food-borne pathogens. The growing demand for fresh, minimally processed foods by consumers had led to the need for natural food preservation methods such as the use of natural antimicrobials and combination with other hurdles, without adverse effects on the consumer or the food itself. Additionally, natural antimicrobials can be used alone or in combination with other non-thermal technologies, but it has also been demonstrated that microorganisms can display tolerance to these factors when they are applied in sub-lethal concentrations. Thus, it may be useful to assess whether pre-treatments with natural antimicrobials like green tea extract,

* Corresponding author: Dr. Rosa J. Jagus
Departamento de Ingeniería Química, Pabellón de Industrias, Ciudad Universitaria (1428), Buenos Aires, Argentina.
E.mail: rjagus@di.fcen.uba.ar
Phone/FAX: 5411-4576-3240

Microgard^{TM} or nisin, would enhance sensitivity or resistance of food-borne bacteria to different stress conditions, such as acidity, heat, hydrogen peroxide or salt addition, which are commonly applied in food-processing methods. For this purpose different techniques were used in order to evaluate possible membrane damages produced by the combined treatments. The results of this study revealed that the bacterial reduction when applying stress conditions was improved in most of the cases after exposure to natural antimicrobials in liquid cheese whey. Synergistic effects are important to minimize the concentration of additives required to achieve a particular antibacterial effect without adversely affecting the sensorial acceptability. This knowledge about the interactions between natural antimicrobials and stressors has important implications for the food industry when considering which food preservation regime is the best to adopt.

INTRODUCTION

Liquid cheese whey (LCW) is a source of food ingredients of high nutritive value. It contains more than half of the solids of the original whole milk, including whey protein (20% of total protein) and most of the lactose, minerals and water-soluble vitamins (González-Martínez et al. 2002; Atra et al. 2005). LCW is generally processed by ultra filtration and spray drying to produce whey powder and protein concentrates. These products are used as ingredients in food formulas (Henning et al. 2006). However, the concentrated fluid obtained from the membrane process could be used directly in the formulation of foods if it maintains a proper microbial stability, avoiding the drying process which can be rather expensive and affects the nutritional and functional properties of LCW (Peters 2005). LCW will then undergo many preservation and storage processes associated to the kind of food into which it was incorporated.

LCW is a fresh, non-fermented dairy product, with a high pH (>6.0) and moisture content. Thus, it is prone to rapid bacterial deterioration, especially under abusive storage temperatures (Hough et al., 1999). Illness caused due to the consumption of contaminated foods has a wide economic and public helath impact worlwide. Historically there have been outbreaks of infection associated with the consumption of cheese, and the predominant organisms responsible have included *Salmonella, Listeria monocytogenes*, virulent strains of *Escherichia coli*, and *Staphylococcus aureus* (De Buyser et al., 2001; Zottola & Smith, 1991). Detailed investigations have shown that the sources of contamination were raw milk or post-pasteurization contamination with organisms from manufacturing environments. Additionally, these microorganisms have been known to display tolerance to factors commonly used to control bacterial growth in foods, such as acidity, low temperatures and sodium chloride addition (Campbell et al. 2004; Greenacre & Brocklehurst 2006; Gandhi & Chikindas 2007).

Listeria spp. are very tough microorganisms and their ability to tolerate many adverse conditions in different foodstuff has been reported (Ghandi & Chikindas, 2007; Sergelidis & Abrahim, 2009). They can adapt and survive in salty environments by accumulation of compatible solutes, which is rather simple since food constitutes a rich source of compatible solutes and their precursors. Also, the acid tolerance response observed in *Listeria spp.* is of particular concern during food processing, representing important implications for the manufacturing of salty and acidic food products like soft cheeses (Shabala et al. 2006; Skandamis et al. 2008). Furthermore, *Listeria monocytogenes* is capable of growing at refrigeration temperature and, although the initial number of this foodborne pathogen may be low in a given product, they can increase during refrigerated storage and thus pose a risk to public health. Thus, it is important to ensure low *Listeria* counts from the beginning of the shelf-life. *Listeria innocua*, a non-pathogenic species, may be used as a biological indicator for *L. monocytogenes* because of its similar response to physical, chemical or thermal treatments (Kamat & Nair, 1996).

Staphylococcus aureus is commonly found in milk and dairy products, particularly in cheeses made either from raw or pasteurized milk, due to it being among the most important etiological agents of bovine mastitis and because it is extensively carried by food industry workers (Jørgensen et al. 2005). *S. aureus* is also one of the main agents of food intoxication caused by milk and dairy product consumption in different countries (Andre´et al., 2008), and is still one of the leading causes of foodborne illness worldwide and the second most commonly reported cause in the United States (Jablonski & Bohach, 2001).

Salmonella spp. were linked to outbreaks associated with the consumption of Cheddar cheese (D'Aoust et al., 1985) and pasteurized milk (Ryan et al., 1987). Since they encounter many diverse and extreme environments, they have developed responses to combat stresses such as extremes of pH, salt and reactive oxygen intermediates (Foster & Spector, 1991).

Escherichia coli is a normal inhabitant of the intestinal tract of humans and warm-blooded animals. Although usually harmless, various *E. coli* strains have acquired genetic determinants (virulence genes) rendering them pathogenic. These pathogens are responsible for three main clinical infections: enteric and diarrhea diseases, urinary tract infections, and sepsis and meningitis. Thus, on the basis of conventional microbiological diagnostics in food control, the absent of *E. coli* without any other characterization of isolated strains is required. On the other hand, the emergence of *Escherichia coli* pathogenic strains is partly due to its increase acid resistance (Leyer et al., 1995) and its ability to grow in a wide temperature range (between 10°C and 43°C).

The use of natural antimicrobial compounds from a wide variety of natural sources is being explored as a means of improving the safety and stability of several foods, while maintaining an image of natural, high quality, healthy products (Gould, 1997). These nontraditional preservation techniques are being developed to satisfy consumer demand with regard to nutritional and sensory aspects of food. Additionally, many of the natural food antimicrobial systems also demonstrated a multifunctional physiological advantage and are emerging as value-added ingredients in various food products (Tiwari et al., 2009).

Natural antimicrobials in food preservation can be used alone or in combination with other technologies in order to reduce the level of each preservation method in what is called the `hurdle effect´. However, many authors have alerted us about the existence of cross-protection responses against multiple stress conditions, since exposure of the pathogen to one kind of sub-lethal stress can confer cross-protection to another lethal stress (O'Byrne and Booth 2002). Thus, it may be useful to assess the response of food borne bacteria to new natural antimicrobials and their subsequent susceptibility or resistance to different stress conditions commonly encountered during food processing and preservation.

In this chapter, the action of natural antimicrobials like nisin, Microgard™ and green tea extract in the conservation of liquid cheese whey is presented. Additionally, the assessment of antagonistic or synergistic responses of microorganisms treated with natural antimicrobials and subsequently subjected to different stress conditions is discussed.

NATURAL ANTIMICROBIALS

Green Tea

Tea (*Camellia sinensis*) is the second most common beverage in the world next to water. Green tea extracts have shown many health benefits, including the prevention of various types of cancer and cardiovascular diseases (Scalbert et al. 2005). These health protective effects are often attributed to the tea polyphenols, in particular, the catechins (He et al. 2006). In terms of the antimicrobial activity, several food-borne microorganisms such as *E. coli*, *S. typhimurium*, *L. monocytogenes*, *S. aureus* and *Campylobacter jejuni* have been reported to be inhibited by tea components (Kim et al. 2004; Si et al. 2006).

Since the antimicrobial effect of green tea polyphenols is concentration dependant (Si, et al., 2006), total polyphenol content of 1% and 3% extracts of commercial green tea was determined using the Folin-Ciocalteau method as described by Singleton and Rossi (1965). The results, expressed as gallic acid equivalents, indicated polyphenol content values of 2087 ± 39 and 4978 ± 122 (mg/L) for 1% and 3% extracts, respectively. Thus, as reported in previous studies (von Staszewski & Jagus, 2007), the antimicrobial activity of green tea depends on the concentration of the extract. However, as can be seen in Figure 1, different microorganisms showed different sensitivity and behavior along the storage at 20 ºC when evaluated in LCW (8 % w/v solid content, pH 5.5, prepared from WPC 35%, provided by Milkaut S.A.).

Although *L. innocua* (CIP 8011, CCMA 29, Facultad de Farmacia y Bioquímica, Buenos Aires, Argentina) treated with 1% green tea extract showed a diminished growth rate, this concentration did not present a significant difference with control at the end of storage (Figure 1A). On the other hand, the 3% green tea extract exerted a bacteriostatic effect against *L. innocua* reaching a difference of 6.5 log cycles less than the control at day 4 of storage. Kim et al. (2004) obtained similar results for *L. monocytogenes* treated with 10% water-soluble green tea extract in brain heart infusion at 35ºC.

The inhibition of 1% green tea extract against *Staphylococcus aureus* (ATCC 6538P) was especially interesting as it completely suppressed the growth of this microorganism (Figure 1B). Wu et al. (2007) also obtained a bacteriostatic activity for 1% green tea extract against *S. aureus* in nutrient broth. Additionally, the 3% extracts showed more than one log cycle reduction of the initial number of bacteria, reaching a difference of 4.8 log cycles with the control at the end of storage. Hamilton-Miller and Shah (1999) and Stapleton et al. (2004) reported that *S. aureus* treated with green tea components form multicellular aggregates with engrossed cell walls and aberrant septa. This severe disorganization of the cell wall architecture, observed by electron microscopy studies, suggests that tea components interfere with peptidoglycan synthesis and compromises the process of cell separation. Additionally, polyphenols can modulate and reduce the halotolerance of *S. aureus* and its resistance to several antibiotics, indicating damage of the cell wall by direct binding of catechins to peptidoglycan (Zhao et al., 2001; Stapleton et al., 2004; Stapleton et al., 2006).

In the case of *Salmonella typhimurium* (ATCC 13311), it was observed that 1% green tea extract did not produce any inhibitory effect during storage, showing a similar behaviour to control (Figure 1C), while the 3% extract showed a slightly inhibitory effect, with a slower growth rate than the control but with similar levels at the end of storage. Si et al. (2006)

obtained a 58% inhibition of *S. typhimurium* after 6 h incubation, using a commercial green tea extract in a 500mg/L concentration.

Similarly, *Escherichia coli* (ATCC 8739) was inhibited very weakly by both extract concentrations studied (Figure 1D). Chou et al. (1999) also noticed a slight inhibition (<30%) of *E. coli* when treated with green tea extract (1%), while Wu et al. (2007) observed no antimicrobial activity at all in the treatment of this microorganism.

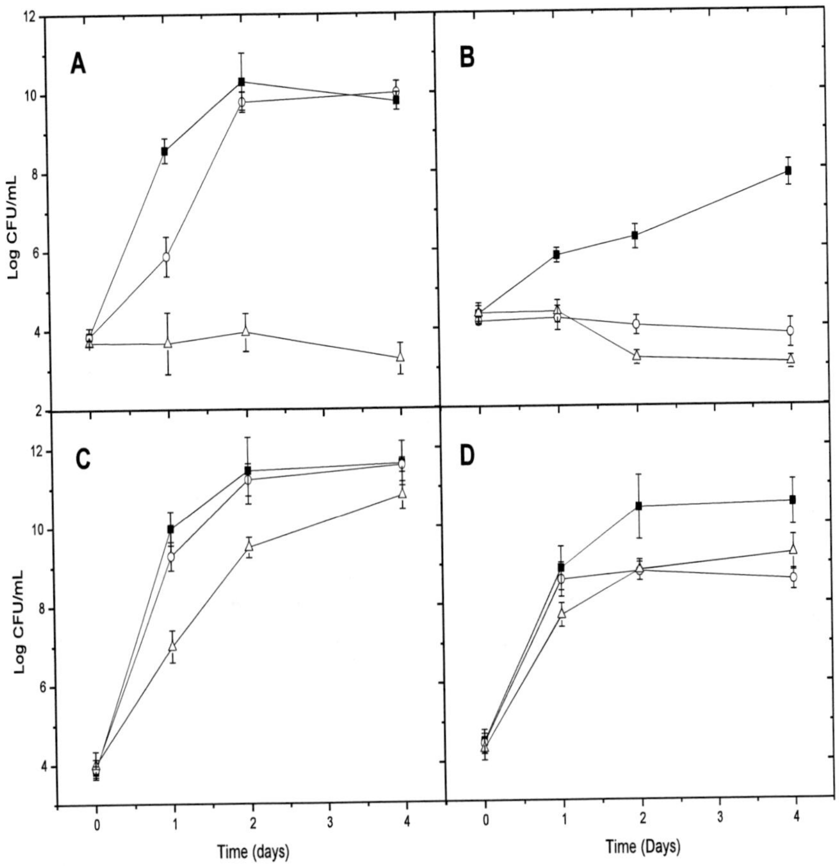

Figure 1: Effectiveness of commercial green tea extract against (A) *Listeria innocua*, (B) *Staphylococcus aureus*, (C) *Salmonella typhimurium* and (D) *Escherichia coli* in WPC. (■) Control; (○) 1% green tea extract; (Δ) 3% green tea extract. Error bars indicate standard deviation.

Collectively, the response obtained for the different microorganisms confirmed what was reported by Kim et al. (2004) and Si et al. (2006), who found that green tea extracts were more potent to *S. aureus* and *L. monocytogenes* than to *Salmonella Spp.* and *E. coli* 0157:H7. According to the data presented, it seams that the growth of gram positive bacteria was inhibited by green tea extracts, while gram negative bacteria were less affected. This was consistent with the results of previous reports, which suggested higher susceptibility of food-borne gram positive bacteria to tea polyphenols compared with gram negative bacteria (Taguri et al., 2004). Susceptibility of bacterial strains to green tea extract has been shown to be related to differences in cell wall components (Ikigai et al., 1993; Zhao et al., 2001). These authors proposed that one of the reasons for Gram negative bacteria being more resistant to tea polyphenols than Gram positive is the presence of a strong negative charge conferred by lipopolysaccharide on the surface of Gram negative bacteria. Thus, the outer membrane of gram negative microorganisms acts as a barrier due to the repulsion between tea components and the cell surface.

Although many authors have carried out studies on the antibacterial mechanism of action of green tea extracts (Caturla et al., 2003; Zhang & Rock, 2004; Gradišar et al., 2007), the exact mechanism of action is not clear. It has been suggested that tea components can inhibit nucleic acid synthesis and affect enzyme activity, particularly those associated with energy production. Protein and lipid synthesis were also affected but to a lesser extend. Additionally, polyphenols could interfere with membrane function (electron transport, nutrient uptake, protein, nucleic acid synthesis and enzyme activity), and interact with membrane proteins, causing deformation in structure and functionality. Ikigai et al. (1993) reported that catechins primarily act on and damage bacterial membrane by perturbing the lipid bilayers, possibly by directly penetrating them and disrupting the barrier function. Alternatively, catechins may change the membrane fluidity and cause membrane fusion, a process that results in leakage of intracellular material and cell aggregation. Indeed, Caturla et al. (2003) demonstrated that galloylated catechins partitioned very efficiently into biological membranes and promoted the formation of detergent-resistant structures or domains within the phospholipids palisade by increasing lipid order. These highly compacted membrane domains affect enzyme activity and proteins functions in membranes, and thereby affect membrane function regulation. In addition, they also showed that galloylated catechins produced leakage from *E. coli*-isolated membrane, reinforcing the global antibacterial activity of these compounds. On the other hand, Si et al. (2006), through studies of electron microscopy, showed that catechins altered bacterial cell morphology, which may have resulted from disturbed cell division. Their observation and results demonstrated that the cell division may have been the target of tea polyphenols because all affected cells showed some degree of incomplete division.

Microgard™

Microgard™ are bacteriocin–like inhibitory products (Danisco, Copenahagen, Denmark) obtained by fermentation of grade A skim milk (Microgard™ 100 and 300) or dextrose (Microgard™ 200) with *Propionibacterium shermanii* or specific *Lactococci*. Their active compounds include diacetyl as well as lactic, propionic and acetic acid and other undefined low molecular weight inhibitors, that can inhibit *Pseudomonas, Salmonella* and *Yersina* and certain fungi (Al-Zoreky et al. 1991). The U.S. Food and Drug Administration has considered propionibacterium metabolites as GRAS (CFR184.1081) and approved Microgard™ products for use in cottage cheese.

The behaviour of *L. innocua* treated with Microgard™ 300 in liquid cheese whey during storage at 20 °C was reported in previous work (von Staszewski & Jagus, 2008). There it has been demonstrated that *L. innocua* showed a pattern similar to the control untreated samples when the concentration of Microgard™ 300 was less than 3% (Figure 2A). With higher concentrations (3% and 5%) a slight delay in the growth rate was achieved, but then restored by the end of storage. These results are in accordance with Zuckerman and Ben Abraham (2002) who reported that the treatment with Microgard™ alone in poultry samples did not inhibit growth of natural *L. monocytogenes.* Also, Degnan et al. (1994) evaluated Microgard™ products as antilisterial agent in crabmeat and observed no appreciable effect on *L. monocytogenes* counts after 6 days compared with the control.

Similarly, *Escherichia coli* presented a very low susceptibility to Microgard™ 100 in liquid cheese whey (Figure 2D). This behavior has also been reported by many authors (Al-Zoreki et al. 1991; Dave et al. 2003; Lemay et al. 2002), who observed that this antimicrobial had a slight or no effect on *E. coli* evaluated in different food systems. Additionally, Dave et al. (2003) observed that Microgard™ products had very little or no effect at 0.25 and 0.5% levels in a meat system but they observed some inhibition of *E. coli* O157:H7 and *L. monocytogenes* in hamburger following treatment with 1% Microgard™ 200 and 300 respectively. However, the Microgard™ products were not sufficiently effective in hamburger when it is heavily (>10^5 CFU/g) contaminated with pathogenic bacteria.

On the other hand, Mugliaroli et al. (2007), who investigated the effectiveness of Microgard™ 100 and 300 against *Staphylococcus aureus* and *Salmonella typhimurium*, respectively, observed a bacteriostatic effect along 4 days of storage in liquid cheese whey at 20 °C (Figure 2B and 2C). Treatments with lower concentration than 5% allowed a slight growth of *S. aureus*, while *S. typhimurium* was completely inhibited by 1%. Similar results were also observed by Al-Zoreky et al. (1991) for *S. typhimurium.*

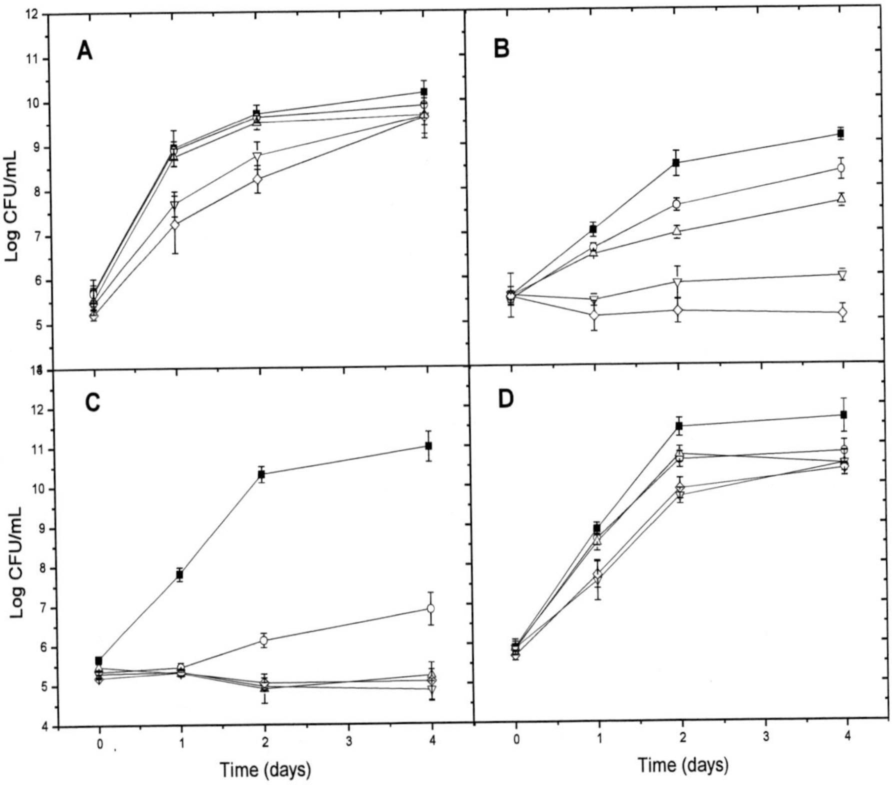

Figure 2: Effectiveness of Microgard™ 300 against (A) *Listeria innocua* and (B) *Staphylococcus aureus*; and Microgard™ 100 against (C) *Salmonella typhimurium* and (D) *Escherichia coli* in WPC. (■) Control; (○) 0.5% MG; (Δ) 1% MG; (∇) 3% MG; (◊) 5% MG. Error bars indicate standard deviation.

Since no great inhibition of *E. coli* or *L. innocua* was achieved with Microgard^{TM} alone, it seams that this antimicrobial offers no advantage when used singly. However, a combination with other techniques, such as other natural antimicrobials (von Staszewski & Jagus, 2008) or ozone application (Jhala et al., 2002) along with Microgard^{TM}, may provide a synergistic effect to make our food safe.

Nisin

Nisin is an antimicrobial peptide, considered GRAS (CFR184.1538) and produced by strains of *Lactococcus lactis* subsp. *lactis* that effectively inhibits Gram-positive bacteria including *L. monocytogenes* and *Bacillus cereus* and also the outgrowth of spores of Bacilli and Clostridia (Cleveland et al. 2001). It is the first antimicrobial peptide with a ''generally recognized as safe'' status in the United States for use in processed cheese (Food & Drug Administration, 1998). In addition, its use in various food products is allowed in several countries (Delves-Broughton, 1990).

Nisin increases the permeability of the membrane by pore formation, resulting in rapid efflux of essential intracellular small molecules (Breukink et al. 1997; Deegan et al. 2006). This phenomenon is mediated by the ability of nisin to bind lipid II, a peptidoglycan precursor of the bacterial cell wall (Bauer & Dicks, 2005). However, is well known that the presence of an outer membrane in Gram-negative bacteria can protect the organism from the antimicrobial action of nisin. For this reason, only *L. innocua* and *S. aureus* were evaluated with nisin in this work.

Figure 3 shows the reduction of *L. innocua* and *S. aureus* initial numbers by the application of increasing concentrations of nisin (prepared from Nisaplin®, Danisco, Copenhagen, Denmark). As can be seen, *L. innocua* resulted to be more sensitive to this antimicrobial than *S. aureus*, showing maximum reductions of 5.8 and 2.8, respectively, with the highest level of nisin tested (500 IU/mL). In fact, Hamama et al. (2002), who studied the fate of enterotoxigenic *S. aureus* in the presence of nisin-producing *Lactococcus lactis* strain during manufacture of fresh cheese, observed that the nisin-producing lactococci are of little help to prevent *S. aureus* growth and subsequent formation of enterotoxin, particularly when the initial milk contamination with the pathogen is relatively high (e.g. 10^5 CFU/mL). Also, Millette et al. (2007) demonstrated that a concentration of 1000 IU/mL of nisin was necessary to reduce by 1.86 log CFU/cm^2 the level of *S. aureus* on ground beef steak, and when a lower concentration of nisin was used (500 IU/mL), a lesser reduction of *S. aureus* population was noticed.

Some of this information was previously published (von Staszewski and Jagus, 2008) and it was concluded that nisin applied alone in liquid cheese whey, had an immediate antilisterial effect and reduced initial counts by 5.3 and 6.7 logs approximately, when 100 and 200 IU/mL were applied. Also Gallo et al. (2007) investigated the influence of pH and storage temperature of the cheese whey in the antilisterial activity of this antimicrobial. They observed that the effectiveness of nisin in controlling the growth of *L. innocua* in LCW was more pronounced at 7°C than at 20°C and as the pH decreased from 6.5 to 5.5.

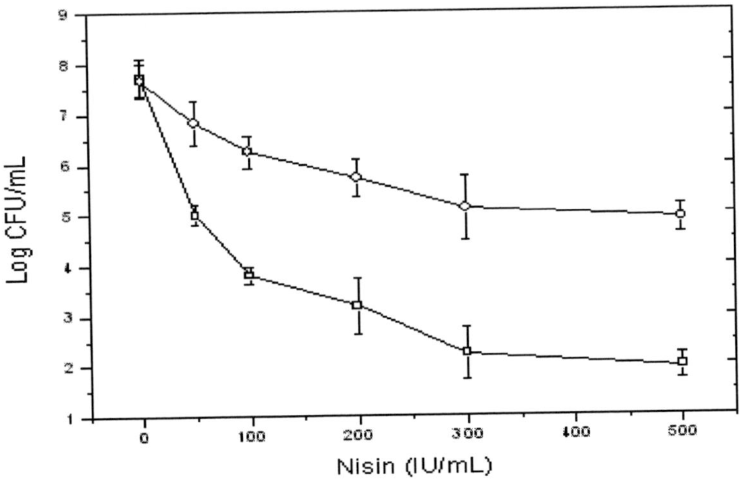

Figure 3: Effectiveness of nisin against (□) *Listeria innocua* and (○) *Staphylococcus aureus* at different nisin concentrations. Error bars indicate standard deviation.

However, it was observed that growth was thereafter restored, showing a temperature-dependant lag phase, but reaching the microbial level of the untreated samples by the end of storage (Figure 4). In fact, at higher temperatures (20 or 25 °C) this event was observed immediately, while a lag phase of 10 and 4 days was observed for storage temperatures of 7°C and 12°C, respectively. Previous reports by other authors also contain evidence consistent with the observed regrowth of nisin treated samples. Schillinger et al. (2001) found that 3000 IU/mL of nisin were required to cause a significant reduction in *Listeria* viable counts in tofu, resulting in a short-term inhibitory effect. The decrease in viable count was followed by rapid regrowth of survivors to nisin, during storage of tofu at 10 °C. After 24h, *Listeria* had already achieved a cell density which nearly compensated the initial reduction caused by nisin.

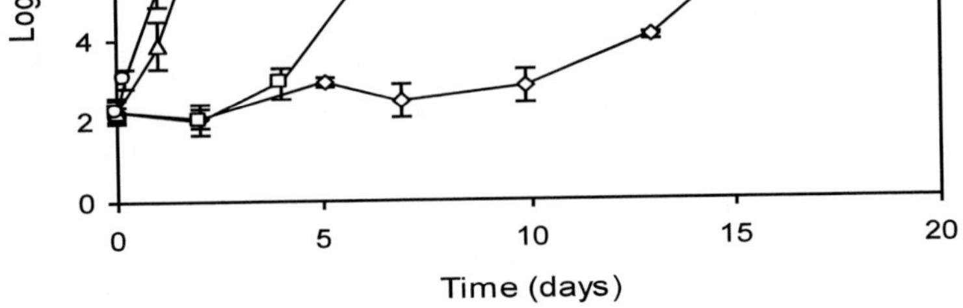

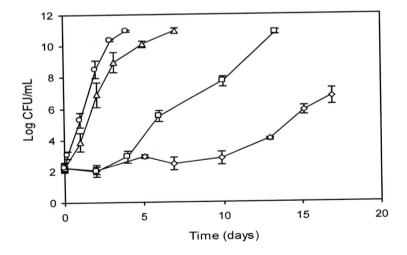

Figure 4: Growth of *L. innocua* treated with nisin (300 IU/mL) and stored in liquid cheese whey (8% w/v) at different temperatures. (◊) 7°C; (□) 12°C; (Δ) 20°C; (○) 25°C. Error bars indicate standard deviation.

Several factors may contribute to the ineffectiveness of nisin to inhibit or provide sufficient prevention of *L. innocua* growth for longer periods of time. Some of them could be due to an insufficient concentration of the antimicrobial, necessary to kill all cells, or unspecific binding of nisin to whey components which may cause partial inactivation of the bacteriocin (Henning et al., 1986). Another potential reason may have been an increased resistance to nisin of some of the inoculated bacteria, which may have been intrinsic or induced during storage (Samelis et al., 2005). Indeed, resistant mutants may arise when nisin is used as an antimicrobial (Chi-Zhang et al., 2004; De Martinis et al., 1997; Ukuku & Shelef, 1997).

STRESS CONDITIONS

The factors used in this study to perform stress assays pretend to simulate some of the conditions present in food processing and preservation. The best known method to destroy vegetative microorganisms is the thermal processing, with temperatures varying from 60ºC to 100ºC. During this treatment, a large amount of energy is transferred to the food with the consequent unwanted reactions that negatively affects the organoleptic and nutritional aspects of food (Sergedilis & Abrahim, 2009). Thus, a pretreatment with a natural antimicrobial could accelerate the thermal inactivation of a given microorganism, allowing a reduction of process time or temperature. The use of salt to lower the water activity is one of the most popular methods of food preservation used by the food industry. In the same way, reducing pH by lactic acid, which is a weak organic acid that has the capacity to penetrate the bacterial cell membrane and affects its intracellular pH, is always a good strategy when the food product allowed it. Finally, the antimicrobial action of hydrogen peroxide (H_2O_2) stems from its ability to form reactive oxygen species, which can damage DNA, enzymes, membrane constituents and could lead to lysis of microorganisms. In many countries, H_2O_2 has been accepted as a food additive for controlling microbial growth in stored milk before cheese-making (EU Risk Assessment Report, 2003) and as sterilizing agent of plastic packaging material used in aseptic systems.

In this work, LCW was inoculated with *Listeria innocua* and treated with the natural antimicrobials while stored for 48 h at 20ºC. After this time, treated and untreated (control) samples were properly diluted in order to obtain a similar number of bacteria (10^8 CFU/mL) in all samples. Then, stress assays were performed by incubating for 60 min. 1.0 ml aliquots of the samples into 9.0 ml of phosphate buffer (pH 7.0, 0.1M) as follows: (a) *heat-shock*, the inoculated buffer was incubated at 52°C; (b) *saline,* buffer containing 10% (w/v) NaCl (Anedra®, Argentina); (c) *oxidative*, buffer containing 0.05% (v/v) H_2O_2 (Cicarelli, Argentina); (d) *acidic,* buffer containing lactic acid (Anedra®, Argentina) to reach pH 4.00. Except for heat-shock, the other stress assays were all incubated at 20°C. After standing the 60 min. the samples were diluted with peptone water 0.1% (w/v) and microbial counts were made on plates of tryptic soy agar enriched with 0.6% yeast extract. Each experiment was performed in duplicate.

Listeria innocua

The response of *Listeria innocua* to the heat stress in the control samples did not show a significant reduction (P>0.05), while the saline and acidic treatments produced only slight bactericidal effects. Bearns and Girard (1958) observed that *Listeria monocytogenes* are very heat resistant microorganisms because they recover after a heat treatment at 61.5 ºC for 35 min, which is consistent with the results exposed in this work. The oxidative stress was the most effective in reducing the number of microorganisms (Table 1).

Little differences were obtained when compared the response of *L. innocua* to different stresses with and without green tea extract. The exception was the consecutive treatment of green tea and lactic acid, which showed an additional reduction of 5.3 log cycles as compared with the acidic condition alone. This indicates that the presence of tea components enhances the antimicrobial activity of this organic acid. Apostolidis et al. (2008) obtained similar results when evaluating the inhibition of *Listeria monocytogenes* by oregano and cranberry phenolic compounds in combination with sodium lactate. They observed that this combined treatment had the best inhibitory effect both in broth and cooked ground beef systems. Additionally, they proposed a mechanism in which phenolic compounds can stack themselves on the plasma membrane causing changes in membrane fluidity and destabilization, resulting in partial membrane disruption. This could allow other small phenolic compounds and lactic acid to enter the cytosol and act on specific enzymes involved in key energy pathways.

Table 1. Effect of green tea extract (3%), Microgard™ 300 (5%) or nisin (300 IU/mL) treatments on the reduction (-Log N/No) of *Listeria innocua* (mean±SD) by different stress conditions in liquid cheese whey

	Heat	Saline	Oxidative	Acidic
Control	0.3 ± 0.2	1.1 ± 0.3	4.1 ± 0.3	1.2 ± 0.3
Green tea extract	1.1 ± 0.1	2.0 ± 0.1	3.5 ± 0.3	6.5 ± 0.3
Microgard™	3.7 ± 0.1	2.0 ± 0.2	4.0 ± 0.0	1.5 ± 0.4
Nisin	4.7 ± 0.2	4.7 ± 0.3	5.0 ± 0.1	5.4 ± 0.2

Although Microgard™ products include organic acids, which could induce resistance to the subsequent stresses as reported by Phan-Thanh et al. (2000), the results obtained in this work does not indicate the presence of cross-protection effects. This antimicrobial is a fermented product that includes inhibitory metabolites besides the acids, which probably enhance the microbial sensitivity. The heat stress applied to *L. innocua* after the Microgard™ 300 (5%) treatment resulted in a synergistic effect with a 3.7 log cycle reduction of *L. innocua*. On the contrary, this antimicrobial did not modify in a relevant degree the bacterial ability to tolerate sodium chloride, hydrogen peroxide or lactic acid stresses (Table 1).

L. innocua present in LCW loses part of its survival abilities when treated previously with nisin. When cells of *L. innocua* were exposed to the double challenge of nisin and stress conditions, a markedly different profile of inactivation was observed (Table 1).This antimicrobial acts by increasing the permeability of bacterial membrane and this was reflected in the disability of *Listeria* cells to support the external changes in osmolarity, temperature, pH and the presence of hydrogen peroxide. Heat, saline as well as acidic stresses, led to a remarkable increase in the susceptibility of this microorganism, resulting in 4.4, 3.6 and 4.2 log cycles reductions more than the samples with no antimicrobial.

On the other hand, nisin was less effective in increasing the vulnerability to oxidative stress, considering the high efficacy of the stress alone.

Transmission electron microscopy (TEM, Philips EM 301) studies have been conducted in an attempted to understand the mechanisms of single and consecutive treatments of nisin and heat, saline and acidic stresses on *Listeria innocua* cultivated in LCW. Micrographs of *L. innocua* cells in LWC (control) showed uniform shapes with smooth intact membranes, without detachment or plasmolysis and a normal progression of septal formation in the process of cell division (Figure 5a). They also exhibit an evenly distributed cytoplasm within the cell. The nisin-treated cells showed a significant increase in surface roughness, appearance of craters in the cell wall and partial cell wall detachment from the plasma membrane (Figure 5b). Weeks et al. (2006) evaluated *L. monocytogenes* cells exposed to 10-min nisin incubation. They found that nisin caused differential permeabilization of the cell membrane throughout the growth cycle. This was manifested as differences in cell viability and observable changes in membrane integrity when examined using TEM.

L. innocua cells treated thermally did not show significant morphological changes as can be seen in Figure 5c. However, damage to the cell membranes was observed only when heat stress was applied after nisin treatment. TEM micrographs revealed that this combined treatment caused the rupture of the cell wall and detachment from cytoplasmic membrane (Figure 5d).

Saline treated cells demonstrate irregular shapes with cytoplasm-sparse areas within the cell. Membrane structure is disorganized with partial disintegration of cell membrane and incipient plasmolysis is also visible (Figure 5e). *L. innocua* cells treated with nisin and stressed by NaCl presented dramatic plasmolysis, loss of cell wall and fragmented membranes (Figure 5f). As a result, broken or lysed cell membranes produce the leakage of cytoplasmic contents, which explains the important reduction observed in the plate count method (Table 1).

L. innocua cells exposed to lactic acid in LCW showed meaningful damages in cell cytoplasm without apparent changes on the membrane (Figure 5g). This fact is in agreement with the mechanism of action of organic acids, since it cause inhibition of essential metabolic reactions, stress on intracellular pH homeostasis and accumulation of toxic anions. Similar TEM results were observed by Raybaudi-Massilia et al. (2008) when evaluated the action of malic acid (0.6% v/v) of *L. monocytogenes* in melon juice. The cells treated with lactic acid in the presence of nisin showed plasmolysis and a disorganized cell cytoplasm (Figure 5h). Also *Listeria* cells appear to have experienced cell wall disruption or with the consequent leakage of intracellular material. These observations corroborate the synergistic activity of nisin and lactic acid observed in Table 1.

Collectively, this study shows that nisin causes disruption of the cell membrane and enhances the action of subsequent stresses. These observations corroborate the results based on plate counts.

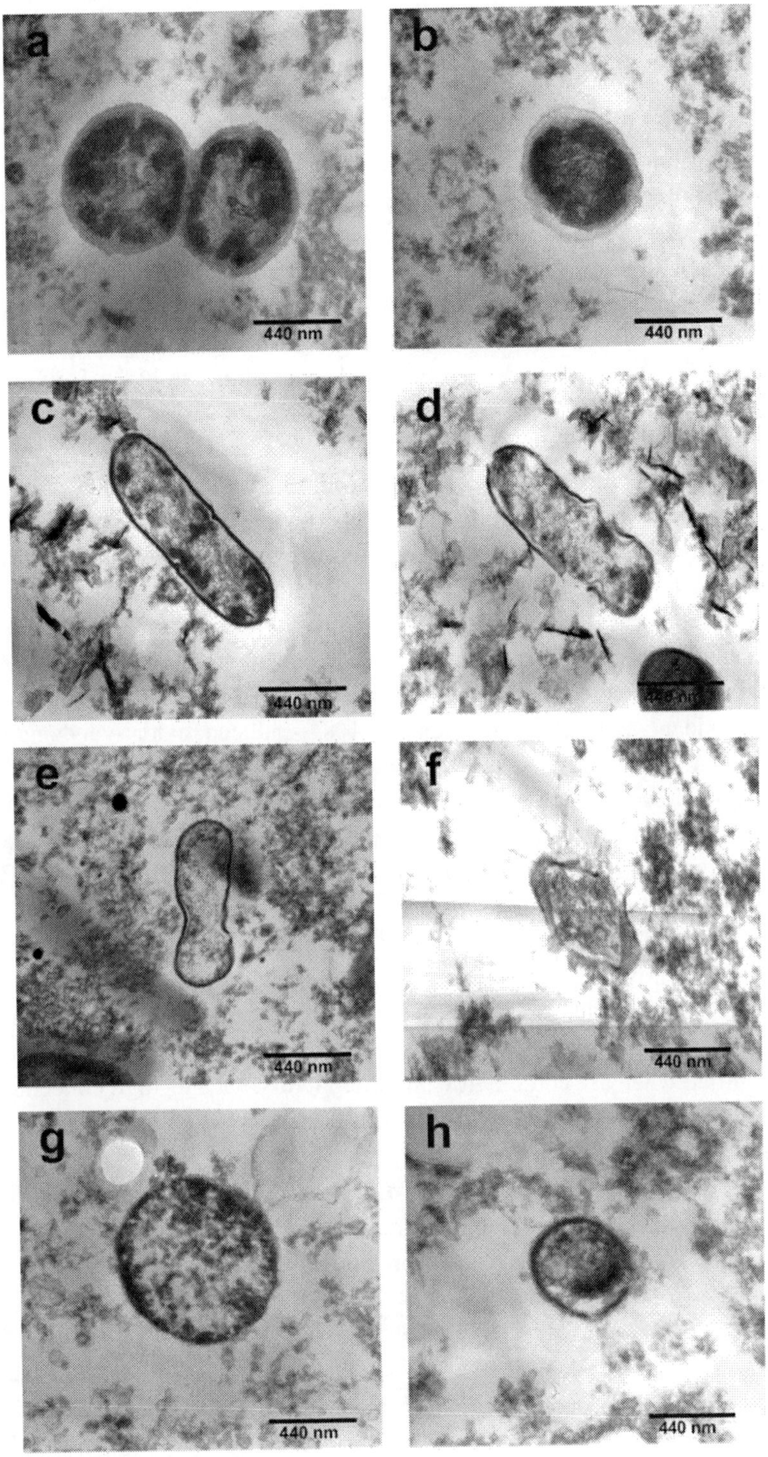

Figure 5. Transmission electron microscopy (TEM) micrographs of *Listeria innocua* cells (magnification x46000) with the following treatments: (a) control; (b) nisin; (c) heat; (d) nisin + heat; (e) saline; (f) nisin + saline; (g) acidic; (h) nisin + acidic.

Staphylococcus aureus

Staphylococcus aureus present in LCW, subjected to heat, oxidative and acidic conditions showed considerable reductions of more than 1.5 log cycles (Table 2). The less effective condition was the saline treatment which reduced less than one order, corroborating the well known ability of this microorganism to tolerate high sodium chloride concentrations (O´Byrne & Booth, 2002).

The treatment of this microorganism with green tea extract followed by heat or saline stresses showed reductions of 3.6 and 2.0 log cycles higher than the stress conditions alone, respectively. Bikels-Goshen et al. (2010) alerted about the increased resistance of *S. aureus* adapted to green tea catechin (EGCG) to heat and other antimicrobials. However, the main difference with our work that may explains the contradictory results lays on the use of very low concentrations of EGCG (0.002%) and on the lesser time (10 min) of thermal treatment. This non-inhibitory concentration allows the synthesis of stress proteins or heat shock proteins, which would help the microorganism to cope with or alleviate the heat stress. On the other hand, the effectiveness of green tea in increasing the vulnerability to sodium chloride are in agreement with the results presented by Stapleton et al. (2006), who informed that epicatechin gallate, present in green tea, can suppress staphylococcal growth in the presence of salt. Green tea extract was, however, ineffective in enhancing the vulnerability to hydrogen peroxide, while the response to lactic acid was not improved at all (Table 2). Since green tea is a potent antioxidant and its activity against H_2O_2-induced oxidative damage has been demonstrated (Yang et al. 2007), it is expected that the combined treatment with hydrogen peroxide results in a lesser reduction in the number of microorganisms. In fact, this event was observed for all the microorganisms evaluated and indicated that green tea constituents such as flavonoids, anthocyanins, or other phenolic compounds may degrade the H_2O_2 before exerting much bactericidal effect.

Although, heat and hydrogen peroxide stresses did not have a significant improvement when applying consecutively after Microgard™ 300 treatment, the saline and acidic stresses produced 3.1 and 1.5 log cycles reduction higher than the stresses applied alone, respectively. The presence of organic acids in Microgard™ 300 may exert a deleterious effect on *S. aureus* that can be amplified when are combined with other factors, such as high concentration of salt or even other organic acids (Charlier et al., 2009).

Table 2. Effect of green tea extract (3%), Microgard™300 (5%) or nisin (300 IU/mL) treatments on the reduction (-Log N/No) of *Staphylococcus aureus* (mean±SD) by different stress conditions in liquid cheese whey

	Heat	Saline	Oxidative	Acidic
Control	2.8 ± 0.1	0.7 ± 0.2	1.6 ± 0.1	2.0 ± 0.2
Green tea extract	6.4 ± 0.3	2.7 ± 0.2	1.2 ± 0.1	2.2 ± 0.2
Microgard™	3.1 ± 0.2	3.8 ± 0.1	1.8 ± 0.2	3.5 ± 0.1
Nisin	4.0 ± 0.5	3.5 ± 0.2	1.3 ± 0.3	2.8 ± 0.2

The *S. aureus* cells treated with nisin and subsequently subjected to heat or sodium chloride produced 1.2 and 2.8 log cycles reductions higher than stress conditions alone, respectively (Table 2). This effect is probably due, as explain above (see *L. innocua* section), to nisin capacity to increase the bacterial membrane permeability. Additionally, it has been reported a slightly higher reduction in the level of *S. aureus* during the production of fresh cheese when a nisin-producing lactococcus was used as lactic starter (Hamama et al., 2002). These authors concluded that as the pH was similar (≤ 4.5) in all samples, the accentuated decrease of *S. aureus* might only be due to the effect of the accumulated nisin in the cheese. In the present work similar results were obtained when a lactic acid stress was applied to LCW treated with nisin. On the other hand, the application of H_2O_2 in the presence of nisin offers neither additional advantages nor antagonistic effects.

Since these natural antimicrobials can suppress the halotolerant nature of *S. aureus* present in LCW, they could be attractive candidates for the preservation of salt-containing foodstuffs when staphylococcal contamination is a particular problem.

Salmonella typhimurium

Salmonella typhimurium exposed to different stresses, revealed a great susceptibility to heat and oxidative conditions (Table 3). Indeed, Ukuku et al. (2004) obtained similar results when assessing the effect of hot water and hydrogen peroxide treatments on the survival of *Salmonella Spp.* in whole and fresh-cut cantaloupe. However, it appeared to be less affected by saline stress and no effect could be noticed for acid stress. This insensitivity to lactic acid was also reported by Eswaranandam et al. (2004), who observed that this organic acid incorporated into soy protein films (2.6%, pH 3.35) did not produce any clear zone of inhibition and no log cycle reductions were noticed when *Salmonella* was inoculated directly onto the film disc with 0.9% lactic acid (pH 4.55).

This microorganism showed no changes in bacterial reductions when applied heat and acidic stresses after green tea extract treatment and unfortunately, saline and oxidative stresses showed lower reductions of *S. typhimurium* than the stress conditions itselves, resulting in a slightly protective effect (Table 3). These results indicate that green tea treatment is not convenient as a pre-treatment to control *S. typhimurium* growth.

Table 3: Effect of Microgard™100 or green tea extract treatments on the reduction (-Log N/No) of *Salmonella typhimurium* (mean±SD) by different stress conditions in liquid cheese whey

	Heat	Saline	Oxidative	Acidic
Control	6.8 ± 0.1	2.1 ± 0.2	6.2 ± 0.1	0.0 ± 0.1
Green tea extract	7.1 ± 0.2	0.7 ± 0.0	5.2 ± 0.2	0.0 ± 0.1
Microgard™	7.9 ± 0.3	3.1 ± 0.2	5.7 ± 0.1	1.6 ± 0.1

Microgard™ 100 induced a moderate increase vulnerability of *S. typhimurium* to heat, saline and acidic stresses, reaching reductions of 1.1, 1.0 and 1.6 log cycles higher than the stresses alone. As it said, Microgard™100 includes different organic acids, which are known to cause sublethal injury in gram-negative bacteria. Alakomi et al. (2000) reported that these acids can cause lipopolysaccharide release of the gram-negative bacteria outer membrane, particularly of *S. typhimurium*; thus, Microgard^{TM} may act as an enhancer of the effects of different stresses. In fact, Greenacre and Brocklehurst (2006) demonstrated that lactic acid-adapted cells of *S. typhimurium* showed an increased vulnerability to the toxicity of NaCl (2.5M). They informed a reduction of 3 log cycles for the acid adapted cells against the 2 log cycles for the control cells, both challenged to an osmotic stress of $a_w = 0.91$. These authors also noted that different organic acids can produce different responses to subsequent stresses and that a mixture of organic acids, as Microgard^{TM} represents, may result more efficient because of the acids' differing abilities to enter the cell. On the other hand, different responses of acid adapted cells have to be with different cell membrane composition and integrity, which differs from one microorganism to another.

Cell surface hydrofobicity is determined by many structures and biomolecules such as proteins, polysaccharides and lipids present in bacterial surface. A modification in the number of cells that can adhere to the solvent is an indication that some of these structures have changed. It has been suggested, as a simple model, that cell surface hydrophobicity increases as a result of the exposure of lipid molecules caused by rapid structural disturbance of the outer membrane (Tsuchido et al., 1985). Thus, in order to evaluate the cell surface hydrofobicity of cells exposed to natural antimicrobials, the microbial adhesion to solvent (MATS) test were performed as described by Rosenberg et al. (1980). As can be seen in (Figure 6), hydrofobicity assay showed that Microgard^{TM} increased in 8% the affinity to solvent of *S. typhimurium*, while no change in the cell surface of this microorganism was observed after applying green tea extract. The cell surface change produced by Microgard^{TM} could reflect the higher vulnerability of these microorganisms to subsequent stress conditions.

Escherichia coli

When stress conditions were applied without adding previously an antimicrobial, *Escherichia coli* showed important log cycle reductions with heat and H_2O_2, while saline and acidic conditions were less effective in controlling the microorganism growth (Table 4). It has been reported that heat treatment at 55 °C of *E. coli* cells induces blebbing and vesiculation of the outer membrane from the cells, thereby causing the release of proteins and lipopolysaccharide from the cell surface and producing a substantial destruction of the outer membrane structure (Katsui et al., 1982). On the other hand, Yokoigawa et al. (1999), who studied the survival differences between pathogenic and non-pathogenic strains of *E. coli*, observed that this microorganism can grow and survive in many acidic and saline foods.

Table 4: Effect of Microgard^{TM}100 or green tea extract treatments on the reduction (-log N/No) of *Escherichia coli* (mean±SD) by different stress conditions in liquid cheese whey

	Heat	Saline	Oxidative	Acidic
Control	5.8 ± 0.0	2.5 ± 0.0	5.4 ± 0.1	1.1 ± 0.1
Green tea extract	6.2 ± 0.1	3.0 ± 0.2	3.7 ± 0.3	2.6 ± 0.0
Microgard^{TM}	7.1 ± 0.1	3.8 ± 0.1	7.8 ± 0.2	3.4 ± 0.1

E. coli cells treated with green tea and then exposed to the subsequent stresses resulted a little more sensitive to heat, NaCl and lactic acid. Additionally, *E. coli* showed an increased (8%) in its hydrofobicity when treated with green tea extract (Figure 6). Probably, the antimicrobial may debilitate the outer membrane of these bacteria, thus, they become more unprotected and vulnerable to extreme conditions. Cho et al. (2007) identified unique changes in saturated and unsaturated fatty acids of the cell membrane of *E. coli* treated with green tea polyphenols, whereas scanning electron microscopic analysis demonstrated the presence of perforations and irregular rod forms with wrinkled surfaces in cells treated with lethal concentrations of tea polyphenols.

Microgard™ 100 treatment produced an increase vulnerability to all of the consecutive treatments comparing with those stresses without antimicrobial (Table 4). This is very interesting since this antimicrobial had a slight or no effect on *E. coli* (see section Microgard™). However, although a reduction in cell number is not significant some kind of injury on the integrity of microorganism cell should occur in order to explain the higher effect of subsequent stresses. Indeed, a great change (22%) in the cell surface hydrofobicity of *E. coli* was observed after applying Microgard™ (Figure 6) and, as a consequence, a possible correlation between changes in cell surface by the action of Microgard™ and a greater activity of stresses in the consecutive treatment could be established.

Another biochemical modification that may also took place is the reduction of catalase and superoxide dismutase activities on injured *E. coli* cells, which may explain the toxic effects of a relatively low peroxide level (McDonald et al., 1983).

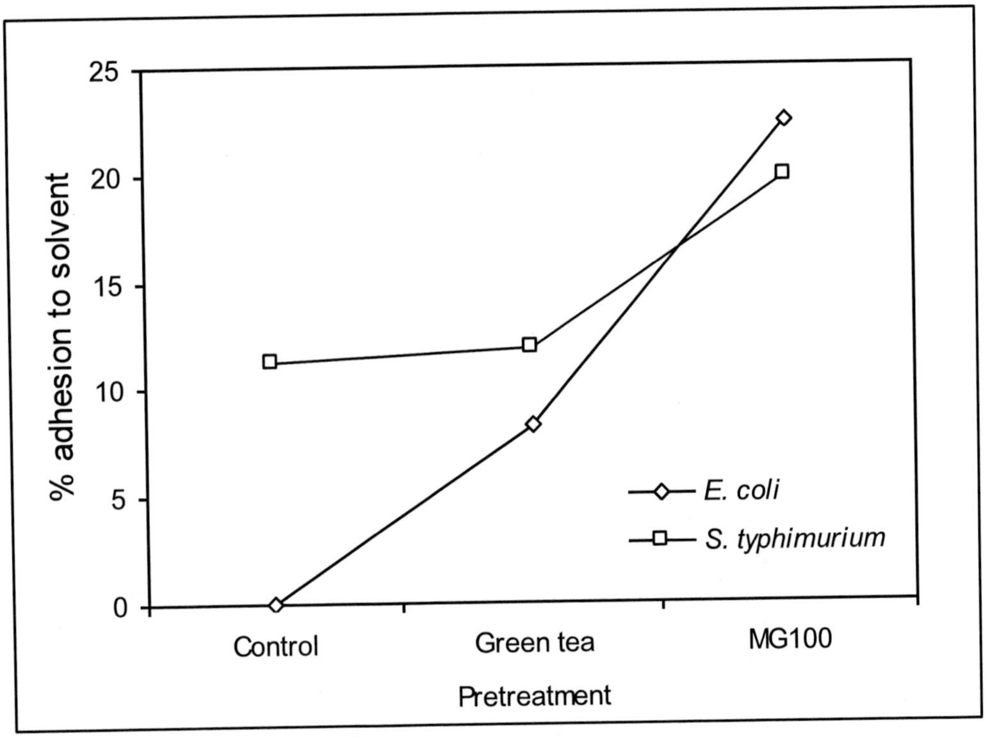

Figure 6. Influence of natural antimicrobial treatments on the surface hydrofobicity of Gram negative microorganisms.

Conclusion

Data exposed in the present chapter showed that the response to stress conditions of a given microorganism, previously treated with an antimicrobial, varies with the microorganism and the type of stress. However, in most of the cases the natural antimicrobial improved the performance of the stress applied in liquid cheese whey. Synergistic effects would have to be exploited to minimize the concentration of additives required to achieve a particular antibacterial effect without adversely affecting the sensorial acceptability. This knowledge about the interactions between natural antimicrobials and stressors has important implications for the food industry when considering which food preservation regime is the best to adopt.

References

Alakomi, H.L., Skyttä E, Saarela M, Mattila-Sandholm T, Latva-Kala K & Helander, I.M. (2000). Lactic acid permeabilizes Gram-negative bacteria by disrupting the outer membrane. *Applied and Environmental Microbiology*, 66, 2001-2005.

Al-Zoreky, N., Ayres, J.W. & Sandine, W.E. (1991). Antimicrobial activity of Microgard^{TM} against food spoilage and pathogenic microorganisms. *Journal of Dairy Science*, 74, 758-763.

Andre′, M.C., Hidalgo Campos, M.R., Borges, L.J., Kipnis, A., Pimenta, F.C. & Bisol Serafini, A. (2008). Comparison of *Staphylococcus aureus* isolates from food handlers, raw bovine milk and Minas Frescal cheese by antibiogram and pulsed-field gel electrophoresis following SmaI digestion. *Food Control*, 19, 200–207.

Apostolidis, E., Kwon, Y.I., Shetty, K. (2008). Inhibition of *Listeria monocytogenes* by oregano, cranberry and sodium lactate combination in broth and cooked ground beef systems and likely mode of action through proline metabolism. *International Journal of Food Microbiology*, 128, 317-324.

Atra, R., Vatai, G., Bekassy-Molnar, E. & Balint, A. (2005). Investigation of ultra-and nanofiltration for utilization of whey protein and lactose. *Journal of Food Engineering*, 67, 325-332.

Bauer, R. & Dicks, L.M.T. (2005). Mode of action of lipid II-targeting lantibiotics. *International Journal of Food Microbiology*, 101, 201-216.

Bearns, R. E. & Girard, K. F. (1958). The effects of pasteurization on *Listeria monocytogenes*. *Canadian Journal of Microbiology*, 4, 55-61.

Bikels-Goshen, T., Landau, E., Saguy, S., & Shapira, R. (2010). Staphylococcal strains adapted to epigallocatechin gallate (EGCG) show reduced susceptibility to vancomycin, oxacillin and ampicillin, increased heat tolerance, and altered cell morphology. *International Journal of Food Microbiology*, doi:10.1016/j.ijfoodmicro.2010.01.011.

Breukink, E., Van Kraaij, C., Demel, R. A., Siezen, R.J., Kuipers, O.P. & De Kruijff, B. (1997). The C-terminal region of nisin is responsible for the initial interaction of nisin with the target membrane. *Biochemistry*, 36, 6968-6976.

Campbell, J., Bang, W., Isonhood, J., Gerard, P.D. & Drake, M.A. (2004). Effects of salt, acid, and MSG on cold storage survival and subsequent acid tolerance of *Escherichia coli* 0157:H7. *Food Microbiology*, 21, 727-735.

Caturla, N., Vera-Samper, E., Villalaín, J., Mateo, C. R., & Micol, V. (2003). The relationship between the antioxidant and the antibacterial properties of galloylated catechins and the structure of phospholipid model membranes. *Free Radical Biology & Medicine*, 34, 648-662.

Charlier, C., Cretenet, M., Even, S., & Le Loir, Y. (2009). Interactions between *Staphylococcus aureus* and lactic acid bacteria: An old story with new perspectives. *International Journal of Food Microbiology*, 131, 30-39.

Chi-Zhang, Y., Yam, K. L., & Chikindas, M. L. (2004). Effective control of *Listeria monocytogenes* by combination of nisin formulated and slowly released into broth system. *International Journal of Food Microbiology*, 90, 15–22.

Cho, Y. S., Schiller, N. L., Kahng, H. Y. & Oh, K. H. (2007). Cellular responses and proteomic analysis of *Escherichia coli* exposed to green tea polyphenols. *Current Microbiology*, 55, 501-506.

Chou, C-C., Lin, L-L., & Chung, K-T. (1999). Antimicrobial activity of tea as affected by the degree of fermentation and manufacturing season. *International Journal of Food Microbiology*, 48, 125-130.

Cleveland, J., Montville, T.J., Nes, I.F. & Chikindas, M.L. (2001) Bacteriocins: safe, natural antimicrobials for food preservation. *International Journal of Food Microbiology*, 71, 1-20.

Connor, D.E. & Kotrola, J.S. (1995). Growth and survival of *Escherichia coli* O157:H7 under acidic conditions. *Applied and Environmental Microbiology*, 61, 382-385

D'Aoust, J.Y., Warburton, D. W., & Sewell, A. M., (1985). *Salmonella typhimurium* phage-type 10 from Cheddar cheese implicated in a major Canadian foodborne outbreak. *Journal of Food Protection*, 48, 1062–1066.

Dave, R.I., Sharma, P., Julson, J., Muthukumarappan, K. & Henning, D.R. (2003). Effectiveness of Microgard® in controlling *Escherichia coli* 0157:H7 and *Listeria monocytogenes*. *Journal of Food Science and Technology*, 40, 262-266.

De Buyser, M.L., Dufour, B., Maire, M., & Lafarge, V. (2001). Implication of milk and milk products in food-borne diseases in France and in different industrialised countries. *International Journal of Food Microbiology*, 67, 1–17.

De Martinis, E. C. P., Crandall, A. D., Mazzotta, A. S., & Montville, T. J. (1997). Influence of pH, salt and temperature on nisin resistance of *Listeria monocytogenes*. *Journal of Food Protection*, 60, 420–423.

Deegan, L., Cotter, P.D., Colin, H. & Ross, P. (2006). Bacteriocins: Biological tools for bio-preservation and shelf-life extension. *International Dairy Journal*, 16, 1058-1071.

Degnan, A. J., Kaspar, C. W., Otwell, W. S., Tamplin, M. L., & Luchansky, J. B. (1994). Evaluation of lactic acid bacterium fermentation products and food-grade chemicals to control *Listeria monocytogenes* in blue crab (*Callinectes sapidus*) meat. *Applied and Environmental Microbiology*, 60, 3198-3203.

Delves-Broughton, J. (1990). Review: Nisin and its application as a food preservative. *Journal of the Society of Dairy Technology*, 43, 73–76.

Eswaranandam, S., Hettiarachchy, N. S., & Johnson, M. G. (2004). Antimicrobial activity of citric, lactic, malic, or tartaric acids and nisin-incorporated soy protein film against *Listeria monocytogenes*, *Escherichia coli* 0157:H7, and *Salmonella gaminara*. *Journal of Food Science*, 69, 79-84.

European Union Risk Assessment Report (2003). Hydrogen peroxide (Vol. 38, pp. 1–258).

Ferruzzi, M.G. & Green, R.J. (2006). Analysis of catechins from milk-tea beverages by enzyme assisted extraction followed by high performance liquid chromatography. *Food Chemistry*, 99, 484-491.

Food and Drug Administration (2006) *Code of Federal Regulations*. 21CFR184.1081.

Food and Drug Administration (2006) *Code of Federal Regulations* 21CFR184.1538.

Foster, J.W. (1991). Salmonella acid shock proteins are required for the adaptative acid tolerance response. *Journal of Bacteriology*, 173, 6896-6902

Gallo, L.I., Pilosof, A.M.R., Jagus, R.J. (2007). Effective control of *Listeria innocua* by combination of nisin, ph and low temperature in liquid cheese whey. *Food Control,* 18, 1086-1092.

Gandhi, M. & Chikindas, M.L. (2007). *Listeria*: A foodborne pathogen that knows how to survive. *International Journal of Food Microbiology*, 113, 1-15.

González-Martínez, C., Becerra, M., Cháfer, M., Albors, A., Carot, J.M. & Chiralt, A, (2002). Influence of substituting milk powder for whey powder on yoghurt quality. *Trends in Food science and Technology*, 13, 334-240.

Gould, G. W. (1997). Methods of preservation and extension of shelf life. *International Journal of Food Microbiology*, 33 (1), 51-64.

Govaris, A., Koidis, P. & Papatheodorou, K. (2001). The fate of *Escherichia coli* O157:H7 in Myzithra, Anthotyros, and Manouri whey cheeses during storage at 2°C and 12°C. *Food Microbiology*, 18, 565-570.

Gradišar, H., Pristovšek, P., Plaper, A., & Jerala, R. (2007). Green tea catechins inhibit bacterial DNA gyrase by interaction with its ATP binding site. *Journal of medicinal Chemistry*, 50, 264-271.

Greenacre, E.J. & Brocklehurst, T.F. (2006). The acetic acid tolerance response induces cross-protection to salt stress in *Salmonella typhimurium. International Journal of Food Microbiology*, 112, 62-65.

Hamama, A., Hankouri, N. E., & Ayadi, M. E. (2002). Fate of enterotoxigenic *Staphylococcus aureus* in the presence of nisin-producing *Lactococcus lactis* strain during manufacture of Jben, a Moroccan traditional fresh cheese. *International Dairy Journal*, 12, 933–938.

Hamilton-Miller, J. M. T., & Shah, S. (1999). Disorganization of cell division of methicillin-resistant *Staphylococcus aureus* by a component of tea (*Camellia sinensis*): a study by electron microscopy. *FEMS Microbiology Letters*, 176, 463-469.

He, Q., Shi, B. & Yao, K. (2006). Interactions of gallotannins with proteins, amino acids, phospholipids and sugars. *Food Chemistry*, 95, 1178-1182.

Henning, S., Metz, R., & Hammes, W. P. (1986). Studies of the mode of action of nisin. *International Journal of Food Microbiology*, 33, 121–134.

Henning, D.R., Baer, R.J., Hassan, A.N. & Dave, R. (2006). Major advances in concentrated and dry milk products, cheese, and fat-based spreads. *Journal of Dairy Science,* 89, 1179-1188.

Hough, G., Puglieso, M. L., Sanchez, R., & Da Silva, O. M. (1999). Sensory and microbiological shelf-life of a commercial Ricotta cheese. *Journal of Dairy Science*, 82, 454-459.

Ikigai, H., Nakae, T., Hara, Y., & Shimamura, T. (1993). Bactericidal catechins damage the lipid bilayer. *Biochimica et Biophysica Acta*, 1147, 132-136.

Jablonski, L. M. & Bohach, G. (2001). Food Microbiology: Fundamentals and Frontiers. Washington, DC: ASM Press.

Jhala, R., Muthukumarappan, K., Julson, J. L., Dave, R. I., and Mahapatra, A. K. (2002). Synergistic effects of ozone and Microgard 300 for controlling *L. Monocytogenes* in ready-to-eat cooked and cured ham. Am. Soc. Agric. Eng. (ASAE), Paper No. 026143. ASAE, St Joseph, MI.

Jørgensen, H. J., Mørk, T. and Rørvik, L. M. (2005). The Occurrence of *Staphylococcus aureus* on a Farm with Small-Scale Production of Raw Milk Cheese. *Journal of Dairy Science,* 88, 3810–3817.

Kamat, A.S. & Nair, P.M. (1996). Identification of *Listeria innocua* as a biological indicator for inactivation of *Listeria monocytogenes* by some meat processing treatments. *Lebensmittel-Wissenschaft und-Technologie*, 29, 714-720.

Katsui, N., Tsuchido, T., Hiramatsu, R., Fujikawa, S., Takano, M. & Shibasaki, I. (1982). Heat-induced blebbing and vesiculation of the outer membrane of *Escherichia coli. Journal of Bacteriology*, 151, 1523-1531.

Kim, S., Ruengwilysup, C. & Fung, D.Y.C. (2004). Antibacterial effect of water-soluble tea extracts on foodborne pathogens in laboratory medium and in a food model. *Journal of Food Protection*, 67, 2608-2612.

Lemay, M.J., Choquette, J., Delaquis, P.J., Garièpy, C., Rodrigue, N. & Saucier, L. (2002). Antimicrobial effect of natural preservatives in a cooked and acidified chicken meat model. *International Journal of Food Microbiology*, 78, 217-226.

Leyer, G.J., Wang, J.J. & Johnson, E.A. (1995). Acid adaptation of *Escherichia coli* O157:H7 increases survival in acidic foods. *Applied and Environmental Microbiology*, 61, 3752-3755.

McDonald, L. C., Hackney, C. R. & Ray, B. (1983). Enhanced recovery of injured *Escherichia coli* by compounds that degrade hydrogen peroxide or block its formation. *Applied and Environmental Microbiology*, 45, 360-365.

Millette, M., Le Tien, C., Smoragiewicz, W., & Lacroix, M. (2007). Inhibition of *Staphylococcus aureus* on beef by nisin-containing modified alginate films and beads. *Food Control*, 18, 878–884

Mugliaroli, S., von Staszewski, M. & Jagus, R.J. (2007). Efecto combinado de antimicrobianos naturales sobre el desarrollo de *Staphylococcus aureus* Y *Salmonella typhimurium* en concentrado proteico de suero líquido (CPSL). In proceedings from XI Congreso Argentino de Ciencia y Tecnología de Alimentos, Bs. As., Argentina. Area 17, Vol: Tecnologías emergentes de preservación de alimentos.

O'Byrne, C.P. & Booth, I.R. (2002). Osmoregulation and its importance to food-borne microorganisms. *International Journal of Food Microbiology*, 74, 203-216.

Peters, R.H. (2005). Economic aspects of cheese making as influenced by whey processing options. *International Dairy Journal*, 15, 537-545.

Phan-Thanh, L., Mahouin, F. & Aligé, S. (2000). Acid responses of *Listeria monocytogenes. International Journal of Food Microbiology*, 55, 121-126.

Raybaudi-Massilia, R.M., Mosqueda-Melgar, J. & Martín-Belloso, O. (2008). Antimicrobial activity of malic acid against *Listeria monocytogenes*, *Salmonella Enteritidis* and *Escherichia coli* O157:H7 in apple, pear and melon juices. *Food Control, 20* (2), 105-112.

Rosenberg, M., Gutnick, D. & Rosenberg, E. (1980). Adherence of bacteria to hydrocarbon: A simple method for measuring cell surface hydrophobicity. *FEMS Microbiology Letters*, 9, 29–33.

Ryan, C.A., Nickels, N.T., Hargrett-Bean, N.T., Potter, M.E., Endo, T., Mayer, L., Langkop, C.W., Gibson, L., McDonald, R.C., Kenney, R.T., Puhr, N.D., McDonnell, P.J., Martin, R.J.,

Cohen, M.L., Blake, P.A. (1987). Massive outbreak of anitmicrobial-resistant salmonellosis traced to pasteurized milk. *Journal of American Medical Association*, 258, 3269–3274.

Samelis, J., Bedie, G. K., Sofos, J. N., Belk, K. E., Scanga, J. A., & Smith, G. C. (2005). Combination of nisin with organic acids or salts to control Listeria monocytogenes on sliced pork bologna stored at 4 ºC in vacuum packages. *Lebensmittel Wissennschaft und Technologie*, 38, 21–28.

Scalbert, A., Manach, C., Morand, C. & Rémésy, C. (2005). Dietary polyphenols and the prevention of diseases. *Critical Review of Food Science and Nutrition,* 45, 287-306.

Schillinger, U., Becker, B., Vignolo, G., & Holzapfel, W. H. (2001) Efficacy of nisin in combination with protective cultures against *Listeria monocytogenes* Scott A in tofu. *International Journal of Food Microbiology*, 71, 159–168.

Sergelidis, D., & Abrahim, A. (2009). Adaptive response of *Listeria monocytogenes* to heat and its impact on food safety. *Food Control, 20*, 1-10.

Shabala, L., Mcmeekin, T., Budde, B.B. & Siegumfeldt, H. (2006). *Listeria innocua* and *Lactobacillus delbrueckii* subsp. *Bulgaricus* employ different strategies to cope with acid stress. *International Journal of Food Microbiology*, 110, 1-7.

Si, W., Gong, J., Tsao, R., Kalab, M., Yang, R. & Yin, Y. (2006). Bioassay-guided purification and identification of antimicrobial components in chinese green tea extract. *Journal of Chromatography*, 1125, 204-210.

Singleton, V. L., & Rossi, J. A. (1965). Colorimetry of total phenolics with phosphomolybdic-phosphotungstic acid reagents. *American Journal of Enology and Viticulture*, 16, 144-158.

Skandamis, P. N., Yoon, Y., Stopforth, J. D., Kendall, P. A., & Sofos, J. N. (2008). Heat and acid tolerance of *Listeria monocytogenes* after exposure to single and multiple sublethal stresses. *Food Microbiology, 25,* 294-303.

Stapleton, P.D., Shah, S., Hamilton-Miller, J.M.T., Hara, Y., Nagaoka, Y., Kumagai, A., Uesato, S. & Taylor, P.W. (2004). Anti-*Staphylococcus aureus* activity and oxacillin resistance modulating capacity of 3-O-acyl-catechins. *International Journal of Antimicrobial Agents*, 24, 374-380.

Stapleton, P.D., Gettert, J. & Taylor, P.W. (2006). Epicatechin gallate, a component of green tea, reduces halotolerance in *Staphylococcus aureus*. *International Journal of Food Microbiology*, 111, 276-279.

Taguri, T., Tanaka, T. & Kouno, I. (2004). Antimicrobial activity of 10 different plant polyphenols against bacteria causing food-borne disease. *Biological Pharmaceutical Bulletin*, 27, 1965-1969.

Tiwari, B. K., Valdramidis, V. P., O´Donnell, C. P., Muthukumarappan, K., Bourke, P., & Cullen, P. J. (2009). Application of natural antimicrobials for food preservation. *Journal of Agricultural and Food Chemistry*, 57, 5987-6000.

Tsuchido, T., Katsui, N., Takeuchi, A., Takano, M. & Shibasaki, I. (1985). Destruction of the outer membrane permeability barrier of *Escherichia coli* by heat treatment. *Applied and Environmental Microbiology*, 50, 363-371.

Ukuku, D. O., & Shelef, L. A. (1997). Sensitivity of six strains of *Listeria monocytogenes* to nisin. *Journal of Food Protection*, 60, 867–869.

Ukuku, D. O., Pilizota, V., & Sapers, G. M. (2004). Effect of hot water and hydrogen peroxide treatments on survival of *Salmonella* and microbial quality of whole and fresh-cut cantaloupe. *Journal of Food Protection*, 67, 432-437.

von Staszewski, M. & Jagus, R. J. (2007). Efecto antimicrobiano de distintas variedades de té verde argentino en suero de queso líquido. *Alimentos, Ciencia e Ingeniería,* 16 (3), 305-307.

von Staszewski, M & Jagus, R.J. (2008). Natural antimicrobials: effect of Microgard™ and nisin against *Listeria innocua* in liquid cheese whey. *International Dairy Journal*, 18, 255-259.

Weeks, M.E., Nebe von Caron, G., James, D.C., Smales, C.M. & Robinson, G.K. (2006). Monitoring changes in nisin susceptibility of *Listeria monocytogenes* Scott A as an indicator of growth phase using FACS. *Journal of Microbiological Methods, 66, 43-55.*

Wu, S.C., Yen, G.C., Wang, B.S., Chiu, C.K., Yen, W.J., Chang, L.W., & Duh, P.D. (2007). Antimutagenic and antimicrobial activities of pu-erh tea. *LWT*, 40, 506-512.

Yang, Z., Tu, Y., Xia, H., Jie, G., Chen, X. & He, P. (2007). Suppression of free-radicals and protection against H_2O_2-induced oxidative damage in HPF-1 cell by oxidized phenolic compounds present in black tea. *Food Chemistry*, 105, 1349-1356.

Yokoigawa, K., Takikawa, A. & Kawai, H. (1999). Difference between *Escherichia coli* O157:H7 and non-pathogenic *E. coli*: survival and growth in seasonings. *Journal of Bioscience and Bioengineering*, 88, 574-576.

Zhang, Y.M., & Rock, C. O. (2004). Evaluation of epigallocatechin gallate and related plant polyphenols as inhibitors of the FabG and FabI reductases of bacterial type II fatty-acid synthase. *The Journal of Biological Chemistry*, 279, 30994-31001.

Zhao, W-H., Hu, Z.Q., Okubo, S., Hara, Y., & Shimamura, T. (2001). Mechanism of synergy between epigallocatechin gallate and β-lactams against methicillin-resistant *Staphylococcus* aureus. Antimicrobial Agents and Chemotherapy, 45, 1737-1742.

Zottola, E. A., & Smith, L. B. (1991). Pathogens in cheese. *Food Microbiology*, 8, 171–182.

Zuckerman, H. & Ben Abraham, R. (2002). Control of growth of L. monocytogenes in fresh salmon using Microgard™ and nisin. *Lebensmittel-Wissenschaft und Technologie,* 35, 543-548.

In: Focus on Food Engineering
Editor: Robert J. Shreck

ISBN: 978-1-61209-598-1

Chapter 9

CHANGES IN RHEOLOGICAL PROPERTIES OF HARD CHEESE DURING ITS AGEING

***Libor Severa*[1], *Jan Trnka*[2], *Jaroslav Buchar*[1], *Pavla Stoklasová*[2], and *Šárka Nedomová*[1]**

[1] Mendel University in Brno, Zemědělská 1, 613 00 Brno, Czech Republic
[2] Institute of Thermomechanics, Czech Academy of Science, Dolejškova 5, 182 00, Praha 8, Czech Republic

ABSTRACT

This study is aimed at characterizing the non-linear viscoelastic behavior of hard cheese during ageing. Edam block (corresponding to the Gouda cheese) and Brie were used as examples. The rheological properties of these cheeses were evaluated using different experimental techniques:

a) The compression tests, including stress relaxation, were performed. The material data from monotonic compression and stress relaxation tests were used for creation of a constitutive equation describing the cheese viscoleastic properties.
b) Viscoelastic properties were also obtained from indentation tests.

In the next step, the behavior of the tested cheeses under dynamic loading was studied during ageing processes. The block of tested cheese was loaded by the impact of an aluminum bar. The force between the bar and cheese was recorded. The surface displacements, as well as surface velocities, were obtained at different points from the position of bar impact. The laser vibrometers were used to record the displacements. Response functions were evaluated both in the time and frequency domains. It was found that the degree of the cheese ageing process is characterized well by the reduction of the surface displacement maximum. This process of ageing is also by the maximum of the impact force. The spectral analysis of the response functions revealed that there was a dominant frequency, which depends only on the degree of the cheese ageing process. The developed method represents a promising procedure for the continuous recording of cheese ageing.

1. INTRODUCTION

Cheese ageing is a complex process involving many physicochemical changes such as a change in pH, a progressive breakdown of the proteins to smaller polypeptides and the gradual accumulation of amino acids (Foxet et al., 1993). Cheese texture may also vary with a change in the physical state of the fats that are already present in the cheese (Dufour et al., 2000; Watkinson et al., 1997). These changes can be described e.g. by the rheological properties of cheese. Rheological characterization of cheese is important as a means of determining body and texture characteristics and also for examining how these parameters are affected by composition, processing techniques and storage conditions (Konstance and Holsinger, 1992).

The most common method employed in studying the mechanical properties of cheeses is the uniaxial force-compression: a constant rate of compression is applied to the sample and the resulting stress is continuously recorded. The method is suitable for the study of the rheological parameters of cheese on ageing and has been employed in French cheeses (Antoniou, et al. 2000), Cheddar cheese (Hort et al., 1997), Parmigiano Reggiano cheese (Noel et al., 1996), Swiss-type cheese (Bachmann et al., 1999) and Gouda cheese (Spangler et al., 1990). This test also enables the characterization of the viscoelastic properties of soft solid foods and other biological materials in terms of the stress relaxation. Del Nobile et al. (2007) used this procedure to characterize a variety of products. In stress relaxation tests, a constant strain is applied, and the stress required in maintaining the deformation is measured as a function of time. When a stress relaxation test is performed, different behaviors can be observed: ideal elastic materials do not relax whereas ideal viscous materials instantaneously show a relaxation. Viscoelastic solids gradually relax and reach an equilibrium stress greater than 0, whereas for viscoelastic fluids, instead, the residual stress vanishes to zero (Steffe, 1996). This method has been widely used for the study of the cheese viscoelastic properties (Sadowska et al., 2009). These tests are used as standards for conventional structural materials because the measured forces and displacements can be converted into the stress-strain properties using simple theories, but they can be tedious and difficult when applied to soft foods because of the need to prepare a specimen of specific size and shape. This problem can be solved when using the indentation tests.

Tests based on the indentation technique are popular in food texture evaluation because they do not need samples with strict shape requirements (Ozkan et al., 2002). However, because of the nonuniformity in the strain distribution, only limited theories that relate the indentation force–displacement response to the stress-strain properties are available (e.g., Sneddon, 1965; Sakai, 2002).

At present, many non-destructive techniques such as small displacement probes, vibrating rheometers, near infrared spectroscopy (NIR), computer vision, biosensors, ultrasonic analysis and sonic measurements are emerging (Blazquez et al., 2006; Davie et al., 1996; Mulet et al., 1999; Ortiz et al., 2001).

One of these methods is the acoustic impulse–response technique. This procedure enables the non-destructive measurement of firmness when the food is excited by being struck with a probe and the frequency spectrum from the recorded acoustic signal is obtained. This technique is widely used for the evaluation of the quality of fruits, hen's eggs and some other products (Cho et al., 2000; Diezma-Iglesias et al., 2004; De Belie et al., 2000a, 2000b;

Diezma-Iglesias et al., 2006; Zude et al., 2006). This procedure was also used for cheese texture assessment (Conde et al., 2008).

The objective of this study was to determine the influence of ageing time on changes in the viscoelastic properties of two types of cheese. The compression, relaxation, indentation and impact response methods were used.

2. Material and Experimental Technique

Two types of cheese produced in the Czech Republic were used. The experiments were first carried out on blocks of Certified Origin EDAM cheese. Semi-hard cheese called either Eidamský blok or Eidamská cihla have been made in the Czech Republic since the 1920's. Their technological scheme is similar to the production of Gouda cheese (Kněz, 1960).

Commercially produced Edam cheeses (29 cm x 10 cm x 10 cm blocks) were manufactured by a company located in Jihlava. The cheese was produced on January 27, 2008. It was stipulated to the suppliers that cheese should be of typical quality and standard properties. Samples of cheese arrived to testing laboratory on February 4, 2008. After receiving the experimental material, it was stored at 12 °C. The blocks of cheese were tested at 43rd (March 10, 2008), 61st (March 28, 2008), 79th (April 15, 2008), and 108th (May 14, 2008) day after the production. The content of fat, moisture, salt and pH of the cheese were evaluated by common procedures. In our experiments Edam cheeses with fat content of 30 and 45 % were used.

The next experiments were carried out on blocks of Certified Origin Brie cheese, manufactured by a company located in Czech-Moravia Highland. The cheese has a form of cylinder (diameter about 230 mm, height above 35mm). An example of the Brie cheese block is shown in Figure 1.

Figure 1. Photo of the Brie cheese block. L and T represent longitudinal and transversal direction of loading.

The blocks were matured in the chambers, where relative humidity and temperature were maintained according to the company procedures. The blocks of cheese were tested in the following periods after production: one week (February 5, 2009), three weeks (February 19, 2009), four weeks (February 26, 2009), five weeks (March 5, 2009), six weeks (March 12, 2009), seven weeks (March 19, 2009), and eight weeks (March 26, 2009).

The cheese specimens were tested in compression, relaxation and indentation using the TIRATEST universal testing machine (Germany). Following experiments have been performed:

a) The specimens of Edam cheese (in form of cylinder with height of 10 mm and diameter of 20 mm) were compressed at following cross-head speeds: 1, 10, 100, and 1000 mm/min.
b) The relaxation tests were performed also on TIRATEST testing machine. Each sample was compressed to different height at the cross-head speed of 10 mm/min. A certain level of the loading force corresponds to each height. The compression anvil was held in fixed position for 300 seconds to maintain a constant strain. Time-dependent changes in stresses were recorded twice in a second, using a computer program. The relaxation test data represent average from five measurements.
c) The indentation tests on Edam cheese were performed at five constant speeds of 1, 5, 10, 20, and 100 mm/min using spherical indenter 10-mm diameter, $D=2R$, respectively.
d) The indentation tests on Brie cheese were performed at constant speed of 20 mm/min using both spherical indenter 10-mm diameters, $D=2R$, and bar indenter (6 mm in diameter), respectively.

The impact tests were carried out using an impact device specially designed and built for cheese measurements. The experimental set up is schematically shown in Figure 2. The impact set-up consisted of a free-falling cylindrical bar (6 mm in diameter, 200 mm in height – made from aluminum alloy). The bar was instrumented by strain gauges. Such instrumentation enables to record the time history of force at the interface between cheese and bar. Surface displacement as well as surface velocity were measured using the laser-vibrometer at following distances from the point of the bar impact 30, 45, 60, 75, 90, 105, 120, and 135 mm.

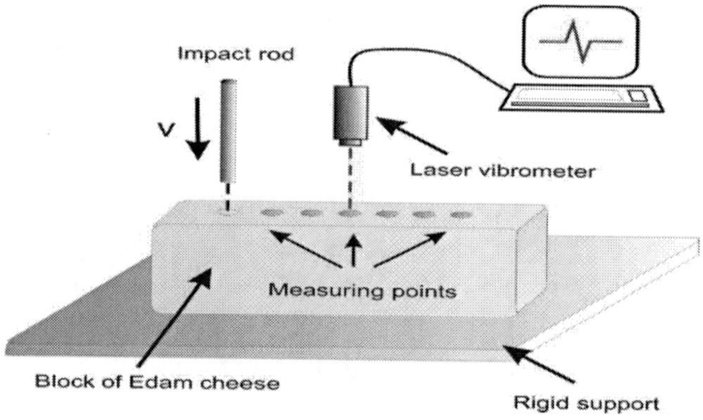

Figure 2. Schematic of the experimental device. (The same experimental arrangement has been used for the testing of the Brie cheese).

3. Results and Discussion

3.1. Edam Cheese Properties at Compression

In Figure 3, examples of the experimental records of true stress – true strains are displayed.

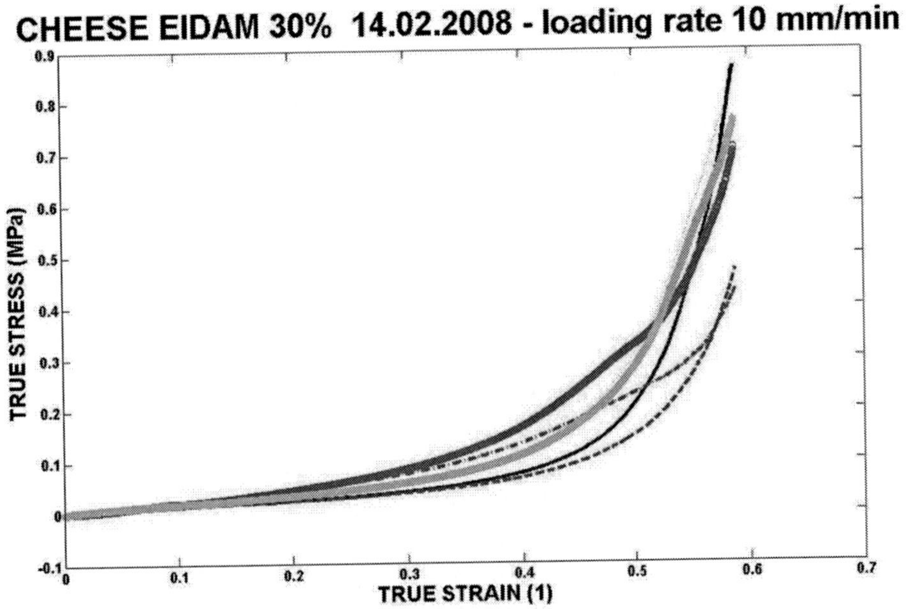

Figure 3. Experimental records of the stress – strain curves.

One can see that there is a scatter of data. Average curves can be fitted by the function:

$$\sigma = Ae^{B\varepsilon} \tag{1}$$

where σ and ε are the true strain and the true stress, respectively (Mancini et al., 1999). A and B are the constants and have to be regarded as fitting parameters. This function describes the stress strain curves obtained for Edam cheese in all stages of its maturity.

The data enable calculation of the elastic modulus E as the tangent to the stress strain curve at the origin:

$$E = \frac{\partial \sigma}{\partial \varepsilon}\Big|\varepsilon = 0 \qquad E = AB \tag{2}$$

The values of the elastic moduli for a loading rate of 10 mm/min are shown in Figure 4. It can be seen that the elastic modulus decreases with the time of the cheese ageing (as expected). The main changes occur in the beginning of the ageing. Elastic modulus also depends on the fat content. The increase of the fat content leads to decrease of the elastic modulus. The compression deformation properties described by the stress – strain curves are also loading rate dependent. The increase in the loading rate leads to the increase in the stress level in all values of the strain. This phenomenon is shown in Figure 5. Similar results were achieved for all tested specimens of the Edam cheese.

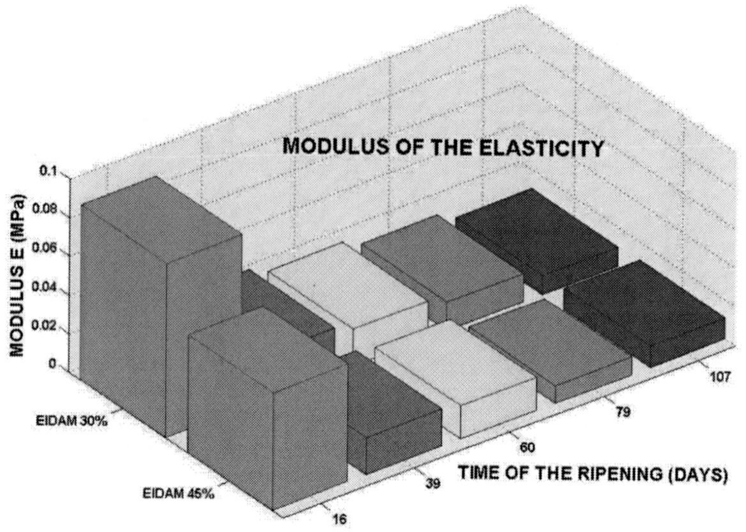

Figure 4. Elastic moduli of the tested cheeses.

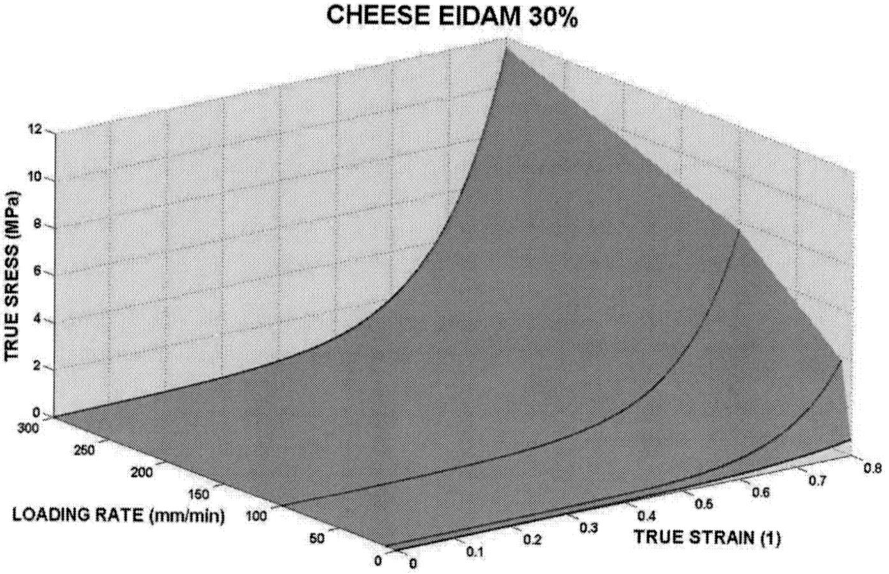

Figure 5. Influence of the loading rate on the stress strain dependence. Cheese tested on February 16, 2008.

The effect of loading rate on the stress – strain curve is the evidence of cheese viscoelastic behavior. A common mechanical test to characterize the viscoelastic properties of the soft solid foods and other biological materials is a stress relaxation. Example of the experimental record of the relaxation curve is shown in Figure 6.

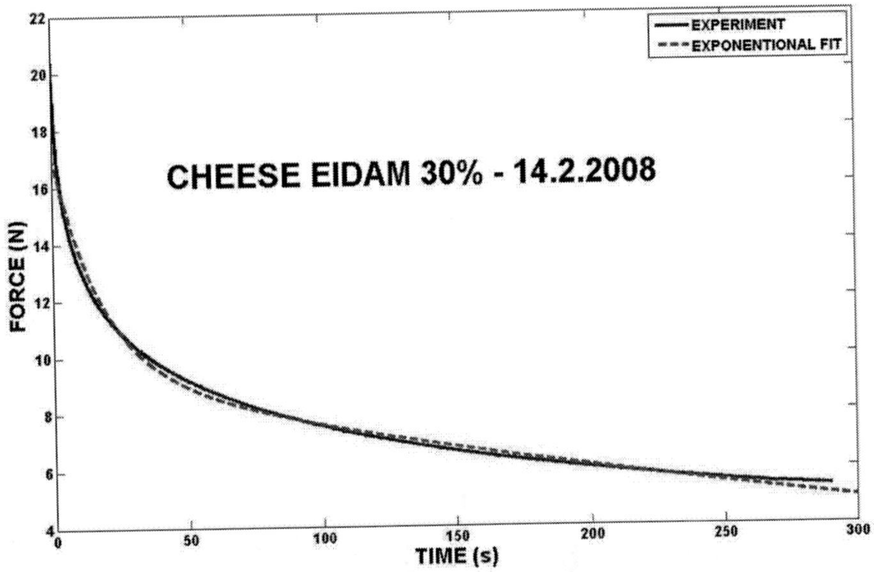

Figure 6. Relaxation curve of the cheese specimen. Loading rate 10 mm/min.

The same characteristics were typical for all stress relaxation curves obtained for all specimens tested at the different stages of their maturity and for all values of the preloading force F_o – see example in Figure 7.

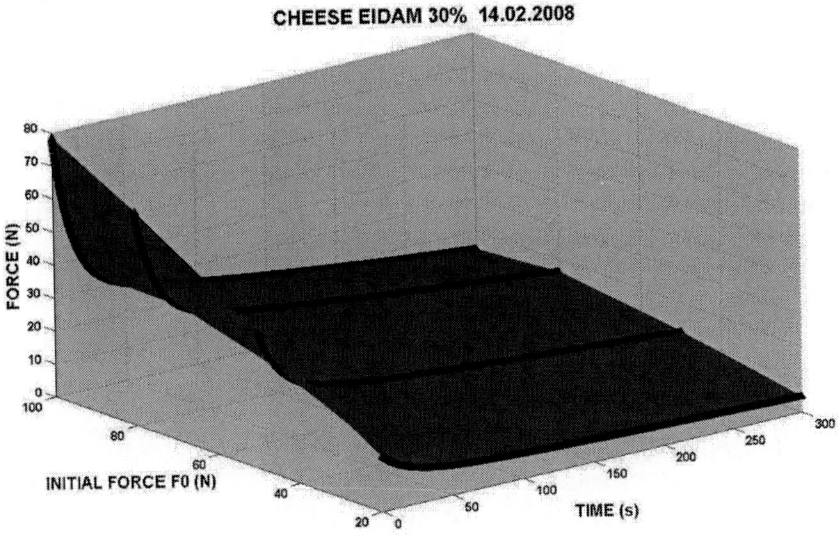

Figure 7. Stress relaxation curves obtained for the different values of the preloading force F_o.

The analysis of the records revealed that data can be fitted by the function:

$$F = ae^{-bt} + ce^{-dt} + F_1 \tag{3}$$

The coefficient of Pearson's linear correlation between model and experimental points was higher than 0.997. Example of the influence of cheese ageing time is shown in Figure 8.

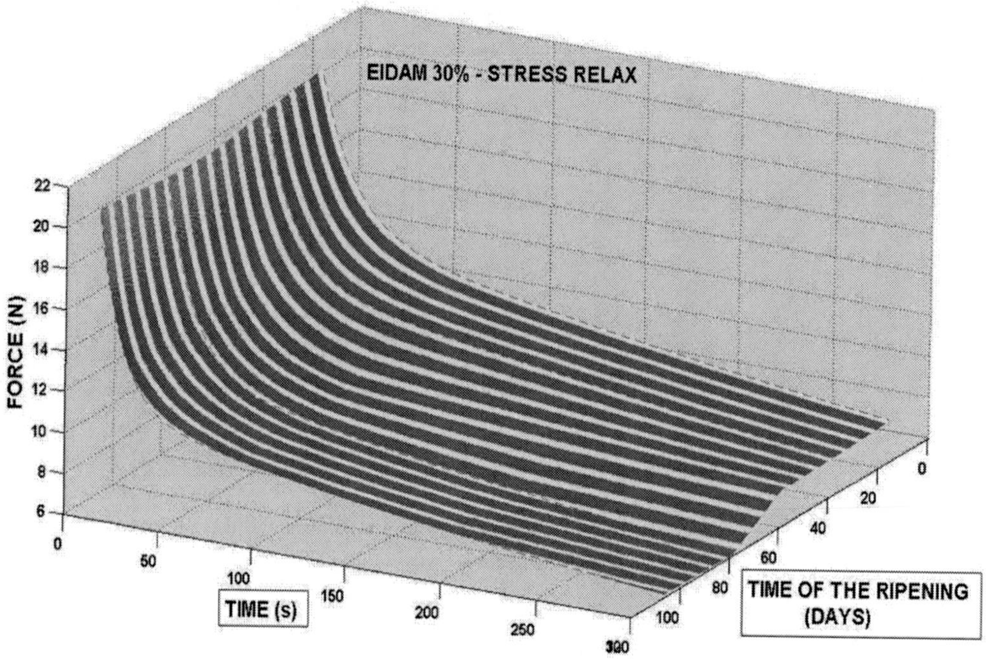

Figure 8: Stress relaxation in the Edam cheese. $F_o = 20$ N.

In order to interpret the experimental data a generalized Maxwell model can be used. The model contains n of Maxwell elements and a spring in parallel; each element consisting of dashpot in series (Bock et al., 1989; Waananen and Okos, 1992; Watts and Bilanski, 1991). The generalized Maxwell model can be written as follows:

$$\sigma(t) = \sum_{i=1}^{i=n} C_i e^{-\frac{t}{\tau_i}} + \sigma_o, \tag{4}$$

where σ is the stress which is given by

$$\sigma(t) = \frac{F(t)}{S}, \tag{5}$$

where S is the cross section of the specimen. Owing to Eq. (2), our model involves a parallel coupling of a Hooke's body and two Maxwell's bodies. The stress–relaxation behavior of cheese can be than described as

$$E(t) = \frac{\sigma(t)}{\varepsilon_o} = E_1 e^{-\frac{t}{\tau_1}} + E_2 e^{-\frac{t}{\tau_2}} + E_o, \tag{6}$$

where ε_o is the strain corresponding to the force F_o. The parameter τ is usually expressed as:

$$\tau = \frac{\eta}{E}, \tag{7}$$

where η is the viscosity. The Eq. (6) can be written as

$$E(t) = E_1 e^{-\frac{E_1}{\eta_1}t} + E_2 e^{-\frac{E_2}{\eta_2}t} + E_o \tag{8}$$

The sum of the values of three elastic moduli (E_o+E_1+E_2) in the model can be used as an indicator of elasticity, while the sum of the values of two viscous moduli (η_1+η_2) – as an indicator of viscosity.

Parameters of Eq. (8) are given in the Tables 1- 5.

Table 1a. EDAM cheese 30%

STRAIN	E_1 (MPa)	η_1 (MPas)	E_2 (MPa)	η_2 (MPas)	E_o (MPa)	E_o+E_1+E_2 (MPa)	η_1+η_2 (MPas)
0.2269	0.1071	1.9764	0.1352	62.9217	0.0438	0.2860	64.8981
0.2649	0.2363	3.8181	0.2578	121.1701	0.0853	0.5795	124.9882
0.2649	0.3861	5.3676	0.3667	166.7669	0.1434	0.8962	172.1345
0.2649	0.4255	5.5953	0.2941	111.9624	0.1709	0.8905	117.5577

February 14, 2008.

Table 1b. EDAM cheese 45 %

STRAIN	E_1 (MPa)	η_1 (MPas)	E_2 (MPa)	η_2 (MPas)	E_o (MPa)	E_o+E_1+E_2 (MPa)	η_1+η_2 (MPas)
0.27	0.07	0.52	0.06	6.372	0.06		6.898
46	30	59	89	3	82	0.2101	2
0.29	0.23	3.36	0.20	78.62	0.06		81.98
25	90	26	22	49	88	0.5100	75
0.38	0.31	3.50	0.20	72.48	0.06		75.98
55	11	38	28	10	71	0.5810	48
0.26	0.53	6.20	0.44	156.7	0.19		162.9
84	50	65	05	059	92	1.1747	124

February 14, 2008.

Table 2a. EDAM cheese 30 %

STRAIN	E_1 (MPa)	η_1 (MPas)	E_2 (MPa)	η_2 (MPas)	E_o (MPa)	$E_o+E_1+E_2$ (MPa)	$\eta_1+\eta_2$ (MPas)
0.4378	0.0499	0.3369	0.0462	4.3155	0.0365	0.1326	4.6524
0.4378	0.1262	0.7032	0.0965	7.7996	0.0736	0.2962	8.5028
0.4559	0.2387	1.3058	0.1657	12.4432	0.1035	0.5079	13.7490
0.5491	0.2112	0.8835	0.1264	7.8767	0.0725	0.4100	8.7603

March 10, 2008.

Table 2b. EDAM cheese 45 %

STRAIN	E_1 (MPa)	η_1 (MPas)	E_2 (MPa)	η_2 (MPas)	E_o (MPa)	$E_o+E_1+E_2$ (MPa)	$\eta_1+\eta_2$ (MPas)
0.5247	0.0395	3.2795	0.0466	0.3043	0.0255	0.1116	3.5838
0.5357	0.1228	0.6752	0.0885	6.7082	0.0539	0.2652	7.3834
0.6120	0.2243	0.5821	0.0884	0.1427	0.0427	0.3554	0.7248
0.4078	0.4363	1.1923	0.1655	8.6191	0.0782	0.6800	9.8115

March 10, 2008.

Table 3a. EDAM cheese 30 %

STRAIN	E_1 (MPa)	η_1 (MPas)	E_2 (MPa)	η_2 (MPas)	E_o (MPa)	$E_o+E_1+E_2$ (MPa)	$\eta_1+\eta_2$ (MPas)
0.5772	0.0523	0.8201	0.0377	31.1998	0.0215	0.1115	32.0198
0.6392	0.1242	1.6239	0.0694	18.7973	0.0454	0.2390	20.4212
0.5753	0.2384	1.8699	0.1103	30.0210	0.0545	0.4032	31.8909
0.6819	0.2664	1.6333	0.0970	22.6843	0.0480	0.4114	24.3176

March 28, 2008.

Table 3b. EDAM cheese 45%

STRAIN	E_1 (MPa)	η_1 (MPas)	E_2 (MPa)	η_2 (MPas)	E_o (MPa)	$E_o+E_1+E_2$ (MPa)	$\eta_1+\eta_2$ (MPas)
0.6901	0.0260	0.2289	0.0588	48.4460	0.0194	0.1042	48.6749
0.6964	0.1284	1.2983	0.0441	18.2580	0.0415	0.2139	19.5563
0.7617	0.1748	1.6487	0.0712	17.1748	0.0343	0.2802	18.8235
0.4999	0.3490	2.0735	0.1268	27.2071	0.0638	0.5396	29.2806

March 28, 2008.

The Figure 9 illustrates changes in the model values of viscoelastic modulus *E(t)* (Eq. (8)). This Figure presents the courses obtained for different values of the initial strains. The sum values of viscosity and damping were determined based on transformations of selected coefficients in the semi-empirical Maxwell model. It was found that the courses of the viscoelastic moduli were not affected by the value of this parameter. This fact approves that cheese behaves as linear viscoelastic solid.

Table 4a. EDAM cheese 30 %

STRAIN	E_1 (MPa)	η_1 (MPas)	E_2 (MPa)	η_2 (MPas)	E_o (MPa)	$E_o+E_1+E_2$ (MPa)	$\eta_1+\eta_2$ (MPas)
0.6487	0.0450	0.6346	0.0345	11.5881	0.0182	0.0977	12.2227
0.7156	0.1895	1.5472	0.0337	9.1226	0.0049	0.2282	10.6698
0.6365	0.2080	1.7903	0.0930	0.2506	0.0493	0.3503	2.0408
0.7500	0.1884	1.3547	0.0796	20.5623	0.0691	0.3371	21.9169

April 16, 2008.

Table 4b. EDAM cheese 45 %

STRAIN	E_1 (MPa)	η_1 (MPas)	E_2 (MPa)	η_2 (MPas)	E_o (MPa)	$E_o+E_1+E_2$ (MPa)	$\eta_1+\eta_2$ (MPas)
0.7749	0.0421	0.4514	0.0236	6.0197	0.0159	0.0816	6.4711
0.7789	0.1063	0.7819	0.0406	9.1581	0.0250	0.1719	9.9400
0.8385	0.1819	0.9546	0.0538	12.4516	0.0351	0.2709	13.4062
0.5471	0.3016	1.6442	0.1150	1.7934	0.0844	0.5009	3.4376

April 16, 2008.

Table 5a. EDAM cheese 30 %

STRAIN	E_1 (MPa)	η_1 (MPas)	E_2 (MPa)	η_2 (MPas)	E_o (MPa)	$E_o+E_1+E_2$ (MPa)	$\eta_1+\eta_2$ (MPas)
0.6827	0.0441	0.5453	0.0326	10.5788	0.0266	0.1032	11.1241
0.7519	0.1141	0.8423	0.0500	13.1306	0.0066	0.1707	13.9730
0.6656	0.2034	1.0293	0.0793	20.9875	0.0740	0.3567	22.0168
0.7824	0.1789	0.8682	0.0662	17.4108	0.0883	0.3334	18.2790

May 13, 2008.

Table 5b. EDAM cheese 45%

STRAIN	E_1 (MPa)	η_1 (MPas)	E_2 (MPa)	η_2 (MPas)	E_o (MPa)	$E_o+E_1+E_2$ (MPa)	$\eta_1+\eta_2$ (MPas)
0.8151	0.0396	0.3913	0.0228	7.2656	0.0107	0.0731	7.6568
0.8180	0.1000	0.4878	0.0381	9.6721	0.0209	0.1590	10.1599
0.8750	0.1444	0.6280	0.0580	15.1168	0.0301	0.2325	15.7449
0.5695	0.2801	1.4248	0.1138	28.9843	0.0677	0.4616	30.4091

May 13, 2008.

Development of the viscoelastic modulus during cheese ageing is displayed in Figure 10. If we compare Figure10 with Figure 7 a clear difference can be seen. The relaxation curves shown in Figure 7 are less sensitive to the cheese maturity than the viscoelastic modulus. Viscoelastic moduli lie well above the moduli obtained from the stress-strain curves. The elasticity of the cheese given by the sum of moduli $E_0+E_1+E_2$ decreases with ageing time – see Figure 11. There is a very small (if not negligible) difference between cheeses with different fat content. This result was expected. The same tendency is exhibited also in case of cheese viscosity, $(\eta_1+\eta_2)$, see Figure 12.

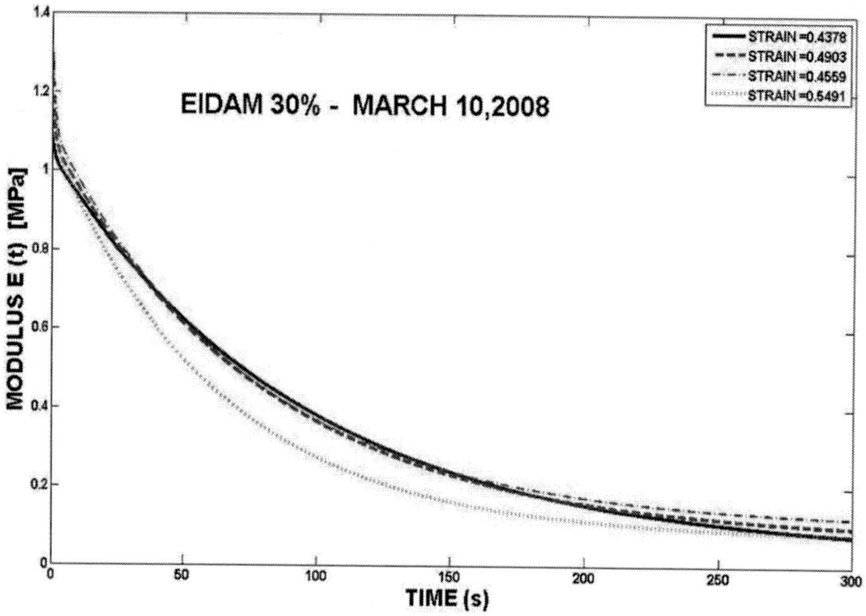

Figure 9: Time history of the viscoelastic moduli.

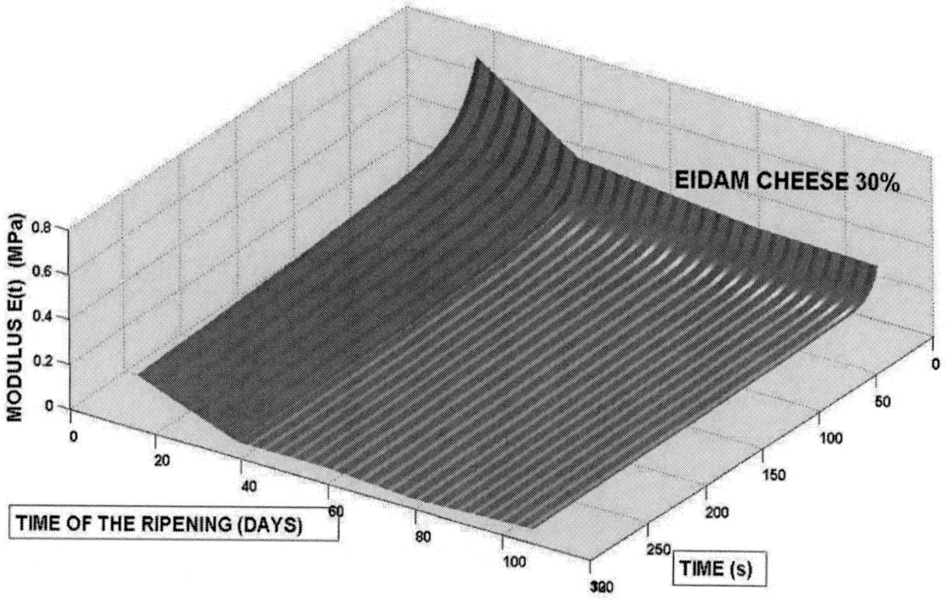

Figure 10. Dependence of the viscoelastic modulus *E(t)* on the ageing time.

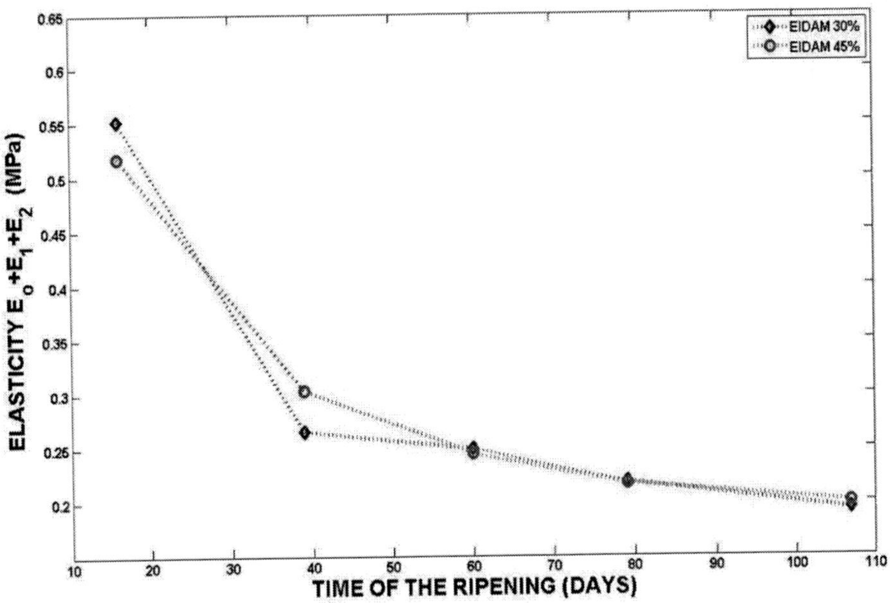

Figure 11. Elasticity of the tested cheeses.

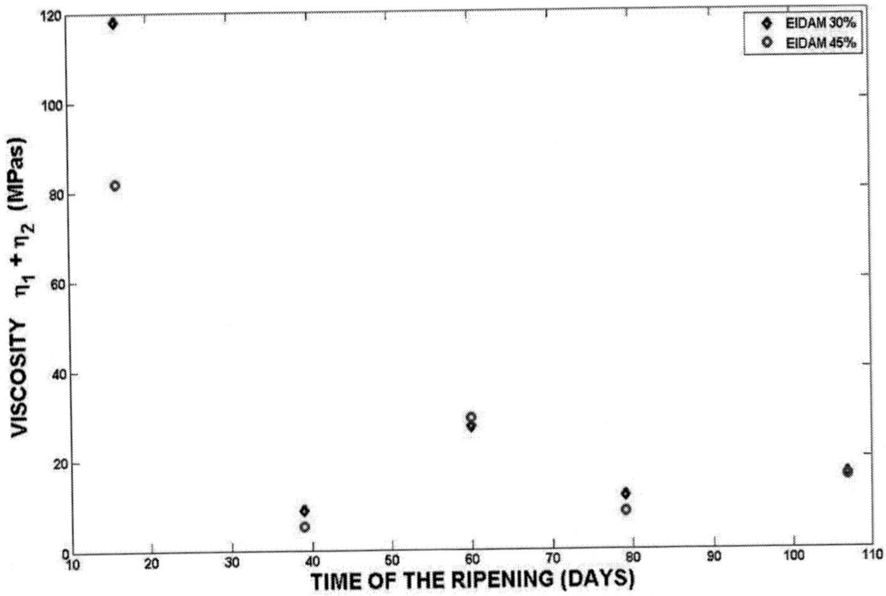

Figure 12. Viscosity of the tested cheeses.

The relaxation test together with the compression test enable creation of constitutive equation of the viscoelastic materials. On the other hand they can be tedious and difficult when applied to soft foods because of the need to prepare specimen of specific size and shape. This problem can be solved by using the indentation tests.

3.2. Indentation Test of Edam Cheese

Tests based on the indentation technique are popular in food texture evaluation because they do not need samples with strict shape requirements (Ozkan et al., 2002). However, because of the nonuniformity in the strain distribution, only limited theories that relate the indentation force–displacement response to the stress-strain properties are available (e.g., Sneddon, 1965; Sakai, 2002). As a result, indentation force-displacement measurements for many foods are interpreted empirically (Anand and Scanlon, 2002). However, there is a need for converting data from existing "empirical" tests into the fundamental material properties so that sensory assessment of foods can be improved (Bourne, 1994).

This part of study was focused on determination of Edam cheese viscoelastic properties from the indentation test. Spherical indenter, which simulates the actions of a cheese grader when "thumbing" a cheese was used. Axisymmetric indentation tests were performed at five constant speeds of 1, 5, 10, 20, and 100 mm/min using spherical indenter (10-mm diameter, $D=2R$).

In Figure 13 an example of the experimental record of indentation force P versus penetration depth h is displayed. The detail analysis shows that experimental data can be fitted by the function:

$$P(t) = ah^3 + bh^2 + ch + d\,, \tag{9}$$

where parameters a, b, c, and d are given in the Tables 6 – 9.

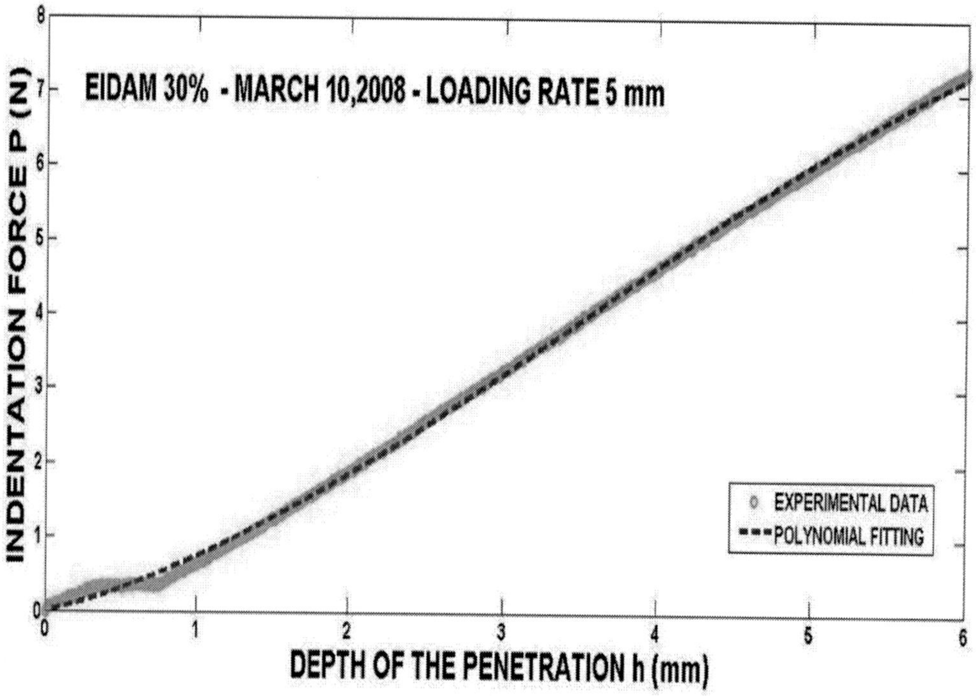

Figure 13. The dependence of the indentation force on the penetration depth.

Table 6. Parameters of the polynomial fitting of *P(h)*

Fat Content	Loading Rate	*a*	*b*	*c*	*d*	R^2
30%	1	-0.02100	0.2204	0.3068	-0.03257	0.9910
	5	-0.02479	0.2680	0.4851	0.00574	0.9950
	10	-0.02239	0.2452	0.8220	-0.11910	0.9960
	20	-0.02056	0.3164	0.1554	0.00667	9.9940
	100	-0.03702	0.4474	0.5287	0	0.9998
45%	1	-0.01017	0.1045	0.1886	0	0.9987
	5	-0.01860	0.1875	0.3156	-0.00997	0.9991
	10	-0.01108	0.1249	0.2305	0.04396	0.9993
	20	-0.01402	0.1590	0.2893	0.04491	0.9940
	100	-0.01197	0.1917	0.1644	0.09416	0.9998

Edam cheese tested on March 10, 2008.

Table 7. Parameters of the polynomial fitting of *P(h)*

Fat Content	Loading Rate	*a*	*b*	*c*	*d*	R^2
30%	1	0.003958	-0.03751	0.8984	-0.14460	0.9985
	5	0.001974	-0.005166	1.1690	0.02030	0.9989
	10	-0.008963	0.1349	0.06969	-0.52850	0.9963
	20	-0.005421	0.1006	0.6557	-0.02214	0.9987
	100	0.0009304	-0.009417	0.5961	-1.96300	0.9999
45%	1	0.0009304	-0.009417	0.5961	0.01533	0.9999
	5	0.006919	-0.06142	0.9293	-0.05719	0.9990
	10	-0.03209	0.4288	-0.5979	0.11700	0.9959
	20	-0.004775	0.04302	0.6046	-0.0166	0.9959
	100	-0.02967	0.3846	-0.4066	0.02172	0.996

Edam cheese tested on March 28, 2008.

Table 8. Parameters of the polynomial fitting of *P(h)*

Fat Content	Loading Rate	*a*	*b*	*c*	*d*	R^2
30%	1	-0.005655	0.04701	0.7981	-0.04141	0.9999
	5	-0.002492	0.01076	1.276	0.07194	0.9999
	10	-0.01005	0.1009	1.068	-0.09581	0.9999
	20	-0.01162	0.1525	0.6751	0.2491	0.9997
	100	-0.01222	0.1304	1.037	-0.1939	0.9999

Table 8. (Continued)

Fat Content	Loading Rate	*a*	*b*	*c*	*d*	R^2
45%	1	-0.003466	0.01606	0.5809	-0.05697	0.9998
	5	-0.003058	0.01024	0.8089	0.1068	0.9999
	10	-0.003507	0.02419	0.5933	-0.01446	0.9999
	20	-0.002595	0.01733	0.6883	0.03484	0.9999
	100	-0.002359	0.01266	1.404	0.1578	0.9999

Cheese tested on April 16, 2008.

Table 9. Parameters of the polynomial fitting of the *P(h)*

Fat Content	Loading Rate	*a*	*b*	*c*	*d*	R^2
30%	1	-0.00245	0.01858	0.4407	-0.00708	0.9998
	5	-0.00734	0.07599	0.5494	-0.09191	0.9999
	10	-0.00453	0.02904	0.9724	-0.028	0.9999
	20	-0.00453	0.01174	0.9724	0.09387	0.9995
	100	-0.00886	0.0794	1.069	-0.3153	0.9999
45%	1	-0.00882	0.07382	0.5018	-0.08236	0.9998
	5	-0.02033	0.1931	0.3686	0.1565	0.9995
	10	-0.02314	0.1591	0.9182	-0.02164	0.9999
	20	-0.01602	0.081	1.228	0.08917	0.9999
	100	-0.01521	0.1209	1.232	-0.07458	0.9999

Cheese tested on May 14, 2008.

The indentation force increases with the decrease of the fat content – see example in Figure 14.

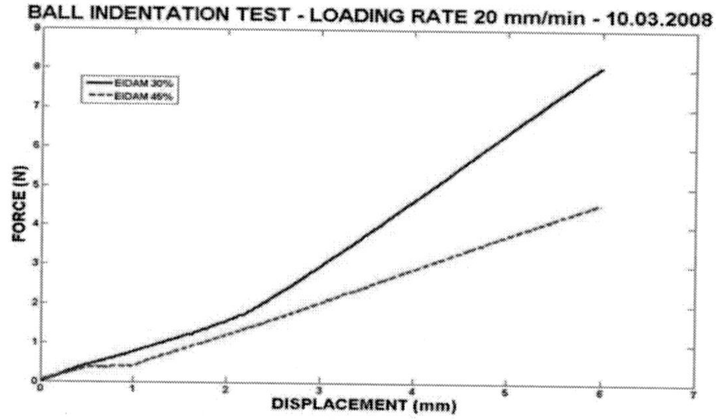

Figure 14. Influence of the fat content on indentation force *P*.

Indentation force increases with loading rate – see example in Figure 15. The influence of ageing time is shown in Figs. 16 – 20. It is evident that indentation force exhibits a decrease with time of cheese ageing for all loading rates there. There are some exceptions in the early stages of the ageing. In order to explain this effect, additional experiments are needed.

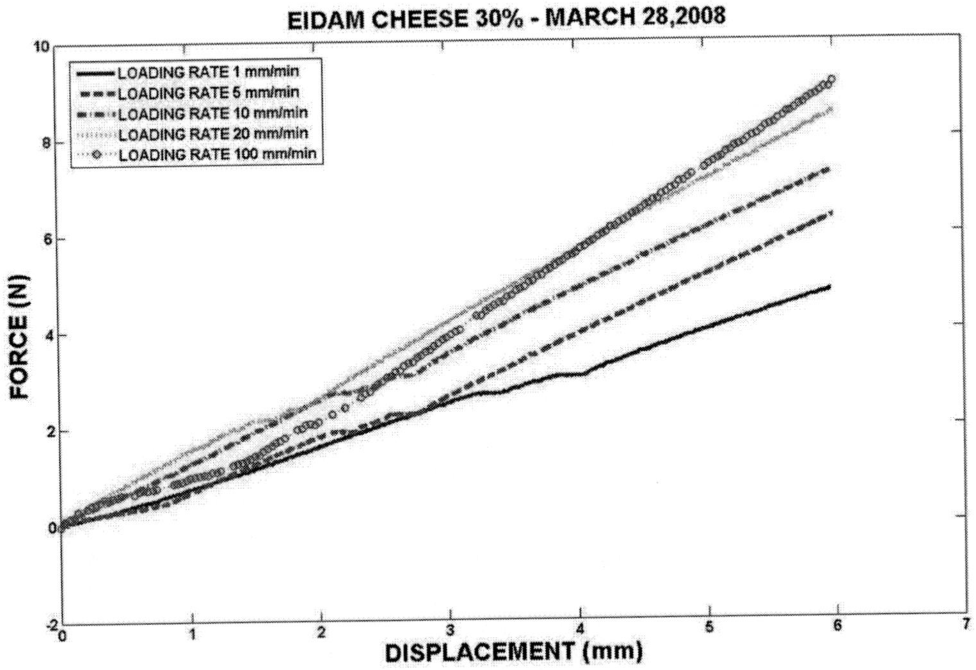

Figure 15. Influence of the loading rate on the indentation force.

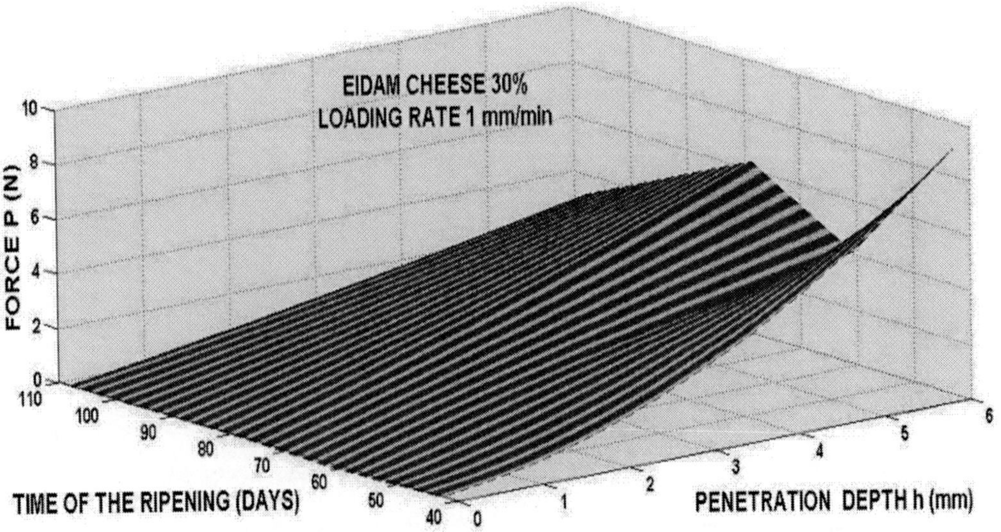

Figure 16. Influence of ageing time on indentation force.

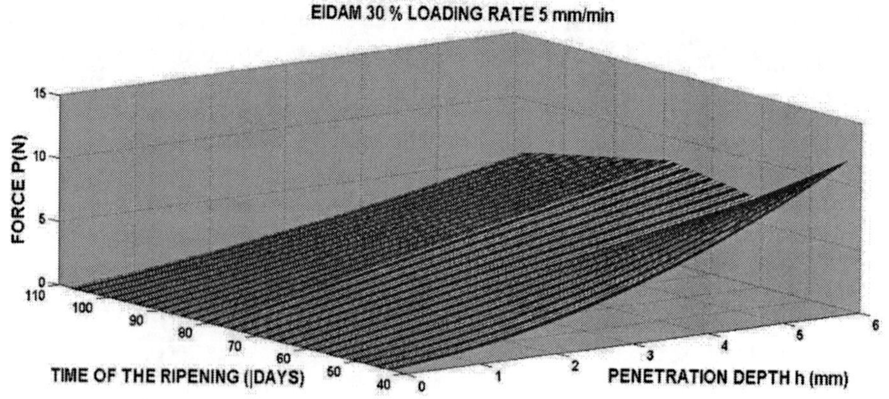

Figure 17. Influence of ageing time on indentation force.

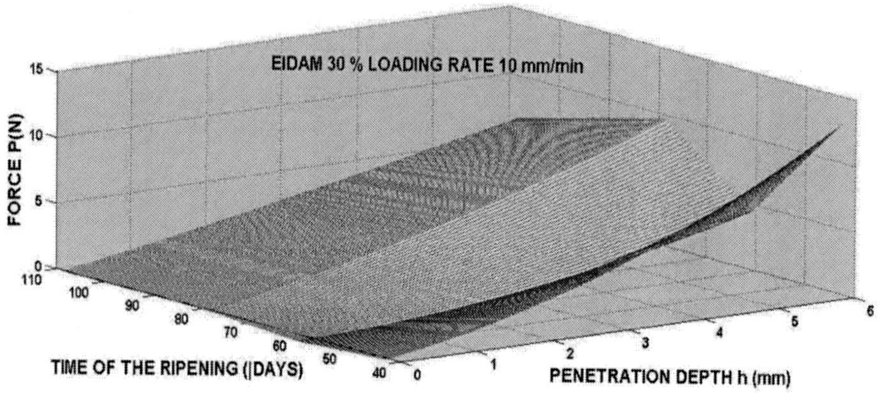

Figure 18. Influence of ageing time on indentation force.

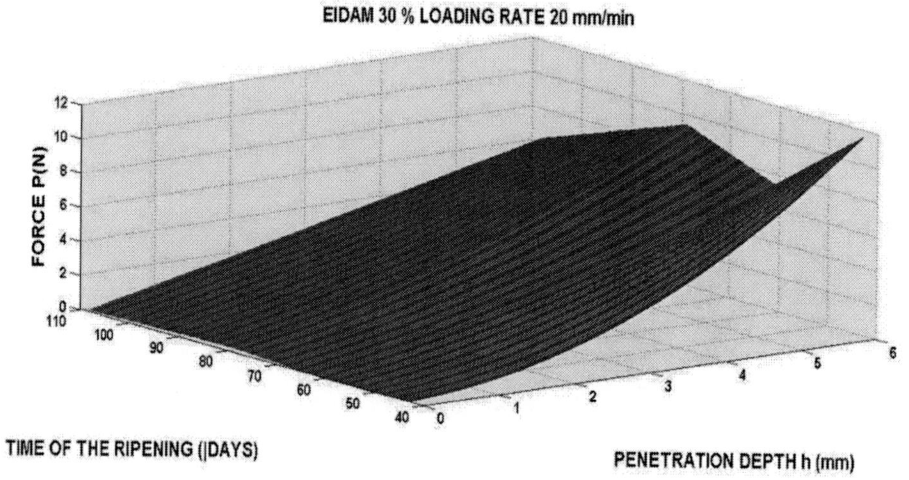

Figure 19. Influence of ageing time on indentation force.

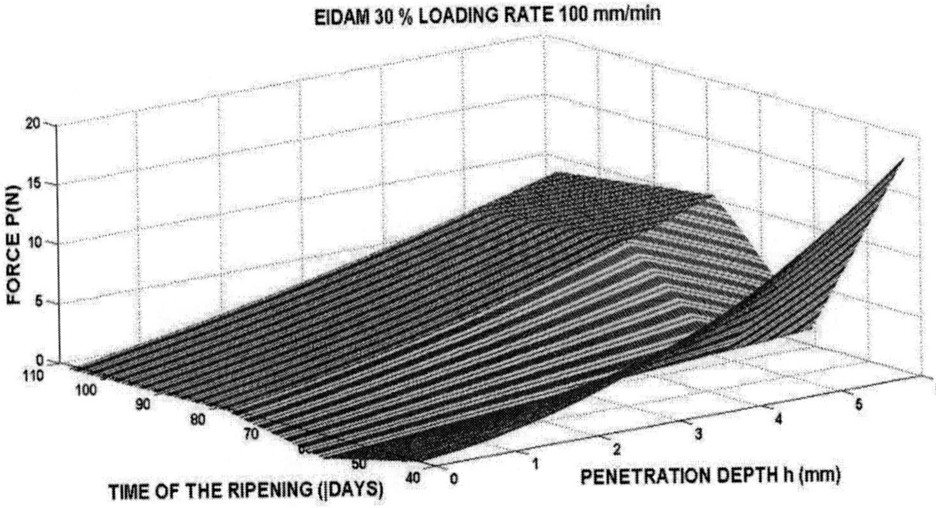

Figure 20. Influence of ageing time on indentation force.

These data can be consequently used for evaluation of the cheese viscoelastic properties. There are many problems connected with the evaluation of the indentation test results. The contact stress and strain in indentation problems, even for an elastic contact, are highly concentrated in the contact region, where extremely inhomogeneous deformations are developed. Such a complex mechanical fields complicate possibility of description of the constitutive relations of the applied force P to the internal stress, as well as the penetration depth h to the adjoining strains. In order to overcome these difficulties Meyer's principle of geometrical similarity can be used (Tabor, 1953). This procedure introduces the mean contact pressure σ, given by the ratio of the indentation load P to the projected contact area A_c, $\sigma = P/A_c$ .For a spherical indenter, the representative stress is given by:

$$\sigma = \frac{P}{\pi r^2} \approx \frac{\gamma_s P}{2\pi R h} , \qquad (10)$$

where R is the radius of the spherical indenter, h represents the total penetration depth which is related to the contact depth h_c by $h = \gamma_c h_c$ – see Figure 21.

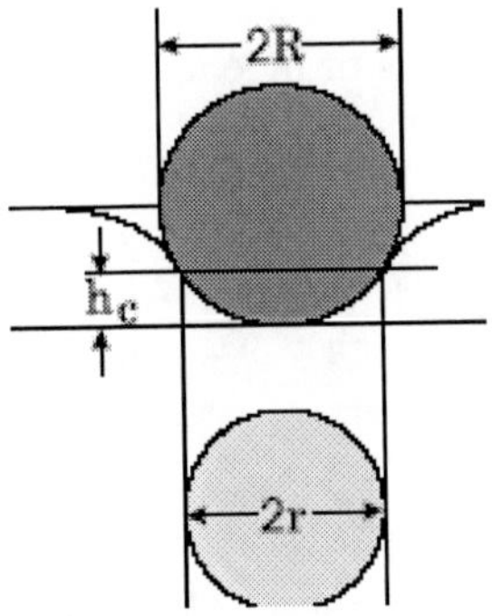

Figure 21. Schematic of the indentation geometry.

The representative strain is defined as:

$$d\varepsilon = k_s \frac{dr}{R}, \tag{11}$$

where dr is the infinitesimal increment of the circle radius of contact r – see Figure 21. The frontal coefficient k_s is an indenter constant which can be obtained from the elastic solution (Sneddon, 1965). Its value is $4/\pi$. The value of γ is 2.

Let us consider an indenter pressed into contact with a linear viscoelastic body. During indentation loading both penetration indenter and its contact area growing with time. The viscoelastic properties are described by the relaxation modulus $E(t)$ instead of the Young modulus E for purely elastic materials. The stress increment is than expressed as

$$d\sigma(t) = \frac{1}{(1-\nu^2)} E(t-t')d\varepsilon(t') \tag{12}$$

with assumption that Poisson ratio ν of viscoelastic material is independent on time, for simplicity. The complete solution was performed by Sakai (2002).

For a constant rate of penetration v_0, the indentation force P can be expressed as:

$$P(t) = \frac{\pi\sqrt{R}}{1-\nu^2}\frac{k_s}{2}\left(\frac{2}{\gamma_s}\right)^{\frac{3}{2}} \int_0^t E(t-t')t'^{\frac{1}{2}}dt' \tag{13}$$

The use of the Laplace transform leads to the solution for the stress relaxation modulus $E(t)$:

$$E(t) = \frac{4(1-\nu^2)}{\sqrt{R}k_s}\left(\frac{\gamma_s}{2\pi v_o}\right)^{\frac{3}{2}} L^{-1}\left[s^{\frac{3}{2}}P(s)\right], \tag{14}$$

where $P(s)$ is the Laplace transform of indentation load $P(t)$ with the transform variable s. This transform is defined as:

$$L[P(t)] = P(s) = \int_0^\infty P(t)e^{-ts}dt \tag{15}$$

And the inverse transform is given as:

$$L^{-1}[P(s)] = \frac{1}{2\pi i}\int_{c-i\infty}^{c+i\infty} P(s)e^{st}\,ds \tag{16}$$

The evaluation of this transform is very easy when using appropriate software (e.g. MATLAB). When this procedure is applied to the experimental data obtained in the previous chapter, a substitution $h = v_o t$, where v_o is the cross-head speed. The Laplace transform of the function given by the Eq. (9) is:

$$P(s) = \frac{6av_o^3}{s^4} + \frac{2bv_o^2}{s^3} + \frac{cv_o}{s^2} + \frac{d}{s} \tag{17}$$

In order to evaluate the function given by Eq.(14) the inverse transformation of the function $P(s)*s^{(3/2)}$ must be performed. The use of MATLAB software leads to:

$$L^{-1}\left[s^{\frac{3}{2}}P(s)\right] = 6av_o^3 L^{-1}\left(s^{-\frac{5}{2}}\right) + 2bv_o^2 L^{-1}\left(s^{-\frac{3}{2}}\right) + \frac{cv_o}{\sqrt{\pi t}} + dL^{-1}\left(s^{\frac{1}{2}}\right) \tag{18}$$

It follows that this transform has no analytical form and can be evaluated only numerically. In Figure 22, the time histories of the stress relaxation (viscoelastic) moduli are displayed.

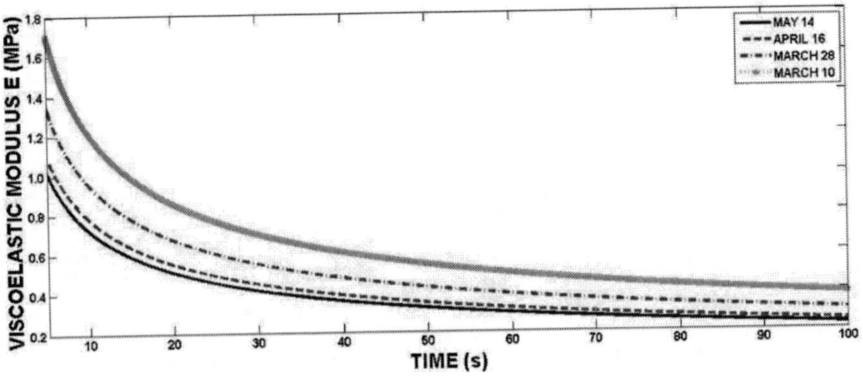

Figure 22. The influence of ageing time on viscoelastic moduli. Edam cheese – 30%.

The same dependence was obtained for the cheese with 45% fat content. It was shown that these moduli are independent on the loading rate. It means that moduli represent viscoelastic properties. The values of viscoelastic moduli are nearly the same as the moduli obtained from the relaxation tests. Both procedures thus lead to the same results. The indentation test was used for the testing of Brie cheese. This cheese is much softer than Edam cheese. It means that preparation of the specimen for the compression tests is very difficult if not impossible.

3.3. Indentation Test of the Brie Cheese

The development of the Brie cheese structure during its ageing is characterized by the growth of a crumb and by many voids and holes – see examples in the Figs. 23 – 24.

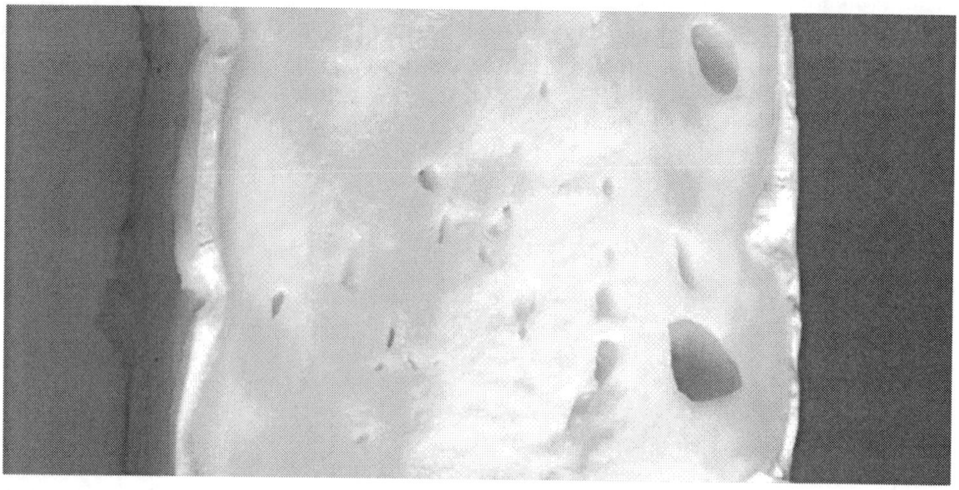

Figure 23. Cross section of the Brie cheese. Three weeks of ageing.

Figure 24. Cross section of the Brie cheese. Six weeks of the ageing.

It is obvious that preparation of the specimen for the compression or relaxation test is impossible. Interpretation of the indentation tests must be based on observation of the cheese structure because these tests may be strongly affected by presence of the holes.

Axisymmetric indentation tests were performed at constant speed of 20 mm/min using both spherical indenter, 10-mm in diameter, $D=2R$, and bar indenter, 6 mm in diameter, respectively. The measurement was performed in two directions shown in Figure 1 (L and T).

Examples of the force – displacements (penetration) records are displayed in Figures 25a-d.

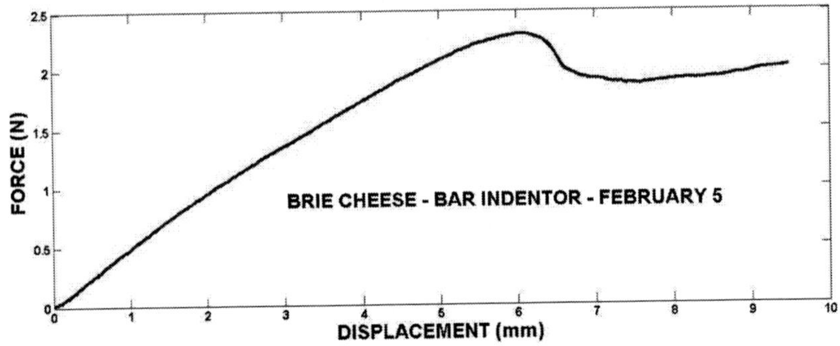

a)

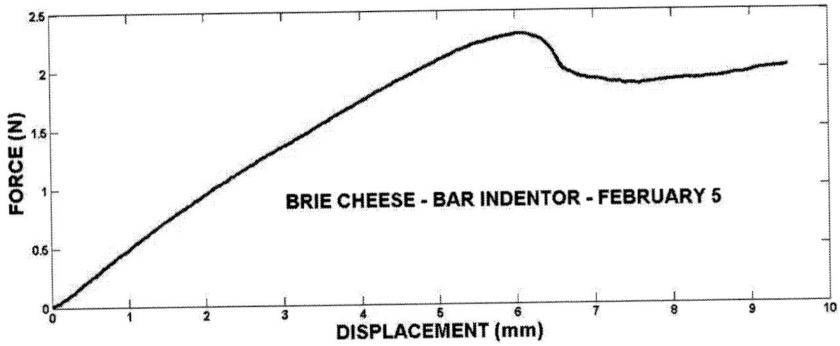

b)

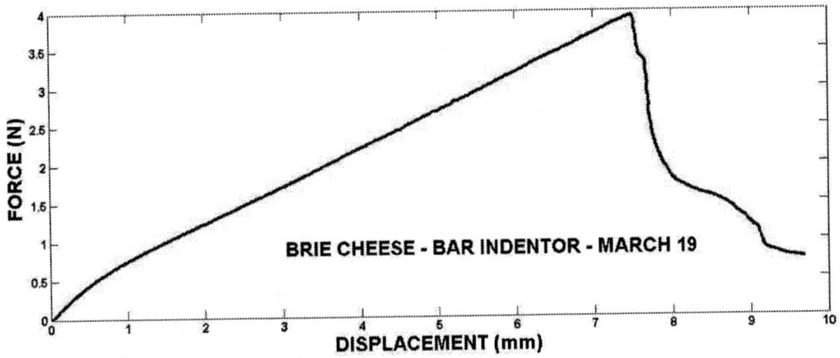

c)

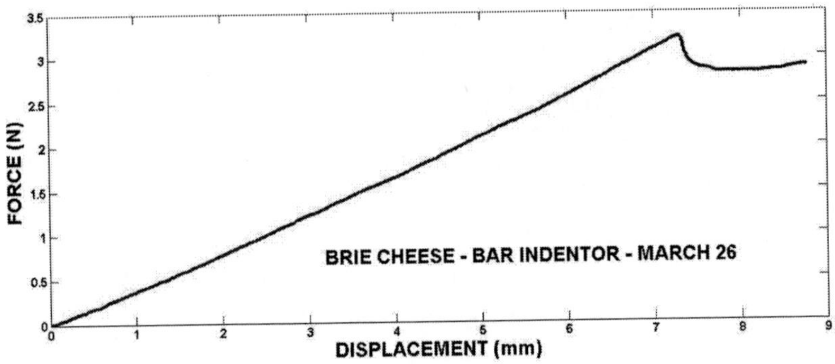

d)

Figure 25 a-d. Force-displacement records.

The force increases up to its maximum then it falls and remains nearly constant. The same qualitative features were found during the ball indentation. The observed features of the force development during indenter penetration can be explained as a consequence of development of the cheese structure. The force maximum corresponds to the moment of the crumb breakage. The scatter in the experimental data can be connected with occurrence of many voids in the cheese. Following parameters were evaluated from the experimental record of indentation force F versus depth of the penetration x:

- Breaking force (N) corresponding to force at the major failure event. It was considered as empirical measure of the crumb strength;
- Displacement, x_{max}, at fracture.
- Work, W (J) corresponding to the area under the force F – displacement x curve until the breaking event occurred. This parameter was used as empirical index of toughness.

Its value is given by the relation:

$$W = \int_{0}^{x_{\max}} F(x)dx \tag{19}$$

Twenty experiments were performed for every period of cheese ageing (10 experiments in L direction and 10 experiments in the T direction). Influence of ageing time on the breaking force (maximum of the force) during bar penetration into the cheese is shown in Figure 26.

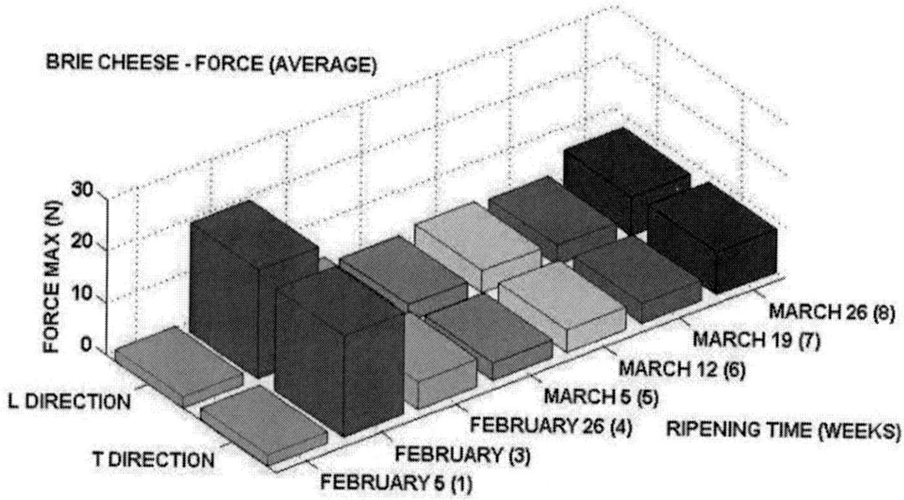

Figure 26. Bar indentation. Influence of ageing time on the maximum of the force.

It is obvious that breaking force of the cheese crumb increases up to certain maximum and then decreases. There are no statistically significant differences between the forces measured in the L and T directions. The same conclusions are valid for the forces found during the ball penetration – see Figure 27.

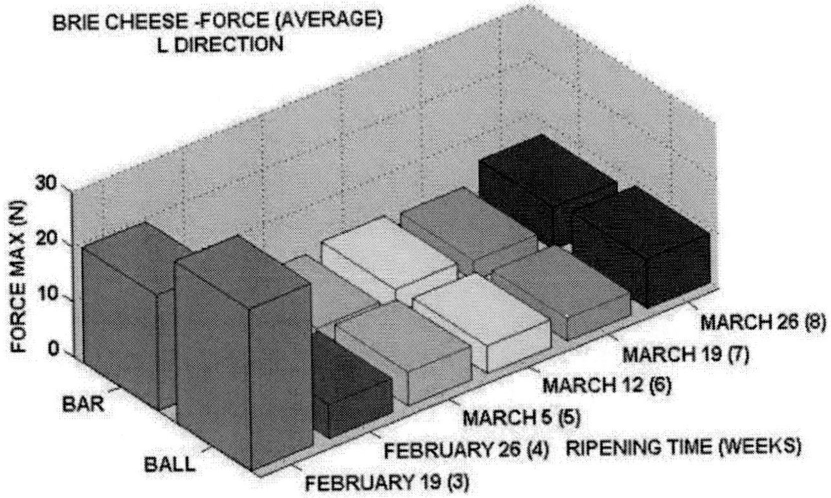

Figure 27. Ball and bar indentation.

The values of the forces are higher for the ball penetration. Displacement at the force maximum is displayed in Figure 28.

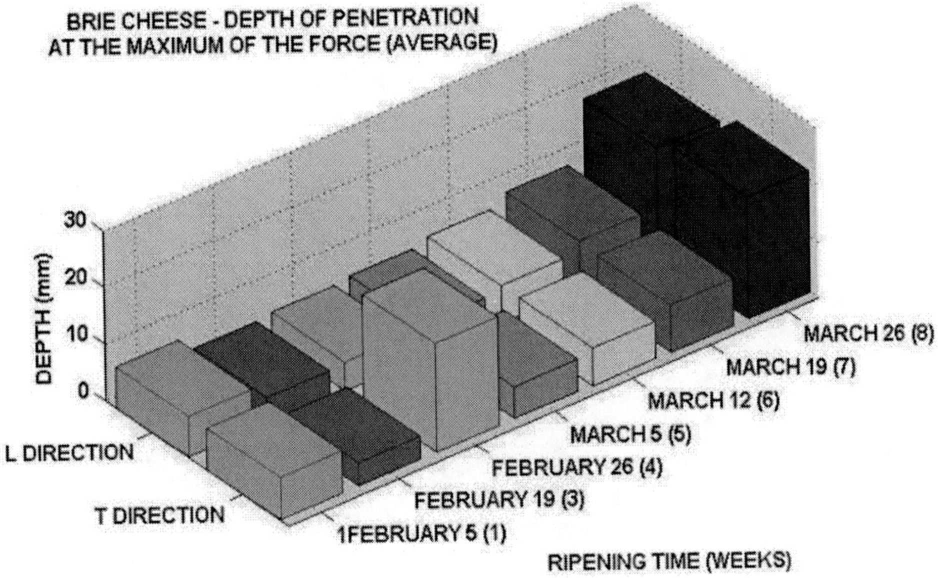

Figure 28. Displacement of the bar indenter at the force maximum.

Difference between the results obtained from bar and ball indentation tests is illustrated in Figure 29.

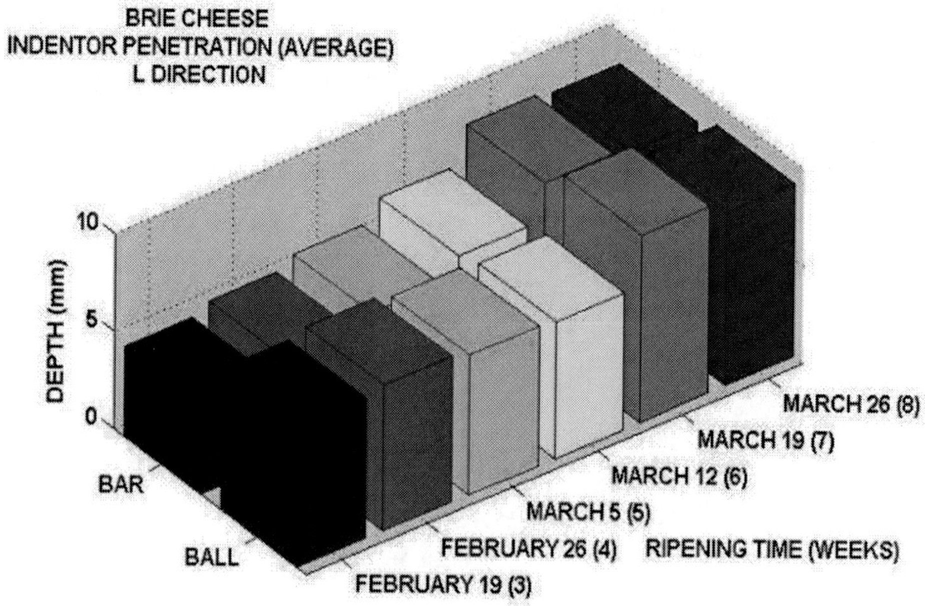

Figure 29. Comparison between depths of the ball and bar indenters.

Displacement increases with ageing time. The values of work W correspond to the values of breaking force. The values of these parameters are displayed in Figure 30 and Figure 31.

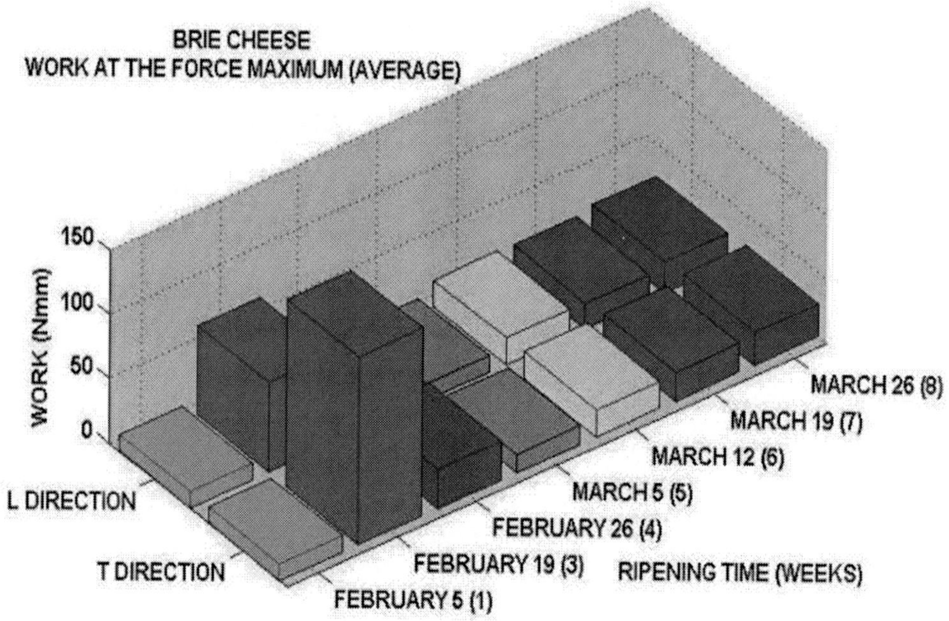

Figure 30. Changes in the average values of work, W, given by Eq. (9).

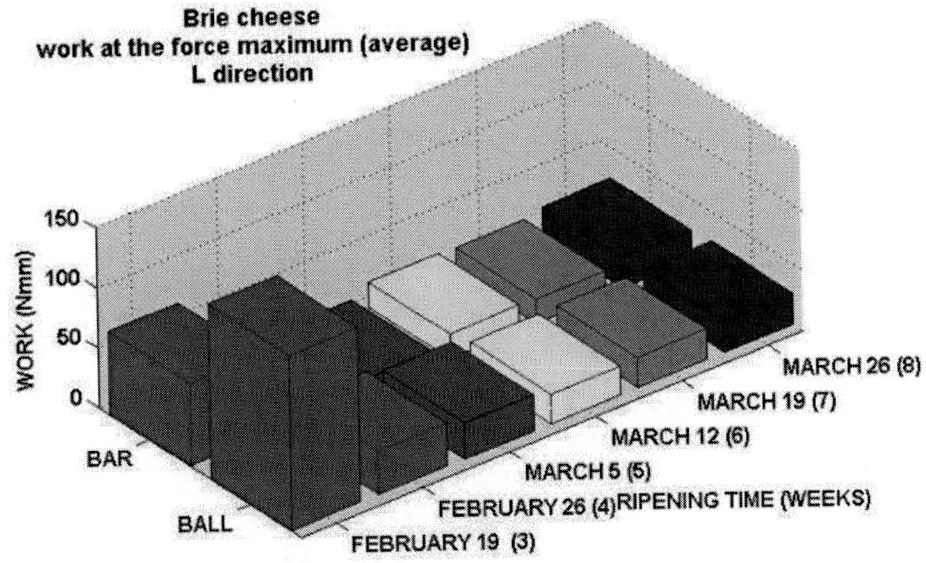

Figure 31. Work performed during penetration of the bar and ball indenters.

Obtained data enable description of the development of the Brie cheese rheological properties during its ageing. Used parameters are semi-empirical. The true description of these properties must be expressed in terms of constitutive model. It is well known that cheese behaves as a viscoelastic solid. The procedure how to obtain parameters of this model from the indentation test results was described in the previous paragraph 3.2. Viscoelastic modulus *E(t)* is given (Sakai et al., 2002) as :

$$E(t) = \frac{4(1-\nu^2)}{\sqrt{R}k_s}\left(\frac{\gamma_s}{2\pi v_o}\right)^{\frac{3}{2}} L^{-1}\left[s^{\frac{3}{2}}P(s)\right], \tag{20}$$

where R is the ball radius, ν is the Poisson's ratio, $k_s = 4/\pi$, $\gamma_s = 2$, v_o is the indenter speed and $P(s)$ is the Laplace transform of indentation load $P(t)$ with the transform variable s.

Analysis of the force F – depth of the penetration h up to the depth at the force maximum x_{max} revealed that this dependence can be fitted by the polynome:

$$P(t) = ah^3 + bh^2 + ch + d, \tag{21}$$

where $h = v_o t$. The use of MATLAB software enables to express the inverse Laplace transform as :

$$L^{-1}\left[s^{\frac{3}{2}}P(s)\right] = 6av_o^3 L^{-1}\left(s^{-\frac{5}{2}}\right) + 2bv_o^2 L^{-1}\left(s^{-\frac{3}{2}}\right) + \frac{cv_o}{\sqrt{\pi t}} + dL^{-1}\left(s^{\frac{1}{2}}\right) \tag{22}$$

The numerical procedure outlined e.g. in (Brančík, 1998) was used. Results of the numerical computations are shown in Figure 32. The computation was performed for $\nu = 0.45$. The results are not very sensitive to changes in the value of this quantity.

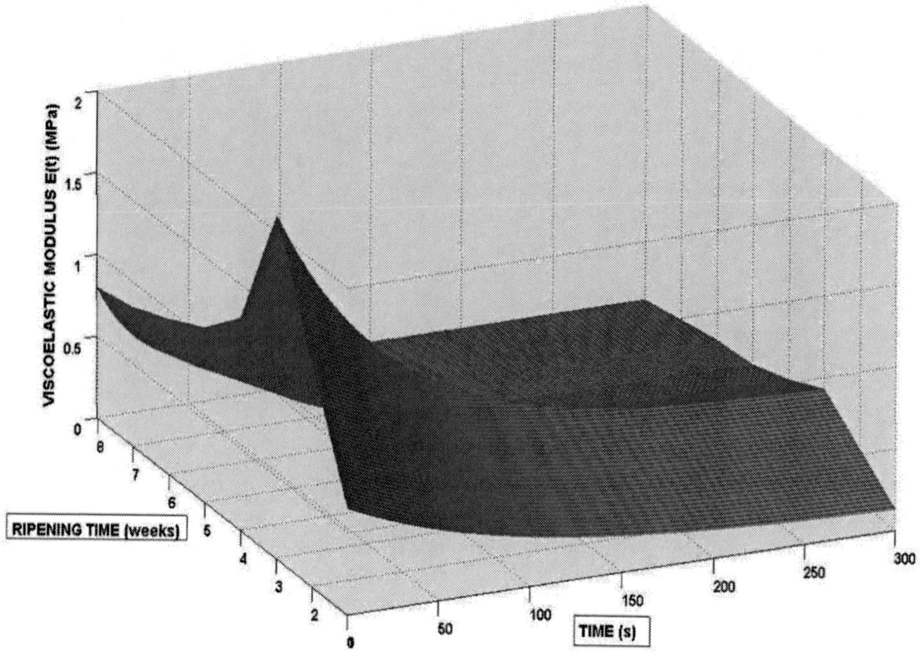

Figure 32. Influence of cheese ageing time on its viscoelastic modulus.

The viscoelastic modulus decreases with time. Intensity of this decrease increases with ageing time. Instantaneous value of E follows the conclusions obtained from the results of indentation tests. The value of the modulus describes the initial response of the cheese to the loading. During longer period, the modulus decreases to asymptotic value. It is obvious that this modulus, i.e. $E = E(t \to \infty)$ is affected by time of cheese ageing significantly less than the value $E(t=0)$. It means that degree of cheese maturity plays dominant role namely for the short time loading (impact etc.).

The mechanical response of Brie cheese is significantly affected by the development of cheese crumb. Strength of this crumb is much higher than strength of the inner part of the cheese. The hardness of this surface layer reaches its maximum in the third week of ageing. After this period, the strength rapidly decreases. The higher resistance against indenter penetration was observed for the ball indenter.

Viscoelastic properties were described by the viscoelastic (relaxation) modulus. This modulus decreases with time. It was shown that these moduli reflect influence of ageing time. Time dependence of this modulus suggest that influence ageing time on the rheological cheese behavior is significant, namely for the short time loading. In case of long-term loading process, this influence is less expressive.

3.4. Response of the Edam Cheese to Low Velocity Impact

The blocks of Edam cheese were tested using the experimental device shown in Figure 2. Velocity of the bar impact was kept constant, 1.2 m/s. The response functions (surface displacements and/or surface velocities, respectively) were evaluated both in the time and frequency domains. These two approaches are presented separately.

3.4.1. Time Domain

An example of the experimental record force versus time is shown in Figure 33. It can be seen that there is a very good reproducibility of the experiments. The function force, *F*, – time, *t*, can be represented by three parameters:

- maximum value of the force
- time of the maximum force achieving
- time of the pulse *F(t)* duration

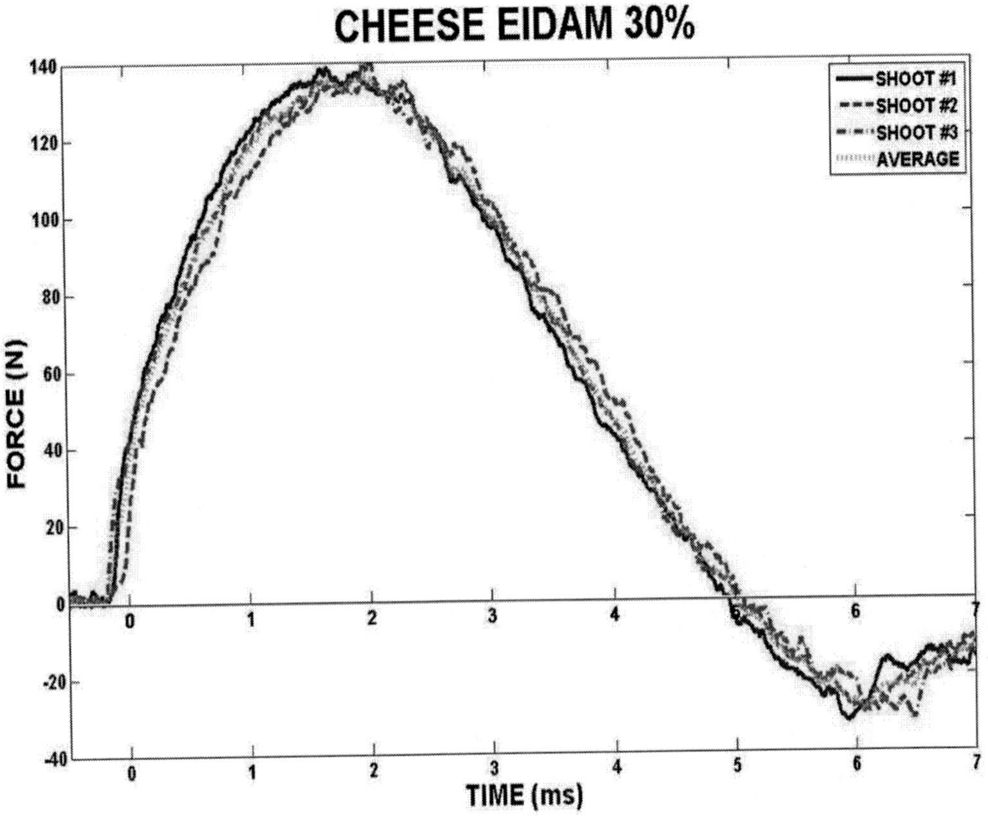

Figure 33. Time history of the force at the contact between cheese block and striking bar.

The maximum values of the loading forces are plotted in Figure 34. Each point represents an average value from three measurements.

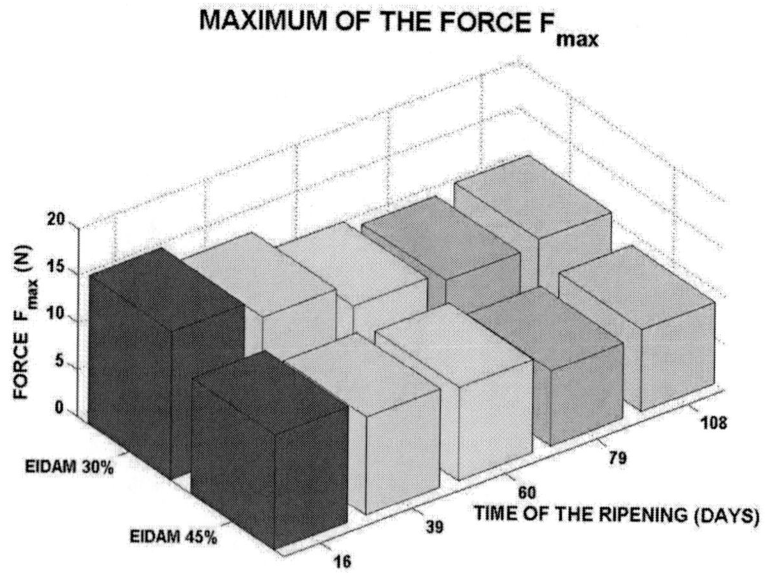

Figure 34. Maximum values of the loading forces versus cheese ageing time.

It is obvious that maximum value of the loading force decreases with the time of ageing. Its value also decreases with increase of fat content. Response function of the cheese block to impact loading is described by the time histories of surface displacement and/or by the surface velocity. These quantities correspond to the stress wave, which originates at the moment of cheese loading by the striking bar. An example of the surface displacement development in the increasing distance from the place of the bar-cheese contact is shown in Figure 35.

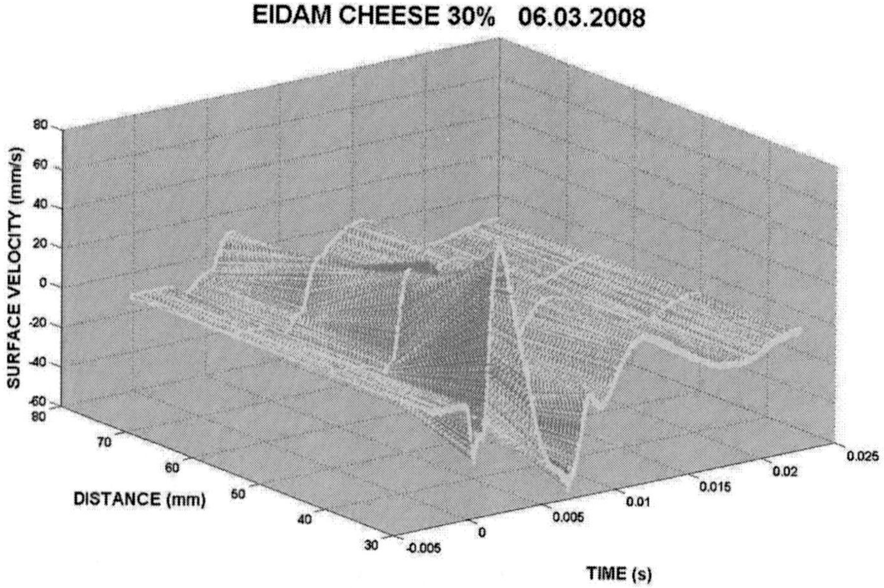

Figure 35. Surface displacement at the different points from the bar – cheese contact.

The surface displacement is highly attenuated in the direction of stress wave propagation. This behavior is typical for wave propagation in the non-linear viscoelastic materials. The same features were exhibited in case of surface velocity versus time functions – see Figure 36. Attenuation of the maximum value of surface displacement in the direction of stress wave propagation is shown in Figure 37.

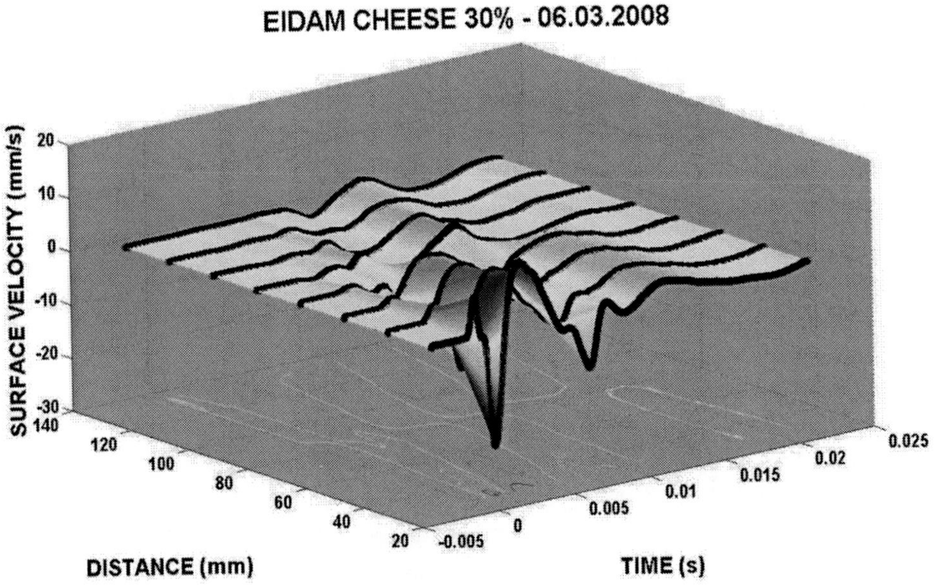

Figure 36. Surface velocity at different points from the bar-cheese contact.

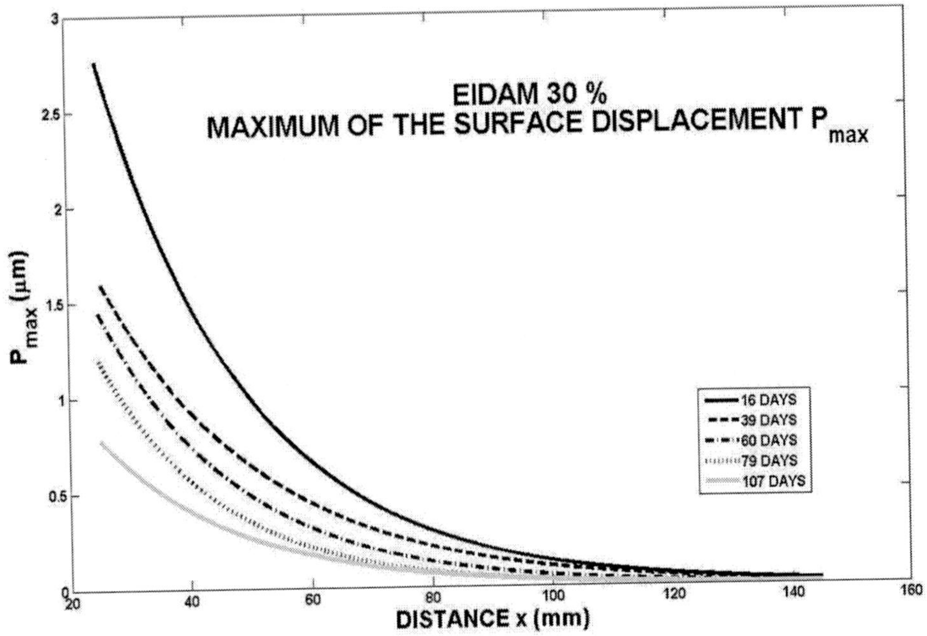

Figure 37. Maximum values of the surface displacements versus distance in stress wave propagation.

The maximum decreases with ageing time. The maximum also decreases with increasing fat content. The maximum of surface velocity also decreases in the direction of surface wave propagation – see Figure 38.

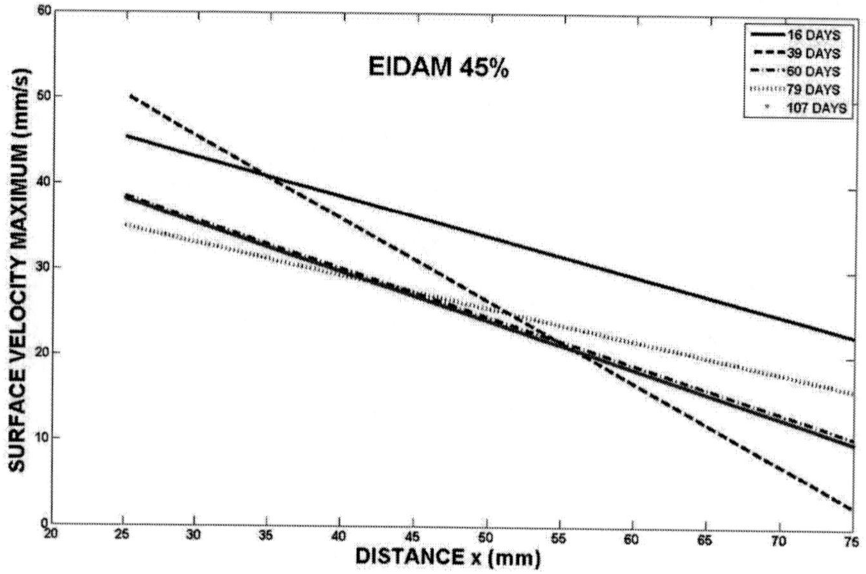

Figure 38. Maximum values of the surface velocity versus distance in stress wave propagation.

It can be seen that namely time histories of the impact forces and surface displacement are meaningful for description of the main textural changes during Edam cheese maturation.

3.4.2. Frequency Domain

Analysis in the frequency domain is based on Fourier transform (see e.g. Marchant, 2003). The function *f(t)* in the time domain is substituted by its spectral function *F(ω)*, where *ω* denotes the angular frequency. Transform into the frequency domain is a complex valued function, that is, with magnitude and phase. The fast Fourier technique (FFT) was used for the evaluation of the magnitude and phase. This algorithm is a part of the MATLAB software. An example of the spectral function amplitude is shown in Figure 39. This function is characterized by a peak value at relatively low frequency. For the whole spectrum, the momentum M_0 (Eq. (22), the momentum M_1 (Eq. (23) the central frequency *CF* (Eq. 24) and the variance *Var* (Eq. 25) were calculated (Oppenhein and Schafer, 1989).

$$M_o = \sum F(\omega)\Delta\omega \tag{23}$$

$$M_1 = \sum F(\omega)\omega\Delta\omega \tag{24}$$

$$CF = \frac{M_1}{M_o} \tag{25}$$

$$Var = \frac{\sum(\omega - CF)F(\omega)}{\sum F(\omega)} \tag{26}$$

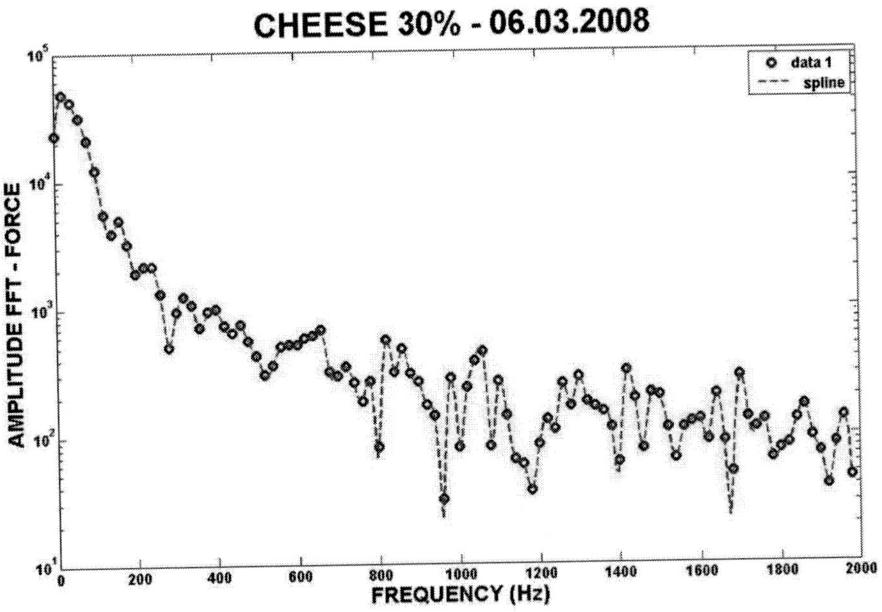

Figure 39. Spectral function amplitude versus frequency function.

Values of the central frequency are displayed in Figure 40.

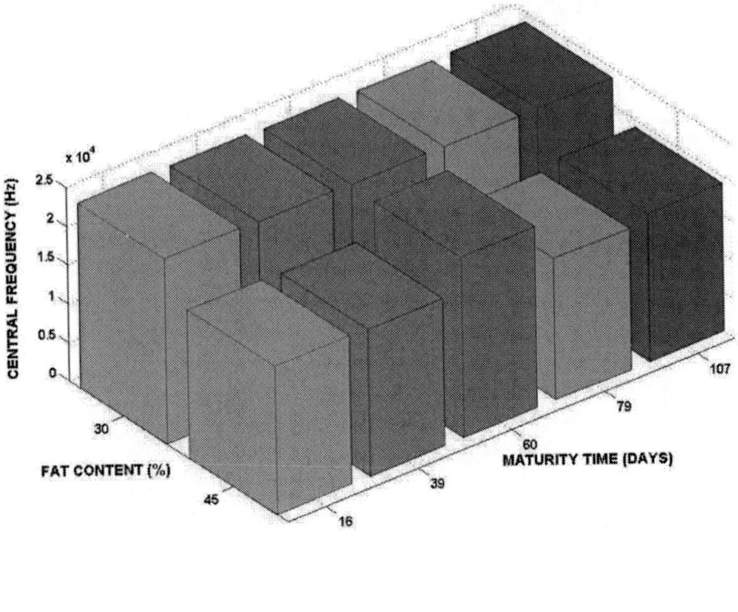

Figure 40. Central frequencies.

The central frequency decreases with increasing fat content. Its dependence on the ageing time is not very clear. Further information is included in the spectral function of the surface displacement versus time function. Frequency dependence of the amplitude changes in the direction of stress wave propagation is shown in Figure 41.

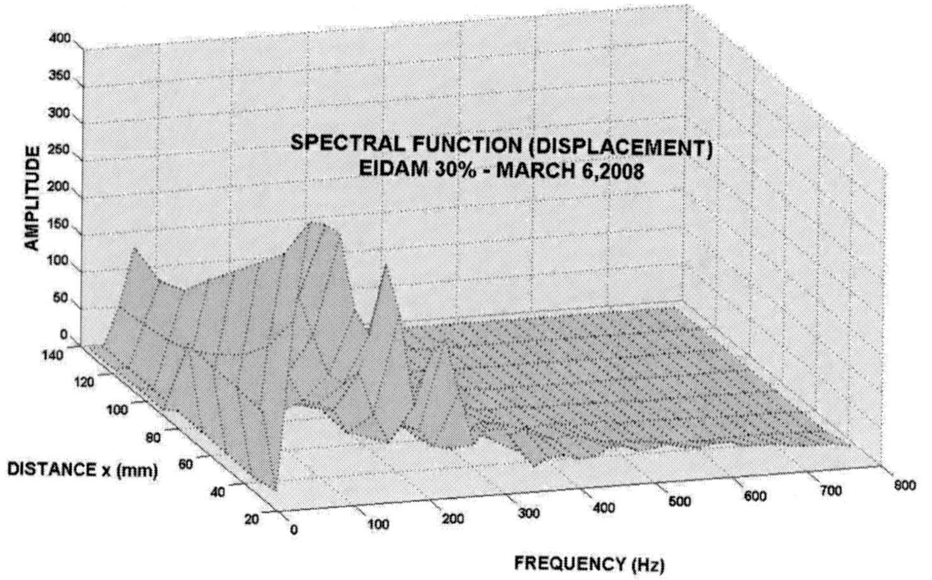

Figure 41. Development of the spectral function amplitude in the direction of stress wave propagation.

The amplitude versus frequency function is characterized by a maximum. Corresponding frequency is usually denoted as a dominant frequency. Analysis of the data led to the conclusion that dominant frequency did not depend on the distance in the direction of stress wave propagation. Its value was strongly dependent on ageing time – see Figure 42.

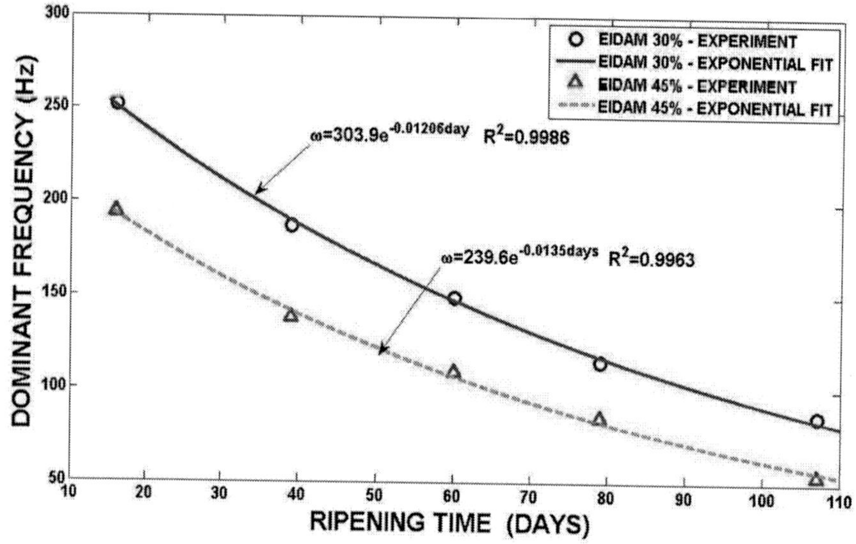

Figure 42. Effect of cheese ageing on the dominant frequency.

It seems that dominant frequency is a very convenient parameter for describing the degree of cheese maturation. It is sufficient to evaluate its value at one point on the cheese surface.

3.5. Response of Brie Cheese to Low Velocity Impact

The Brie cheese was tested using the same experimental technique as shown in Figure 2. The view on this experimental arrangement is shown in Figure 43.

Figure 43. Photo of the experimental arrangement for Brie cheese testing at low velocity impact.

Brie cheese was tested by the impact of the instrumented aluminum bar falling from the heights of 65 and 150 mm. The force at the interface between the cheese surface and the bar was recorded. At the distances x = 30, 45, 60, 75, 90, and 105 mm displacements of the cheese surfaces were measured using the laser vibrometers. These measurements were performed in the L and T direction - see Figure1. Experimental records were evaluated both in time and frequency domain.

3.5.1. Time Domain

Experimental records of force versus time are shown in Figure 44. It can be seen that there is a very good reproducibility of the experiments. The maximum values of the impact force increase with the height of the bar fall, i.e. with the impact velocity. This increase is shown in the Figs. 45 and 46.

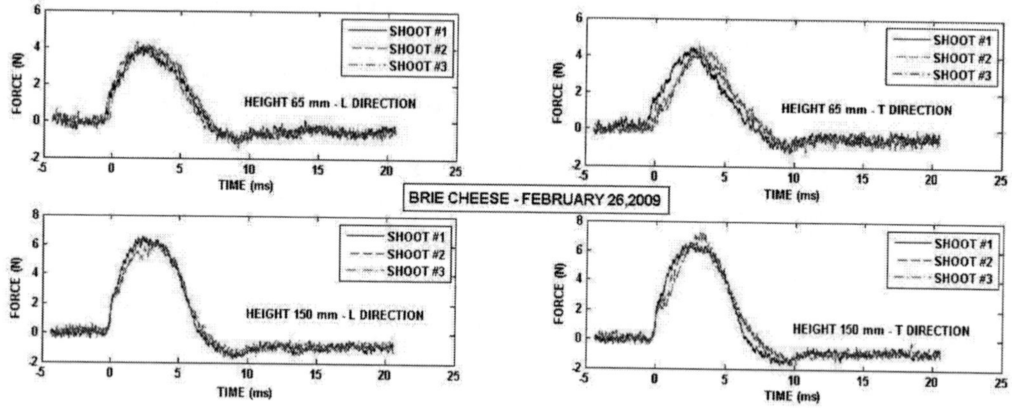

Figure 44. Experimental records of the loading force.

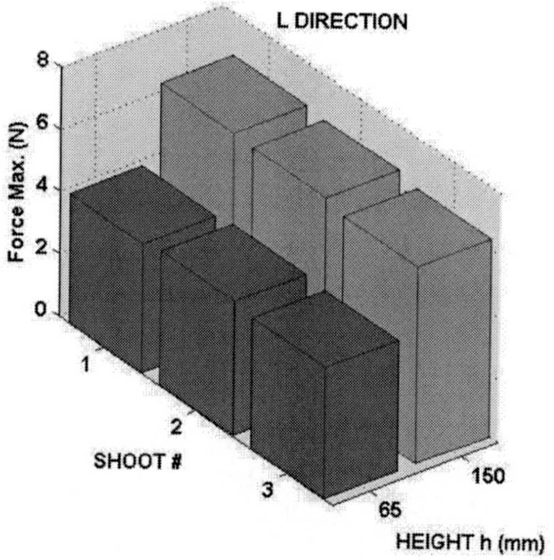

Cheese tested on February 2009.

Figure 45. Influence of the impact velocity on the maximum value of the force.

As it was mentioned in the beginning of the paragraph 3.4.1., the shape of the force F – time t dependence is further described by following parameters:

- time of the maximum force achieving, t_I
- time of the pulse $F(t)$ duration, λ.

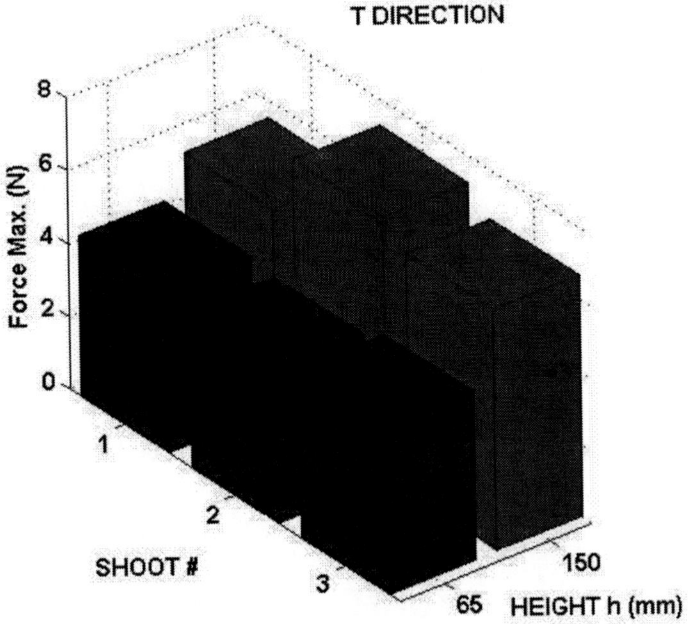

Cheese tested on February 2009.

Figure 46. Influence of the impact velocity on the maximum value of the force.

It was found that these parameters are independent on impact velocity. The parameters are different in the directions L and T - see Figs. 47 and 48.

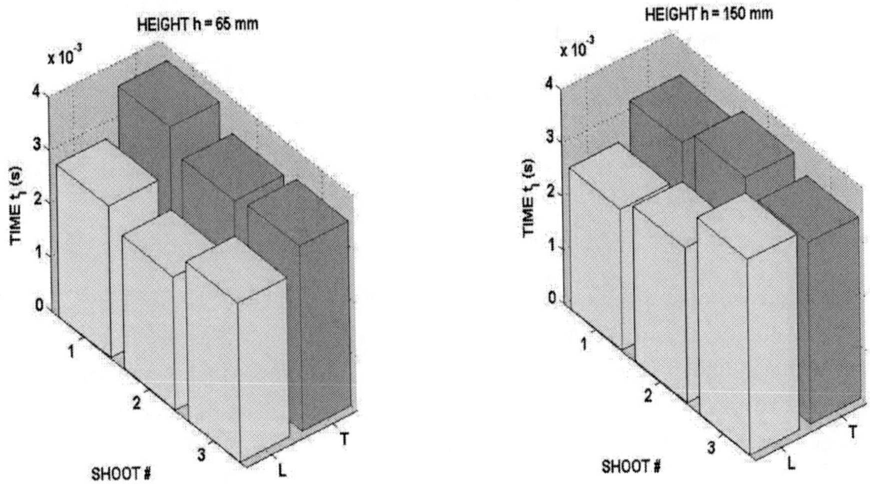

Figure 47. The values of time t_I. Cheese tested on February 26, 2009.

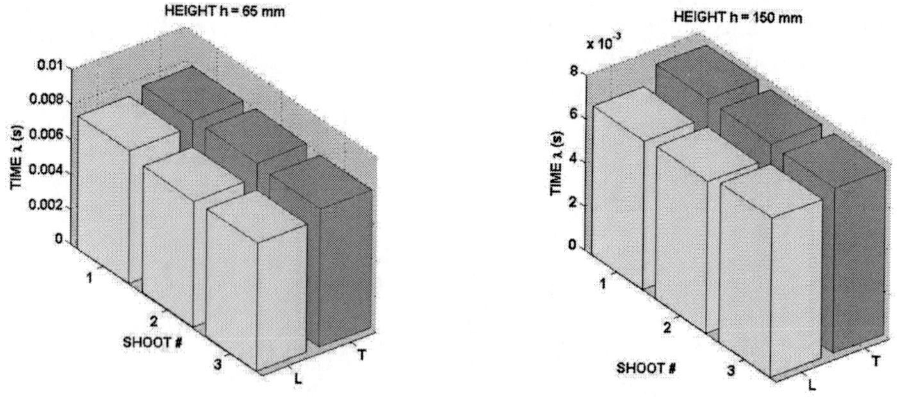

Figure 48. The values of time λ. Cheese tested on February 26, 2009.

These parameters can be used for evaluation of the effect of cheese ageing duration. Dependence of the force maximum value on the ageing time is shown in Figure 49.

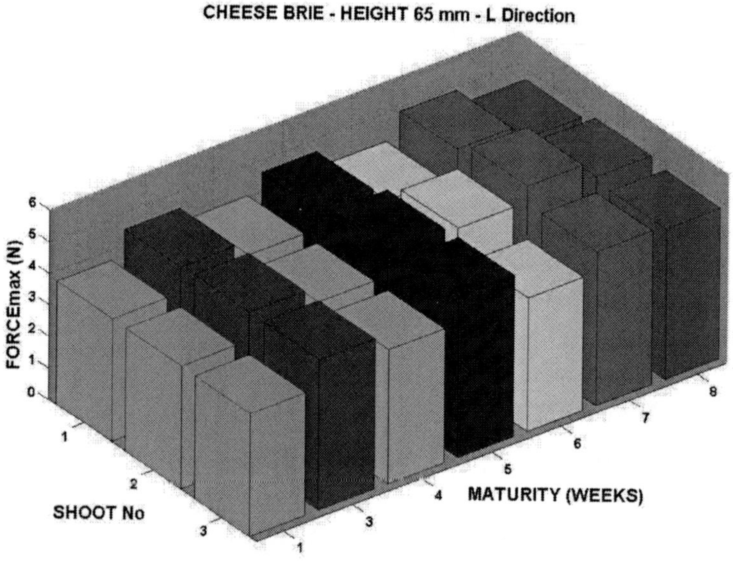

Figure 49. Peak values of the impact forces in different stages of the cheese maturity.

Certain increase of the force maximum in the third week of cheese ageing can be seen, similarly as in the case of indentation test - see subchapter 3.3. Further development of this maximum is nearly independent on the ageing time. The same conclusions were obtained for the height of 150 mm and for the both directions, L and T. The parameters t_I and λ exhibit independence on the cheese ageing time.

Response of the cheese to impact loading is further described by the time histories of the surface displacement and/or by the surface velocity. The shape of the displacement versus time dependence is mainly influenced by distance from the impact point. An example of the development of the surface displacement in the increasing distance from the place of the bar –

cheese contact is shown in Figure 50. The displacement decreases with this distance. It corresponds to the wave attenuation. The qualitative features of the surface displacement time profiles are similar for different impact velocities and for both directions L and T. Example of the surface displacement changes with the time of the cheese ageing is illustrated in Figure 51.

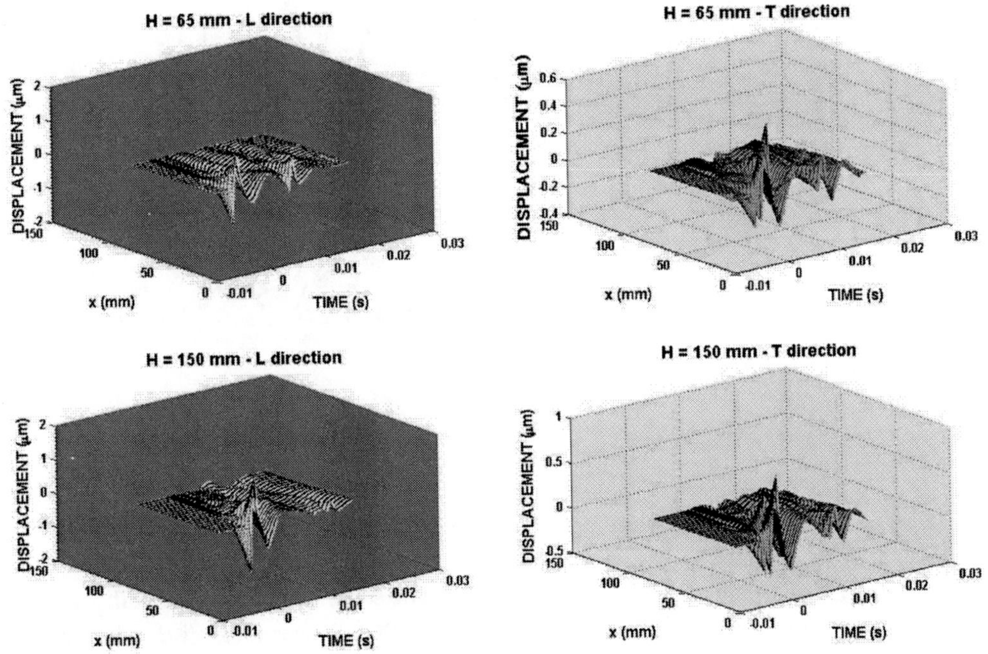

Figure 50. Development of the surface displacement in the different points from the place of the bar – cheese contact.

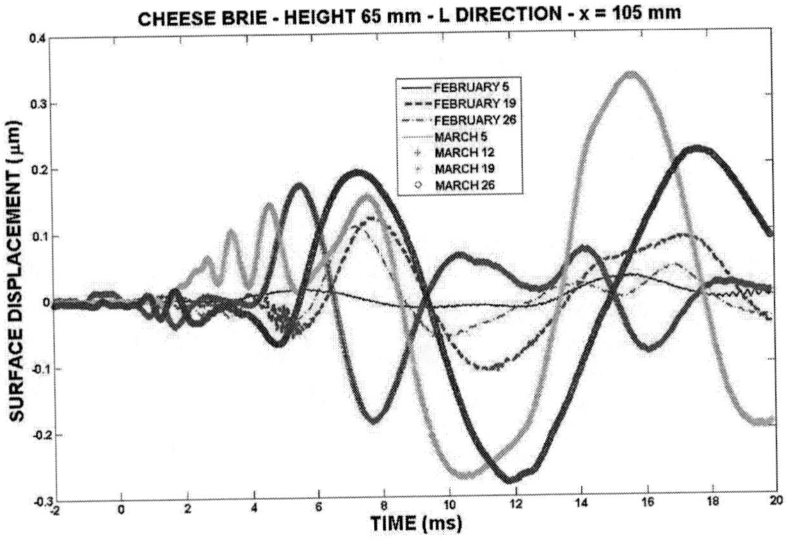

Figure 51. Influence of cheese ageing time on the surface displacement time profile.

The displacement vs. time function is characterized by many oscillations. Dependence of the surface displacement maximum on the cheese ageing time is shown in Figure 52.

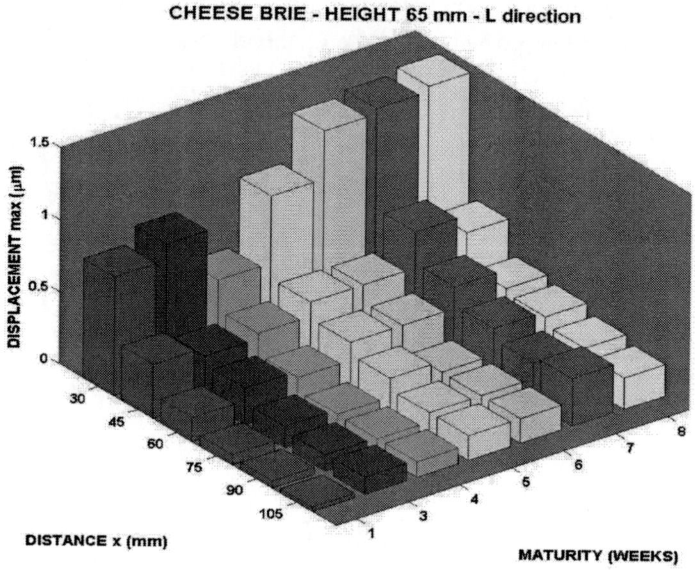

Figure 52. Maximum of the surface displacement in the different stages of the cheese maturity.

It can be seen that this maximum increases with the cheese's ageing time. This effect is probably a consequence of the cheese crumb development. This crumb exhibits more pronounced elasticity than the inner part of cheese. The same conclusions remain valid for all performed experiments. This behavior is opposite than that observed for Edam cheese.

3.5.2 Frequency Domain

The frequency response of the Brie cheese to impact loading was described using the same quantities as the response of the Edam cheese. Frequency dependence of the spectral function of the force is displayed in Figure 53. This amplitude is independent on cheese ageing time. There is a maximum (at certain frequency), which is constant. The same features exhibit the spectral functions of all forces. There is no dependency on wave propagation (L,T).

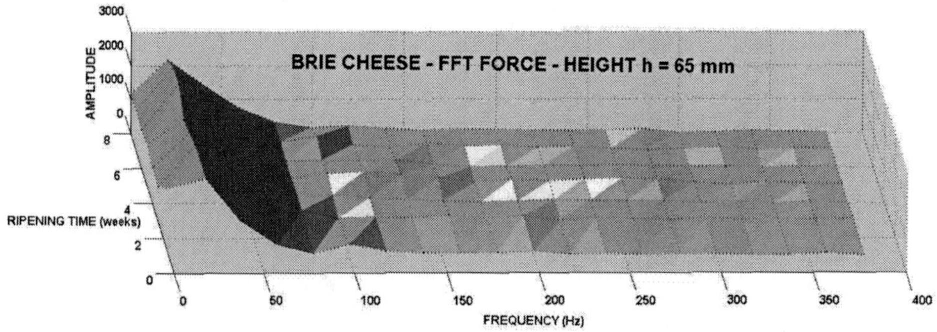

Figure 53. Amplitude of the spectral function. Impact force.

The development of the frequency dependence of the spectral function amplitude of the surface displacement with the time of the cheese ageing is illustrated in the Figs. 54 – 56.

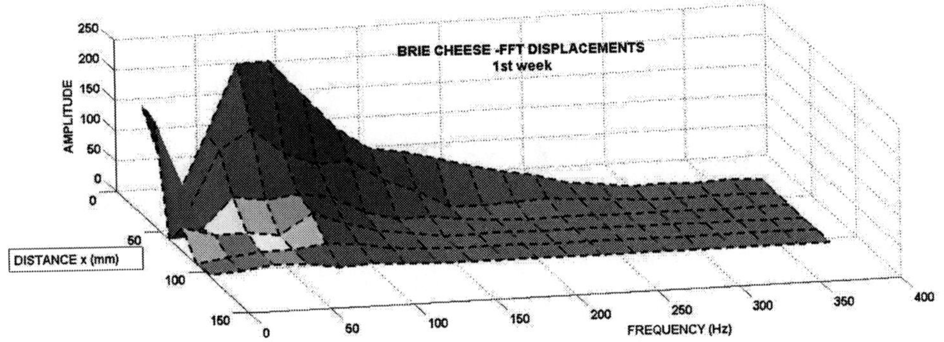

Figure 54. Amplitude of the spectral function. Surface displacements.

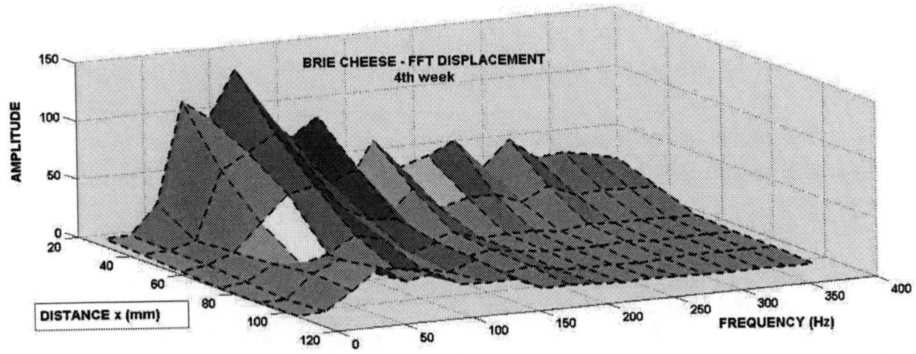

Figure 55. Amplitude of the spectral function. Surface displacements.

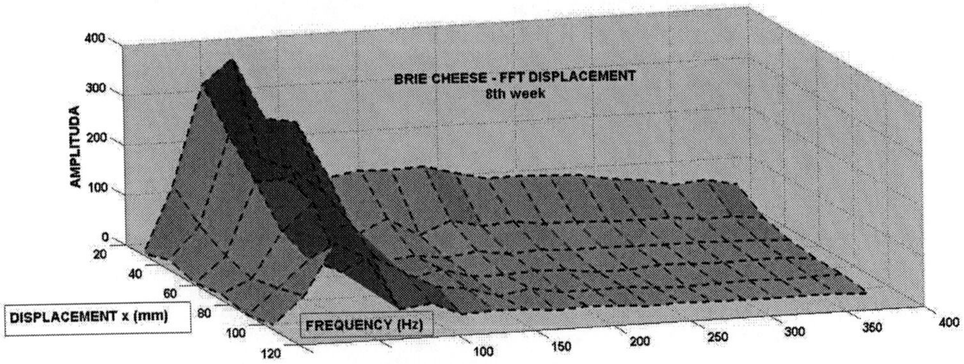

Figure 56. Amplitude of the spectral function. Surface displacements.

There is a maximum of this amplitude at certain frequency - so called dominant frequency ω_c. This frequency corresponds to particular resonant frequency. Its values are displayed in Figure 57.

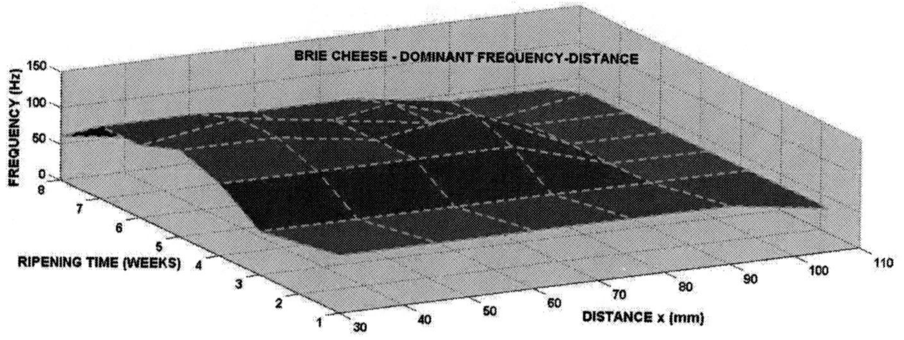

Figure 57. Dominant frequency.

The value of the dominant frequency is not a single valued function of the ageing time as it has been shown for the Edam cheese. This frequency decreases with distance from the loading point.

Another parameter, which may be very useful, is the transfer function amplitude (see 3.4.2 for a detail definition). The values of this parameter are displayed in Figure 58.

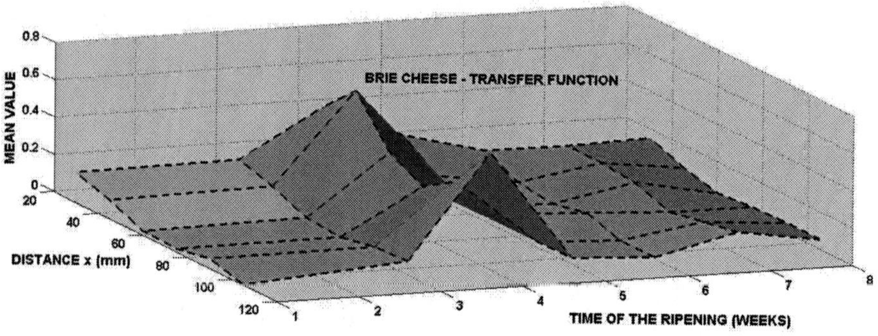

Figure 58. Amplitude of the transfer function.

The maximum of this amplitude occurs in the third week after the cheese production. Previous results showed that in the same period the hardness of the cheese crumb achieved its maximum.

4. CONCLUSION

The chapter summarizes the results on mechanical behavior of Edam and Brie cheese during their ageing. These results were obtained using three main experimental procedures:

- Compression test (Edam only)
- Indentation test
- Low velocity impact test.

The following results concerning tested cheeses were discovered:

- The response of the Edam cheese to the compression loading is sensitive to loading rate. The level of the stress decreases with the increase in the fat content. Elastic moduli obtained from the compression test decreased with time of cheese ageing.
- Relaxation tests show on the influence of fat content and especially time of cheese ageing.
- Evaluation of the relaxation tests was performed in terms of semi-empirical Maxwell model. Obtained results show that tested types of cheese behave like linear viscoelastic body. Elasticity as well as viscosity of the tested cheeses fall down during the cheese ageing. There is nearly no difference in the behavior of cheeses with the different fat content.
- The indentation test performed on Edam cheese was interpreted in terms of theory of indentation of viscoelastic materials. It was shown that this approach enables obtaining a general expression for evaluation of the relaxation modulus, which describes the viscoelastic properties. These moduli reflect influence of ageing time. They are also independent on loading rate. It means that these moduli can be used for evaluation of cheese maturity degree. The indentation test thus represents the proper tool for non-destructive testing of cheeses.
- The mechanical response of Brie cheese to the indentation by the two types of the indenters was studied. It was found that qualitative features of this response are independent on the shape of the indenter (ball or bar). The indentation test reflects observed development of the cheese structure. The mechanical response of the Brie cheese is significantly affected by the development of the cheese crumb. The strength of this crumb is much higher than strength of the inner part of the cheese. The hardness of this surface layer reaches its maximum in the third week of ageing. After this period, the strength rapidly decreases. The higher resistance against indenter penetration was observed for the ball indenter.
- The experimental results were interpreted in terms of the theory of viscoelasticity by the same way as in the case of Edam cheese. The viscoelastic properties were described by the viscoelastic (relaxation) modulus. Dependence of this modulus on time suggest that influence of ageing time on the rheological cheese behavior is significant namely for the short time loading. This influence is less expressive for long-term loading process.
- This presented chapter summarizes the results of an extensive study focused on the Edam and Brie cheeses' responses to non-destructive impact. An experimental arrangement was developed to perform non-destructive impact loading and response measurements on block of cheese in a single operation.
- In the case of Edam cheese it was found that both loading force as well as response functions enable description of texture changes during the cheese ageing. Softening of the cheese during its ageing is well described by the decrease of loading force

maximum. The same conclusions were obtained from the time histories of the surface displacements. The valuable information was received using the frequency analysis of obtained functions. Namely, the dominant frequency represents a very promising tool for the cheese maturation characterization. Non-destructive impact tests could be useful for prediction of textural characteristics of Edam cheese and therefore estimating the degree of cheese maturity.

- Response of the Brie cheese to the low velocity impact loading is given namely by the properties of its surface layer (crumb). Main parameters of this response enable identifying the time at which the crumb achieved the maximum of its hardness. The description of the cheese texture at longer time using these parameters is difficult if not impossible.

A comparison of the single experimental methods suggests that the most effective one is probably the indentation test. This test is applicable to all types of soft cheese and its interpretation can lead to exact evaluation of the cheese viscoelastic properties. The acoustic methods are useful namely for description of the cheese surface layers behavior.

Acknowledgments

The research has been supported by the Grant Agency of the Czech Academy of Sciences under Contract No. IAA201990701

References

Anand, A. & Scanlon, M. G. (2002). Dimensional effects on the prediction of texture-related mechanical properties of foods by indentation. *Transactions of the ASAE, 45*, 1045-1050.

Antoniou, K. D., Petridis, D., Raphaelides, S., Ben Omar, Z., & Kesteloot, R. (2000). Texture assessment of French cheeses. *Journal of Food Science, 65,* 168-172.

Bachmann, H. P., Butikofer, U., & Meyer, J. (1999). Prediction of flavon and texture development in Swiss-type cheese. *Lebensmittel-Wissenschaft und Technologie, 32,* 284-289.

Blazquez, C., Downey, G., O'Callaghan, D., Howard, V., Delahunty, C., & Sheehan, E. (2006). Modelling of sensory and instrumental texture parameters in processed cheese by near infrared reflectance spectroscopy. *Journal of Dairy Research, 73,* 58-69.

Bock, R. G., Puri, V. M., & Manbeck, H. B. (1989). Modeling stress relaxation response of wheat en mase using the triaxial test. *Transactions of the ASAE, 32,* 1701-1708.

Bourne, M. C. (1994). Converting from empirical to rheological tests on foods - it's a matter of time. *Cereal Foods World, 39,* 37-39.

Brančík, L. (1998). The fast computing method of numerical inversion of Laplace Transforms using FFT algorithm. In: *Proc. of 5th EDS '98.* Int. Conf., Brno, Czech Republic, June 1998, pp. 97-100.

Cho, H. K., Choi, W. K., & Paek, J. H. (2000). Detection of surface cracks in shell eggs by acoustic impulse method. *Transactions of the ASAE, 43*, 1921-1926.

Conde, T., Cárcel, J. A., García-Pérez, J. V., & Benedit, J. (2007). Non-destructive analysis of Manchego cheese texture using impact force–deformation and acoustic impulse–response techniques. *Journal of Food Engineering, 82,* 238-245.

Davie, I. J., Banks, N. H., Jeffery, P. B., Studman, C. J., & Kay, P. (1996). Non-destructive measurement of kiwifruit firmness. *New Zealand Journal of Crop and Horticultural Science, 24,* 151-157.

De Belie, N., Schotte, S., Coucke, P., & De Baerdemaeker, J. (2000a). Development of an automated monitoring device to quantify changes in firmness of apples during storage. *Postharvest Biology and Technology, 18,* 1-8.

De Belie, N., Schotte, S., Lammertyn, J., Nicolai, B., & De Baerdemaeker, J. (2000b). Firmness changes of pear fruit before and after harvest with the acoustic impulse response technique. *Journal of Agricultural Engineering Research, 77,* 183-191.

Del Nobile, M. A., Chillo, S., Mentana, A., & Baiano, A. (2007). Use of the generalized Maxwell model for describing the stress relaxation behavior of solid-like foods. *Journal of Food Engineering, 78,* 978-983.

Diezma-Iglesias, B., Ruiz-Altisent, M., & Barreiro, P. (2004). Detection of internal quality in seedless watermelon by acoustic impulse response. *Biosystems Engineering, 88,* 221-230.

Diezma-Iglesias, B., Valero, C., García-Ramos, F. J., & Ruiz-Altisent, M. (2006). Monitoring of firmness evolution of peaches during storage by combining acoustic and impact methods. *Journal of Food Engineering, 77,* 926–935.

Dufour, E., Mazerolles, G., Devaux, M. F., Duboz, G., Duployer, M. H., & Mouhous, R. N., (2000). Phase transition of triglycerides during semi-hard cheese ripening. *International Dairy Journal, 10,* 81-93.

Fox, P. F., Law, J., McSweeney, P. L. H., & Wallace, J. (1993). Biochemistry of cheese ripening. In P. F. Fox (Ed.), *Cheese, chemistry, physics and microbiology* (2nd ed., pp. 389-438). London, UK: Chapman and Hall.

Hort, J., Grys, G., & Woodman, J. (1997). The relationships between the chemical, rheological and textural properties of Cheddar cheese. *Lait, 77,* 587-600.

Kněz ,V. (1960). *Cheese production* (in Czech). SNTL, Prague, 380 p.

Konstance, R. P. & Holsinger, V. H. (1992). Development of rheological test methods for cheese. *Journal of Food Technology, 46,* 105-109.

Mancini, M., Moresi, M., & Rancini, R. (1999). Mechanical properties of alginate gels: empirical characterization. *Journal of Food Engineering, 39,* 369-378.

Marchant, B. P. (2003). Time - Frequency Analysis for Biosystems Engineering. *Biosystems Engineering, 85,* 261-281.

Mulet, A., Benedito, J., Bon, J., & Sanjuan, N. (1999). Low intensity ultrasonics in food technology. *Food Science and Technology International, 5,* 285-297.

Noel, Y., Zannoni, M., & Hunter, E. A. (1996). Texture of Parmigiano Reggiano cheese: Statistical relationships between rheological and sensory variants. *Lait, 76,* 243-254.

Oppenhein, A. V. & Schafer, R. W. (1989). *Discrete-time signal processing.* New Jersey: Prentice Hall International, Inc.

Ortiz, C., Barreiro, P., Correa, E., Riquelme, F., & Ruiz-Altisent, M. (2001). Non-destructive identification of woolly peaches using impact response and near-infrared spectroscopy. *Journal of Agricultural Engineering Research, 78,* 281-289.

Ozkan, N., Xin, H., & Chen, X. D. (2002). Application of a depth sensing indentation hardness test to evaluate the mechanical properties of food materials. *J. Food Sci., 67,* 1814-1820.

Sadowska, J., Białobrzewski, I., Jelinski, T., & Markowski, M. (2009). Effect of fat content and storage time on the rheological properties of Dutch-type cheese. *Journal of Food Engineering, 94,* 254-259.

Sakai, M. (2002). Time-dependent viscoelastic relation between load and penetration for an axisymmetric indenter. *Phil. Mag. A, 82*, 1841-1849.

Sneddon, I. N. (1965). The relation between load and penetration in the axisymmetric Boussinesq problem for a punch of arbitrary profile. *Int. J. Engng. Sci., 3*, 47-57.

Spangler, P. L., Jensen, L. A., Amundson, C. H., Olson, N. F., & Hill, C. G. (1990). Gouda cheese made from ultrafiltrated milk: effects of concentration factor, rennet concentration and coagulation temperature. *Journal of Dairy Science, 73,* 1420-1428.

Steffe, J. F. (1996). Rheological Methods. In *Food Process Engineering* (2nd ed., 428 p.), East Lansing, MI: Freeman Press.

Tabor, D. (1951). The Hardness of Metals. Clarendon, Oxford (Chapter 2).

Waananen, K. M. & Okos, M. R. (1992). Stress-relaxation properties of yellow-dent corn kernels under uniaxial loading. *Transactions of the ASAE, 35,* 1249-1258.

Watkinson, P., Boston, G., Campanella, O., Coker, C., Johnston, K., & Luckman, M. (1997). Rheological properties and maturation of New Zealand Cheddar cheese. *Lait, 77,* 109-120.

Watts, K. C. & Bilanski, W. K. (1991). Stress relaxation of alfalfa under constant displacement. *Transactions of the ASAE, 34,* 2491-2498.

Zude, M., Herold, B., Roger, J. M., Bellon-Maurel, V., & Landahl, S. (2006). Non-destructive tests on the prediction of apple fruit flesh firmness and soluble solids content on tree and in shelf life. *Journal of Food Engineering, 77,* 254-260.

In: Focus on Food Engineering
Editor: Robert J. Shreck

ISBN: 978-1-61209-598-1

Chapter 10

FOCUSING ON LAMB RENNET PASTE: COMBINING TRADITION AND INNOVATION IN CHEESE PRODUCTION

Antonella Santillo* and Marzia Albenzio
Department of Production and Innovation in Mediterranean Agriculture and Food Systems (PrIME). University of Foggia, Italy

ABSTRACT

This paper reviews the factors affecting the rennet paste composition and highlights **strengths and weaknesses in the** production of this type of coagulant and in its use for cheesemaking. Rennet paste is used almost exclusively in the manufacture of PDO cheese from ovine and goat milk and of some pasta filata cheeses as Caciocavallo.

New strategies for improving enzymatic and hygienic quality of rennet paste are presented as the incorporation of selected probiotic strains into rennet. Innovative lamb rennet paste containing probiotic is able to transfer microbioal cells into the curd matrix during the milk coagulation step. The study of ovine cheese produced using lamb rennet paste containing *Lactobacillus acidophilus* and a mix of *Bifodobacterium lactis* and *Bifidobacterium longum* evidenced enhanced nutritional and health properties and accelerated ripening process in terms of proteolysis and lipolysis, with cheese maintaining acceptable organoleptic characteristics. In particular, *Lb. acidophilus* and bifidobacteria evidenced the ability to liberate *9c,11t*-CLA, *9t,11t*-CLA, and *10t,12c*-CLa in the ovine cheese as an outcome of the peculiar metabolic pathway associated to microbial strains. The current consumer market of dairy products requires a continuous effort in terms of quality and innovation; dairy compartment offers the chance of developing innovative products beginning from traditional cheeses which exert ameliorated healthfulness and beneficial properties.

Key Words: lamb rennet paste, probiotic, ovine cheese, cheese ripening, functional cheese.

---

* Corresponding author. Tel. +39-0881-589326 Fax ++39-0881-589331
E-mail address: a.santillo@unifg (A. Santillo)

Introduction

The essential step in the manufacture of most cheese varieties involves coagulation of the casein component of the milk protein system to form a gel by the use of rennet or coagulant. Rennet composition varies along with several factors such as source (animal species, diet, microbial, and genetic features), physical state (liquid, powder, and paste), and enzymatic composition (chymosin/pepsin ratio, lipolytic enzymes). Rennet paste is used almost exclusively in the manufacture of ovine and goat cheeses from raw milk. In particular, some of the PDO sheep milk cheeses from Southern Europe, such as Idiazabal and Roncal in Spain, Fiore Sardo, Pecorino Romano, Canestrato Pugliese in Italy, and Feta in Greece are produced using lamb rennet paste. Although rennet paste has a complete enzymes outfit, the use of artisanal rennets often entails problems concerning curd formation and final characteristics of the cheese, probably because the rennet activities are not standardized. Therefore many manufacturers are replacing the artisanal lamb rennet by standardized commercial calf rennets (Vicente et al., 2001). The presence of lipolytic enzymes in rennet paste is able to influence greatly taste and flavour of cheese, which is not request in many cheese typology so this could also be a limit for the use of artisanal lamb rennet paste in cheese production. Moreover, poor hygienic control is supposed to be one of the main problems with artisanal production of rennet pastes (Nunez et al., 1991). However, recent study has demonstrated that an adequate artisanal production does not influence rennet hygienic quality (Virto et al., 2003; Santillo et al., 2007b) as well as the hygienic quality of the final cheese (Etayo et al., 2006; Gil et al., 2007; Santillo et al., 2007b).

Nonetheless, there is a renewed interest in promoting the use of lamb rennet paste as an alternative to calf rennet to maintain the authenticity of traditional cheeses as well as to give different flavors to new products (Addis et al., 2005; Santillo et al., 2007a). Indeed, the current consumer market of dairy products requires a continuous effort to improve nutritional features and health aspect of food products. Dairy compartment offers the chance of developing functional products beginning from traditional cheese.

In this review new strategies of improving enzymatic and microbial characteristics of lamb rennet paste are reported. New achievements on the effects of innovative lamb rennet paste on the nutritional and ripening pattern of ovine cheese are also presented.

Enzyme Activities in Rennet Paste

Rennet paste produced from lamb's or kid's abomasum is characterized by the presence of both proteolytic and lipolytic enzymes. Lipases are inactivated in liquid and/or powder calf rennets due to the production process providing a step of acid activation of the proenzymes; thus only rennet paste presents the complete lipolytic and proteolytic enzyme pattern.

The principal proteolytic enzyme in rennet paste is ascribed to chymosin, capable of hydrolyzing k-CN at Phe_{105}-Met_{106} which is the catalytic mechanism of the milk clotting enzymes. Chymosin (EC 3.4.23) belongs to the group of aspartic acid proteinases; the isoelectric point and proteolytic pH optimum is about 4, but it has high specific milk clotting activity at the milk pH. In addition ruminants secrete proteolytic enzymes as pepsin (EC 3.4.23.1) and gastricsin (EC 3.4.23.3); the former has strong general proteolytic activity besides its milk clotting activity. Both chymosin and pepsin are produced as inactive zymogens (i.e. prochymosin and pepsinogen) and secreted by chief cells and, to some extent, by the mucous neck cells in glands of the fundic region of the abomasal mucosa (Andrén, 1992). The activation of zymogens is due to the low pH owing to the presence of HCl secreted by parietal cells in the fundic glands.

Rennet paste contains lipases identified as pregastric esterases (PGEs) (of oral origin) and gastric lipases. The term PGE is the most widely accepted terminology for esterolytic or lipolytic enzymes secreted by oral tissues of mammals (Nelson, 1977). This potent lipolytic enzyme complex is secreted mainly from palatine glands and from tissues in other regions, i.e. root of the tongue and esophageal region, of the oral cavity of young ruminants (Russel et al., 1980). Pancreatic lipase activity is low at birth but subsequently increases. Pregastric and pancreatic secretions are the principal sources of lipolytic enzymes; PGE plays an important role in lipolysis particularly during neonatal life, it preferentially hydrolyzes short chain fatty acids esterified in the sn-3 position, releasing high level of butyric acid (Addis et al., 2008). The presence and the activity of lipases make the rennet paste suitable for the production of certain cheese varieties in which lipolysis is considered to be desirable e.g. Parmesan, Pecorino Romano, Blue cheeses and and pasta filata cheeses (Fox, 2000).

Factors Affecting Enzymatic Composition of Lamb Rennet Paste

In Mediterranean area lamb rennet paste is produced according to protocols which derive from local traditions and vary along with the geographical region. An habitual protocol to manufacture lamb rennet paste provides for the removal of the perivisceral fat, then the whole stomach is salted (18-20% wt/wt) and the samples are ripened at 6°C and 70% R.H. for about 2 months. At the end of ripening abomasa are ground to obtain a paste. The industrial process does not differ substantially from the artisanal one and no activation procedures (e.g. HCl addition) of chymosin are described for artisanal and industrial process (Addis et al., 2008).

A limiting factor in the production of rennet paste with standardized enzymatic and microbial features is ascribed to the high variability of raw abomasa conferred to the production plant. In fact, abomasa come from many local farms and show different characteristics owing to different breeding systems of lamb (i.e. feeding regime, age at slaughtering). Abomasa are generally pooled for colour and size in the attempt to obtain pastes with homogeneous enzymatic composition.

Enzymatic and microbial composition of rennet paste varies along with several factors such as diet of lamb (milk feeding, weaning), age at slaughter, time of fasting prior to slaughtering and abomasa processing technology. Rennet paste produced using abomasa from suckling lambs exhibited an enzymatic and microbial activity reported in Table 1. Rennet showed white colour of the paste and lactobacilli and lactococci cells were detected due to the feeding regime of the lambs based on maternal milk.

Table 1. International Milk Clotting Units (IMCU), enzymatic activities, and microbial cell loads of industrial lamb rennet paste.

IMCU/g	155
Chymosin, %	70
Pepsin, %	30
Lipases, Lipolytic Unit/g	1,020
Mesophilic bacteria, $\log_{10}$ cfu/g	4.68
Mesophilic lactobacilli, $\log_{10}$ cfu/g	1.18
Mesophilic lactococci, $\log_{10}$ cfu/g	2.95

From Santillo & Albenzio, 2008.

Guilloteau et al. (1983) found that chymosin activity in lambs abomasa peaked at 2 d of age, then gradually decreased up to 7 d, and had a new rise till 14 d before decreasing again. Andrén and Bjorck (1986) reported that with continuous suckling the secretion of prochymosin could last at least six months. Casein in the milk is the main factor causing the secretion of prochymosin, thus the abomasa of calves fed a casein diet contain higher chymosin and stable level of pepsin (Zhang et al., 2005). The latter enzyme has been shown to increase regularly with age, weaning having little effect. Pepsin is characterized by a lower ratio of milk clotting activity to proteolytic activity than chymosin thus its relative contribute to proteolysis increases with age (Andrén, 2003). In calves, PGE activity is slightly affected by age and diet (Nelson et al., 1977); Collins et al. (2003) reported that suckling stimulates the secretion of PGE at the base of the tongue, which is carried into the abomasa with the milk.

Santillo et al. (2007b) carried out a survey on lambs subjected to maternal milk (MS) and artificial milk diet (AR) and slaughtered at 20 and 40 day of age with the aim of studying the effects of both diet and age on the rennet enzyme composition. Chymosin and lipase were higher in AR than in MS whereas pepsin displayed and opposite trend in rennet obtained from lambs slaughtered at 20 d. This could be because MS lambs had access to hay and concentrate given to their mothers so solid feed intake led to a depression in chymosin and lipases synthesis. In general, enzyme activities decreased with increased age at slaughter so that no differences emerged between diets in rennet from lambs slaughtered at an older age. Addis et al. (2008) have shown that enzymatic characteristics of lamb rennet paste are also influenced by the slaughtering procedures (Table 2). The authors compared milk *vs* mixed milk-pasture diet and time elapsed between suckling and slaughtering. Apart from diet, lambs slaughtered soon after suckling had almost 100% of chymosin activity which decreased passing from 2 to 12 h after suckling. The opposite trend was observed for pepsin which increased along with time prior to slaughtering. PGE acitivity was found the highest in lambs fed exclusively on milk and slaughtered after suckling whereas grazing partially inhibited the production of PGE in suckling lambs and can enhanced the activity of other lipases.

Table 2. Total milk clotting activity and enzymatic content of lamb paste rennets.

Rennet[1]	A	B	C	D	E	F
Total clotting activity, IMCU/g	94	95	278	263	48	279
Chymosin, %	100	88	77	98	82	76
Pepsin, %	0	12	23	2	18	24
Pregastric lipases activity, LFU/g	6.78	1.6	0.22	0.44	2	0.85

[1]A = lambs fed only milk and slaughtered after suckling; B = lambs fed only milk and slaughtered 2 h after suckling; C = lambs fed only milk and slaughtered 12 h after suckling; D = lambs fed milk and pasture and slaughtered after suckling; E = lambs fed milk and pasture and slaughtered 2 h after suckling; F = lambs fed milk and pasture and slaughtered 12 h after suckling.

Adapted from Addis et al., 2008.

Microbial feed additives facilitate the establishment and maintenance of suitable microbial flora in the gastro intestinal tract of animals. The two most commonly used microbial additives are *Lactobacillus* spp. and *Saccharomyces cerevisae* (Agarwal et al., 2002); the former is primarily responsible for the exclusion of enterotoxygenic bacteria, whereas the latter mainly affects the functioning of rumen (Fuller, 1989). In particular, *Lactobacillus acidophilus* is indigenous to the intestinal tract of humans and animals, and is most often used as feed supplement (Abu-Tarboush et al., 1996). The addition of selected lactic acid bacteria improves the solubility and stability of milk replacer through milk acidification and shows a positive impact on growth performance and health status of lambs. Indeed, the lactobacilli compete with the pathogens for their establishment in the gut and thus control diarrhea (Fuller, 1989).

Although many research (Ellinger et al., 1980; Modler et al., 1990; Hughes and Hoover, 1991; Cruywagen et al., 1995; Lubbadeh et al., 1999) focused on the effect of probiotic on growth performance and health of animals, no studies have been reported on the effects on enzyme composition of rennet paste. Recently, Santillo et al. (2007b) demonstrated that the addition of *Lb. acidophilus* to milk substitute in lambs diet can improve the enzymatic composition of rennet paste as an outcome of enzyme activities brought about by probiotic added to milk substitute (Table 3). In this study lambs were subjected to three different feeding regimes (mother suckling -MS-, artificial rearing -AR-, and artificial rearing with *Lactobacillus acidophilus* supplementation -ARLb-) and slaughtered at 20 d and 40 d. MS lambs were kept with their dams whereas AR and ARLb lambs were removed from their dams 24 to 30 hours after birth and fed on a commercial milk substitute from buckets. *Lb. acidophilus* influenced positively the lactic microflora of rennet, lactobacilli and lactococci resulting in higher cell loads than in the traditional rennet paste probably due to the mutualism relationship between those microbial groups based on the production of metabolites such as peptides and amino acids. In particular, the lactic microflora contributed with indigenous lipases to total lipases activity in the rennet; indeed, LAB possess esterolytic/lipolytic enzymes capable of hydrolyzing a range of esters of FFA, tri-, di-, and monoacyl glyceride substrates (Collins et al., 2003; Santillo et al., 2007b).

Table 3. Effects of feeding regimen and age at slaughter of lambs on International Milk Clotting Units (IMCU), enzymatic activities of artisanal lamb rennet paste.

Item	Age at slaughtering (d)	Feeding regime[1]			SEM	Effects[2]		
		MS	AR	ARLb		Feeding regime	Age at slaughtering	Feeding x Age
Total coagulating activity, IMCU/g	20	189[c]	157[bc]	148[bc]				
	40	58[a]	62[a]	137[b]	14.79	*	***	*
Chymosin, R.U./g	20	679.77[b]	889.80[d]	750.55[d]				
	40	85.93[a]	167.02[a]	454.50[c]	30.43	***	***	***
Pepsin, R.U./g	20	748.84[a]	117.79	148.19				
	40	102.43	136.79	40.06	50.93	***	***	***
Lipases, L.U./g	20	210[a]	2960[b]	4400[b]				
	40	475[a]	990[a]	3240[b]	642	***	NS	*

[a-b-c] Means with different superscripts differ for each item in rows and columns ($P < 0.05$).

[1] MS, mother suckling; AR, artificial rearing; ARLb, artificial rearing with *Lactobacillus acidophilus* supplementation.

[2]NS not significant; *$P<0.05$; ***$P < 0.001$.

From Santillo et al., 2007b.

INNOVATIVE LAMB RENNET PASTE CONTAINING PROBIOTIC

Shah (2007) reviewed the health benefits of food products containing probiotic organisms such as antimicrobial activity and prevention of gastrointestinal infections, effectiveness against diarrhoea, improvement in lactose metabolism, antimutagenic and anticarcinogenic properties, reduction in serum cholesterol, suppression of *Helicobacter pylori* infection, improvement in inflammatory bowel disease, and immune system stimulation. The main probiotic organisms that are currently used worldwide belong to the genera *Lactobacillus* and *Bifidobacterium*.

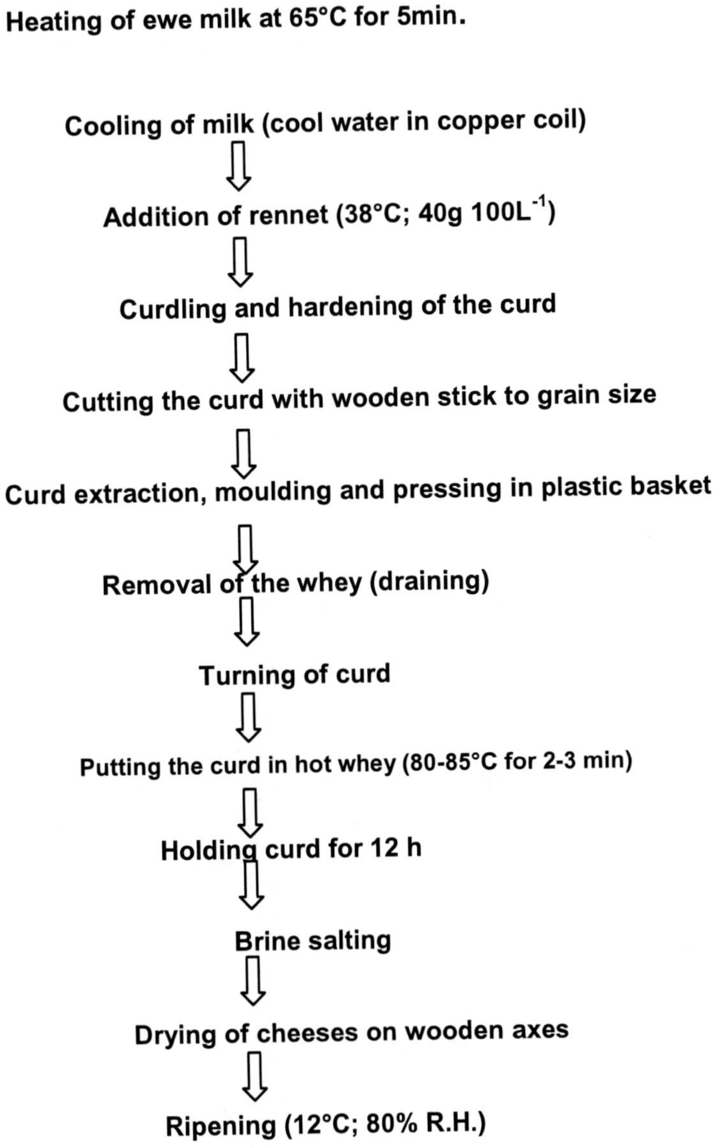

From Santillo, et al., 2007a.

Figure 1. Traditional protocol of Pecorino foggiano cheese production.

The incorporation of probiotic cultures into dairy food is successful when the cultures maintain viability until being consumed and also if the adjunct cultures do not adversely affect the composition, texture, or sensory features of the product (Corbo et al., 2001). The current consumer market focuses on quality and innovation attributes of dairy products. Dairy compartment offers the chance of developing innovative products beginning from traditional cheese which have their own typical features. Addis et al. (2008) reported that many traditional ewe milk cheeses i.e. PDO Idiazabal, PDO Roncal, and PDO Pecorino romano cheese are produced in certain Southern European countries using lamb rennet paste. In such context, the product innovation is bound by the specification of the Production Protocol acknowledged by the National Board or European Community. Together with PDO cheeses many dairy products from ewe and goat milk produced according to traditional protocols are of great value for the agricultural economy in Southern Italy. Pecorino foggiano cheese, a traditional uncooked sheep's milk cheese (Figure 1), can be sold as short-time ripened cheese with a soft texture and a thin yellow rind or as a long-time ripened cheese with a harder texture and a more piquant and intense flavour (Santillo et al., 2007a). In developing a probiotic cheese, it is essential that the technological steps involved in its manufacture do not significantly differ from the narrow specification limits in place for producing traditional cheeses so as ensure that market share is not lost and production costs are not increased significantly (Gomes et al., 1995).

Microflora contained in rennet paste, i.e. lactic acid microflora, is transferred to the milk during cheese production and can contribute to the biochemical pathways in the cheese matrix during ripening. Given that the rennet paste is a carrier of microbial cells in the vat milk and thus in the cheese, Santillo & Albenzio (2008) explored the possibility of using rennet paste containing selected bacteria in the cheese production. This could provide a spin-off for health properties of cheese (i. e. production of probiotic cheese) and for its ripening features (acceleration of ripening process). It has to be considered that traditional cheeses are often made following a manual or semi-manual process scarcely susceptible to modifications of the production process which is often handed down through generations of producers. In this context the use of innovative rennet paste containing selected probiotics could lead to a product innovation without modifying the production process of traditional cheese.

Santillo & Albenzio (2008) tested the effectiveness of the incorporation of probiotic bacteria cultures as *Lactobacillus acidophilus* (LA-5), *Bifidobacterium lactis* (BB-12), and *Bifidobacterium longum* (BB-46; Chr. Hansen, Milan, Italy) into traditional lamb rennet paste at a concentration of 11 $\log_{10}$ cfu/g of rennet to produce Pecorino foggiano cheese verifying that the cell load of added probiotic remained stable at a level 8 $\log_{10}$ cfu/g of cheese until the end of ripening (60 d).

Role of Lamb Rennet Paste in Cheese Proteolysis

Proteolysis is the main process in cheese ripening determining changes in the texture - due to the breakdown of the protein network - and in flavour formation through the release of peptides, free amino acids, and catabolic products. Proteolysis is ruled by several agents in cheese, such as rennet, indigenous milk enzymes, starter and non starter bacteria, and secondary inocula and their enzymes. Each of these agents contribute to proteolysis at different stages of cheese production and ripening.

Primary proteolysis concerns the intact casein chains and is carried out mainly by the coagulant and plasmin activities; during secondary proteolysis generated peptides are further hydrolysed by the proteinases and peptidase of starter and non starter bacteria.

Proteolytic enzymes associated to rennet paste are able to influence the proteolytic pattern of cheese during ripening. When animal rennet is used in internal bacterially ripened cheese varieties, proteolysis of β-casein is less than that of α_s-casein (Fox, 1993). In fact, chymosin shows higher specificity towards α- than β-CN resulting in a more extensive degradation of the former casein fraction in cheese during ripening (Bustamante et al., 2003; Irigoyen et al., 2002; Santillo et al., 2007a).

Proteolysis in cheese is an a useful index of cheese maturity and quality. Among other techniques, electrophoresis allows monitoring hydrolysis of the individual caseins and identification of the peptides formed. Bustamante et al. (2003) and Santillo et al. (2007a) found that cheese produced using lamb rennet paste displays a specific band, named A1 band, in Urea-PAGE of pH 4.6 insoluble nitrogen fraction (Figure 2). The identification of the sequence of this fragment could allow to exclusively associate the presence of this band to the use of lamb rennet paste in cheese production.

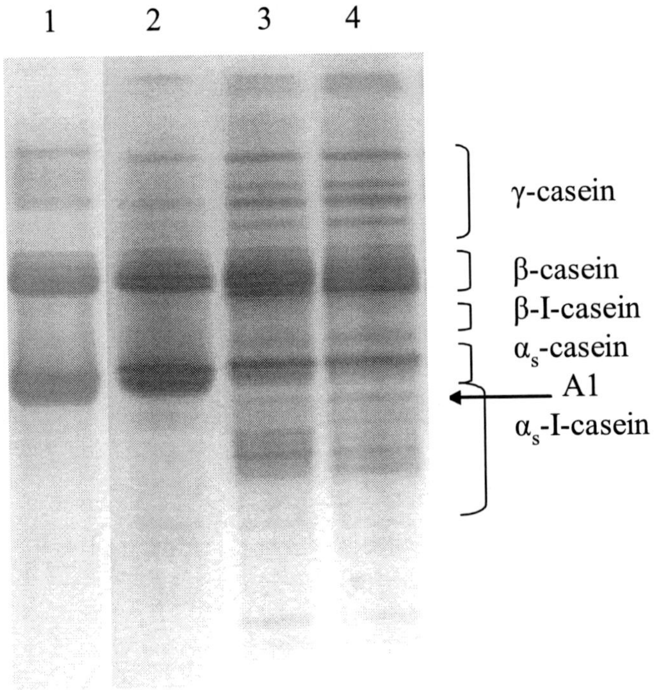

From Santillo et al., 2007a.

Figure 2. Urea polyacrylamide gel electrophoresis of pH 4.6-insoluble nitrogen fraction of Pecorino foggiano cheese made using lamb rennet paste: a representative proteolytic pattern illustrates groups of bands analyzed for quantification. Ovine sodium caseinate (lane 1); Pecorino foggiano after 1d (lane 2), 30 d (lane 3), and 60 d (lane 4) of ripening.

The determination of nitrogen fractions (e.g. Water Soluble Nitrogen, WSN; Non Casein Nitrogen, NCN; Non Protein Nitrogen, NPN; Phostotungstic Acid Soluble Nitrogen, PTASN, Proteose-Peptone, PP) is valuable for assessing the overall extent of proteolysis and the general contribution of each proteolytic agent. In particular, WSN contains numerous small- and medium- sized peptides, free amino acids and their degradation products, organic acid and their salts (McSweeney & Fox, 1997). Pirisi et al. (2007) reported higher WSN in PDO Fiore Sardo ovine cheese produced using industrial lamb rennet (60:40 chymosin/pepsin ratio) than traditional lamb rennet paste (100:0 chymosin/pepsin ratio) suggesting that the greater proteolytic activity in the former cheese was due to the co-presence of pepsin which increases the general proteolytic activity on caseins.

The choice of probiotic strains is critical for their contribution to the proteolytic process in cheese. Although LAB are weakly proteolytic, they possess a proteinase and a wide range of peptidases, which are principally responsible for the formation of small peptides and amino acids in cheese (Fox et al., 2000). El-Soda et al. (1992) showed that the peptide hydrolase system of *Bifidobacteria* spp. was comparable with that of LAB with respect to the presence of general amino-peptidase Pep N and several di-, tri and probably imino-peptidases. In Canestrato pugliese cheese with added bifidobacteria, more pronounced imino-, amino, and dipeptidase activities were found (Corbo et al., 2001). Bergamini et al. (2009a) tested different probiotic cultures for Pategrás Argentino cheese demonstrating that each culture influences the proteolysis differently: *B. lactis* did not impact proteolysis, *Lb. paracasei* showed a minor influence, and *Lb. acidophilus* increased the level of small nitrogen compounds and free amino acids. Pecorino foggiano cheese (Santillo & Albenzio, 2008) produced using traditional rennet or the same rennet containing probiotics shows differences in WSN level ascribed to the proteolytic activity of the probiotic strains. In particular, the proteolytic enzymes brought about by a mix of *B. longum* and *B. lactis* were responsible for higher levels of WSN than in cheese with *Lb. acidophilus* (Figure 3).

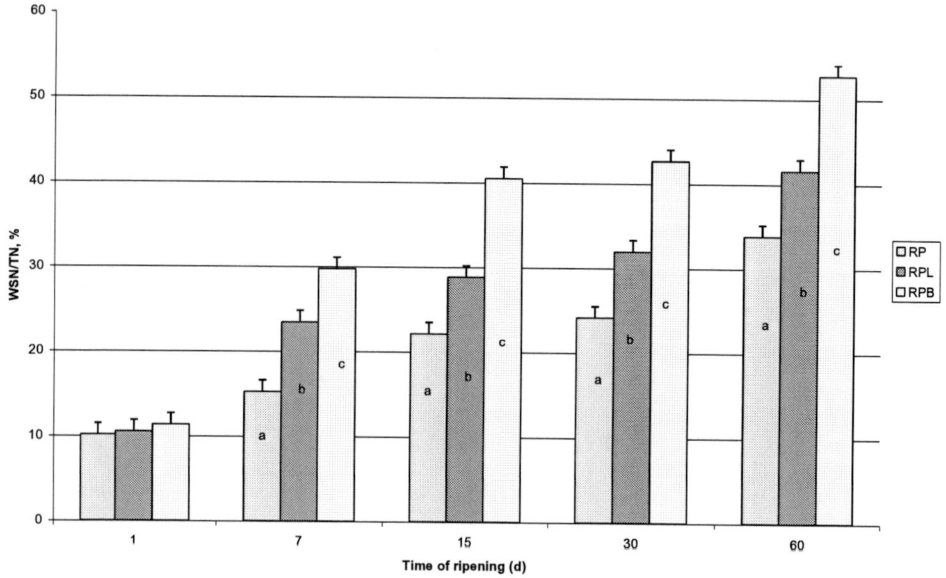

[1]RP= cheese manufactured using traditional lamb rennet paste; RPL= cheese manufactured using lamb rennet paste containing *Lb. acidophilus* (LA-5); RPB= cheese manufactured using lamb rennet paste containing *B. lactis* (BB-12) and *B. longum* (BB-46).

Figure 3. Effect of lamb rennet paste on water soluble nitrogen of ovine cheese during ripening.

Variations in the concentration of free amino acids (FAA) during cheese ripening may be considered as an index of secondary proteolysis. In Idiazabal cheese, Vicente et al. (2001) found that rennet type (artisanal lamb rennet *vs* commercial calf rennet) influenced the free amino acid content, being higher when the cheeses were made with commercial calf rennet. Santillo & Albenzio (2008) compared the level of FAA in cheese produced using lamb rennet paste with and without probiotics and found that the total FAA were higher in the cheeses containing probiotic; thus it can be inferred that the effect of peptidase activities associated to microbial cells is added to the effect of the rennet enzymes on the release of free amino acids during cheese ripening. Analogously to the starter bacteria, probiotic strains liberate different levels of FAA based on their enzyme system and the degree of autolysis in the cheese.

The PCA analysis applied to the FAA content of Pecorino foggiano cheese, obtained with lamb rennet paste containing *B. longum* and *B. lactis*, and *Lb. acidophilus* and with lamb rennet paste without probiotic, during ripening is shown in Figure 4. The probiotic strain added to rennet influences also the type of amino acids; cheeses containing *Lb. acidophilus* and bifidobacteria lay in a well-defined zone along the second principal component of the PCA biplot relevant to the free amino acid composition in Pecorino cheese. In particular, cheeses containing a mix of bifidobacteria were characterized by higher contents of aspartic acid whereas cheeses containing *Lb. acidophilus* showed higher contents of glutamic acid, tyrosine, asparagine, and glutamine. The amino acids freed in the cheese matrix during secondary proteolysis undergo further catabolic reactions which involve decarboxylation, deamintion, transammination, desulfurations leading to the production of a wide array of compounds which contribute to flavour formation such as amines, acids, and thiols. A number of different LAB and other cheese microorganisms have been evaluated for their ability to degrade amino acids to aroma compounds. Interestingly, the evolution of amino acid pattern evidenced that cheese containing *Lb. acidophilus* displayed a peak of free amino acid at 30 d and a subsequent drop passing from 30 to 60 d of ripening. This trend was attributed to the catabolic activity of FAA carried out by *Lb. acidophilus* which was not evaluated in the study. However, a great variety of peptidolitic enzymes, amino peptidase, di- and tripeptidases, and proline-specific peptidases, was observed in *Lb. acidophilus* (Bergamini et al., 2009b), being these enzimatyc activities largely strain dependent (Macedo et al., 2000; Di Cagno et al., 2006).

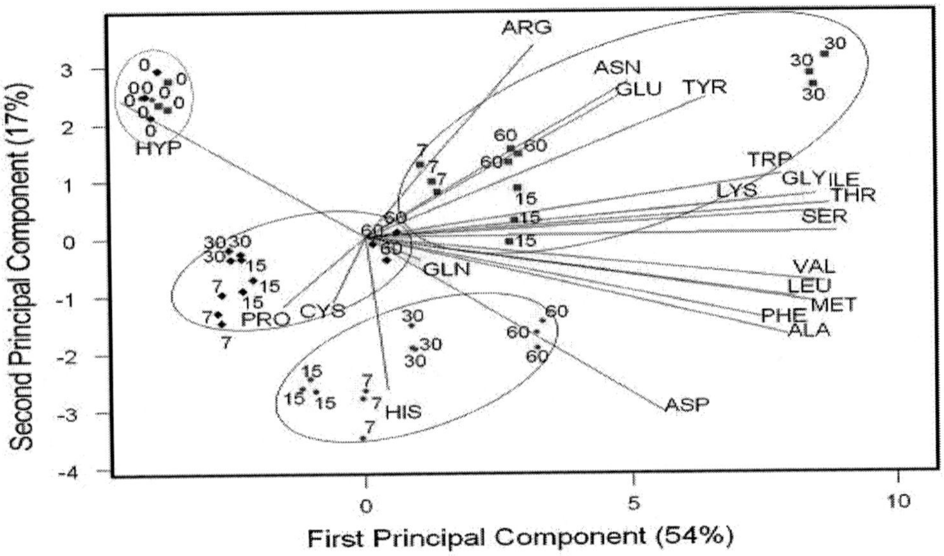

Figure 4. Principal component analysis of the FAA of Pecorino cheese manufactured with different rennet paste at 0 (cheese curds), 7, 15, 30, 60 d of ripening: ● RP-CH, cheese manufactured using traditional lamb rennet paste; ■ RPL-CH, cheese manufactured using lamb rennet paste containing *Lb. acidophilus*; ♦ RPB-CH, cheese manufactured using lamb rennet paste containing *B. lactis* and *B. longum*.

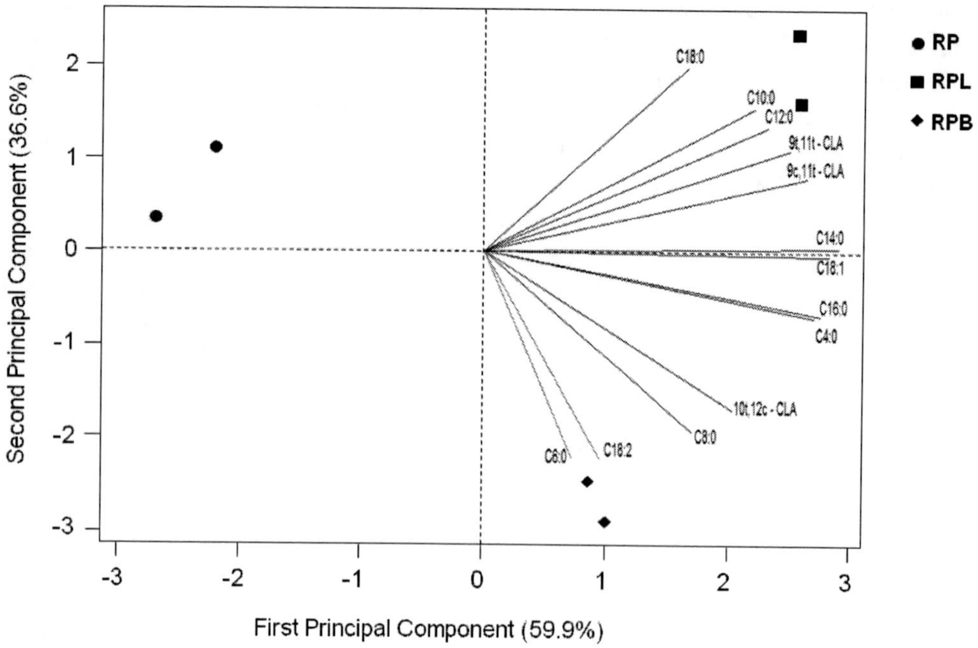

From Santillo et al., 2009.

Figure 5. Principal component analysis of the FFAs and CLA isomers in the sheep milk cream incubated with different rennet paste: ● RP, traditional lamb rennet paste, ■ RPL, lamb rennet paste containing *Lb. acidophilus*; ♦ RPB, lamb rennet paste containing *B. lactis* and *B. longum*.

ROLE OF LAMB RENNET PASTE IN CHEESE LIPOLYSIS

As for proteolysis, lipolysis is another important biochemical event occurring during cheese ripening. This process leads to the release of free fatty acids (FFA) and their catabolites such as methyl ketones, secondary alcohols, alchanes, lactones, and esters, which contribute to cheese flavour. The lipolytic agents in cheese are endogenous milk lipases, rennet pregastric and gastric lipases, and lipolytic enzymes associated to microflora (Collins et al., 2003).

Lipolysis is quite limited in most cheese varieties exceptions are some typology of cheese e.g Provolone, ovine cheese (Gobbetti et al., 2002; Virto et al., 2003; Albenzio et al., 2001). Cheese from ewe milk is characterized by a more intense lipolysis owing to higher level of milk fat and to the extended time of ripening to obtain its peculiar characteristics, i.e. thickness of the rind, hardness, and flavour. Cheese produced from raw milk contains very high level of FFA compared to cheese produced from pasteurized milk owing to the lipases and esterases activities of the milk microflora especially Non Starter Lactic Acid Bacteria (NSLAB) (Albenzio et al., 2001).

Recent studies evidence that improved healthfulness and functional properties of fat fraction in milk and dairy products can be achieved by reducing the content of medium chain saturated fatty acid (MCFA) and enhancing the content of conjugated linoleic acid (CLA), vaccenic acid (VA), ω_3 FA, branched –chain FA, butyric acid, and sfingolipids because of their recognised beneficial properties for human health (Staijns, 2008).

Pregastric lipases associated to rennet paste present a strict selectivity for short n-chain fatty acid in the sn3 position on triglycerides whereas gastric esterases are preferentially active on mono- and di-glycerides and usually hydrolyse medium and long n-chain fatty acid located at the sn1 and sn2 positions of triglycerides (Ha & Lindsay, 1990). Thus the balance of different lipases in rennet paste influences the lipolytic pattern of cheese during ripening, with a major content of short chain fatty acids in correspondence to high content of PGE. In fact, the high flavour intensity comes from short and medium chain FFA, in particular butyric acid which contributes to the cheesy lipolyzed aroma (Pinho et al., 2003).

In general, hard Italian cheeses from ewe milk show a profile of FFA characterized by butyric acid which occurs at the highest concentration caproic, capric, palmitic, and congeners of C18:0 acids (Gobbetti & Di Cagno, 2003). Accordingly, when rennet paste was used for Pecorino cheese production the most abundant FFA were: butyric acid, caproic acid, palmitic acid, oleic acid, and linoleic acid (Santillo et al., 2007b). Moreover, rennet paste containing probiotics influenced the FFA pattern in the cheese owing to the contribute of the lipases associated to the selected strains to the total lipolytic activity of rennet paste (Santillo et al., 2009).

Testing total lipolyitic activity of rennet paste containing probiotic on a natural substrate as the sheep milk substrate is useful to evaluate the FFA profile. Total lipolytic activity detected in the sheep milk cream substrate, after incubation at 37°C for 24 h, was the lowest (2,906±356.6 LU/g) in rennet without probiotics and the highest in rennet paste containing *Lb. acidophilus* (5,957±374.5 LU/g) and a mix of *B. longum* and *B. lactis* (5,654±367.4 LU/g); thus highlighting the ability of these probiotic strains to contribute to total lipase activity in rennet paste (Santillo et al., 2009). Levels of FFA detected in the sheep milk cream substrate were in accordance with the lipolytic activity of rennet, as evidenced by PCA analysis (Figure 5). In particular, butyric acid released by rennet paste containing probiotics, doubled the value of the traditional rennet as an outcome of the peculiar metabolic pathways associated to the strains. Moreover, cheese obtained using rennet paste containing *Lb. acidophilus* showed the highest levels of C10:0, C12:0, C18:0 whereas cheese obtained using rennet paste containing bifidobacteria reported the highest concentration of C6:0, C8:0, and C18:2, with the latter FFA being about 80% higher in this cheese than in cheese containing *Lb. acidophilus*.

Cokley et al. (2003) in *in vitro* studies assessed the ability of a range of lactobacilli, lactococci, bifidobacteria, and pediococci to produce CLA from free linoleic acid; a range of bifidobacteria strains tested exhibited considerable CLA biosynthetic ability. *B. lactis* showed a good percentage of conversion of linoleic acid to CLA. In MRS broth and skim milk Alonso et al. (2003) verified also that some strains of *Lactobacillus acidophilus* are able to produce CLA from free linoleic acid. Sheep milk cream incubated with rennet paste containing *Lb. acidophilus* exhibited also the highest contents of *9c,11t-* and *9t,11t*-CLA whereas the same substrate incubated with rennet paste containing *B. lactis* and *B. longum* displayed the highest levels of *10t,12c*-CLA as evidenced by the high positive loadings with the first principal component in the PCA biplot (Figure 6).

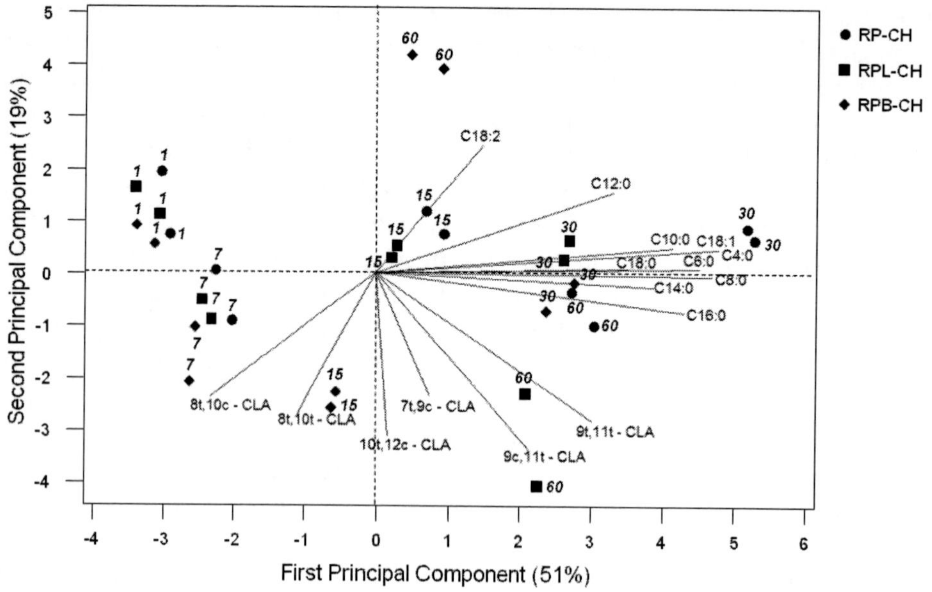

From Santillo et al., 2009.

Figure 6. Principal component analysis of the FFAs and CLA isomers of Pecorino cheese manufactured with different rennet paste at 1, 7, 15, 30, and 60 d of ripening: ● RP-CH, cheese manufactured using traditional lamb rennet paste; ■ RPL-CH, cheese manufactured using lamb rennet paste containing *Lb. acidophilus*; ♦ RPB-CH, cheese manufactured using lamb rennet paste containing *B. lactis* and *B. longum*.

Although the content of CLA in most cheese varieties is documented (Jiang et al., 1997; Zlatanos et al., 2002; Seçkin et al., 2005) few studies (Santillo et al., 2007b; 2009) reported the CLA content in ewe cheese containing probiotic. Free fatty acid and CLA profile in cheese (Figure 6) produced using rennet paste containing probiotics followed closely the one illustrated for sheep milk cream substrate so that the study of the lipolytic activity of rennet in a natural substrate is predictive of the free fatty acid profile in cheese. Challenging the probiotic added to rennet paste for the ability to synthesize short chain fatty acid and CLA offers the opportunity of improving health-promoting functional cheese with the benefits of ameliorated fat fraction, enriched CLA and probiotic bacteria.

Influence of Lamb Rennet Paste on Rheological Parameters and Preference and Acceptance test of Cheese

Cheese texture may be defined as a composite sensory attribute resulting from a combination of physical properties and perceived by the senses of sight, touch, and hearing (Pinho et al., 2004). While these attributes are manifested during cheese consumption, mechanical properties of cheese are determined by the application of a fixed stress or strain, i.e. compression, shearing, or cutting, to a sample of cheese under defined experimental conditions. These properties are related to the composition, structure, and strength of the attractions between the structural elements of the cheese. The behaviour of cheese subjected to fixed stress can be described by hardness, firmness, springiness, adhesiveness, cohesiveness, gumminess, and chewiness.

Rheological parameters play an important role in the quality of cheese being closely related to the intensity of the ripening process, in terms of both proteolysis and lipolysis, and to the definition of the cheese structure.

Park (2007) and Revilla et al. (2007) reported a high correlation between the level of proteolysis and texture in the ewes cheese, evidencing the involvement of αs_1-I-CN in cheese texture. Values of pH also play an important role in cheese texture; as the pH of cheese curds decreases, there is a concomitant loss of colloidal calcium phosphate from the casein submicelles with a progressive dissociation of the submicelles into smaller casein aggregates at a pH value below 5.5 (Lebecque et al., 2001). Lower pH value in cheese facilitates a major casein dissociation; the disaggregation exposes a larger surface area of proteins to proteinases and leads to an increase in enzyme-substrate interaction (Upreti et al., 2006).

Cheeses produced using traditional rennet paste or rennet paste containing probiotic displayed different rheological parameters as a consequence of the degree of casein breakdown observed during ripening (Santillo & Albenzio, 2008). *B. longum* and *B. lactis* showed a drop in cheese pH as well as a greater proteolysis as an outcome of the metabolic activity of bifidobacteria. These events resulted in lower values of hardness, cohesiveness, springiness, gumminess, and chewiness than those found in cheese obtained with traditional rennet and rennet containing *Lb. acidophilus* (Table 4).

It has been shown that high acidity, protein, and total solid contents generally make the cheese harder and less easily deformed (Kehagias et al., 1995). The adjunct probiotic bacteria in lamb rennet paste yielded higher cheese moisture probably due to the grater production of lactic acid; this could lead to the coagulum to loose more calcium into the whey and result in a lower ability of the curd to contract and thus expel water in subsequent cheese aging (Jimenez-Marquez et al., 2005).

Although textural evaluation by instrumental method is standardized and more reproducible, it is necessary to evaluate the consumer acceptance of food products, especially when innovations are introduced. Addition of lactobacilli in cheese has been associated with an increased proteolysis and intensification of flavour as well as proteolytic enzymes produced by certain probiotic adjunct were also found to degrade bitter peptides (Lynch et al., 1999; Lane & Fox, 1999; Ong et al., 2007).

Excessive proteolysis and lipolysis could result in off-flavors because high concentrations of bitter peptides and volatile FFA, respectively, influence cheese flavor either directly or as precursors for other compounds (Broadbent et al., 2002; Pinho et al., 2004).

Ong et al. (2007) found a positive and significant correlation between the scores of bitterness and the level of water soluble nitrogen in Cheddar cheese containing different strains of probiotic. Addition of bifidobacteria to Gouda (Gomes et al., 1995) and Cottage (Blanchette et al., 1996) cheeses had also a negative effect on cheese flavour, and resulted in a reduced acceptability with respect to the traditional cheeses. The evaluation of the consumers degree of liking is useful when the possibility of product innovation is challenged. Santillo & Albenzio (2008) reported that Pecorino cheese produced using lamb rennet paste containing probiotic did not differ from traditional cheese in preference and acceptance test performed on non trained consumers panel. The maintaince of organoleptic characteristics similar to conventional cheese without disappointing the consumer expectation is of great importance and make new products attractive for commercial exploitation.

Table 4. Rheological parameters of ovine cheese manufactured using traditional lamb rennet paste and lamb rennet paste containing probiotic.

		Lamb rennet paste[1]				Effect		
Parameter	Ripening time, d	RP	RPL	RPB	SEM	Rennet	Time	Rennet x Time
Hardness, N	60	35.57[b]	37.51[b]	25.32[a]	3.66	*	*	NS
Cohesiveness	60	0.18[b]	0.15[b]	0.06[a]	0.02	***	NS	NS
Springiness, mm	60	7.89[c]	7.13[b]	5.1[a]	0.14	**	*	*
Gumminess, N	60	0.58[b]	0.57[b]	0.19[a]	0.1	***	*	NS
Chewiness, Nxmm	60	10.03[b]	10.11[b]	1.45[a]	0.46	***	***	***

[a-b]Means with different superscripts differ for each item in rows ($P < 0.05$).

[1]RP= cheese manufactured using traditional lamb rennet paste; RPL= cheese manufactured using lamb rennet paste containing *Lb. acidophilus* (LA-5); RPB= cheese manufactured using lamb rennet paste containing *B. lactis* (BB-12) and *B. longum* (BB-46).

NS not significant;* $P < 0.05$.

From Santillo & Albenzio, 2008.

Table 5. Acceptance of ovine cheese manufactured using traditional lamb rennet paste and lamb rennet paste containing probiotic.

	Lamb rennet paste[1]				
Parameter	RP	RPL	RPB	SEM	Effect
Colour	6.78[b]	6.06[a]	6.78[b]	0.23	*
Smell	6.59	6.31	6.84	0.24	NS
Taste	6.25	6.81	6.31	0.27	NS

[a-b]Means with different superscripts differ for each item in rows ($P < 0.05$).

[1]RP= cheese manufactured using traditional lamb rennet paste; RPL= cheese manufactured using lamb rennet paste containing *Lb. acidophilus* (LA-5); RPB= cheese manufactured using lamb rennet paste containing *B. lactis* (BB-12) and *B. longum* (BB-46).

NS not significant;* $P < 0.05$.

From Santillo & Albenzio, 2008.

CONCLUSION

Many traditional dairy products from ewe and goat milk are of great value for the agricultural economy in the Mediterranean area. In particular, the production protocols of traditional ovine cheese provide for the use of lamb rennet paste able to impart typical features to the cheese.

Dairy compartment offers the chance of developing innovative products beginning from the traditional cheeses; in the light of this, the use of rennet containing probiotic is a suitable strategy for innovation in traditional ovine cheese without modification of the production procedures. This could offer several advantages: firstly, the possibility to produce new probiotic version of traditional cheese; secondly, to accelerate the ripening process in terms of both proteolysis and lipolysis due to the metabolic activity of the incorporated probiotic; lastly, to improve healthfulness and functional properties of dairy products.

The essential requirement for the commercial exploitation of these new dairy products is the maintenance of organoleptic characteristics of traditional cheese in order to meet the consumer expectation. In fact, dairy products from ewe milk are preferentially consumed by costumers who are particularly fond and familiar with the typical sheepy aroma.

REFERENCES

Abu-Tarboush H.M, Al-Saiady M.Y. and Keir-El Din H. (1996). Evaluation of diet containing lactobacilli on performance, fecal coliform and lactobacilli of young dairy calves. *Animal Feed Science and Technology*, 57: 39-49.

Addis M., Pirisi A., Di Salvo R., Podda F. & Piredda G. (2005). The influence of the enzymatic composition of lamb rennet paste on some properties of experimentally produced PDO Fiore Sardo cheese. *International Dairy Journal*, 15: 1271-1278.

Addis M., Piredda G. & Pirisi A. (2008). The use of lamb rennet paste in traditional sheep milk cheese production. *Small Ruminant Research*, 79: 2-10.

Agarwal N., Kamra D.N., Chaudhary L.C., Agarwal I., Sahoo A. & Pathak N.N. (2002). Microbial status and rumen enzyme profile of cross bred calves fed on different microbial feed additives. *Letters in Applied Microbiology*, 34: 329-336.

Albenzio M., Corbo M.R., Rehman S.U., Fox P.F., De Angelis M., Corsetti A., Sevi A. & M. Gobbetti. (2001). Microbiological and biochemical characteristics of Canestrato Pugliese cheese made from raw milk, pasteurized milk, or by heating the curd in hot whey. *International Journal of Food Microbiology*, 67: 35-48.

Alonso L., Cuesta E.P. & Gilliland S.E. (2003). Production of free conjugated linoleic acid by *Lactobacillus acidophilus* and *Lactobacillus casei* of human intestinal origin. *Journal of Dairy Science*, 86: 1941-1946.

Andren A. (1992). Production of prochimosin, pepsinogen and progastricsin, and their cellular and intracellular localization in bovine abomasal mucosa. *Scandinavian Journal of Clinical and Laboratory Investigation*, 52: 59-64.

Andren A. (2003). Rennets and coagulant. Bodmin, UK: MPG Books Ltd. Pp 281-286.

Andrén A. & Bjorck L. (1986). Milk-feeding maintains the prochymosin production in cells of bovine abomasal mucosa. *Acta Physiologica Scandinavica*, 126: 419-427.

Bergamini C.V., Hynes E.R., Candioti M.C. & Zalazar C.A. (2009a). Multivariate analyses of proteolysis patterns differentiate the impact of six strains of probiotic bacteria on a semi-hard cheese. *Journal of Dairy Science*, 92: 2455-2467.

Bergamini C.V., Hynes E.R., Palma S.B., Sabbag N.G. & Zalazar C.A. (2009b). Proteolytic activity of the three probiotic strains in semi-hard cheese as single and mixed cultures: *Lactobacillus acidophilus*, *Lactobcillus paracasei*, and *Bifidabacterium lactis*. International Dairy Journal 19: 467-475.

Blanchette, L., Roy D., Bèlanger G. & Gauthier S. F. (1996). Production of Cottage cheese using dressing fermented by bifidobacteria. Journal of Dairy Science 79: 8-15.

Broadbent J.R., Barness M., Brennand C., Strickland M., Huhouck K., Johnson M.E, & Steele J.L. (2002). Contribution of *Lactococcus lactis* cell envelope proteinase specificity to peptide accumulation and bitterness in reduced- fat Cheddar chees. *Applied and Environmental Micriobiology*, 68:1778-1785.

Bustamante, M., Virto M., Aramburu I., Barron L. J. R., Pérez-Elortondo F. J., Albisu M. & de Renobales M. (2003). Lamb rennet paste in ovine cheese (Idiazabal) manufacture. Proteolysis and relationship between analytical and sensory parameters. *International Dairy Journal*, 13: 547-557.

Coakley M., Ross R.P., Nordgren M., Fitzgerald G., Devery R. & Stanton C. (2003). Conjugated linoleic acid biosynthesis by huma-derived *Bifidobacterium* species. *Journal of Applied Microbiology*, 94: 138-145.

Collins, Y. F., MacSweeney P. L. H. & Wilkinson M. G. (2003). Lipolysis and free fatty acid catabolism in cheese: a review of current knowledge. International Dairy Journal 13: 841-866.

Corbo M. R., Albenzio M., De Angelis M., Sevi A. & Gobbetti M. (2001). Microbiological and biochemical properties of Canestrato pugliese hard cheese supplemented with Bifidobacteria. *Journal of Dairy Science*, 84: 551-561.

Cruywagen CW, Jordaan I & Venter L. (1995). Effect of Lactobacillus acidophilus supplementation of milk replacer on preweaning performance of calves. *Journal of Dairy Science*, 79: 483-486.

Di Cagno R., Quinto M. & Corsetti A. (2006). Assessing the proteolytic and lipolytic activities of single strains of mesophilic lactobacilli as adjunct cultures using a Caciotta cheese model system. *International Dairy Journal*, 16: 119-130.

Ellinger D.K., Muller L.D. & Glantz P.J. (1980). Influence of feeding fermented colostrum and Lactobacillus acidophilus on fecal flora of dairy calves. Journal of Dairy Science 63: 478-482.

El Soda M., Macedo A. & Olson N. F. (1992). The peptide hydrolase system of *Bifidobacterium* species. Milchwissenschaft 47: 87-90.

Etayo, I.; Pérez-Elortondo, F.J.; Gil, P.; Albisu, M.; Virto, M.; Conde, S.; Rodríguez Barrón, L.J.; Nájera, A.I.; Gómez-Hidalgo, M.E.; Delgado, C.; Guerra, A. & de Renobales. M. (2006). Hygienic quality, lipolysis and sensory properties of Spanish Protected Designation of Origin ewe's milk heeses manufactured with lamb rennet paste. Lait, 86, 415-434.

Fox P. F. (1993). Cheese: An overview. London: Chapman & Hall. Pp. 1-36.

Fox P.F., Guinee T.P., Cogan T.M. & McSweeney P.L.H. (2000). Fundamentals of cheese science. Gaithersburg: Aspen Publishers Inc. pp. 236-281.

Fuller R. (1989). A review: probiotics in man and animals. *Journal of Applied Bacteriology*, 66: 365-378.

Gil P.F., Conde, S., Albisu, M., Pérez-Elortondo, F.G., Etayo, I., Virto, M. & de Renobales M. (2007). Hygienic quality of ewes' milk cheeses manufactured with artisan-produced lamb rennet pastes. *Journal of Dairy Research*, 74: 329-335.

Gobbetti M, Morea M, Baruzzi F, Corbo MR, Matarante A, Considine T, Di Cagno R, Guinee T & Fox PF. 2002 Microbiological, compositional, biochemical and textural characterisation of Caciocavallo Pugliese cheese during ripening. *International Dairy Journal* **12** 511-523

Gobbetti M. & Di Cagno R. (2003). Hard Italian cheeses. Bodmin, UK: MPG Books Ltd. Pp 378-385.

Gomes A. M. P., Malcata F. X., Klaver F. A. M. & Grande H. J. (1995). Incorporation and survival of *bifidobacterium* spp. Strain Bo and *Lactobacillus acidophilus* strain Ki in a cheese product. Netherland Milk Dairy Journal 49: 71-95.

Guilloteau P, Corring T, Garnot P & Martin P. (1983) Effect of age and weaning on enzyme activities of abomasum and pancreas of the lamb. *Journal of Dairy Science* 66 2373–2385.

Ha K. J. & Lindsay R. C. (1990). Method for the quantitative analysis of volatile free and total branched-chain fatty acids in cheese and milk fat. *Journal of Dairy Science*, 73: 1988-1999.

Hughes D.B. & Hoover D.G. (1991). Bifidobacteria-their potential for use in American dairy products. *Food Technology*, 45: 75.

Irigoyen A., Izco J.M., Ibanez F.C. & Torre P. (2002). Influence of calf or lamb rennet on the physicochemical, proteolytic, and sensory characteristics of an ewe's-milk cheese. International Dairy Journal 12: 27-34.

Jiang J., Björck L. & Fondén R. (1997). Conjucated linoleic acid in Swedish dairy products with speciale reference to the manufacture of hard cheeses. *International Dairy Journal*, 7. 863-867.

Jimenez-Marquez S. A., Thibault J. & Lacroix C. (2005). Prediction of moisture in cheese of commercial production using neural networks. *International Dairy Journal*, 15: 1156-1174.

Kehagias C., Koulouris S., Samona A., Malliou S and Koumoutsos G. (1995). Effect of various starter on the quality of cheese in brine. *Food Microbiology*, 12: 413-419.

Lane C.M. & Fox P.F. (1996). Contribution of starter and adjunct lactobacilli to proteolysis in Cheddar cheese during ripening. *International Dairy Journal*, 6: 715-728.

Lebecque A., Laguet A., Devaux M. F. & Dufour E. (2001). Delineation of the texture of Salers cheese by sensory analysis and physical methods. *Lait*, 81:609–623.

Lynch C.M., Muir D.D., Banks J.M., McSweeney P.L.H. & Fox P.F. (1999). Influence of adjunct cultures *Lactobacllus paracasei* ssp. *paracasei* or *Lactobacillus plantarum* on Cheddar cheese ripening. Journal of Dairy Science 82: 1618-1628.

Lubbadeh W., Haddadin M.S.Y., Al-Tamimi M.A. & Robinson R.K. (1999). Effect on the cholesterol content of fresh lamb of supplementing the feed of Awassi ewes and lambs with Lactobacillus acidophilus. *Meat Science*, 52: 381-385

Macedo A.C., Vieira M., Pocas R. & Malcata X.F. (2000). Peptide hydrolase system of lactic acid bacteria isolated from Serra da Estrela Cheese. *International Dairy Journal*, 10:769-774.

Modler H.W., McKeller R.C. & Yaguchi M. (1990). Bifidobacteria and bifidogenic factors. *Canadian Institute of Food Science and Technology Journal*, 23:29.

McSweeney P.L.H. & Fox P.F. (1997). Chemical methods for the characterization of proteolysis in cheese during ripening. Lait 77: 41-76.

Nelson, J.H., Jensen R.G. & Pitas R.E. (1977). Pregastric esterase and other oral lipase-a review. Journal of Dairy Science 60: 327-362.

Nunez, M., B. F. Del Pozo, M. A. R. Marin, P. Gaya, & M. Medina. (1991). Effect of vegetable and animal rennet on chemical, microbiological, rheological and sensory characteristics of La Serena cheese. *Journal of Dairy Research*, 58: 511-519.

Ong L., Henriksson A. & Shah N.P. (2007). Chemical analysis and sensory evaluation of Cheddar cheese produced with *Lactobacillus acidophilus, Lb. casei, Lb. paracasei* or *Bifidobacterium* sp. International Dairy Journal 17: 937-945.

Park Y. W. (2007). Rheological characteristics of goat and sheep milk. *Small Ruminant Research*, 68: 73-87.

Pinho O., Ferreira I.M.P.L.V.O. & Ferreira M. A. (2003). Quantification of short-chain free fatty acids in "Terrincho" ewe cheese: intravarietal comparison. Journal of Dairy Science 86:3102-3109.

Pinho O., Mendez E., Halves M.M. & Ferreira I.M.P.L.V.O. (2004). Chemical, physical, and sensorial characteristics of "Terrincho" ewe cheese: changes during ripening and intravarietal comparison. *Journal of Dairy Science*, 87:249-257.

Pirisi A., Pinna G., Addis M., Piredda G., Mauriello R., De Pascale S., Caira S., Mamone G., Ferranti P., Addeo F. & Chianese L. (2007). Relationship between the enzymatic composition of lamb rennet paste and proteolytic, lipolytic pattern and texture of PDO Fiore Sardo ovine cheese. *International Dairy Journal*, 17: 143-156.

Revilla I., Rodríguez-Nogales J.M. & Vivar-Quintana A.M. (2007). Proteolysis and texture of hard ewes' milk cheese durino ripening as affected by somatic cell count. *Journal of Dairy research*, 74: 127-136.

Russell R.W., Caruolo E.V. & Wise G.H. (1980). Effects of pregastric sterase on utilization of the whole milk by preruminant calves. Journal of Dairy Science 63: 1114-1122.

Santillo, A., Caroprese M., Marino R., Muscio A., Sevi A. & Albenzio M. (2007a). Influence of lamb rennet paste on composition and proteolysis during ripening of Pecorino Foggiano cheese. *International Dairy Journal*, 17: 535-546.

Santillo, A., Quinto M., Dentico M., Muscio A., Sevi A. & Albenzio M. (2007b). Rennet Paste from Lambs Fed a Milk Substitute Supplemented with *Lactobacillus acidophilus*: Effects on Lipolysis in Ovine Cheese. Journal of Dairy Science 90: 3134-3142.

Santillo, A. & Albenzio M. (2008). Influence of lamb rennet paste containing probiotic on proteolysis and rheological properties of Pecorino cheese. *Journal of Dairy Science*, 91: 1733-1742.

Santillo A., Albenzio M., Quinto M., Caroprese M., Marino R. & Sevi A. (2009). Probiotic in Lamb Rennet Paste Enhances Rennet Lipolytic Activity, and CLA and Linoleic Acid Content in Pecorino Cheese. *Journal of Dairy Science*, 92: 1330-1337.

Seçkin A.K., Gursoy o., Kinik O. & Akbulut N. (2005). Conjugated linoleic acid (CLA) concentration, fatty acid compositiona and cholesterol of some Turkish dairy products. *LWT*, 38: 909-915.

Shah N.P. (2007). Functional cultures and health benefits. International Dairy Journal 17:1262-1277.

Staijns J.M. (2008). Dairy products and health: focus on their constituent or on the matrix? *International Dairy Journal*, 18: 425-435.

Upreti P., Metzger L.E. & Hayes K. D. (2006). Influence of calcium and phosphorus, lactose, and salt-to-moisture ratio on Cheddar cheese quality: proteolysis during ripening. *Journal of Dairy Science*, 89: 444-453.

Vicente M.S., Ibáñez F.C., Barcina Y. & Barron L.J.R. (2001). Changes in the free amino acid content during ripening of Idiazabal cheese: influence of starter and rennet type. Food Chemistry 72: 309-317.

Virto M., Chavarri F., Bustamante M. A., Barron Aramburu M., Vicente M. S., Perez-Elortando F. J., Albisu M. & de Renobales M. (2003). Lamb rennet paste in ovine cheese manifacture. Lypolisis and flavour. *International Dairy Journal*, 13: 391-399.

Zhang F., Chen J., & Yang F., Li L. (2005). Effects of age and suckling on chymosin and pepsin activities in abomasums of goat kids. *International Journal of Dairy Technology*, 58: 115-118.

Zlatanos S., Laskaridis K., Feist C. & Sagredos A. (2002). CLA content and fatty acid composition of Greek Feta and hard cheeses. *Food Chemistry*, 78: 471-477.

In: Focus on Food Engineering
Editor: Robert J. Shreck

ISBN: 978-1-61209-598-1

Chapter 11

EFFECT OF DIFFERENT FOOD PRESERVATION TREATMENTS ON ENZYME ACTIVITY, MECHANICAL BEHAVIOR AND/OR COLOR OF VEGETAL TISSUES

Lía N. Gerschenson[1,2,*], Ana M. Rojas[1,2], Marina F. de Escalada Pla[1,2] and Maria Emilia Latorre[1,3]

[1]Departamento de Industrias,
Facultad de Ciencias Exactas y Naturales (FCEN), Universidad de Buenos Aires (UBA) Ciudad Universitaria. Intendente Güiraldes 2620, (1428) Ciudad Autónoma de Buenos Aires, Argentina.
[2]Research Member of the National Scientific and Technical Research Council of Argentina (CONICET).
[3]Fellow of CONICET

ABSTRACT

Texture and color are major quality attributes of plant-based foods. Texture lies in a "mechanical unit" whose components are cell wall, cellular membrane or plasmalemma and middle lamella. Color lies in the presence of pigments compartmentalized into the cells, which selectively absorb certain wavelengths of light while reflecting others. The application of different stress factors for food preservation purposes can alter enzymatic activity and can affect texture and color, compromising consumer acceptability of resultant food products.

In this chapter, the effect on the activity of some enzymes, textural behavior and color changes, of the osmotic treatment of cucumber (*Cucumis sativus* L.) and butternut (*Cucurbita moschata*, Duch. ex. Poiret) or of the gamma irradiation of red beet (*Beta vulgaris* L. var. *conditiva*) or of the blanching of kiwifruit (*Actinidia deliciosa*, A. Chev.), is analyzed.

* Corresponding author: Lía Noemí Gerschenson
e-mail: lia@di.fcen.uba.ar
[1] Phone: 54 – 11 – 4576-3366 / 3397
Fax number: 54 – 11 – 4576-3366

INTRODUCTION

Cell wall (CW) and middle lamella (ML) work together controlling the way by which vegetal tissues undergo mechanical deformation and failure during mastication (Waldron et al., 1997). The relative resistance of these two entities determines the perception of juiciness or mealiness (Szczesniak and Ilker, 1988) but it is also affected by cell adhesion which can be attributed to different chemical compounds.

Texture is one of the most important quality attributes of fruits and vegetables. Moreover, the texture of biological materials is strongly influenced by its underlying tissue and cellular structure (Aguilera and Stanley, 1998). The textural quality of plant materials is generally negatively affected by processing operations. For example, vegetal tissues are usually blanched in order to inactivate enzymes and extend shelf-life; this procedure, however, decreases tissue firmness due to losses in the structural integrity of cell walls, middle lamellae and cellular membranes (Stanley et al., 1995). Osmotic dehydration is a commonly used technique for the concentration of solid foods, and has been extensively applied to the partial dehydration of fruits and vegetables (Raoult et al., 1989). Low-dose gamma irradiation, used as a post-harvest and pre-shipment treatment, and in combination with other processes, has shown to be a promising technique for extending the shelf-life of fruits and for their preservation. The decimal reduction time (D-values) is the amount of radiation energy required to inactivate 90% of specific pathogens; for *Escherichia coli* O157:H7, on fresh-cut vegetables, the D-values are mostly between 0.12 and 0.20 kGy (Foley et al., 2004; Niemira et al., 2002; Niemira and Fan 2006) and the application of ionizing radiation for technological purposes can affect product texture.

Color lies in the presence of pigments compartmentalized into the cells, which selectively absorb certain wavelengths of light while reflecting others. In general, these pigments are associated with vitamin value or antioxidant capacity of vegetal tissue. Along with texture, it is one of the principal factors for quality appraisal by consumers. All biological pigments selectively absorb certain wavelengths of light while reflecting others. The light absorbed may be used by the plant to power chemical reactions, while the reflected wavelengths of light determine the color the pigment will appear to the eye. Betalains of red beet are water-soluble nitrogen-containing pigments derived from tyrosine that are found only in a limited number of plants, specifically in plants belonging to the order Caryophyllales and in the fungal genera *Amanita* and *Hygrocybe* (Gandia-Herrero et al., 2005). Interest in betalains has grown after discovering their antiradical activity (Strack et al., 2003), and they are widely used as additives in the food industry because of their natural colorant properties and absence of toxicity, even at high concentrations (Schwartz et al., 1980). Chlorophylls (a and b) constitute the photosystems I and II with other photopigments (Lehninger, 1972); they are bound to specific proteins and embedded into the thylacoid membranes in chloroplasts and are responsible for the color of many plant-based foods. Factors producing membrane injury can also alter the chloroplast integrity and its function into the cell, determining that free chlorophylls, which are extremely unstable, easily degrade to pheophytin. In the case of squashes and pumpkins, they are an important food source of carotenoids which are partially responsible for their color (Azevedo-Meleiro and Rodriguez-Amaya, 2007).

Different enzymes are present in vegetal tissues, for example polyphenol oxidase (PPO), peroxidase (POX), pectin methylesterase (PME), polygalacturonase (PG). Polyphenol oxidase

(PPO) and peroxidase (POX) are used for controlling the efficiency of the blanching process. PPO is the key enzyme in melanin biosynthesis and in the enzymatic browning of fruits and vegetables. The role of PPO in the secondary metabolism of plants still remains unclear, but its implication in betalain biosynthesis has been proposed. PPO is a copper enzyme that catalyzes two different reactions using molecular oxygen: the hydroxylation of monophenols to o-diphenols (monophenolase activity) and the oxidation of the o-diphenols to o-quinones (diphenolase activity; Sánchez-Ferrer et al., 1995).

Peroxidase (POX) involves a group of enzymes known to play a very crucial role in free radicals scavenging within plant systems (Regalado, Garcia-Almadarez and Duarte-Vazques, 2004), being also involved in various developmental and metabolic processes. Plants contain two classes of POX: the intracellular Class I, and the Class III, which is secreted into the cell wall or the surrounding medium, being only present in land plants, as an adaptation to the terrestrial life in the presence of elevated oxygen concentrations. The exchange of electrons and protons is produced by the Fe(III)-protoporphyrin IX (*heme*) group of the Class III of POX. In its standard peroxidative cycle, the enzyme catalyzes the reduction of H_2O_2 by taking electrons from donor molecules such as phenolic compounds, lignin precursors, auxin or secondary metabolites. It is considered that POX plays an important role in processes such as lignification (Egley et al., 1983) as well as the insolubilization of pectin-extensin complexes (Jackson et al., 1999).

Pectin methylesterases (PME) are ubiquitous enzymes that modify the degree of methyl esterification of pectins, which are major components of plant cell walls (Pelloux et al., 2007). Such changes in pectin structure are associated with changes in cellular adhesion, plasticity, pH and ionic contents of the cell wall and influence plant development and stress responses. The PME activity is linked to chemical changes in the cell wall-middle lamella structure occurring during the thermal treatment of the vegetal tissues (Mc Feeters et al., 1985), affecting their mechanical behavior.

The normal role of polygalacturonase (PG) is to hydrolyze pectins during fruit ripening, which leads to the softening of the fruit. The PG in cucumbers is an exo-splitting enzyme such as in carrots and peaches. It exhibits the highest affinity for large substrate molecules to which it cleaves most rapidly and the mechanism is not random cleavage but rather a specific hydrolysis of terminal linkages (Pressey and Avants, 1975). These certain linkages susceptible to exo-polygalacturonase action might be critical bonds in the cell wall of fruit. Polygalacturonase (PG) from the mesocarp tissue of mango fruit at different stages of ripening was extracted and purified by Chaimanee (1992) and the increase in both endo- and exo-polygalacturonase activities correlated well with the increase of ripeness.

EXPERIMENTAL

Enzyme Activity

Peroxidase (POX)

The procedure of Fúster et al. (1994) was followed to extract soluble and total POX from tissue. The samples were homogenized in an ice-cooled Omni Mixer for 5 min at 4°C with a 50 mol m^{-3} phosphate buffer solution (pH 7.2) to extract a soluble fraction of this enzyme.

The homogenate was immediately divided into two fractions and solid NaCl was added to one of them up to reach a final 10^3 mol m^{-3} NaCl concentration in the suspension. This portion was submitted again to homogenization with an Omni Mixer for 7 min at 4°C, obtaining a complete dissolution of the NaCl and also the extraction of ionically bound POX. The supernatant obtained after the centrifugation (4°C) of this suspension volume, was used to determine the total POX activity. The other fraction was used to evaluate soluble POX. The enzyme activity was assayed as described by Marangoni et al. (1995), using guaiacol as substrate and expressing the activity as the change in absorbance (470 nm) or UAb min^{-1} mg^{-1} protein.

Polyphenol oxidase (PPO)

The PPO activity was evaluated according to Coseteng and Lee (1987) and Xuan et al. (2008). The powder was suspended in a 10^2 mol m^{-3} phosphate buffer solution (pH = 6.0) (1 g powder per 5 ml buffer) and 10^3 mol m^{-3} NaCl solution for 45 min at 5-7 °C. The suspension was then centrifuged (Model 5804R, Eppendorf AG, Hamburg, Germany) at 9000 rpm for 10 min (7°C). The supernatant was assayed for the total POX activity, at 25 °C by using 40 mol m^{-3} pyrocatechol (Merck, Buenos Aires, Argentina) in a 10 mol m^{-3} phosphate buffer solution (pH= 7.0). The activity was evaluated following the change in absorbance at 420 nm with a spectrophotometer (SpectroSC model, LaboMed Inc., Culver City, CA, USA). One unit of activity was defined as the change in absorbance (420 nm) or UAb min^{-1} mg^{-1} protein.

Polygalacturonase (PG)

The samples were homogenized in an ice-cooled Omni Mixer for 5 min at 4°C with a 50 g kg^{-1} NaCl solution (Patel y Phaff, 1960) containing 5 mol m^{-3} of sodium tetrationate as a protease inhibitor. Enzyme activity was measured by adding 0.1 ml of filtrated homogenate to 0.2 ml of 100 mol m^{-3} sodium acetate (pH 4.5) and 0.2 ml of 150 mol m^{-3} NaCl solutions. The blanks were prepared replacing 0.1 ml of filtrated homogenate with the same volume of 100 mol m^{-3} sodium acetate buffer (pH 4.5). The reaction was initiated by adding 0.5 ml of 10 kg m^{-3} polygalacturonic acid (Sigma grade III) adjusted to pH 4.5. After 1 hour of incubation at 37°C, the enzymatic reaction was stopped by an addition of sodium carbonate solution and reaction mixtures were analyzed for reducing groups by the arsenomolybdate method (Nelson, 1944). The PG activity was expressed as mg galacturonic acid reducing group equivalents min^{-1} mg^{-1} protein (Marangoni et al., 1995).

Pectin methylesterase (PME)

The PME activity was extracted and determined according to Hagerman and Austin (1986). Each sample was extracted into an ice-cooled Omni Mixer (4°C) with a 88 kg m^{-3} NaCl solution containing 20 g kg^{-1} polyvinylpirrolidone (PVP) as a scavenger of phenolic compounds. The homogenate was centrifugated for 10 min at 20000 g (5°C). The supernatant was separated, its pH was adjusted to 7.5 and 30-100 μl were finally used for the colorimetric procedure. This assay used 0.1 kg m^{-3} bromothymol blue solution (prepared in a 3 mol m^{-3}, pH 7.5, potassium phosphate buffer) as a colorimetric indicator of the decrease of pH that occurs due to the hydrolytic activity of PME. The enzyme activity was assayed at 25°C and mentioned as the change in absorbance (620 nm) or UAb min^{-1} mg^{-1} protein.

Protein Concentration

The protein was determined in each enzymatic extract by Lowry assay using bovine serum albumin as a standard (Lowry et al., 1951).

Color Evaluation

The color of samples was assessed by using a colorimeter (Minolta Co. Ltd., Osaka, Japan) with illuminant D65 and 10°observer angle. Samples were placed onto a white tile and values of the International Commission on Illumination (CIE) color space coordinates $L^* a^* b^*$ values were acquired, being L^* the lightness, a^* the grade of greenness/redness and b^* the grade of blueness/yellowness.

Mechanical Behavior

Both compression and relaxation tests were performed using an Instron Testing Machine (Instron Corp, Canton, MA, USA) provided with a 5000 N transductor and a 30 mm in diameter upper steel plate.

In the compression test, force (F)-deformation (L) curves were recorded.

For the relaxation test, the specimens were compressed in a range of 10-25% of deformation. The deformations applied were the lowest ones that allowed to record Instron Machine response without macroscopic tissue failure, when the different materials were compressed. At the preset deformation level, the crosshead was stopped and the relaxation force [F(t)] was recorded for at least 10 min. Curves of F as a function of time were fitted to the generalized Maxwell model (Nussinovitch et al., 1989; Peleg and Calzada, 1976):

$$F(t) = F_\infty + \sum_{i=1}^{n} F_i \exp(-t/\tau_i)$$

and the different parameters were obtained: force at infinite time of relaxation (F_∞) , initial relaxation force (F_i) and relaxation time (τ_i) for each Maxwell body.

Statistical Analysis

Significant differences ($p<0.05$) among treatments were tested through analysis of variance (ANOVA) followed by pairwise multiple comparisons evaluated by Tukey's significant difference test (Sokal and Rohlf 1969). Nonlinear regression analysis was applied for modeling relaxation curves (Statgraphic Statistical package, version 2.6, STSC, 1987, Rockville, MD, USA).

RESULTS AND DISCUSSION

Cucumber (*Cucumis sativus* L.)

The effect of equilibration into 0, 190 or 320 mol m^{-3} buffered (20 mol m^{-3} potassium phosphate, pH 6.8) solutions of polyethylene glycol 400 (PEG) was studied to simulate the stress supported when tissues are contacted during industrial processing, with a hypotonic, isotonic or hypertonic solution, respectively. Peroxidase (POX) and poligalacturonase (PG) activities and texture were evaluated for the different samples to conclude about the effect of treatments on cucumber (Sajnin et al., 2003).

Enzymes

As can be seen in **Table 1**, the total POX and PG activities, in general, increased significantly ($p<0.05$) for cucumber mesocarp tissue equilibrated in 0, 190 and 320 mol m^{-3} PEG solutions where the tissue was in a hypotonic, isotonic and incipient plasmolyzed condition, respectively.

Miesle et al. (1991) determined that peroxidase activity increased in response to the ripening (a normal senescent process) of highbush blueberry fruit and its maximum activity, when expressed on a fresh-weight basis, was coincident with berry softening. The same researchers reported that POX might have various functions related to fruit ripening not only with reference to cell wall synthesis but also in relation to changes in cell wall plasticity. Miller (1986) postulated that POX plays an important role during polysaccharide degradation and fruit softening. It can be observed in **Table 1** that, around 55% of the total POX extracted from raw (unsoaked) tissue was as a soluble fraction and it increased significantly ($p<0.05$) after 36 hours of equilibration, especially in 0 and 190 mol m^{-3} PEG solutions. However, the ionic form of POX did not change significantly after these treatments. According to previously cited literature, it can be thought that the total POX may have increased in response to tissue injury during tissue immersion and equilibration.

Table 1. Enzyme activity for cucumber

Concentration of PEG solutions (mol/m^3)	POX soluble (UAb/min mg)	POX ionically bounded (UAb/min mg)	POX total (UAb/min mg)	PG. 10^2 (mg GalA/ min mg)
0 (hypotonic)	134 ± 7^{A}	29 ± 4^{B}	163 ± 9^{D}	1.13 ± 0.11^{E}
190 (isotonic)	145 ± 7^{A}	56 ± 7^{C}	201 ± 6	1.30 ± 0.40^{E}
320 (hypertonic)	98 ± 2	49 ± 5^{C}	147 ± 4^{D}	1.30 ± 0.20^{E}
Raw tissue	58 ± 2	39 ± 6^{BC}	97 ± 1	0.64 ± 0.10

Adapted from Sajnin et al. (2003). POX: peroxidase. PG: polygalacturonase.
Average and standard deviation are shown (n=3). Same letters per column indicate the absence of significant differences ($p>0.05$).

The PG in cucumbers is an exo-splitting enzyme (Pressey and Avants, 1975). It exhibits the highest affinity for large substrate molecules to which it cleaves most rapidly and the mechanism is not a random cleavage but rather an specific hydrolysis of terminal linkages.

These linkages susceptible to exo-polygalacturonase action might be critical bonds in the cell wall of fruit. It has been reported that PG exhibited very large increases in activity in cucumber tissues in response to storage at 25° or 38°C after mechanical stress of these fruits (Miller et al., 1987). As pectic substances are involved in textural characteristics of cucumbers, PG could play a major role in texture expression (Buescher et al., 1979). As can be observed in the table, the PG increased significantly ($p<0.05$) in all equilibrated tissues in comparison with raw (unsoaked) cucumber. The higher value of the PG activity might be a response to immersion-equilibration and this increased activity might produce tissue alteration due to the PG capacity for cell wall pectin hydrolysis.

Mechanical Behavior

For uniaxial compression assays, tissue failure could only be detected for raw (unsoaked) cucumber tissue, allowing to conclude that immersion affected *per se* cucumber mesocarp tissue. Alteration of the shape of the force-deformation curves recorded under compression, for all the equilibrated tissue samples compared to those obtained for raw tissue, might be the signal of cell-cell adhesion weakness. Calcium leakage from tissue due to calcium absence in PEG solutions may promote dissociation of Ca-pectin complexes in cell wall and middle lamella. As these complexes are essential for tissue integrity in the case of cucumber tissue, softening by partial cell separation might have altered tissue behavior under compression (Sajnin et al., 2003).

Table 2. Mechanical parameters for cucumber

Concentration of PEG solutions (mol/m^3)	**$\Delta F/\Delta L$ (N/m)**	**F_1 (N)**	**τ_1(s)**	**F_2 (N)**	**τ_2(s)**	**F_3 (N)**	**τ_3(s)**	**F_∞ (N)**
0 (hypotonic)	26647±4431	17.8±1.2	661±40	12.2±1.2	84±6	13.3±1.2	4±0	21.5±2.3
190 (isotonic)	5808±1257	1.8±0.1	167±10	2.9±0.2	16±1	4.3±0.3	$2±0^A$	1.9±0.1
320(hypertonic)	ND	0.4±0.0	121±9	0.9±0.1	4±0	---	---	0.05±0.00
Raw tissue	16347±4251	3.9±0.5	329±21	3.7±0.2	24±3	5.4±0.6	$2±0^A$	11.9±0.1

Adapted from Sajnin et al. (2003). $\Delta F/\Delta L$ calculated from a tangent drawn in the force-deformation initial period; the other parameters were calculated from relaxation curve (deformation: 10 %) with equation : $F(t) = F_\infty + \sum [F_i \exp(- t / \tau_i)]$ and all are informed on the basis of average and standard deviation (n=8). Same letters per column indicate the absence of significant differences ($p>0.05$). ND: not detectable.

As can be seen in **Table 2**, the tangent of the initial period of the force-deformation curve ($\Delta F/\Delta L$) decreased as PEG molarity increased. Tissue equilibrated in PEG concentrations higher than 190 mol m^{-3} (isotonic condition) did not show $\Delta F/\Delta L$. The value of the ratio $\Delta F/\Delta L$ for tissue equilibrated in the solution without PEG was higher than the one obtained from tissue equilibrated in the other solutions and raw tissue; the degree of cell bursting that occurred might have been low or compensated by the high turgor of non-burst cells. This behavior was different from that observed for succulent fruits like melon and kiwifruit (Sajnin et al., 1999; Stadelmann et al., 1966) which showed a decrease in $\Delta F/\Delta L$ for tissue equilibrated in solutions without osmotica.

The relaxation curves fitted significantly ($p<0.001$) to a discrete Maxwellian model of, at most three elements with an independent Hookean spring, for raw tissue and tissue equilibrated with solutions of PEG concentrations lower than 320 mol m^{-3} (**Table 2**). The independent spring in parallel to all Maxwell elements represented the residual relaxation force (F_∞). One Maxwell element was lost for tissue equilibrated in 320 mol m^{-3} PEG solutions revealing the damage caused to tissue structure by hypertonic solutions. Residual relaxation force (F_∞) was lower for the tissue equilibrated in an isotonic solution than in raw tissue, confirming the effect of immersion *per se*.

Although sometimes not significantly ($p>0.05$), relaxation times (τ_i) of each Maxwell element, in general, increased with turgor pressure (**Table 2**). Consequently, cucumber mesocarp tissue that swelled after equilibration in 0 mol m^{-3} PEG solutions showed the highest τ_i values. Relaxation times (τ_i), which are also called the characteristic time of a Maxwell fluid, can be thought as the time it takes a body to be stretched out when deformed (Rao, 1992). A higher τ_i value obtained as a consequence of equilibration might tell about a viscoelastic material whose viscous component flowed slower than the one belonging to the corresponding element of the raw sample, after material deformation under an external force applied. Probably, higher turgor produced an increased intercellular contact retarding relaxation. Residual relaxation force (F_∞) as well as the pre-exponential factors decreased with the PEG concentration increase, denoting a decrease in turgor pressure.

Butternut (*Cucurbita moschata* Duch. ex. Poiret)

Turgor pressure of the raw tissue was adjusted by immersion in hypotonic (0 mol m^{-3}), isotonic (250 mol m^{-3}) and hypertonic (570 mol m^{-3}) buffered (20 mol m^{-3} potassium phosphate, pH 6.8) solutions of polyethylene glycol 400 (PEG) with the object to observe the response of enzyme activity and mechanical behavior, in order to understand changes that occur when turgor pressure is modified during preservation, as a consequence of immersion or osmotic treatment (de Escalada Pla et al., 2005 and 2006).

Enzymes

The activity of enzymes related to the CW was evaluated in tissue samples. The enzyme PME removes esterified methyl groups non-randomly so that blocks of contiguous ionized galacturonate residues are generated in the pectin chain. These acidic blocks are involved in the formation of interpolymer Ca-bridged junction zones (Carpita and Gibeaut, 1993; Fry, 1986). As can be observed in **Table 3**, this enzyme was only detectable in raw (unsoaked) pumpkin tissue.

The enzyme PG exhibits the highest affinity for large substrate molecules of polygalacturonans to which it cleaves most rapidly producing specific hydrolysis of terminal linkages. As can be observed in **Table 3**, PG exhibited the same activity for raw and 0 mol m^{-3} -PEG equilibrated butternut tissues, but a significant decrease was observed for isotonic and plasmolyzed tissues. Consequently, the activation or the *de novo* synthesis of PME and PG was not found after the equilibration of butternut and, in general, there was a decrease in PME and PG with an immersion and/or change of turgor pressure.

The total POX increased twofold after soaking the pumpkin tissue (**Table 3**). Probably, equilibration of samples of cut tissue in buffered-PEG solutions promoted a change in cells and/or CW-ML structure related to the POX increase. Although the total POX activity was not different among the treated tissues, important differences were observed when it was distinguished between soluble and ionically-bound POX fractions. It is known that soluble forms are cytoplasmic and they are involved in the metabolic control of ripening and senescence of fruits whereas ionically-bound POX is, in general, encountered in CW and ML (Ingham et al., 1998; Miesle et al., 1991). When the CW was submitted to maximal stretching as it occurred in 0 mol m^{-3}-PEG equilibrated tissue, ionically-bound POX showed the highest activity (**Table 3**). POX was probably catalyzing specific reactions in the CW-ML related to CW stretching. POX can produce cross-linking of phenolic compounds (formation of diferuloyl bridges and lignification) or a cross-linking of hydroxyproline-rich glycoproteins (HRGPs). It is likely that POX acted producing cross-linking of extensin, an HRGP located in the cell wall of mature butternut tissue. The formation of extensin network by oxidative-cross-linking has also been reported as occurring in response to tissue damage, providing architectural strength to the CW (Iraki et al., 1989). Bradley et al. (1992) reported that extensin-cross linking begins in a few minutes after vegetable stress. As a consequence, the increase of ionically bound POX might reflect, in our case, the response of cut/immersed/equilibrated tissue to the stress imposed.

As can be seen in **Table 3**, ionically-bound POX and soluble POX showed an enhanced activity with respect to raw tissue for samples equilibrated in an isotonic solution. In plasmolyzed tissue, the vacuole and all the protoplast shrank due to water loss and the increase in the POX activity occurred only at the expense of the soluble fraction. The increase of soluble fractions in isotonically equilibrated and plasmolyzed tissues with respect to raw tissue may be a biochemical indicator of metabolic alteration due to tissue cutting and immersion as well as to plasmolysis, when this corresponds. This increase was not observed in swelled-tissue; probably, the excess of POX synthesized into the cytoplasm was exported to the CW for helping to stretch.

Table 3. Enzyme activity for butternut

Concentration of PEG solutions (mol/m^3)	PME $.10^2$ (UAb/min mg)	PG.10^4 (mg GalA/ min mg)	POX soluble (UAb/min mg)	POX ionically bounded (UAb/min mg)	POX total (UAb/min mg)
0 (hypotonic)[1]	ND	1.25±0.05^A	7.8±0.7^C	20.0±0.2	27.8±2.0^E
250 (isotonic)[1]	ND	0.5±0.0^B	15.9±0.9	16.0±1.0	31.9±3.0^E
570(hypertonic)	ND	0.3±0.0^B	18.0±0.8	10.0±0.9^D	28.1±2.0^E
Raw tissue	4.69±0.30	1.15±0.05^A	7.0±0.8^C	9.4±0.7^D	16.4±1.0

[1]Adapted from de Escalada Pla et al. (2005). PME: pectin methylesterase. PG: polygalacturonase. POX: peroxidase. Average and standard deviation are shown (n=6). Same letters per column indicate the absence of significant differences (p>0.05). ND: not detectable.

Mechanical Behavior

Peaks of bioyield failure were observed during compression for tissue equilibrated under all osmotic conditions and their presence was associated with CW-integrity. The presence of bioyield failure peaks seem to indicate that the CW was integer although it could be more or less stretched depending on the turgor pressure of the cell content or on the hydrostatic pressure that the membrane (MB) and tonoplast exerted on the CW (Pitt, 1982; Stadelmann,

1966). According to different authors (Pitt, 1982; Sajnin et al., 2003), the existence of failure forces in compression curves of tissue, reveals that the middle lamella (ML) is stronger than the cell wall (CW), fact that avoids cell debonding under compression. It can be concluded that the CW of pumpkin tissue had a high resistance for avoiding plasmoptysis (or cell bursting) when the tissue was in the buffered solution without PEG. The resistance of this butternut tissue to bursting might be ascribed to the development of oxidative cross-linking of extensin catalyzed by POX as a response to tissue stress after cut and immersion, in addition to the natural resistance of the cellulose framework.

The ratio of force to deformation at bioyield failure (F_{by}/L_{by}) was calculated for the tissue equilibrated with all the different solutions assayed and is shown in **Table 4**. There was no difference between this ratio for raw (unsoaked) and isotonically equilibrated pumpkin tissues. Consequently, tissue damage promoted by the immersion *per se* was not detected by this parameter. Tissue samples equilibrated in PEG concentrations of 570 mol m^{-3} (hypertonic condition) showed the lowest ratio values and those equilibrated in the 0 mol m^{-3} PEG system showed the highest ones.

Table 4. Mechanical parameters for butternut

Concentration of PEG solutions (mol/m^3)	F_{by}/L_{by}(N/m)	F_1 (N)	τ_1(s)	F_2 (N)	τ_2(s)	F_∞(N)
0 (hypotonic)	312±23^A	32.5±2.1^B	200±8^C	35.0±3.5^D	9±1^E	276.1±3.1^G
250 (isotonic)	298±22^A	52±7	266±38	63.0±7.1	15±2^F	227.5±12.3
570(hypertonic)	148±6	77±14	114±17	231±21	10±5	28.0±3.5
Raw tissue	300±20^A	35.0±3.5^B	195±36^C	38.5±3.4^D	13±4^{EF}	273±5^G

Adapted from de Escalada Pla et al. (2006). F_{by}/L_{by} is the ratio of Force to deformation at the bioyield point; the other parameters were calculated from relaxation curve (deformation: 14-25 %) with equation: $F(t) = F_\infty + \sum [F_i \exp(-t/\tau_i)]$, and are informed on the basis of average and standard deviation (n=6). Same letters per column indicate the absence of significant differences (p>0.05).

For plasmolyzed tissues, an initial lag in the compression-force was observed. Cell void volumes might have developed after MB retraction from the CW due to water loss during plasmolysis. The appearance of cell void volumes in these tissues was probably responsible for the initial increase in deformation accompanied by insignificant force-values during compression. As a consequence, the plasmolyzed tissues showed a more plastic behavior.

The relaxation test allowed to evaluate the mechanical response of the undamaged tissue-structures. The relaxation force, as a function of time, was recorded, in general, at 14% constant deformation. For plasmolyzed tissues it was a must to apply higher constant deformation (25%) in order to obtain detectable forces, as a consequence of the damage suffered by the tissue during equilibration with the PEG solution. Force change with time could be significantly ($p<0.05$) adjusted to the generalized Maxwell model (two Maxwell elements and an isolated spring) and the normalized relaxation parameters (the force after infinite time of relaxation, $F\infty$; the pre-exponential forces corresponding to the relaxation of each Maxwell viscoelastic body, F_i and the relaxation times, τ_i of them) were obtained (**Table 4**). As can be seen in the table, higher values of $F\infty$ were associated with tissue equilibrated in 0 and 250 mol m^{-3}-PEG. Plasmolysed tissue showed a very low residual force at an infinite time of relaxation.

For tissue equilibrated in higher PEG concentrations, the mechanical behavior might be ascribed to the viscoelastic components, showing the highest values of Fi associated to the Maxwell body that shows the lowest relaxation time (τ_2). The need for two viscoelastic elements suggests that different structural elements contributed to each unit. Thus, probably, viscoelastic elements 1 and 2 might reflect the relaxation properties of hemicelluloses and polyuronides, respectively (Sakurai and Nevins, 1993).

The $F\infty$-value of raw (unsoaked) tissue was greater than the one of isotonically equilibrated tissue which also showed higher values of Fi (**Table 4**). Thus, immersion *per se* produced tissue damage that affected the structure. This effect was not detected through the firmness parameter, as previously stated.

Red Beet (*Beta Vulgaris* L. var. *conditiva*)

The effect of low doses of gamma radiation (1 and 2 kGy) on peroxidase (POX), polyphenol oxidase (PPO) activities, as well as on the changes in color and the mechanical behavior of fresh-cut red beet root were analyzed, with the purpose of understanding the influence of the processing on tissue characteristics (Latorre et al., 2009).

Enzymes

The increase in H_2O_2 like the one occurred after γ-irradiation can be controlled in plants by the peroxidases (POX), through its hydroxylic and standard peroxidative cycles. POX can be considered as bifunctional enzymes that can oxidize various substrates in the presence of H_2O_2 but also produce ROS like $^{\bullet}OH$, which is implicated in the scission of polysaccharides such as pectin and xyloglucans of the cell wall, to accomplish the natural process of cell elongation (Fry, 1998). As can be observed in **Table 5**, the POX activity increased significantly ($p<0.05$) with the γ-irradiation dose.

Concerning PPO (tirosinase), it can be observed in **Table 5** that there was no change in the PPO activity with 1 kGy irradiation while it increased significantly ($p<0.05$) for 2 kGy. PPO catalizes reactions involved in tyrosine-betaxanthin to betanidin conversion (Gandía-Herrero et al., 2005).

Table 5.Enzyme activity and color parameters for gamma irradiated red beet

Irradiation dosis (kGy)	a^*	b^*	L^*	POX (UAb/min mg)	PPO (UAb/min mg)
0	40 ± 4^A	17 ± 5^B	10 ± 3^C	8.81±0.81	1.20 ± 0.04^F
1	38 ± 4^A	15 ± 4^B	8 ± 3^C	12.36±0.19	1.23 ± 0.04^F
2	40 ± 1^A	17 ± 2^B	9.9 ± 0.9^C	14.32±0.14	1.53±0.04

Adapted from Latorre et al. (2009). a^*, b^*, L^*: color parameters (n=10). POX: peroxidase; PPO: polyphenoloxidase (n=3). Average and standard deviation are shown. Same letters per column indicate the absence of significant differences ($p>0.05$).

Color

No significant changes ($p>0.05$) in color could be observed between samples, from the comparison of all parameters evaluated (**Table 5**). The parameter "a*" always showed values in the red area of the spectrum.

The non significant ($p>0.05$) change in color parameters observed for this tissue due to irradiation confirms that the increment in the PPO activity developed at 2 kGy was not mainly deviated to melanin biosynthesis and enzymatic browning in red beet tissue herein assayed, but to betacyanin-betaxanthin synthesis. It is suggested that higher PPO levels may be, in part, ascribed to the increased necessity for tyrosinase activity in order to compensate with higher synthesis, the increased consumption of betacyanin and betaxanthin by the generated •OH and other ROS compounds. The proposed detention of the oxidation process (Gandía-Herrero et al., 2005) mediated by tyrosinase in the biosynthetic scheme of betalamic acid formation, implies the natural existence of a reducing agent in the raw (living) tissue, like L-(+)-ascorbic acid.

Mechanical Behavior

Red beet tissue showed a bioyield failure. As can be observed in **Table 6**, irradiation did not increase the ratio of force and deformation at the bioyield (Fby/Lby).

Relaxation data obtained at a constant compressive deformation from non- fractured tissue samples showed the change in the relaxation force with time for the different treatments; these data fitted to a mechanical model constituted by two Maxwell elements and a free spring. The residual relaxation force (F_∞) increased significantly after tissue irradiation (**Table 6**) showing the rise in tissue elasticity probably due to an increase of cross-link density (Holst et al., 2006). This rise can be, in part, ascribed to the insolubilization of the extensin-pectin network in the cell wall as well as to the higher calcium cross-linked pectins in the cell wall-middle llamella, as a consequence of higher POX-activity in response to increasing H_2O_2 production. The formation of a less transient structure, which probably involves covalent bonds derived from oxidative processes mediated by H_2O_2-POX activity (Fry, 1986) determined the higher elasticity observed for the irradiated tissues.

Table 6. Mechanical parameters for gamma irradiated red beet

Irradiation dosis (kGy)	$(F_{by}/L_{by}).10^{-3}$ (N/m)	F_1 (N)	τ_1(s)	F_2 (N)	τ_2(s)	F_∞ (N)
0	11 ± 2^A	5.5±2.9	692 ± 182^B	1.1±0.2	22±5	4.7±0.9
1	12 ± 2^A	6.8±2.1	730 ± 247^B	1.7 ± 0.2^C	28 ± 4^D	10.2 ± 2.1^E
2	12 ± 2^A	5.4±1.8	606 ± 224^B	1.8 ± 0.4^C	28 ± 7^D	12.8 ± 2.2^E

Adapted from Latorre et al. (2009). F_{by}/L_{by} is the ratio of force to deformation at the bioyield point; the other parameters were calculated from relaxation curve (deformation: 10 %) with equation : $F(t) = F_\infty + \sum [F_i \exp(- t / \tau_i)]$, and all are informed on the basis of average and standard deviation (n=6). Same letters per column indicate the absence of significant differences ($p>0.05$).

Kiwifruit (*Actinidia deliciosa* A. Chev.)

The effect of steam blanching on enzyme activity, fruit texture and color was studied with the purpose of understanding the effect of blanching process on this tissue (Llano et al., 2003).

Enzymes

In industrial practice, the efficiency of vegetable blanching (e.g. peas) is tested through POX inactivation since, in general, it is one of the most heat resistant enzymes and it is associated with oxidation determining off-flavors, low nutritive values and other damages. POX enzymes are ubiquitous, occurring in all higher plants. POX is found in most plant tissues and has been proposed to have various functions related to fruit ripening, including cell wall synthesis, changes in cell wall plasticity, lignification, degradation of indole-3-acetic acid and anthocyanin breakdown. Peroxidase activity is also induced by mechanical stress as encountered during handling and processing (Miesle et al., 1991; Miller et al., 1987). Fuster et al. (1994) found that kiwifruit ripening (a normal senescent process that affects cell membrane fluidity among other characteristics) diminished ionically bound form of POX while the soluble fraction of this enzyme increased.

In **Table 7,** POX activity can be observed when the blanching of halves of kiwifruit was performed through contact with water vapor at atmospheric pressure (99.8 ºC) and for 0-8 min. No difference was observed for each blanching time between the activity of the total and soluble POX fractions and, as a consequence, only total POX is reported. The activity of the total POX diminished abruptly for 5 minutes of heat treatment, when the tissue temperature ranged between 48° (center) and 90°C (periphery), showing the need for heat treatment times greater than 3 min for effective blanching of halves of kiwifruit.

The PME activity is usually linked to chemical changes into the cell wall-middle lamella structure occurring during thermal treatment of vegetal tissues (McFeeters et al., 1985). This enzyme shows optimal activity around 55-62°C. Some authors reported enhanced activity around 65°C. As can be seen in **Table 7**, the enzyme activity decreased rapidly from 1 min of blanching and on. Consequently, the probability of demethylation of pectin by the nucleophilic attack of this enzyme on the methoxyl groups, decreased during heat treatment.

Color

The green color of outer pericarp tissue of kiwifruit halves stayed up to 3 min of blanching, without visual evidence of browning. From 5 min of heating on, tissue became yellow-brown and a constant a^* value of -4 was measured in the Hunter Lab colorimeter (**Table 7**). These results were probably due to chlorophyll degradation and consequently, pheophytin formation as it was determined by Robertson (1985) in kiwifruit slices heated at 100°C for 5 min. Chlorophylls are bound to specific proteins and embedded in the thylacoid membranes in chloroplasts. Factors producing membrane injury can also alter the chloroplast integrity determining that free chlorophylls easily degrade to pheophytin.

Table 7. Enzyme activity and color parameters for blanched kiwifruit

Time of blanching (min)	a^*	POX.10^3 (UAb/min mg)	PME.10^3 (UAb/min mg)
0	-8.9 ± 0.2^A	5.82 ± 0.08	2.00 ± 0.10
1	-8.4 ± 0.2^A	4.90 ± 0.10	0.70 ± 0.03^C
3	-6.2 ± 0.2	5.47 ± 0.07	0.95 ± 0.03
5	-4.1 ± 0.1^B	1.08 ± 0.02	0.67 ± 0.02^C
8	-3.9 ± 0.1^B	ND	0.029 ± 0.001

Adapted from Llano et al. (2003). a^*: color parameter. POX: peroxidase. PME: pectin methylesterase. Average and standard deviation are shown (n=3). Same letters per column indicate the absence of significant differences ($p>0.05$). ND: not detectable.

Mechanical Behavior

In **Table 8**, the change of the ratio of failure force to failure deformation (F_f/L_f) with heating time (beginning at 1 min) can be observed; the progressive and uninterrupted decrease in this ratio attained a 50% decayment after 3 min of heating. The F_f/L_f value determined might then be understood as a global measurement of the integrity of the cell and of the strength of the tissue.

The adjustment to mechanical models of the relaxation curves obtained was assayed. Raw and 1-3 min heated outer pericarp tissues could be described by a generalized Maxwell mechanical array (Peleg and Normand, 1983) containing three elements (**Table 8**). F_∞ is the force after 10 minutes of relaxation (residual relaxation force) which is due to the presence of structures that perform as a single spring which is in parallel to a group of Maxwell bodies in the generalized Maxwell mechanical array. The independent spring and one or two Maxwell elements were lost, respectively, for 5 and 8 min of heating as can be observed in **Table 8**. These changes might be the expression, in the mechanical model, of an increase or predominance of the viscous component in the behavior of the tissue due to membrane damage with a resultant decompartimentalization of intracellular fluids. Residual relaxation force (F_∞) determined after 5 min of heating was not important (**Table 8**). It is interesting to remark that the loss of POX activity and greenness of the outer pericarp kiwifruit halves increased considerably after 5 min of heating as previously stated.

Table 8. Mechanical parameters for blanched kiwifruit

Time of blanching (min)	(F_f/L_f). 10^{-2} (N/m)	F_1(N)	τ_1(s)	F_2(N)	τ_2(s)	F_3(N)	τ_3(s)	F_∞ (N)
0	87 ± 7	1.9 ± 0.5^B	3 ± 1^C	1.3 ± 0.2^D	38 ± 7^E	1.5 ± 0.3^F	344 ± 47^G	2.1 ± 0.3^H
1	63 ± 6	2.1 ± 0.5^B	3 ± 1^C	1.4 ± 0.2^D	35 ± 11^E	1.9 ± 0.3^F	368 ± 143^G	2.3 ± 0.4^H
3	40 ± 6^A	2.0 ± 0.1^B	4 ± 1^C	1.6 ± 0.1^D	49 ± 6^E	1.6 ± 0.4^F	405 ± 43^G	1.4 ± 0.2
5	30 ± 8^A	1.8 ± 0.5^B	14 ± 3	1.2 ± 0.3^D	227 ± 51	--	--	0.4 ± 0.1
8	25 ± 10^A	0.3 ± 0.1	201 ± 97	--	--	--	--	0.1 ± 0.1

Adapted from Llano et al. (2003). F_f/L_f is the ratio of force to deformation at failure; the other parameters were calculated from relaxation curve (deformation: 12 %) with equation: $F(t) = F_\infty + \sum [F_i \exp(-t/\tau_i)]$, and all are informed on the basis of average and standard deviation (n=10). Same letters per column indicate the absence of significant differences ($p>0.05$).

CONCLUSION

When vegetal tissues were submitted, for technological purposes, to immersion in hypotonic, isotonic or hypertonic solutions or to exposure to gamma radiation, texture and/or color changed, affecting product final quality. The texture was altered because of changes in tissue integrity due to mechanical stress or degradation of different components. In the case of color, decompartimentalization of pigments because of tissue injury and pigment degradation were the causes for observed changes. As a response to these changes, some enzymes tended to increase their activity as in the case of the increase of peroxidase to help restoring tissue mechanical characteristics in the case of cucumber and butternut immersed in different solutions or of red beet treated with ionizing radiation. In this last case, also polyphenoloxidase activity increased to restore initial color quality.

In the case of blanching, heat transfer to the kiwifruit tissue, for a certain period, decreased the peroxidase and pectin methylesterase activity. In this process, enzymes inactivation is the desired effect. Because of the damage suffered by the tissue there was no immediate metabolic response to the occurred changes in texture and color although studies along storage must be performed to clarify if later on the response can be activated.

It can be concluded that the different studied treatments and pre-treatments affected the color and texture of vegetal tissues. The increase in enzyme activity due to these treatments is a response of tissue to the suffered injury and changes, tending to restore original conditions of the affected characteristics. This response depends on the tissue treated and on the possibility of reaction that remains after the suffered injury.

ACKNOWLEDGMENTS

The authors acknowledge the financial support of the National Agency of Scientific and Technological Promotion of Argentina (ANPCyT), the National Scientific and Technical Research Council of Argentina (CONICET) and the University of Buenos Aires (UBA).

REFERENCES

Aguilera, J.M., & Stanley, D.W. (1998). Microstructural Principles of Food Processing Engineering. 2nd edition. Gaithersburg, MD: Aspen Publishers, Inc.

Azevedo-Meleiro, C.H. & Rodriguez-Amaya, D.B. (2007). Qualitative and Quantitative Differences in Carotenoid Composition among Cucurbita moschata, Cucurbita maxima, and Cucurbita pepo. *Journal of Agricultural and Food Chemistry*, 55, 4027-4033.

Bradley, D.J., Kjellbom, P. & Lamb, C.J. (1992). Elicitor-and wound-induced oxidative cross-linking of proline-rich plant cell wall protein: a novel, rapid defense response. *Cell*, 70,21-35.

Buescher,R. W., Hudson, J. M. & Adams, J, R. (1979). Inhibition of polygalacturonase softening of cucumber pickles by calcium chloride. *Journal of Food Science,* 44(6), 1786-1787 (1979).

Carpita, N.C.& Gibeaut, D.M. (1993). Structural models of primary cell walls in flowering plants: consistency of molecular structure with the physical properties of the walls during growth. *The Plant Journal*, 3(1), 1-30.

Chaimanee, P. (1992). Changes in cell wall hydrolase polygalacturonase during ripening in mango fruit. *Acta Horticulturae* (ISHS), 321, 804-810.

Coseteng, M.Y. & Lee, C.Y. (1987). Changes in Apple Polyphenol Concentrations in Relation to degree of Browning. *Journal of Food Science*, 52 (4), 985-989.

de Escalada Pla, M., Ponce, N.M., Wider, E., Stortz, C.A., Rojas, A.M. & Gerschenson, L.N. (2005). Chemical and biochemical changes of pumpkin (Cucumis moschata, Duch.) tissue in relation to osmotic stress. *Journal of the Science of Food and Agriculture*, 85(11), 1852-1860.

de Escalada Pla, M., Delbón, M., Rojas, A.M. y Gerschenson, L.N. (2006). Effect of immersion and turgor pressure change on mechanical properties of pumpkin (*Cucumis moschata*, Duch.)". *Journal of the Science of Food and Agriculture*, 86(15), 2628-2637.

Egley, E., Pau, 1 R.N. Jr, Vaughn, K.C. & Duke, S.O. (1983). Role of peroxidase in the development of water-impermeable seed coats in *Sida spinosa* L. Plant, 157 (3), 157-224.

Foley, D., Euper, M., Caporaso, F. & Prakash, A. (2004). Irradiation and chlorination effectively reduces Escherichia coli O157:H7 inoculated on cilantro (Coriandrum sativum) without negatively affecting quality. *Journal of Food Protection*, 67(10), 2092-2098.

Fry, S. C. (1986). Cross-linking of matrix polymers in the growing cell walls of angiosperms. Annual Review of Plant Physiology, 37,165-186.

Fry, S.C. (1998).Oxidative scission of plant cell wall polysaccharides by ascorbate-induced hydroxyl radicals. Biochemical Journal, 332, 507–515.

Fúster, C., Prestamo, G. & Cano M. P. (1994). Drip loss, peroxidase and sensory changes in kiwi fruit slices during frozen storage. Journal of the Science of Food and Agriculture, 64, 23-29.

Gandia-Herrero, F., Escribano, J. & Garcia-Carmona, F. (2005). Betaxanthins as substrates for tyrosinase. An Approach to the Role of Tyrosinase in the Biosynthetic Pathway of Betalains. *Plant Physiology*, 138, 421–432.

Hagerman, A.E. & Austin, P.J. (1986). Continuous spectrophotometric assay for plant pectin methylesterase. *Journal of Agriculture and Food Chemistry*, 34(3), 440-444.

Holst, P.S., Kjøniksen, A.L., Bu, H., Sande, S.A. & Nyström, B. (2006). Rheological properties of pH-induced association and gelation of pectin. *Polymer Bulletin*, 56(2-3), 239-246.

Ingham, L.M., Parker, M.L. & Waldron, K.W. (1998). Peroxidase: changes in soluble and bound forms during maturation and ripening of apples. *Physiology Plantarum*, 102, 93-100.

Iraki, N.M., Bressan, R.A., Hasegawa, P.M. & Carpita, N.C. (1989). Alteration of the physical and chemical structure of the primary cell wall of growth-limited plant cells adapted to osmotic stress. *Plant Physiology*, 91,39-47.

Jackson, P., Paulo, S., Brownleader, M.D., Freire, P. & Ricardo, C.P.P. (1999). An extensin peroxidase is associated with white-light inhibition of lupin (*Lupinus albus*) hypocotyl growth. *Australian Journal of Plant Physiology*, 26, 313–326.

Llano, K., Haedo, A., Gerschenson, L.N. & Rojas, A.M. (2003). Mechanical and biochemical response of kiwifruit tissue to steam blanching. Food Research International, 36(8), 767-775.

Latorre, M.E., Narvaiz, P., Rojas, A.M., & Gerschenson, L.N. (2009). Effects of gamma irradiation on biochemical and physico-chemical parameters of fresh-cut red beet (*Beta vulgaris* L. var.*conditiva*) root. Journal of Food Engineering. In press. Available online 28 December 2009. doi:10.1016/j.jfoodeng.2009.12.024.

Lehninger, A.L. (1972). Bioquímica. Las bases moleculares de la estructura y función celular. Barcelona, España: Omega Ediciones S.A.

Lowry, O, H, Rosebrough, N. J., Farr, A. L. & Randall, R. J. (1951). Protein measurement with the Folin phenol reagent. *Journal of Biological Chemistry*, 193, 265-275.

Marangoni, A, G, Jackman, R. L. & Stanley, D. W. (1995). Chilling-associated softening of tomato fruit is related to increased pectin mehylesterase activity. *Journal of Food Science*, 60(6), 1277-1281.

McFeeters, R.F., Fleming, H.P. & Thompson, R.L. (1985). Pectinesterase activity, pectin methylation, and texture changes during storage of blanched cucumber slices. *Journal of Food Science*, 50, 201-205, 219.

Miesle, T. J., Proctor, A. & Lagrimini, L. M. (1991). Peroxidase activity, isoenzymes and tissue localization in developing highbush blueberry fruit. *Journal of the American Society of Horticultural Science*, 116(5), 827-830.

Miller, T. J. (1986). Oxidation of cell wall polysaccharides by hydrogen peroxide: a potential mechanism for cell wall breakdown in plants. Biochemical and Biophysical Research Communications, 141, 238-244 (1986).

Miller, A.R., Dalmasso, J.P. & Kretchman, D.W. (1987). Mechanical stress, storage time, and temperature influence cell wall-degrading enzymes, firmness, and ethylene production by cucumbers. *Journal of the American Society of Horticultural Science*, 112, 666-671.

Nelson, N. (1944). A photometric adaptation of the Somogyi method for the determination of glucose. *Journal of Biological Chemistry*, 153, 375-380.

Niemira, B.A., Sommers, C.H. & Fan, X. (2002). Suspending lettuce type influences recoverability and radiation sensitivity of Escherichia coli O157:H7. *Journal of Food Protection*, 65(9), 1388-93.

Niemira, B.A., & Fan, X. (2006). Low dose irradiation of fresh and fresh-cut produce: safety, sensory, and shelf life. Chapter 10. In: C.A. Sommers, & X. Fan X. (Eds.), Food Irradiation. Research and Technology. (pp 169-184). Oxford, UK: Blackwell Publishing and IFT Press.

Nussinovitch, A., Peleg, M. & Normand, M.D. (1989). A modified Maxwell and a non-exponential model for the characterization of the stress-relaxation of agar and alginate gels. *Journal of Food Science*, 54, 1013-1016 (1989).

Patel, D. S. & Phaff, H. J. (1960). Studies on the purification of tomato polygalacturonase. *Food Research*, 25, 37-46.

Peleg, M, & Calzada, J.F. (1976). Stress relaxation of deformed fruits and vegetables. *Journal of Food Science*, 41, 1325-1329.

Peleg, M. & Normand, M.D. (1983). Comparison of two methods for stress relaxation data presentation of solid foods. *Rheologica Acta*, 22, 108-113.

Pelloux, J., Rustérucci, C. & Mellerowicz, E.J. (2007). New insights into pectin methylesterase structure and function. *Trends in Plant Science*, 12 (6), 267-277.

Pitt, R.E. (1982). Models for the rheology and statistical strength of uniformly stressed vegetable tissue. *Transactions of the American Society of Agricultural Engineers*, 25(6) 1776-1784.

Pressey, R. & Avants, J. K. (1975). Cucumber polygalacturonase. *Journal of Food Science*, 40, 937-939.

Rao, V, N, M. (1992). Classification, description and measurement of viscoelastic properties of solid foods. In: M.A. Rao, & J.F. Steffe (Eds.). Viscoelastic properties of foods (pp. 3-47). New York, NY: Elsevier Applied Science.

Raoult, A.L., Lafont, F., Rios, G., & Guilbert, S. (1989) .Osmotic dehydration. In: Mujumdar, A. & Roques, M. (Eds.), Drying 89. (pp-487-495). NY: Hemisphere Publishing Corporation.

Regalado, C., Garcia-Almendarez, B. E. & Duarte-Vazquez, M. A. (2004). Biotechnological applications of peroxidases. Phytochemical Reviews, 3(1-2), 243-256.

Robertson, G.L. (1985). Changes in the chlorophyll and pheophytin concentrations of kiwifruit during processing and storage. *Food Chemistry*, 17(1), 25-32.

Sajnín, C., Gerschenson, L. N. & Rojas, A. M.(1999). Turgor pressure in vegetable tissues: comparison of the performance of incipient plasmolysis technique using mannitol and polyethylene glycol. Food Research International, 32, 531-537.

Sajnín, C., Gamba, G., Gerschenson, L.N. & Rojas, A.M. (2003). Textural, histological and biochemical changes in cucumber (Cucumis *sativus* L.) due to immersion and variations in turgor pressure. *Journal of the Science of Food and Agriculture*, 83(7), 731-740.

Sakurai, N. & Nevins, D.J. (1993).Changes in physical properties and cell wall polysaccharides of tomato (*Lycopersicum esculentum*) pericarp tissues. *Physiology Plantarum*, 89, 681-686.

Sánchez-Ferrer, A., Rodríguez-López, J.N., García-Cánovas, F. & García-Carmona, F. (1995). Tyrosinase: a comprehensive review of its mechanism. *Biochimical Biophysical Acta*, 1247, 1-11.

Schwartz, S. J., von Elbe, J. H., Jackman, R. L. & Smith, J. L. (1980). Quantitative determination of individual betacyanin pigments by high performance liquid chromatography. Journal of Agricultural and Food Chemistry, 28, 540-543.

Sokal, R.R. & Rohlf, F.J. (1969). Biometry. The principles and practice of statistics in biological research. San Francisco, CA: W. H. Freeman and Co. Publisher.

Stadelmann, E. (1966). Evaluation of turgidity, plasmolysis and deplasmolysis of plant cells. In: D.M. Prescott (Eds.). Methods in Cell Physiology (volume 11, pp. 143-216). New York, NY: Academic Press.

Stanley, D.,Bourne, M.,Stone, A. & Wismer, W. (1995). Low temperature blanching effects on chemistry, firmness and structure of canned green beans and carrots. *Journal of Food Science*, 60(2), 327-333.

Strack, D., Vogt, T. & Schliemann, W. (2003). Recent advances in betalain research. *Phytochemistry*, 62(3), 247-269.

Szczesniak, A. S. and Ilker, R. (1988). The meaning of textural characteristics-juiciness in plant foodstuffs. Journal of Texture Studies, 19, 61-78.

Waldron, K, W, Ng, A,, Parker, M. L. & Parr, A. J. (1997). Ferulic acid dehydrodimers in the cell walls of *Beta vulgaris* and their possible role in texture. *Journal of the Sciencxe of Food and Agriculture*, 74, 221-228.

Xuan, L., Yanxiang, G., Xiaoting, P., Bi,n Y., Honggao, X. & Jian Z. (2008). Inactivation of peroxidase and polyphenol oxidase in red beet (*Beta vulgaris* L.) extract with high pressure carbon dioxide. *Innovative Food Science and Emerging Technology*, 9, 24-31.

INDEX

A

B

C

D

E

F

G

H

M

Q

R

S

T

U

V

W

X

Y

Z